Astrobiolo~

Understanding Life in the

CHARLES S. COCKELL

School of Physics and Astronomy, University of Edinburgh, UK

WILEY Blackwell

Library of Congress Cataloging-in-Publication Data applied for.

A catalogue record for this book is available from the British Library.

Wiley also publishes its books in a variety of electronic formats. Some content that appears in print may not be available in electronic books.

Cover image: Getty Images - 177854272

Set in 9/11pt, TimesLTStd by SPi Global, Chennai, India.

2   2017

Astrobiology

Contents

Acknowledgements

This textbook was no small undertaking. It was a lot of fun to write, but it could not have been done without help from many people. I'm grateful to colleagues for reading through chapters and providing general and specific feedback on their content, readability and style. They include Michael Benton, Martin Brasier, John Brucato, Alastair Bruce, Casey Bryce, Mark Claire, Claire Cousins, Dave Deamer, Duncan Forgan, Mark Fox-Powell, Jesse Harrison, Harriet Jones, Sun Kwok, Hanna Landenmark, Andy Lawrence, Nicola McLoughlin, Sami Mikhail, Christine Moissl-Eichinger, Sam Payler, John Raven, Ken Rice, Alan Schwartz, Eddie Schwieterman, Rebecca Siddall, Max Van Wyk de Vries. I'd like to thank Sam Payler and Mark Fox-Powell for helping with the glossary.

I would like to give special thanks to all the students of the 2014/2015 Astrobiology course at the University of Edinburgh (PHYS08051) who reviewed the completed chapters and critiqued them as part of their course work. Ultimately the arbiters of this textbook are their peers around the world who use it, so having the class engage in a critical review of the final draft was an essential step in ensuring that the material is accessible and presented in a coherent way. In the process I hope they got a useful experience in critiquing scientific writing. They are, in no particular order: Gordon Amour, Danica Ariyadasa-saez, Ewen Borland, Adam Brown, Craig Brownhill, Andrew Campbell, Joanna Cassidy, Tom Cheong-Macleod, Olivia Chingara, Lily Chubb, Joseph Ciccone, Scott Cocker, Charlotte Davie, Anna de Graaff, Katie Garrett, Aidan Gibbs, Megan Gouw, Lachlan Greig, Darren Hall, Alistair Hudson, Tom Hughes, Claudia Hulmes, Daniel Innes, Isobel Irvine, Munro Kellock, Yoon-Joo Kim, Kyle Laing, Fiona Lavroff, Cara Lynch, Cameron MacFarlane, Thomas McAulay, Patrick McKenna, Connor Minnis, Shivam Mishra, Andrew Mobberley, Andrew Moffat, Edward Morgan, Gillies Munro, Kathryn Murie, Valdemar Nordstrom, Elena Papachristou, Hamish Strachan, Emily Payne, Agniete Pocyte, John Pritchard, Andrew Reilly, Gary Robertson, Francesco Sarandrea, Luciana Sinpetru, Neringa Siugzdinyte, Tomas Soltinsky, Calum Strain, Amy Teasdale, Best Thitasiri, Max Van Wyk de Vries, Anthony Wang, Andrew Williams, Craig Young, Hannah Adamson, Emma Andrews, Corrigan Appleton, Cameron Atkinson, Taylor Baird, James Bell, David Bradley, Chris Brash, David Brown, Duncan Bruce, Holly Bryce, Rhiannon Cajipe-Stansfield, Steph Campbell, Georgia Clark, Josh Clark, Alex Clements, Rachel Conway, David Currie, Jessica Davis, Zak Dean-Stone, Alberta Edwardes, Idenildo Ferreira, Craig Gilbertson, Tom Hardy, Connor Hay, Eilidh Hunter, Ryan Jamieson, Madi Kuhn, Samuel Lees, Summer Lewis, Alex Lyster, Sofie MacDonald, Torquil MacLeod, James Maroulis, Jonny Maxey, Dorothea Meath Baker, Anna Oprandi, Amar Parmar, Ben Pryce, Madeleine Reader, Callum Reid, Josh Reid, Aidan Rocke, Sahl Rowther, Nicolas Rooms, Enn Rusi, Cian Ryan, Agne Semenaite, Oliver Sharpe, Chris Smith, Emma Stam, James Steel, Brinley Terrell, Will Thwaites, Tu thong Tran, Morris Trestman, Laura Turnbull, Manos Tzivakis, Gordon Walker, James Weatherill, Andrew Whittington, Alex Williams.

I would also like to thank the University of Edinburgh and the School of Physics and Astronomy for the opportunity to teach astrobiology. This textbook is structured around the Astrobiology course that I have taught at the university. The feedback of students and discussions with them after lectures and tutorials did much to hone my view of what is useful to undergraduates to learn in developing their interdisciplinary knowledge of science. Further refinements in the content and level have been made based on the feedback and opinions of over 50 000 students who have taken part in our 'Astrobiology and the Search for Extraterrestrial Life' Massive Open Online Course (MOOC).

Some of the images in the book were kindly provided by friends, colleagues and those whom I contacted for them including Lynn Rothschild, Ralf Kaiser, Jonathan Clarke and Victor Tejfel. I'd like to thank the Journal *Astrobiology* for permission to reproduce images. Any errors in the text-book are entirely my responsibility.

I'd like to thank the team at Wiley, including Rachael Ballard, Audrie Tan and Fiona Seymour, for bringing this textbook to fruition. Sue Bowler and Kai Finster are thanked for their thorough review of the textbook before production.

Charles S. Cockell

About the Companion Website

This book is accompanied by a companion website:

www.wiley.com/go/cockell/astrobiology

The website includes:
- A complete astrobiology lecture course comprising slide sets for each of the chapters.
- Exam questions related to the chapters.
- **Astrobiology Periodic Table**
- Other materials.

1

Astrobiology and Life

Learning Outcomes

➤ Understand that astrobiology is concerned with the origin, evolution and distribution of life in the Universe. It investigates life in its cosmic context.
➤ Know about some aspects of the history of astrobiology and how it emerged as a field.
➤ Understand some of the detailed scientific questions that underpin its main lines of enquiry.
➤ Understand some of the complexities in the definition of life and how we can construct useful operational definitions of life.

1.1 About this Textbook

It is not a straightforward task to decide how to structure a book whose objective is to describe life in its cosmic environment – **astrobiology**. This is a formidable task and the fields that must be covered to do justice to this subject, even at an introductory level such as this, include astronomy, biology, chemistry, geosciences, planetary sciences, physics, social sciences and others. I read many books to see how others had approached this problem. One logical way is to start at the beginning. From the Big Bang we arrive at the first generation of stars and then as they die and explode they sow the Universe with the elements needed for life (Figure 1.1). Planets form around new stars. Some are habitable and on at least one of them the origin of life occurs. Life proliferates, transitions from single-celled organisms to animals, all the while

being harassed by extinctions. Ultimately an intelligence develops that sets about detecting planets around distant stars and it puts together a programme to attempt to find other civilisations.

This fine narrative is a logical backdrop to organise the chapters of a textbook. The chronological story has a certain rhythm and emerging complexity about it. It is the approach taken by many astrobiology texts.

However, I decided not to take this approach. This textbook begins with a study of the one data point of life that we know – life on Earth. If we want to investigate how the elements required for life were produced after the Big Bang or why a habitable environment needs certain characteristics, or why certain molecules might have been needed for life to emerge, we need to know about biology first. We need to understand its structure, its requirements and what conditions it can subsist under in order to question how those characteristics were made possible.

We start by looking at the fundamental properties of matter and how those properties underpin the structure of the molecules of life. We then consider how these molecules are assembled into the major components of living cells. With a sound knowledge of the structure of life we then move on to think about how these cells can get the energy they need to grow and reproduce, all the time being mindful of those factors that might be specific to terrestrial life or from which we could learn something about life anywhere.

Supported by our knowledge of living things, we then explore how all life on Earth is related or linked into a tree of life. We investigate what the physical and chemical

Astrobiology: Understanding Life in the Universe, First Edition. Charles S. Cockell.
© 2015 John Wiley & Sons, Ltd. Published 2015 by John Wiley & Sons, Ltd.
Companion Website : www.wiley.com/go/cockell/astrobiology.

Figure 1.1 *Astrobiology seeks to understand the phenomenon of life in its cosmic context. This 'ultra deep field' view imaged by the Hubble Space Telescope includes nearly 10 000 galaxies across the observable Universe in both visible and near infra-red light. The smallest, reddest galaxies are among the youngest, in existence when the Universe was just 800 million years old.* [Source: NASA, ESA, H. Teplitz and M. Rafelski (IPAC/Caltech), A. Koekemoer (STScI), R. Windhorst (Arizona State University), and Z. Levay (STScI)].

limits are to life that might define how diverse or extensive this tree can become in extreme environments at the limits of planetary habitability.

At this point we are now equipped with a solid understanding of the structure, interrelationships and capabilities of the life that we know on the Earth. It's time to put this into its cosmic context.

We will turn to look at how this life might have come about. To start with we return to the beginning of the Universe and investigate how stars and planets form. This distinctly astronomical turn of events in the book is a necessary way to address the question: how did the elements required for life form and where did they form? In particular, we will examine the conditions for the formation of carbon compounds in the Universe.

To begin to understand how the formation of the elements of life, especially carbon, could have led to the origin of life we need to know something about the conditions on the early Earth (we'll be assuming in this book that life originated here, for reasons that will be discussed). We will investigate the environmental characteristics of our planet during the first billion years to understand what sort of

environments and habitats could have existed on the Earth at that time.

The question of how life might have originated in this early environment is our next task. We will consider the chemical reactions and environments in which life could have originated and discuss some of the ideas for the reactions that allowed simple precursors to come together to make the macromolecules of biology – how chemical reactions led to the formation of the first self-replicating cell.

We follow this up by considering the evidence of early life on Earth, when the first organisms emerged and some of the complexities and controversies of the evidence of preserved life in the rock record. These problems make full use of our acquired knowledge about the structure, energy sources and environments in which life can persist.

It's time to take yet another step back and to think about how this first billion years fits into the whole history of our planet. We begin a chapter where we consider how geologists date rocks, order their understanding of the history of the Earth and we discuss some of the major transitions in life that occurred after the first billion years, including the rise of animals.

With an overarching view of the history of the planet, we might be tempted to think that this has all been rather smooth and structured. Unicellular organisms evolved into animals and then intelligence. However, the next two chapters elaborate why this isn't the case. By investigating rises in atmospheric oxygen that have occurred in our planet's past and the occurrence of mass extinctions, we can see that the emergence of life on a planet, and its success over billion year time scales, is fraught with difficulties, including asteroid impacts (Figure 1.2). We will see that life itself is responsible for some of these changes. Are these challenges universal and were the opportunities that presented themselves during the co-evolution of the planet and life universal? This question will be discussed as we progress, but you might like to keep it in mind at any time you are thinking about the history of life on the Earth.

At this stage, we have a fairly complete understanding of planet Earth, its history, its life, its geology. We have, in essence, got to grips with a detailed understanding of the one planet we know that supports life, its characteristics and how life shaped, and was shaped by, its environment. At this point we take this knowledge and expand further to the cosmic context.

In the following chapters we take what we know about the Earth and consider what might define a habitable planet and where in the Universe such environments might exist. Taking a look closer to home – our own Solar System – we investigate how Mars and icy moons

Figure 1.2 *The dinosaurs, these flying reptiles (pterosaurs) and many other forms of life at the end of the Cretaceous are thought to have been driven to extinction by the effects of a large asteroid impact. Therefore, to understand the past history of life on the Earth, we need to investigate our planet's astronomical environment, a key objective of astrobiology (Source: NASA).*

Figure 1.3 *As this artist impression makes clear, the detection of rocky worlds around other stars offers us the possibility of a statistical assessment of how common Earth-like worlds are in the cosmos, an analysis of their diversity, and determining the abundance of detectable life* [Source: NASA/JPL-Caltech/R. Hurt (SSC-Caltech)].

compare to the Earth. Are other planets in our Solar System habitable? We move on from this position to consider the billions of other planets in our Universe by examining the methods used to search for exoplanets, determine their different characteristics (Figure 1.3) and how we might search for life on them.

In the final chapters of the book we consider extraterrestrial intelligence and whether there are any other intelligences in the Universe with which we can communicate. We contemplate the future and fate of our own civilisation.

Each chapter presents a text on a particular aspect of the link between life and cosmos. I have attempted to elaborate some of the principles of astrobiology with respect to each subject area.

You will also notice that the units I use in the textbook are not consistent throughout. For example, growth temperatures of microorganisms are usually discussed in Celsius. Temperatures of planetary surfaces are often expressed in Kelvin. Different scientific fields tend to use different units and, rather than creating complete consistency (which would result in seemingly odd units being used for phenomena where they are not normally used), I have stuck with the normal conventions. These differences highlight the multidisciplinarity of astrobiology.

The chapters also include some boxes that I thought would be useful. I have written boxes that present some points of debate in astrobiology that are worth discussing

with others or contemplating yourself. They are by no means exhaustive and you should use them to encourage other discussions or come up with new questions. Here and there I have written boxes about some of the major facilities that astrobiologists use. There are a vast number of techniques that astrobiologists employ in the laboratory, but some large facilities, coordinated internationally, such as space telescopes, expand the reach of the science significantly. They also give you a flavour of the modern nature of international science.

I often get asked by students, 'What degree do I need to do astrobiology?' Any degree allows a person to explore aspects of astrobiology from different angles. Science should never be closed to inquisitive minds on account of narrow human discipline definitions. So throughout the book I've liberally scattered boxes that contain some personal information about astrobiologists, their original degree areas and what motivates them. There are many fine people in astrobiology and I'd like to emphasise that there is no significance to the astrobiologists not included or included. I chose a selection of colleagues whom I thought would exemplify the variety of disciplines from which astrobiologists come.

I have included some further reading. This was, perhaps, the most difficult task. It is impossible to do justice to all the literature that exists in every field that comprises astrobiology, let alone list all of the main contributions. Instead

I've suggested two or three popular books that relate to each chapter that might provide some enjoyable additional reading. I have also listed a set of papers. They are papers I thought would give a representative sprinkling of just some of the possible avenues an interested reader might pursue and that relate to the main themes and subjects covered in each chapter. Again, no significance should be placed on the omission of many important papers or the inclusion of the ones I listed.

1.2   Astrobiology and Life

It may seem strange to begin a science text book by launching into a discussion underpinned by some history, but to begin this journey into astrobiology, we need to agree on what it is that we are all talking about.

First, what do we mean by the very term 'astrobiology' and where did it come from? Second, throughout this text book you'll find regular mention of the word 'life': 'Life in the Universe', 'Life in extremes', 'Life on Mars', 'Life beyond Earth', 'Life on the early Earth' and so on. With all this talk of life, we should perhaps try to agree exactly what it is we are talking about when we speak of 'life'. I say 'try', because I'll declare from the beginning that the multi-century endeavour to define 'life' is not going to be solved in this textbook (and there may not even be a solution, as we'll shortly discuss). However, we can at least engage in a useful debate about what we might mean by life – we might agree on a working definition that will serve us during this journey into astrobiology. In the following sections, let's explore what modern astrobiology is, review its history and see if we can come up with a working definition of life useful for the rest of the book.

1.3   What is Astrobiology?

So let's begin with 'astrobiology', the subject matter of this textbook. What is this field of science?

We might summarise by saying that astrobiology is the interdisciplinary science that sits at the interface between biological sciences, earth sciences and space sciences – exploring questions that seek to understand the phenomenon of life in its wide universal environment. Although the term was coined in the 1950s, astrobiology

has gained new momentum, spurred on by new efforts in microbiology, chemistry, geosciences, planetary sciences and astronomy.

No science has a fixed, constrained and prescribed set of questions, but most sciences have a general set of questions which can be used to successfully circumscribe what its area of endeavour is, and these can be identified for astrobiology. Astrobiology might be said to address at least four large-scale questions: How did life originate and diversify? How does life co-evolve with a planet? Does life exist beyond the Earth? What is the future of life on the Earth?

Undoubtedly one of the most potent questions in astrobiology is: 'Are we alone in the Universe?' or similar formulations of the question of whether we are the only type of life in the Universe. This is one of the most obvious questions to ask when considering how life fits into its cosmic context. Although this question captures the public imagination, it is just one of many big questions asked by astrobiology. It is, however, reasonable that it is a major question. Discovering whether the Earth is the only planet that harbours an experiment in evolution in a Universe of at least 150 billion galaxies and a galaxy with about 200 billion stars seems like a sensible line of enquiry. These numbers tell us that there are about 10^{22} stars in the Universe, give or take the odd order of magnitude, which we could explore for the presence of habitable worlds. There is no shortage of planetary bodies for the attention of astrobiologists.

Astrobiology recognises that it is difficult to develop a full understanding of life on Earth without understanding its links to the cosmic environment. The Earth seems like a tranquil place (Figure 1.4). However, it is subjected to the vagaries of its astronomical environment. For example, a leading hypothesis for the extinction of the dinosaurs is an asteroid or comet impact about 65 million years ago. This hypothesis underscores the fact that to understand past life on Earth we need to understand how the astronomical environment may have influenced life. Eventually, when the Sun's luminosity increases to a sufficiently high value, the Earth's oceans will boil away and the planet will suffer a runaway greenhouse effect, eventually turning into a Venus-like world (Figure 1.5). Thus, to understand the future of life on Earth, we must also understand our astronomical environment. We need to know how stars are born and die. Investigating the past and future of life on Earth means that we need to look beyond the Earth to get answers.

Figure 1.4 *The one data point we have of a planet that harbours life – the Earth. Astrobiology seeks to understand how the phenomenon of life came about and whether it is unique in the Universe. Here the Earth rises over the lunar landscape in this iconic image taken by Apollo 8 in 1968. The image is sometimes called, 'Earthrise' (Source: NASA).*

We've identified some of the major questions in astrobiology. Let's survey some of the other questions of which these large-scale enquiries are comprised.

First, we start with how life began. Astrobiology is concerned with the origin of life. Questions such as: How did life originate? where did life originate? Was it inevitable? When did it happen? What is the evidence for early life? all encompass this area of research.

Once life did spread across the Earth we wonder what its limits are. If we can find out what the physical and chemical boundaries of life are – the most extreme conditions it can tolerate – we can begin to assess the habitability of other planetary bodies as locations for life. This knowledge also helps us to assess what the impact of human activity and industry might be on the biosphere. Questions that fascinate astrobiologists include: What are the limits of life? How does life survive at physical and chemical extremes? Are these limits universal? What do these limits tell us about habitable conditions or the possible presence of life elsewhere? These probing lines of thought drive us to study life

Figure 1.5 *When the Sun turns into a Red Giant star in several billions years from now, all life on Earth will already have been extinguished and the oceans boiled away. Therefore, to understand the future of life on our planet, we need to know about the evolution of stars. Biology and astronomy are inextricably linked in astrobiology (Source: wikicommons via Fsgregs).*

in extreme environments, from the deep oceans to the freezing wastes of Antarctica.

It is one thing to know about how life originated and what its abilities to live in extreme planetary conditions might be, but how are all the organisms on the Earth related? The diversity of life on the Earth is extraordinary. What unites organisms and what is the relationship between them? Astrobiologists want to better understand the evolutionary links between diverse organisms with questions such as: How is life related? What triggered the appearance of multicellular life? What have been the major catastrophes for life and the effects of mass extinctions?

We are then faced with the profound question of whether this vast evolutionary experiment is unique and whether other planets in the cosmos harbour life. This central question in astrobiology can be summarised with the sub-questions: Is there life elsewhere? If there is, what is it like? If there isn't life elsewhere, why not and what is missing elsewhere that was present on the Earth when life emerged? A rather more specific, but nevertheless very interesting, set of questions can also be identified: Is intelligence inevitable and has it arisen elsewhere? If it has evolved elsewhere, can we communicate with it? What happens if we do?

Finally, of course, we should not forget our own civilisation. Our future is as much part of the evolution of life as any other organisms and this future impels us to ask many questions such as: Will humans leave the Earth permanently? How do we settle other planets? How do we preserve Earth whilst settling space? How will we adapt to space – can society be successfully extended to this environment? These are not so much scientific questions, more technical questions, but they very much bear on the applications of astrobiology to human society. These questions generate direct links between astrobiology and humanities and social sciences as they force us to confront our own place in the cosmos and the story of life.

1.4   History of Astrobiology

Having explored the main questions in astrobiology and summarised them, we might ask ourselves when all this enquiring began. Astrobiology is, from a philosophical standpoint, an ancient science. Greek philosopher, Metrodorus of Chios (fourth century BC; Figure 1.6), a student of Democritus (c. 460–370 BC) (one of the first people to propose the atomic theory of matter) stated: 'It would be strange if a single ear of wheat grew in a

Figure 1.6   *Metrodorus of Chios, ancient Greek philosopher, who wondered about the existence of other worlds* (Source: wikicommons, Keith Schengili-Roberts).

large plain, or there were only one world in the infinite'. The Greeks had a very different view of the Universe than the one we have today. They had no real conception of the planets as rocky bodies or the vast distances to the stars. Indeed, they thought that all the stars were held in the surface of a huge sphere. Metrodorus's statement was, nevertheless, a remarkable perspective on the potential plurality of Earth-like worlds. Metrodorus's view of the world was very different from Plato's (c. 428–348 BC) and Aristotle's (384–322 BC) who asserted the uniqueness of the Earth in the cosmos. Indeed, the idea that the Earth was the centre of the Universe was based on the observation that the stars never moved with respect to one another, which the Greeks incorrectly interpreted to be a result of the fixed position of the Earth rather than the great distances to the stars. Aristotle's view would dominate for many centuries. Until the Enlightenment, the idea that the Earth was a sole inhabited world in the cosmos held its grip on the public view, bolstered by religious doctrine.

In the sixteenth century, the geocentric view of the Universe, which firmly placed the Earth as the centre of the action, was overturned by Nicolaus Copernicus (1473–1543). The Copernican view was the forerunner of newly emerging ideas that stars may be other suns.

In the seventeenth century, more enquiring and inquisitive minds appeared and, with them, new speculations about the place of the Earth in the larger order of things. One of the most astonishing speculations about worlds beyond the Earth was made by Italian astronomer and philosopher, Giordano Bruno (1548–1600; Figure 1.7), who stated in his book *On the Infinite Universe and Worlds*: 'In space there are countless constellations, suns and planets; we see only the suns because they give light; the planets remain invisible, for they are small and dark. There are also numberless earths circling around their suns, no worse and no less than this globe of ours. For no reasonable mind can assume that heavenly bodies that may be far more magnificent than ours would not bear upon them creatures similar or even superior to those upon our human earth'. This was a prescient statement about the possibility of extrasolar planets and a person couldn't do much better today in writing a clear summary of why Earth-like exoplanets are hard to find. Bruno was eventually burned at the stake for a variety of charges, most of which related to him holding beliefs contrary to the Catholic Church concerning the Trinity, Jesus and indiscretions about his views on church ministers. However, one of these charges was explicitly for claiming the plurality of worlds. It is sobering to remember that it was once possible to be executed for discussing the existence of extrasolar planets.

During the Enlightenment, the invention of the telescope allowed scientists to see new moons and planetary bodies. Although one might be forgiven for believing that this would reduce speculation, as more data was available, it had the opposite effect. Armed with new evidence for other worlds, speculation went wild.

Christiaan Huygens (1629–1695), who discovered Saturn's moon, Titan, and invented the pendulum clock, wrote extensively on extraterrestrial life and the habitability of other planets in his book, *Cosmotheoros*, published posthumously in 1698. As well as speculating about astronomers on Venus, he also suggested that other intelligences would understand geometry. About music he said: 'This is a very bold assertion, but it may be true for aught we know, and the inhabitants of the planets may possibly have a greater insight into the theory of music than has yet been discovered among us'.

William Herschel (1738–1822), discoverer of Uranus and infra-red radiation, after observing the strangely

Figure 1.7 *Giordano Bruno, whose speculations about other worlds (the 'plurality of worlds') contributed to his demise (Source: wikicommons).*

circular craters of the Moon speculated about them in preserved manuscripts, in an age when their impact origin was completely unknown: 'By reflecting a little on this subject I am almost convinced that those numberless small Circuses we see on the moon are the works of the Lunarians and may be called their Towns'.

As late as 1909, Percival Lowell (1855–1916), observer of the infamous Martian 'canals' said of Mars in his book, *Mars as the Abode of Life*: 'Every opposition has added to the assurance that the canals are artificial; both by disclosing their peculiarities better and better and by removing generic doubts as to the planet's habitability'.

We could continue with many such quotes (and many other eminent scientists and philosophers were convinced of alien life), but these three are adequate to make two points. First, we would have to wait for the space age and the direct and close-up observation of planetary bodies to truly force astrobiology into an empirical era and, second, these quotes are a warning from the past.

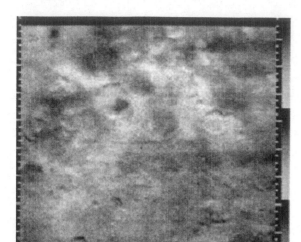

Figure 1.8 *One of the first orbital photographs of Mars, taken by the Mariner 4 craft on 15 July 1965 suggested a dead, desiccated environment unfit for life. The area shown is 262 × 310 km and is a heavily cratered region south of Amazonis Planitia, Mars* (Source: NASA).

The desire to believe in alien life should not trump empirical observation. Life should always be the last explanation after all non-biological explanations have been exhausted.

It was only at the beginning of the space age (Figure 1.8) that the photographic study of planetary surfaces yielded new and more empirically constrained views of the surfaces of other planets. In general, they showed other planets to be devoid of life and this led to a strong retreat from previous optimism. Nevertheless, astrobiology entered into the realms of experimental testing with a range of pioneering experiments and discoveries that would take it from its previous philosophical underpinnings to its present day status as a branch of science.

Laboratory experiments from the 1950s and onwards, simulating conditions on the early Earth and showing the production of amino acids and other building blocks of life, brought the study of the origin of life into the laboratory. The publication of evidence, in the 1980s and onwards, of fossil life on the Earth preserved for more than three billion years turned the search for ancient life

on Earth and the timing of the emergence of life into a scientific quest.

The first experimental search for life on other worlds, undertaken by the robotic Viking biology experiments, landed on Mars in 1976 and gave ambiguous results, but nevertheless demonstrated that we can go to other planets and implement the scientific method in a search for life. Attempts were made in the 1970s to transmit messages to other civilisations with all of its social and ethical implications. Despite the lack of response, the efforts to search for, and communicate with, extraterrestrial intelligence triggered a vigorous discussion about the intersection of astrobiology with social sciences.

The discovery of liquid water oceans in the planetary bodies orbiting in the frigid wastes beyond Mars, such as the moons of Jupiter (Europa, Ganymede) and Saturn (Enceladus) and the discovery of complex organic carbon chemistry on Saturn's moon, Titan, has showed us that we can learn about the habitability of planetary bodies and organic chemistry in surprising places (Figure 1.9). In recent years, the discovery of planets, particularly rocky planets, around other stars (exoplanets) has led to a flourishing of astrobiology and our ability to assess the statistical chances of habitable worlds elsewhere in the Universe.

These experiments and discoveries, from the mid-twentieth century and onwards, set the stage for

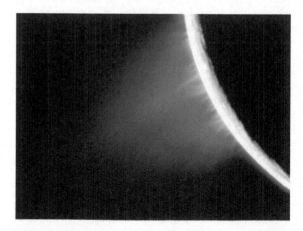

Figure 1.9 *Plumes of water emanating from the south polar region of Saturn's moon, Enceladus, is just one of the many discoveries that have provided an empirical basis with which to test the hypothesis that habitable conditions exist beyond the Earth* (Source: NASA).

Figure 1.11 *Nobel Laureate, Joshua Lederberg, who was at the forefront of United States efforts in exobiology in the twentieth century, at his laboratory in the University of Wisconsin, October 1958.*

Figure 1.10 *Gavriil Tikhov, who wrote an early book called 'Astrobiology' and took a great deal of interest in spectroscopy as a means to look for signatures of extraterrestrial life. Here he observes the spectroscopic signatures of vegetation.*

astrobiology as the truly experimental science that we know today.

Throughout this history different terms have been used to describe the science, which can, if you don't take care, cause much confusion. In the mid-twentieth century, although not the first time the word was used, 'astrobiology' was the title of a 1953 book by Gavriil Tikhov (1875–1960; Figure 1.10). His book explored the possibility of life on other worlds. Tikhov was particularly fascinated by the idea of using the spectroscopy of vegetation to seek vegetation on other planets and even founded a Sector of Astrobotany allied to the Science Academy of Kazakhstan. His methods and ideas were forerunners of the use of spectroscopy to search for biosignatures on exoplanets.

In 1960, Joshua Lederberg (1925–2008; Figure 1.11), a pioneer in bacteriology and molecular biology who won the Nobel Prize for his work on bacterial genetics, used the term **exobiology** to describe the search for life beyond the Earth. Other terms have included cosmobiology, xenobiology and bioastronomy, the latter used by astronomers.

Today, the word astrobiology is used in a wide sense to mean not just the search for life beyond the Earth, but the study of life in its cosmic context in general, including the past history of life on the Earth.

1.5 What is Life?

Attempts to define life are an ancient pursuit. From an astrobiologist's point of view the discussion is rather important since if we want to search for life elsewhere or find it in ancient rocks on Earth, we had better know what we are looking for.

Let's start by saying from the beginning that we probably don't need to accurately define it to progress with our work. We can come up with a working definition of life that encompasses some of its characteristics or its essential processes. For example, life is 'a self-sustained chemical system capable of undergoing Darwinian evolution', is a rather succinct definition by NASA scientist, Gerald Joyce.

Many of these definitions are underpinned by the idea of evolution being a fundamental characteristic of life. Evolution is the process by which variation in a population of organisms, placed under environmental conditions, results in the selection of surviving organisms that pass their traits onto subsequent generations ('natural selection').

There are many other apparent characteristics of life that we could list. For example, life exhibits complex

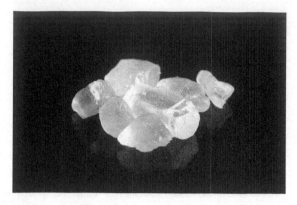

Figure 1.12 *Life grows, but crystals do as well, such as these salt (NaCl) crystals that could grow if placed in a saturated salt solution* (Source: wikicommons, Mark Schellhase).

behaviour and often unpredictable interactions. Life also grows and reproduces, a point recognised by Joyce's definition that life is a 'self-sustained chemical system'. Life also metabolises, which is a process that involves breaking down compounds or changing their form to generate energy and obtain raw materials.

However, the problem with all these characteristics individually is that many non-biological entities exhibit these behaviours. Salt crystals, when exposed to the

appropriate conditions, such as a saturated salt solution, can grow (Figure 1.12).

Computer programs can be made to 'reproduce' in the sense that they can be multiplied and in some cases, even incorporate errors in analogy to biological evolution. Fires, in a rudimentary way, 'metabolise' organic material. They burn organic carbon in oxygen to produce the waste products, carbon dioxide and water. The chemical reaction involved in this process is identical to respiration used to produce energy in animals, the only difference is that the reaction is biochemically controlled in life and uncontrolled in fires (Figure 1.13).

So if we work hard enough, we can find **abiotic** conditions that are similar to biological behaviours. Conversely, we can find biological entities that we might think are alive, but fail to exhibit characteristics that we associate with most life. Viruses, such as influenza viruses, are particles that require a host to replicate, taking over cell machinery to reproduce themselves. As they cannot reproduce on their own, are they alive? In some definitions of life we might include viruses, in some we may not. We can take this argument to extremes. What about a rabbit? Like a virus it too cannot replicate on its own. It requires another rabbit. Is a rabbit dead on its own but a living thing when it is with its mate? Very quickly we arrive at a *reductio ad absurdum* and the discussion goes nowhere.

Figure 1.13 *A wildfire burns organic carbon in oxygen to produce carbon dioxide and water in an identical chemical reaction performed by respiring animals.*

Figure 1.14 *Erwin Schrödinger attempted to define life from a physical perspective.*

Physicist Erwin Schrödinger (1887–1961; Figure 1.14) famously attempted to define life in his seminal book, *What is Life?*, published in 1944 following a series of lectures given at Trinity College, Dublin, in 1943. Quite apart from some fascinating predictions about the nature of the genetic material (that it was an 'aperiodic' crystal – a crystal that lacks long-range order – for which he attempted to estimate the size) he also attempted to get at the nature of life.

Schrödinger recognised that life made 'order from disorder' and, employing the second law of thermodynamics, according to which entropy (a measure of the homogenisation of energy) increases as energy is dissipated in atoms or molecules, Schrödinger explained that life evades the decay to thermodynamic equilibrium by maintaining what he termed 'negative entropy', for example by gathering energy. The phrase 'negative entropy' is rather unwieldy and counter-intuitive. It should be seen more as a popular statement about his views on life than an attempt to define a real physical process. The notion of negative entropy is not limited to life, however. Chemical reactions such as endothermic reactions (reactions that take up energy) are in some sense extracting energy from the surrounding environment to increase order.

Schrödinger's interpretation has a tendency to give the impression that life is 'struggling' against the laws of physics – attempting to maintain order against the ineluctable forces of the Universe that have a tendency to disperse it into disorder. The problem with this view is that it does not explain why life is so successful. If it was such a struggle, why does life seem to be so tenacious and ubiquitous once it got started on Earth?

Another, related, way to view life is to focus less on the organisms themselves, but instead on the process that life is involved with. We can think of life as a process

driving the Universe more efficiently towards disorder than non-biological processes.

To understand this idea, consider my lunch sandwiches. If I place them on a table, and assuming they are not degraded by fungi (which they will be, but this is a thought experiment), it will take a very long time for the energy in their sugar and fat molecules to be released. Indeed, the energy in the sandwich may not be released until it ends up in the Earth's crust from the movements of plate tectonics, heated to great temperatures in the far future when the sugars and fats will be turned into carbon dioxide. However, if I eat the sandwiches, within about an hour or two their contained energy will be released as heat energy in my body, with some portion of it being used to build new molecules. In essence I have accelerated, very greatly, the dissipation of the sandwich into energy. I have enhanced the rate at which the second law of thermodynamics has had its way with the sandwich. Living things represent extraordinary local complexity and organisation, but the process they are engaged in is accelerating the dissipation of energy and the run-down of the Universe. Local complexity in organisms is an inevitable requirement to construct the biological machines necessary for this effect to occur. As the physical universe has a tendency to favour processes that more rapidly dissipate energy, then life is contributing to the processes resulting from the second law of thermodynamics, not fighting it. Seen from this perspective, it is easier to understand why life is successful. It might even be inevitable where organic chemistry allows for it.

But we still haven't *defined* life. One answer to the problem of the definition of life is that life is simply a human word, an artificial definition created by us. It is what philosophers would call 'a non-natural kind', as opposed to a 'natural kind'. The latter term applies to a substance such as gold, whose characteristics can be exactly defined in terms of its physical properties. We can state the molecular mass, melting point and a range of other definitive physical properties of gold that allow for an exact definition of what it is. A good example of a 'non-natural kind' is a 'chair'. If we define it as 'something we can sit on', then does that make my coffee table a chair? And thus we launch into an endless circular discussion about what a chair is. The conversation is rather pointless because ultimately a chair is simply what we define it to be. If that includes coffee tables then so be it. Similarly, maybe life is just a definition that encompasses an interesting segment of all organic chemistry that happens to do certain things, such as reproducing and growing. If we want to include viruses, then so be it; if not, then so be it. Perhaps the crucial point is that we all agree on a definition that we are going to use. Today, astrobiologists tend to think that we can create working definitions that are useful in the search for life elsewhere, such as Joyce's definition quoted earlier.

A working definition of life should not stop us continuing the debate about defining life. Quite apart from refining our understanding of biology, it would be a travesty to destroy some sort of entity elsewhere (if we ever find it) simply because it failed to fit within a narrow definition of life that we have constructed.

1.6   Conclusions

In this introductory chapter we have discussed what astrobiology is. We find it to be an important vehicle to consider life within its cosmic context. It has its own set of scientific questions, such as its most potent: Is there life beyond the Earth? Astrobiology is philosophically an ancient science, but has in recent years gained the empirical knowledge to begin to address questions about life in the Universe, driven in particular by technological advances in space missions. The definition of astrobiology's subject matter, life, has proven to be extremely intractable, but we can produce working definitions of life that allow us to advance our quest to understand the origin, evolution and distribution of life in the Universe.

Further Reading

Books

Catling, D.C. (2014) *A Very Short Introduction to Astrobiology*. Oxford University Press, Oxford.

Crowe, M.J. (2003) *The Extraterrestrial Life Debate 1750–1900*. Dover Publications, New York.

Dick, S.J. (1996) *The Biological Universe*. Cambridge University Press, Cambridge.

Schrödinger, E. (2012) *What is Life?* Cambridge University Press, Cambridge (originally published in 1944).

Papers

Bains, W. (2014) What do we think life is? A simple illustration and its consequences. *International Journal of Astrobiology* 13, 101–111.

Benner, S.A. (2010) Defining life. *Astrobiology* 10, 1021–1030.

Chao, L. (2000) The meaning of life. *BioScience* 50, 245–250.

Chela-Flores, J. (2013) From systems chemistry to systems astrobiology: life in the universe as an emergent phenomenon. *International Journal of Astrobiology* 12, 8–16.

Cleland, C.E., Chyba, C.F. (2002) Defining 'life'. *Origins of Life and Evolution of Biospheres* 32, 387–393.

Des Marais, D.J., Walter, M.R. (1999) Astrobiology: exploring the origins, evolution and distribution of life in the universe. *Annual Reviews of Ecology and Systematics* 30, 397–420.

Machery, E. (2012) Why I stopped worrying about the definition of life … and why you should as well. *Synthese* 185, 145–164.

McKay, C.P. (2004) What is life and how do we search for it on other worlds? *PLoS Biology* 2, e302. doi:10.1371/journal.pbio.0020302.

Nealson, K.H., Tsapin, A., Storrie-Lombardi, M. (2002) Search for life in the Universe: unconventional methods for an unconventional problem. *International Microbiology* 5, 223–230.

Lafleur, L.J. (1941) Astrobiology. *Leaflets of the Astronomical Society of the Pacific* 143, 333–340.

Lederberg, J. (1960) Exobiology: approaches to life beyond the Earth. *Science* 132, 393–400.

2

Matter, the Stuff of Life

Learning Outcomes

➤ Understand the concept of atoms, ions and molecules and their basic structure.
➤ Be able to describe and explain different bonding types: ionic, covalent, metallic, van der Waals' interactions and hydrogen bonding.
➤ Understand and give examples of the role of these bonding types in biological systems.
➤ Understand phase diagrams and the unusual characteristics of the water phase diagram.
➤ Understand how phase diagrams can be used to understand planetary environments and their suitability for liquid water.
➤ Describe extreme states of matter, including plasma and degenerate matter.
➤ Understand some of the interactions between matter and light that are important to astrobiology.
➤ Be able to calculate the energy, frequency and wavelength of light absorbed or emitted for given energy level changes of electrons in atoms and thus the implications for absorption and emission spectroscopy of stars and exoplanets.

2.1 Matter and Life

At the heart of our understanding of physical processes in the Universe, including the physical principles that govern the assembly of life, is the structure of matter. The structure of matter and the interactions of atoms and molecules may not seem to be directly linked to astrobiology. However, to understand how the molecules of life are assembled we need to understand how atoms associate and the types of associations that occur. Ultimately, it is the binding between individual atoms, **ions** and molecules that results in the complex array of structures we call life.

2.2 We are Made of 'Ordinary' Matter

The matter of which stars, planets and life is made is probably less than 5% of all the matter in the Universe. Most of the matter is in more elusive forms such as **Dark Matter**. Dark Matter is not thought to have relevance to life, although of course it is thought to influence the characteristics of the Universe in which life exists. We won't consider it further in this textbook. The small part of the Universe that we call ordinary matter or **baryonic matter** is the material that we will consider in this chapter and it is usually just referred to as 'matter'. It is the matter that makes life.

Matter is constructed of elementary particles which are assembled into atoms. The nucleus of the atom is made of **protons**, which are positively charged, and **neutrons**, which have no charge. They have very small masses (the neutron, 1.675×10^{-27} kg; and the proton, 1.672×10^{-27} kg). These particles are themselves made of the elementary particles, **quarks**. Quarks are in the domain of particle physics and will not be discussed in detail in this book. Around these nuclei are distributed negatively charged **electrons**, creating a complete atom (Figure 2.1).

Astrobiology: Understanding Life in the Universe, First Edition. Charles S. Cockell.
© 2015 John Wiley & Sons, Ltd. Published 2015 by John Wiley & Sons, Ltd.
Companion Website : www.wiley.com/go/cockell/astrobiology.

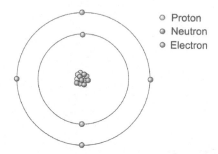

Figure 2.1 *The structure of a typical atom showing the nucleus and electrons. Note that this is not to scale. The electrons are in orbitals about 10 000 times further out than the diameter of the nucleus. Electrons occupy energy levels, depicted here as circles* (Source: Charles Cockell).

2.3 Matter: Its Nucleus

In a given nucleus of any atom there are a total of Z protons with a positive charge and N neutrons of zero charge, where the number of protons constitutes the **atomic number**. The number of protons defines the type of element the atom is. So, for instance, in the biologically important atom, carbon, the number of protons is 6 (which defines it as the element carbon) and the number of neutrons is usually 6. The atom has an atomic number of 6. The **atomic mass number** is the total number of protons and neutrons, which for carbon is usually 12. Protons, because they are positively charged, tend to repel each other. This would cause the nucleus to fall apart. However, the nucleus is held together by the strong nuclear force that operates over very small distances and overwhelms the electrostatic repulsion. In all stable elements there are at

least an equal number of neutrons as protons and in many elements, more neutrons.

2.3.1 Isotopes

Not all atoms of the same element have the same number of neutrons. The value of N can vary. Atoms that have the same number of protons, but different numbers of neutrons are called **isotopes**. As we will see later, isotopes turn out to be very important tools for dating the age of rocks, looking for evidence for life on the early Earth and hunting for signatures of life elsewhere.

Examples of different isotopes are those found in the element carbon, which has three important isotopes (Figure 2.2). Isotopes are often designated by showing the atomic mass number as a superscript on the left-hand side of the element:

Carbon 12 (^{12}C) – 6 neutrons, 6 protons (a stable isotope; which makes up 98.9% of carbon on the Earth).
Carbon 13 (^{13}C) – 7 neutrons, 6 protons (a stable isotope; 1.1% of carbon).
Carbon 14 (^{14}C) – 8 neutrons, 6 protons (an unstable isotope; one part per trillion!).

Carbon 12 (^{12}C) is the most common form of carbon and constitutes more than 98% of carbon in living matter. ^{13}C is rarer but, like ^{12}C, it is a long-lived **stable isotope**. ^{14}C in contrast, with two additional neutrons, is not stable. It is called a radioactive isotope or **radioisotope**. One of the neutrons decays into a proton with the release of an electron and the atom becomes nitrogen 14 (^{14}N). This decay has a half-life of 5730 years. In other words, after 5730 years, half of a sample of ^{14}C will have decayed. The unstable nature of ^{14}C means that it is a small proportion

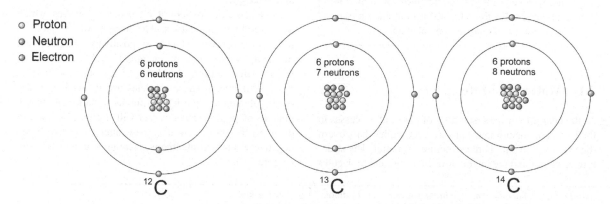

Figure 2.2 *Three isotopes of carbon. They vary in the number of neutrons* (Source: Charles Cockell).

of carbon isotopes but, despite this, it turns out to be enormously useful. As it decays with a known rate, it can be used in dating to determine how old objects are ('carbon dating'). Later in the book we'll see how radioisotopes can be used to put absolute dates on the fossil and geological record of the Earth.

2.4   Electrons, Atoms and Ions

An atomic nucleus is surrounded by electrons. An atom has Z electrons and they orbit at a great distance from the nucleus relative to the diameter of the nucleus itself. A typical atomic nucleus has a diameter of $\sim 10^{-14}$ m, whereas a whole atom has a radius of 2 to 5×10^{-10} m. As we already said that an atom has Z protons, this means that the positive charge on the protons cancels out the negative charge on the Z electrons, making an atom, overall, neutral.

Electrons have something of a split personality. They exhibit particle-like properties and the behaviour of waves, like light. Therefore, to consider electrons orbiting the nucleus in the same sense that a planet orbits a star is not technically correct. Nevertheless, it is a useful analogy and we can consider electrons to occupy orbitals around the nucleus. Each orbital can only take a maximum of two electrons, so as we progress through higher atomic number elements in the periodic table, electrons can be considered to be stacked into sequential orbitals. The orbitals are given letter designations (s, p, d, f, g). The structure of electron orbitals will not be considered in detail in this book, but from the point of view of molecular structure, and the consequences for molecules involved in life, the crucial point to understand is that atoms have a tendency to react until they have full electron orbitals. Once electron orbitals are full then there are no 'extra' or 'missing' electrons to get involved in chemical reactions. This explains the inert and stable characteristics of the noble gases, such as argon and neon, which have full electron shells.

Consider, for example, the sodium atom (Na). It is in Group 1 of the Periodic Table (see Appendix) and has an electron structure written as $1s^2\ 2s^2\ 2p^6\ 3s^1$ (the letters refer to the different orbitals; the superscript shows the number of electrons in each orbital). The sodium has a single electron in orbital 3s. By losing this lone electron the sodium atom becomes more stable because the next shell down is full (the 2p shell is made of three separate shells called the x, y and z shells, each of which have a pair of electrons in them, giving it six electrons in total). However, by losing this electron, the sodium atom gains a net positive charge, as it now has one more proton than electrons. The product of this electron loss, Na^+, is called an **ion**. An ion is an atom that had gained or lost electrons.

Similarly, chlorine, for example, in Group 17 of the Periodic Table, has the electronic structure $1s^2\ 2s^2\ 2p^6\ 3s^2\ 3p^5$. By gaining an electron, the chloride ion (Cl^-) fills its 3p electron orbitals (three sets of orbitals with a maximum of two electrons each) and attains a noble gas electronic configuration, making it more stable.

The tendency of atoms to lose or gain electrons to attain a noble gas electron configuration explains the key features of many bonds that we will look at in the next section.

2.5   Types of Bonding in Matter

With this knowledge of atoms and ions, we now consider how they bond together to construct molecules and ultimately large, complex molecules or **macromolecules**, such as the genetic material **DNA (deoxyribonucleic acid)**, that make up life.

There are five basic types of bonding that hold atoms and molecules together to give rise to the structure of ordinary matter and life. We will consider each of these in this chapter. They are:

- Ionic bonding
- Covalent bonding
- Metallic bonding
- van der Waals' interactions
- Hydrogen bonding.

The first three of these types of bonding are primarily involved in holding atoms together to make molecules, although both covalent and ionic bonds are also involved in holding different parts of biological molecules together, as we shall shortly see. The last two types of bonding are primarily involved in holding molecules together. Let's have a look at some of the features of these bonds and particularly how they relate to life.

2.6   Ionic Bonding

Ionic bonding is the electrostatic force of attraction between positively (+ve) and negatively (–ve) charged ions (primarily between non-metals such as chloride or fluoride ions and metals such as sodium or potassium ions). Most ionic compounds are crystalline solids at room temperature.

Chloride ions
(Cl⁻)   Sodium ions
(Na⁺)

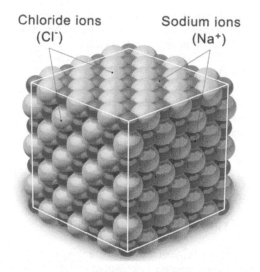

Figure 2.3 *The structure of NaCl showing the alternating sodium and chloride ions* (Source: wikicommons).

The crucial feature of an ionic bond is that each atom either gains or accepts an electron so that the resulting ion has its lowest energy (noble gas-like) configuration. Table salt, NaCl, is a typical example of ionic bonding and you can see its structure in Figure 2.3. In this salt, sodium gives up an electron and chlorine gains this electron so that both ions gain a noble gas configuration, as described in Section 2.4.

In the case of NaCl, the Na^+ and Cl^- ions are attracted to one another, but clearly chloride ions will be repelled from each other, as will sodium ions. The natural configuration they take up is therefore an alternating packed cubic structure. Other similar examples are caesium chloride (CsCl) and sodium fluoride (NaF).

Each +ve ion can attract several −ve ions (and vice versa). Ionic bonds are typically very strong.

We can consider from an energetic (thermodynamic) point of view the stability of these bonds. Let's consider sodium fluoride, NaF. The energy required to break an ionic chemical bond in this structure is about 3×10^{-19} joules (J). We can also calculate the typical thermal energy (the energy in its motion) in a bond at a specified temperature, which is approximated by $\sim k_B T$. k_B is the **Boltzmann constant**, which has a value of 1.381×10^{-23} JK^{-1} and T is the temperature, measured in kelvin. At room temperature (300 K), the value of $k_B T$ is 4.1×10^{-21} J. The ionic bond energy is therefore about 75 times the thermal energy in the bond. In other words, the thermal energy in NaF at room temperature will not cause the ions to break apart.

You need to put in a lot more energy to break apart the ionic bonds, explaining why this compound is stable at room temperature.

2.6.1 Ionic Bonds and Life

Ionic bonds are important for life because they also play a role in holding molecules together.

In Figure 2.4 you can see a typical **protein** chain made up of **amino acids** strung together in a long chain. Some of these amino acids are charged (we will come back to amino acids in Chapter 3).

In proteins, negatively charged amino acids will tend to form an ionic bond with positively charged ones. Figure 2.5 illustrates this bonding.

The negative charge in the aspartic acid, which is one type of amino acid, is attracted to the positive charge in the lysine, another type of amino acid, to form an ionic bond, which helps keep the chains together. This bonding is important in proteins that have functions to perform in cells, for example proteins that are **enzymes** or biological catalysts that play a role in accelerating chemical reactions in life. These ionic bonds help the proteins keep their shape, which is necessary if they are to

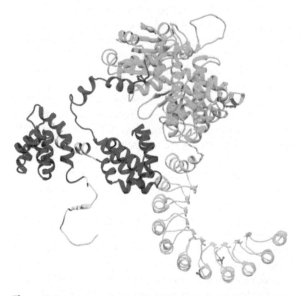

Figure 2.4 *A typical protein. The coloured ribbons and lines depict the chains made of amino acids strung together, a chain that is itself folded together into a complex three-dimensional molecule. This one (called NOD2) is involved in the human immune system* (Source: wikicommons, LPKozlowski).

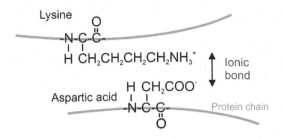

Figure 2.5 *An ionic bond in a protein formed between two amino acids, the positively charged amino acid, lysine, and the negatively charged amino acid, aspartic acid. (Source: Charles Cockell).*

perform their functions correctly. From a very colloquial perspective you can think of ionic bonds as 'bolts' that can help hold together the three-dimensional structure of biological molecules, such as proteins.

2.7 Covalent Bonding

In ionic bonding the central feature of the bond is the *transfer* of an electron from one atom to another to generate two ions. In **covalent bonds** an electron is *shared*. Covalent bonds take place between atoms that are generally close to each other in the Periodic Table and have small difference in **electronegativity** (the tendency of an atom to attract electrons). Similarly to ionic bonds, the sharing of an electron allows the noble gas electron configuration to be attained by both atoms.

The simplest example of a covalent bond is the hydrogen molecule, shown in Figure 2.6. Here each electron in each hydrogen atom is shared with the other, such that together they have a full electron shell (with two electrons).

Covalent bonds are very strong. An example is diamond, a covalent network of carbon. Taking the same logic that we used for ionic bonds, the energy per bond is ~600×10^{-21} J. That's equivalent to 150 times the thermal energy at room temperature ($150 \, k_B T$). In other words, the thermal energy in the bonds of diamond is much lower than the energy needed to break carbon–carbon bonds in diamond. Diamond is very stable at room temperature.

2.7.1 Covalent Bonds and Life

Covalent bonds play a crucial role in biology because they are the bonds that hold atoms together in the vast array of carbon compounds that make life. However, they are also found in specific situations where biologically important molecular structure is required. In analogy to the role of ionic bonds in holding charged amino acids together in protein chains, covalent bonds can form between sulfur-containing amino acids (cysteine and methionine). The sulfur atoms within the amino acids join together and form a **disulfide bridge**. These bonds anchor the structure of the protein, making sure that its three-dimensional shape is maintained.

Figure 2.7 shows covalent bonds holding two protein chains together in a loop.

These bonds are used in some organisms for holding proteins together in extreme conditions. Some microbes that live at very high temperatures have extra covalent bonds in proteins to help maintain their stability, for example in hot springs like the one in Yellowstone National Park shown in Figure 2.8.

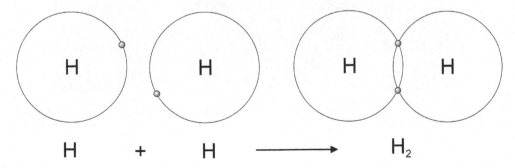

Figure 2.6 *The covalent bond in the hydrogen molecule. The two electrons are shared (Source: Charles Cockell).*

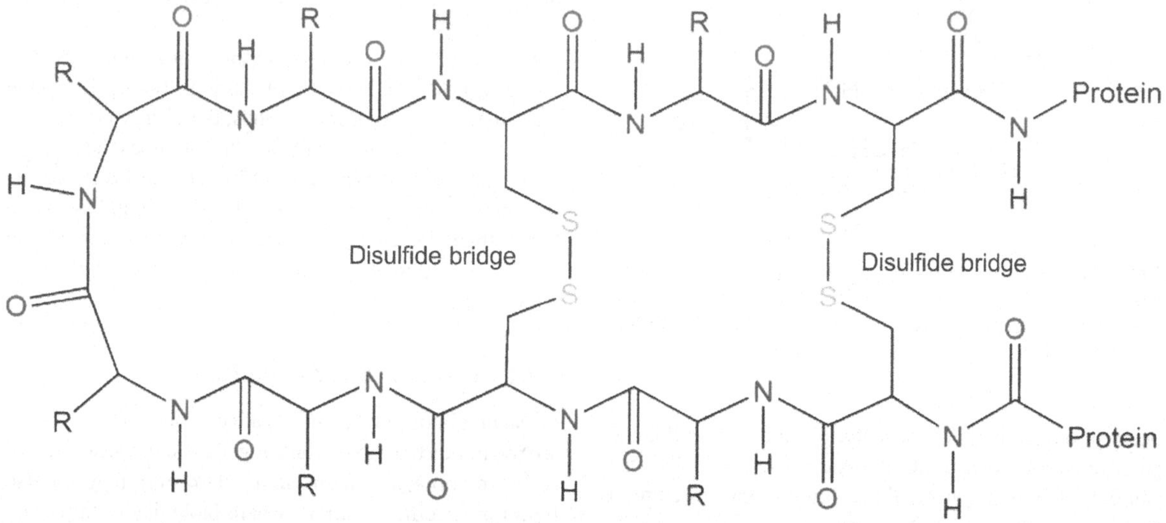

Figure 2.7   *As well as holding atoms together in molecules, covalent bonds link within molecules to provide structure. Here covalent bonds in two disulfide bridges hold together a protein chain made of amino acids in a particular shape. The covalent bonds, as in many similar diagrams, are shown as solid black lines* (Source: Charles Cockell).

Figure 2.8   *Covalent bonds are used in some microbes to stabilise proteins against high temperatures in volcanic pools, such as these in Yellowstone National Park. The microbes are mixed in with the browns and yellows in the spring, colours caused by microbial pigments and minerals such as iron.*

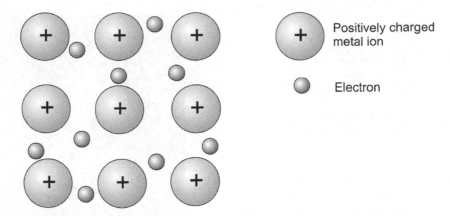

Figure 2.9 *Metallic bonding showing a 'sea' of delocalised electrons around positively charged metal ions* (Source: Charles Cockell).

2.8 Metallic Bonding

Metallic bonding is the type of bonding found in metal elements. It occurs through the electrostatic force of attraction between positively charged ions and their delocalised outer electrons (Figure 2.9). The electrons become detached from the atoms and form a 'sea' between the metal ions. In this way, the atoms, by delocalising the electrons (essentially losing them), achieve a stable noble gas configuration.

Metals form strong bonds. Again, using the reasoning discussed earlier, consider potassium. Its energy per bond is $\sim500\times10^{-21}$ J, which is equivalent to approximately 125 times the thermal energy in the bond at room temperature (~300 K). Many metals are therefore stable at room temperature (e.g. iron, silver).

Metallic bonds are not relevant for life as they are not used by biological systems (although many metal ions are used in enzymes to catalyse chemical reactions or carry out the transfer of electrons, such as copper, iron and magnesium).

2.9 van der Waals' Interactions

We have shown how atoms and molecules can be held together by ionic and covalent bonding. There are two other types of bonding that play a crucial role in holding matter together and play prominent roles in life. These two bonding types are involved in binding between molecules.

One class of interactions between molecules are **van der Waals' interactions** or forces. These interactions are much weaker than covalent and ionic bonds. They are on

Figure 2.10 *The dipoles of two HCl molecules involved in Keesom interactions. They attract each other like tiny bar magnets* (Source: Charles Cockell).

the order of 1% of the strength of a covalent bond. Nevertheless, they are essential in biological systems.

Van der Waals' forces can be divided into three categories.

2.9.1 Dipole–dipole (Keesom) Forces

Molecules have a charge distribution which is never quite even (i.e. it has a directionality or *anisotropy*), which results from the non-even distribution of the electrons and their negative charge. This uneven electron charge distribution results in a small permanent electric *dipole* across the length of the molecule. An example is shown with hydrogen chloride in Figure 2.10. The dipoles attract one another like small bar magnets, resulting in attraction between the molecules.

2.9.2 Dipole–induced Dipole (Debye; Pronounced deh-beye) Forces

van der Waals' interactions can be induced in a molecule that has no charge. In this type of interaction, one of the

Figure 2.11   *An induced dipole in the otherwise uncharged neon is an example of a Debye interaction* (Source: Charles Cockell).

molecules has a charge (HCl), but the other does not, for example here, neon. Nevertheless, a charge can be induced in neon by the presence of the charge in HCl (Figure 2.11), which influences the electron distribution in the neighbouring atom or molecule. As a result the molecules are attracted.

2.9.3   Dispersion Forces

Yet another type of van der Waals' interactions is one in which neutral atoms can attract one another. The charge distribution, even in neutral atoms and molecules, is never quite even. A charge anisotropy will exist on account of the electrons in orbit around the atom, which, at a snap shot in time, will always be slightly unevenly distributed around the atom (Figure 2.12).

This causes a small dipole moment to be established in the atoms or molecules. These **dispersion forces** or London forces are important for attraction between inert gases (Ne, Ar etc.) and covalently bonded molecules (H_2, N_2, CH_4 etc.), but exist between all atoms and molecules to some degree.

2.9.4   van der Waals' Interactions and Life

van der Waals' interactions are involved in the attraction between molecules involved in biological systems, both inside and outside cells. A particularly remarkable example of van der Waals' forces in action in biology is the attachment of a gecko to a wall of glass. Geckos have

Figure 2.12   *Charge imbalance (exaggerated here) around an atom results in a small charge distribution in the atom and creates a dipole* (Source: Charles Cockell).

Figure 2.13   *van der Waals' in action. A gecko attached to a window. The hairs on its feet interact with the surface using van der Waals' interactions.*

many tens of thousands of *setae* – tiny hairs – on their feet, each of which attaches to a wall using van der Waals' interactions between the surface and the tiny projections on the *setae* (Figure 2.13). The combined force is large enough to hold up the gecko and accounts for their ability to attach to a smooth glass surface.

2.10   Hydrogen Bonding

Finally, we come to **hydrogen bonding**, which plays an enormously important role in biological processes. It is found in molecules containing an OH group (a hydroxyl group), including water, ethanol, methanol and numerous other organic molecules.

A hydrogen atom, having one electron, can be covalently bonded to only one atom. However, the hydrogen atom can involve itself in an additional electrostatic bond with a second atom of highly electronegative character such as fluorine or oxygen. This second bond is

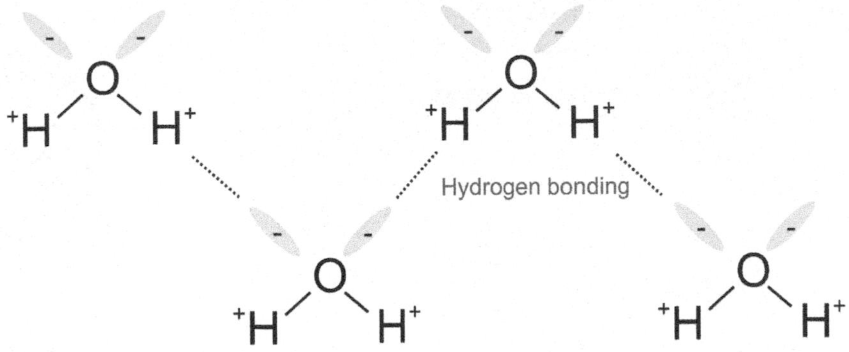

Figure 2.14 *Hydrogen bonding in water ice. The dotted lines show the hydrogen bonding. The green lobes are the electrons on oxygen that take part in the interaction with the hydrogen atom on other water molecules* (Source: Charles Cockell).

a hydrogen bond between two atoms in two different molecules.

How does this work? Let's go through the basic points using water:

- The charge density in a covalent bond such as an OH bond is highly asymmetric and the centre of charge is much closer to the O atom.
- This leaves the H atom behaving like an H^+ (proton) sitting on the surface of an O atom.
- Other electrons on the O atom distribute themselves so as to minimise repulsion. These electrons form lobes of electron density on the opposite side of the O atom to the OH bond.
- Another H_2O molecule orients itself so that its positively charged H is close to these negatively charged electrons.

You can see the result of these interactions much more clearly in Figure 2.14, which shows the hydrogen bonding in water ice depicted in two dimensions.

2.10.1 Hydrogen Bonds and Life

Hydrogen bonding is very prominent in life. In the next chapter we will discuss in more detail the structure of the molecule of information storage in life – **deoxyribonucleic acid**, DNA. For the time being it is useful to understand that DNA is made up of two complementary strands that bind together to form its characteristic double helix shape. These two strands are held together down their middle by hydrogen bonding.

In the diagram in Figure 2.15 you can see the structure of a DNA double helix which has been flattened into a two-dimensional depiction (from its normal three-dimensional helical structure). Along the two edges of the structure you can see the pentagonal **ribose** sugars that make up its backbone linked together with phosphate groups. In the middle are the four **base pairs** that encode the genetic information. They are pointing towards each other, with dotted lines showing the hydrogen bonding. There are two hydrogen bonds for the pairing between adenine (A) and thymine (T) bases and three hydrogen bonds between guanine (G) and cytosine (C).

The hydrogen bonding is just strong enough to hold the strands together so that DNA does not fall apart, but it is just weak enough so that the two strands can be unzipped when they need to be pulled apart for replication of the DNA molecule when cells divide. In Chapter 4 we will investigate the structure and functions of DNA in more detail.

2.11 The Equation of State Describes the Relationship between Different Types of Matter

Having investigated how matter is held together, we now look at types of matter, their interrelationships and some of the important consequences for life.

The matter that makes up the majority of the ordinary matter can be found in different forms. The major forms are solid, liquid, gas and plasma. All of these states of matter are familiar to us. The states of matter are determined by the pressure and temperature regimen under which they exist. This relationship is called the **equation of state**.

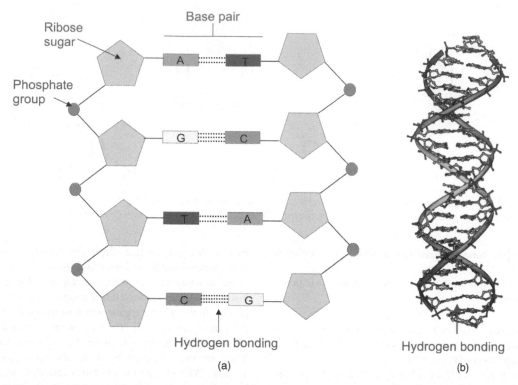

Ribose sugar

Phosphate group

Base pair

Hydrogen bonding

Hydrogen bonding

(a)

(b)

Figure 2.15 *Hydrogen bonding in the molecule, DNA. (a) The dotted lines along the middle of the flattened two-dimensional molecule on the left are the hydrogen bonds that hold the two strands together. On the right (b) is the three-dimensional double helix (Source: wikicommons, Michael Ströck).*

Debate Point: Is the structure of life universal?

Some people say that astrobiologists are narrow-minded and that alien life forms, if they exist, will be constructed in ways unimaginable to us and possibly very differently from life on Earth. Do you agree with this? As you progress through this chapter and later chapters, you might like to consider at what level of hierarchy this statement may or may not be true. Hydrogen bonding, for example, would be expected to be the same anywhere in the Universe as it is determined by the interactions between atoms in the universal Periodic Table. So too with all the forms of bonding explored in this chapter. Surely, therefore, we are on safe ground to say that if life exists elsewhere it would use the same types of bonding to hold its atoms and molecules together? In the next chapter we will look at how molecules in terrestrial life are put together. Are these molecules universal structures? Would we expect cells to be put together using the same basic molecules? We will then progress in a later chapter to look at the evolutionary relationships between whole organisms. Are these universal? Consider at what scale in the structure of life from atoms to communities of organisms you would be confident to say that all life in the Universe would share identical characteristics.

2.12 Phase Diagrams

One very common way to show this equation of state is to focus on just the pressure and temperature (although we could also look at relationships between pressure and volume or temperature and volume, but these are less interesting and give us less information). This depiction of the state of matter as a function of pressure and temperature is called a **phase diagram**. You can see a phase diagram for water in Figure 2.16.

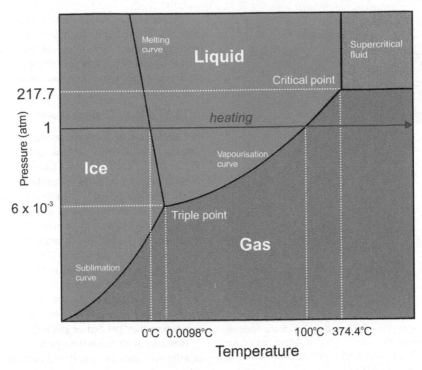

Figure 2.16 *A phase diagram for water. The axes are not drawn to a fixed scaling, but they are drawn to exaggerate values of important features of the diagram* (Source: Charles Cockell).

Let's examine some of the main features of this diagram. On the *x* axis is temperature and on the *y* axis pressure. Follow the red horizontal line from the pressure of 1 atm (our usual experience) shown on the figure from the left to the right.

At atmospheric pressure and low temperatures water is a solid ice. At 0 °C as it is warmed, the water undergoes a phase change as it meets the melting curve and becomes liquid, an experience we are all familiar with when we see ice melting. As we heat it up further, the water undergoes another phase change at 100 °C as it turns into gas in reaching the vapourisation curve. At this point, water boils. If we continued to heat it to very high temperatures (several hundred K), it would turn into a plasma as the electrons are driven off the nuclei (this phase is not shown on the diagram).

There are two features of the diagram to point out. At high temperatures and pressures there is a point called the **critical point**, at which gas and liquid become indistinguishable. Matter in this part of the phase diagram is called a **supercritical fluid**. This state of matter is not used in biological systems, although it has general importance for understanding the behaviour of matter. Some **exoplanets** with high surface pressures and temperatures may even have atmospheres and surfaces covered in supercritical water (Chapter 19).

You will also notice that at pressures lower than atmospheric pressure water boils at lower temperatures than 100 °C. This is consistent with the experience of mountaineers. The higher they go, the lower the temperature at which water boils (making it more difficult to cook vegetables). At the summit of Mount Everest (a height of 8848 m), the boiling point of water is 71 °C. If we continue to reduce the pressure we hit a point on the graph called the **triple point**. This is where all three phases of matter can co-exist. You will notice that in this region of the graph and at lower pressures, if we heat ice it turns directly into gas – it undergoes **sublimation**. There is no liquid phase in this region of the phase diagram.

2.12.1 Matter and Mars

We can get a better understanding of phase diagrams and their importance in astrobiology by looking at another

Figure 2.17 *Ice exposed by the robotic scoop at the Phoenix landing site in the north polar region of Mars. Water ice exposed on day 20 ('sol' 20) had begun to sublime (white arrows) by sol 24* (Source: NASA).

planet – Mars. The image in Figure 2.17 is from the subsurface of the Phoenix landing site on Mars. Phoenix landed on Mars in 2008 in the north polar region at 68.22 °N, 234.25 °E. Water ice was exposed by the robotic arm scoop, which uncovered some of the surface dust. After four days (or four '**sol**', a Martian day) the ice has begun to disappear. It is vapourising. But no liquid water is formed. This tells us the atmospheric pressure on Mars must be at the triple point or lower. In fact, the mean atmospheric pressure on Mars is approximately 6 millibar (mb) [equivalent to 0.0059 atm, or in standard scientific units, 600 pascals (Pa)], near the triple point.

However, Mars hasn't always been like this. There is plenty of evidence that, in the early history of the planet (more than 3.5 billion years ago), there was liquid water on the surface. Valley networks, dried lakes and other features suggest that persistent bodies of liquid water could form on the surface and we will come to this evidence later in the book. This tells us that early in the history of Mars the atmospheric pressure must have been higher to have allowed for the liquid state of water to have existed. Thus, Mars has lost much of its atmosphere since its very early history.

The phase diagram is therefore an elegant way of understanding how the different phases of matter interact. It has great explanatory power as it can tell us about how conditions on planetary scales have changed over time. It allows us to relate geological observations to physical principles.

2.12.2   Phase Diagrams and Life

We might also think about the consequences of the phase diagram for life. Have a look at the 'melting curve' in Figure 2.16 – that is the curve between the solid and liquid regions of water at above the triple point.

The line has a negative gradient – it goes from right to left as we follow it upwards in pressure from the triple point. This is unusual. Most materials have a positive gradient (Figure 2.18). This is caused by the fact that solid water (ice) is less dense than liquid water. For most substances, if you take a point in the liquid portion of the diagram near to the melting curve and imagine pressurising it (i.e. moving it up the *y* axis), it will transition into a solid. In other words, if you pressurise the liquids of most substances they get denser and solidify.

However, with water, if we pressurise its solid phase, the density will increase and it will turn into liquid water (in other words, now take a point on the water phase diagram in Figure 2.16 in the ice region near to the melting curve and move up the *y* axis). It will turn into liquid.

The biological consequences are important because it means that when water freezes it floats on the surface

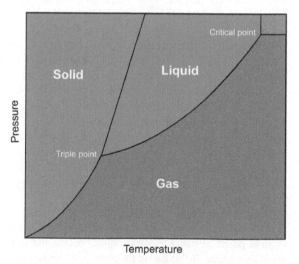

Figure 2.18 *A phase diagram of a 'typical' substance with a positive melting curve gradient* (Source: Charles Cockell).

Figure 2.19 *Fish take advantage of a negative melting curve on the phase diagram of water by surviving under ice (Source: Marsinfomage/flikr).*

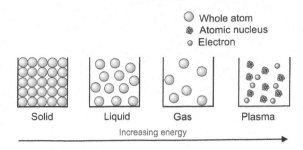

Figure 2.20 *The structure of plasma compared to other states of matter (Source: Charles Cockell).*

of a lake. This allows large multicellular organisms to remain active and alive when the outside temperatures have dropped to below a temperature required to freeze water – the floating ice insulates the water below from freezing (Figure 2.19).

2.13   Other States of Matter

In the previous section we looked at gases, liquids and solids and their interrelations. We explored a few examples of the consequences this can have for planetary sciences and life.

In this section we'll look at some other states of matter. These states do not have any role in the structure of life, but they do have an important role to play in the structure of the Universe, particularly stars, their characteristics and the potential environments around them where planets might exist. They also provide some useful context to the matter that is present in life and show how biology makes use of just a subset of the different states of matter in the Universe.

2.13.1   Plasma

Plasma is an extremely important phase of matter, which makes up about 99% of 'ordinary' matter in the Universe.

Plasma was first discovered by William Crookes (1832–1919), in 1879, but it wasn't called 'plasma' until 1928, when Irving Langmuir (1881–1957) coined the term. It is sometimes called the 'fourth state of matter'. Unlike gases, solids or liquids, plasma has a very large component of ions (Figure 2.20). The electrons in the outer orbitals are stripped away at high temperatures and the result is a collection of ions and electrons, which are no longer bound. As a result of the free electrons, plasma responds strongly to electromagnetic fields, which partly explains the complex patterns it can adopt when exposed to such fields.

Hot plasma is typically at a temperature of thousands of degrees kelvin. An example is gas in the Sun's atmosphere where high temperatures ionise it. Other examples are lightning and the northern lights, Aurora Borealis, in which transient heating of Earth's atmospheric gases ionises atoms. Everyday examples include the glow discharge of a plasma of gas in neon lighting or fluorescent bulbs, which are colder plasmas produced typically at temperatures of ~300–1000 K.

2.13.2   Degenerate Matter

Degenerate matter is a collection of free, non-interacting particles and occurs at very high density or low temperature (Figure 2.21). Degenerate matter is also called a Fermi gas or a degenerate gas, although it behaves very differently from conventional gases. Degenerate matter was first described for a mixture of ions and electrons in 1926 by Ralph Fowler (1889–1944). In electron degenerate matter, electrons are forced to dissociate from parent atoms, generating a material with atomic nuclei and delocalised electrons (in some sense therefore the electron 'sea' of metals could be considered degenerate matter). This stripping away of electrons can be achieved by subjecting a gas to intense pressures. The resulting material has a very high density ($\sim 1000\,\text{kg cm}^{-3}$).

To understand this type of matter it is instructive to move from the atomic and molecular scale, which has

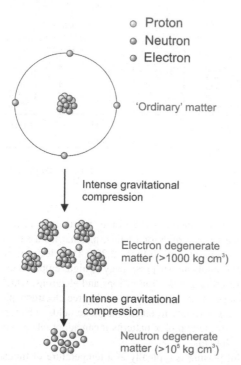

○ Proton
◉ Neutron
◉ Electron

'Ordinary' matter

Intense gravitational compression

Electron degenerate matter (>1000 kg cm³)

Intense gravitational compression

Neutron degenerate matter (>10⁵ kg cm³)

Figure 2.21 *Electron and neutron degenerate matter. In electron degenerate matter the electrons become delocalised from the nuclei of the atoms. In neutron degenerate matter the electrons are forced to combine with protons to form neutrons* (Source: Charles Cockell).

generally attracted our attention so far, to the astronomical scale. It is within astronomical objects that we find this material. This also gives us an opportunity to explore some astronomy.

Electron degenerate matter can be found in **white dwarf stars**. White dwarfs are thought to be the final state of about 97% of the stars in the Galaxy and such a star will be formed from our own Sun when it reaches the end of its life.

Inside a white dwarf, indeed any star, there is a battle going on. On the one hand, gravitational forces have the effect of attempting to collapse the star, but on the other hand gas pressure tends to prevent this gravitational collapse from occurring by pushing outwards. In a white dwarf, this balance is such that the pressures inside the star are sufficient to strip electrons off atomic nuclei and form electron degenerate matter. However, there is an upper limit to the mass of an electron-degenerate object, the **Chandrasekhar limit**, beyond which electron degeneracy pressure cannot support the object against

collapse under its own gravity. The limit is approximately 1.44 times the mass of the Sun (solar masses) for objects with compositions similar to the Sun. If we have higher mass than this, then the star will collapse further.

If we continue compressing matter the energy of the electrons increases to the point where it is energetically favourable for them to combine with protons to produce neutrons (Figure 2.21) and neutron degenerate matter is formed. The density of this material is even greater than electron degenerate matter ($>10^5$ kg cm^{-3}). This is the material from which **neutron stars** are constructed. A neutron star has a diameter on the order of one-thousandth that of a white dwarf. The interior structure of these objects is uncertain, but one model of such a structure is shown in Figure 2.22.

Neutron stars spin very rapidly with their enormous magnetic fields generating beams of radio or light energy that, if pointing in the direction of the Earth, can be detected as **pulsars** and have a frequency between about 5–650 s.

Amazingly, even neutron stars have not escaped the attentions of astronomers and planetary scientists as abodes for life. The popular article by Frank Drake, *Life on a Neutron Star*, published in *Astronomy* in December 1973 has become something of a classic. This was followed by science fiction stories, for example Robert Forward's books, *Starquake* and the *The Dragon's Egg*. These are depictions of the 'cheela', a civilisation of tiny beings that live on the surface of a neutron star under its intense gravity. They intervene to help some hapless humans in orbit around their star who are suffering a malfunction on their spaceship. These ideas are fascinating and thought-provoking. However, neutron stars are unlikely places for life. As the book progresses you can consider some of the factors that might cause you to agree or disagree with this statement.

Neutron degenerate matter is associated with huge gravitational forces. When a neutron star has a close companion, it pulls material to it. This material flies down to the surface of the star and crashes onto the surface, releasing energy. This energy is emitted mostly as X-rays and is modulated with the neutron star spin.

Typical characteristics of a neutron star are that they have a mass of greater than 1.4 solar masses, a radius of 10–80 km and a density of $\sim 10^{11}$ kg cm^{-3}.

There is an upper limit to the mass of a neutron-degenerate object, the Tolman–Oppenheimer–Volkoff limit, which is analogous to the Chandrasekhar limit for white dwarfs. The precise limit is unknown. Above this limit, a neutron star may collapse into a black hole.

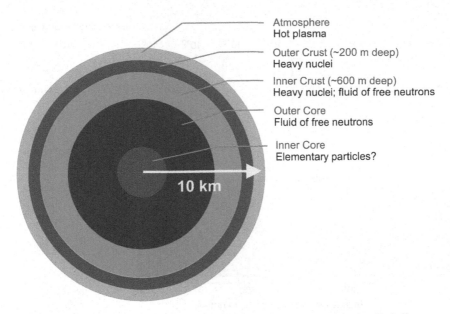

Figure 2.22 A hypothetical structure of a neutron star (Source: Charles Cockell).

The composition of the interior of a black hole is really unknown. Does the interior of black holes contain quark degenerate matter, where the neutrons have themselves disintegrated into their constituent subatomic quarks? Quark degenerate matter has also been suggested to occur in hypothetical 'quark stars'.

At the centre of a black hole is thought to be a singularity or singularity ring where density is infinite. This can be regarded as some of the most extreme matter in the Universe.

2.14   The Interaction between Matter and Light

Matter has many complex interactions with radiation and different regions of the electromagnetic spectrum, from the heating of objects exposed to infra-red to the penetrating nature of X-rays used to image inside objects. However, here we will look at just one important interaction – the absorption and emission of electromagnetic radiation, which is of special importance to astrobiologists. The emission of light tells us about the structure of other stars, galaxies and the interstellar medium. The absorption of light is used to determine the presence of gases in the atmospheres of exoplanets and will be used to search for life. We will discuss this more in Chapter 19.

Light can be understood as a wave with a wavelength, λ (Figure 2.23). The frequency with which the peaks or troughs of those waves pass a given point is the frequency of light, f, and the relationship between these two properties is given by:

$$\lambda = c/f \qquad (2.1)$$

where c is the speed of light.

The wave description of light is quite adequate for many applications, but it fails to explain some basic observations. If we shine light (just a form of energy) at a surface the energy imparted to the electrons can cause them to jump energy levels or be lost entirely. However, we can only induce electrons to be emitted from the atoms at particular wavelengths of light. Furthermore, increasing the intensity of light of an inappropriate frequency does not allow electrons to be emitted. These observations suggest that light comes in packages or **quanta**, some frequencies having enough energy and others not to release the electrons or cause them to change energy levels within an atom. These different energy levels are referred to as quantum levels. The energy difference between these quantum levels can be given by:

$$E = hf \qquad (2.2)$$

where f is the frequency of the light that will allow an electron to jump between the energy levels and h is Planck's constant (6.626×10^{-34} J s^{-1});

or $E = hc/\lambda$ if we want to express this in terms of the wavelength of light.

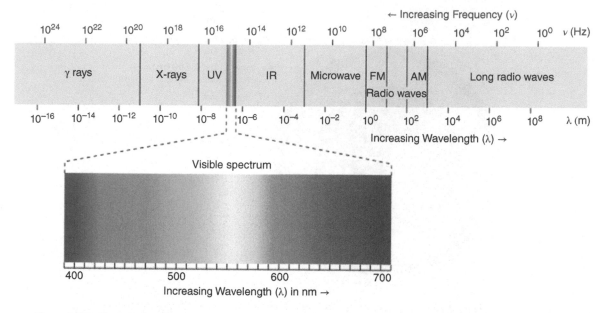

Figure 2.23 *The wavelength and frequency of different types of electromagnetic radiation* (Source: wikicommons).

The energy required to make an electron jump from one energy level to another (resulting in absorption of that energy; Figure 2.24a), say from level n_1 to energy level n_2, is therefore equal to:

$$E(n_1) = E(n_2) + hf \qquad (2.3)$$

Note that the energy in any given energy level is usually given as a negative number (Figure 2.24a). Hence the energy of the smaller integer energy level is subtracted from the larger integer energy level to get the energy difference between them.

This equation also tells us what frequency of photon will be emitted when an electron *drops* an energy level (from n_2 to n_1) and emits its energy as a photon. This is given by $f = [E(n_1) - E(n_2)]/h$. This is conceptually shown in Figure 2.24a as the emission caused when an electron drops an energy level.

We can also calculate the value of E, the energy of a given energy level, from first principles by considering the structure of the atom itself. The value of E in electronvolts (eV) at any energy level can be calculated by the equation given below, which we won't derive here, but will state. It is obtained by thinking about an electron as orbiting the atom much like a planet orbits a star. Although in reality this is not how electrons behave, it is a good enough approximation for deriving some of the basic properties of atoms. This approach to considering electrons is

known as the Rutherford–Bohr model of the atom, after the physicists Ernest Rutherford (1871–1937) and Niels Bohr (1885–1962) who elaborated this model.

The energy of an electron at a given energy level in an atom is given by:

$$E(n) = -Z^2 m_e e^4 k_e /2n^2 h^2 \qquad (2.4)$$

where Z is the atomic number (or number of electrons or protons in the atom), m_e is the mass of an electron, e is the charge of the electron, k_e is Coulomb's constant, n is an integer that defines the energy level and h is Planck's constant.

This approximates to:

$$E(n) = -13.6 Z^2 /n^2 \text{ eV} \qquad (2.5)$$

The value, 13.6, is also known as the **Rydberg unit of energy (Ry)**.

Taking the formula we have just seen (Equation 2.5) we can observe that the energy needed to make an electron jump from one energy level to another (or the energy of a photon given off when an electron drops from one energy level to another) is given by:

$$E = E(n_1) - E(n_2) = Z^2 Ry(1/n_1^2 - 1/n_2^2) \qquad (2.6)$$

where n_1 and n_2 are the energy levels and $n_1 < n_2$.

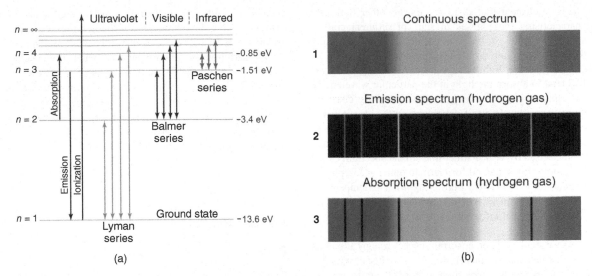

Figure 2.24 *(a) Energy levels in the hydrogen atom and the transition series associated with the movement of electrons between these levels. Also illustrated are transitions that cause emission lines, absorption lines and ionisation, involving the loss of an electron from the ground state. (b) Emission lines (2) and Fraunhofer (absorption) lines (3) in a spectrum, compared to a continuous spectrum (1).*

This expression gives the energy in **electron volts**. The value can be multiplied by 1.602×10^{19} to give the value in joules.

2.14.1 The Special Case of the Hydrogen Atom

For hydrogen the value of Z is 1, which leads us to the general formula for the energy required for an electron to change energy levels as:

$$E = E(n_1) - E(n_2) = Ry(1/n_1^2 - 1/n_2^2) \quad (2.7)$$

The wavelength of light that would be emitted or taken up by a transition between these two energy levels is therefore also given by:

$$\lambda = Ry(1/n_1^2 - 1/n_2^2)/hc \quad (2.8)$$

This equation can also be expressed in a different way, known at the Rydberg formula (for hydrogen):

$$1/\lambda = R_\infty(1/n_1^2 - 1/n_2^2) \quad (2.9)$$

where the Rydberg Constant (R_∞) is equal to Ry/hc and has a value of $1.097 \times 10^7 \text{ m}^{-1}$.

A special case of this equation is when hydrogen is ionised. In this case, the electron is completely removed from the atom $(n_2 = \infty)$ from a given energy level (n). If the electron is in the ground state (i.e. $n_1 = 1$) then

the energy required for ionisation is equal to Ry eV or 13.6 eV.

The different energy levels to be found in hydrogen are illustrated in Figure 2.24a, with examples of a transition that leads to an emission line and one that leads to an absorption line. Also illustrated is the principle of ionisation where an electron has sufficient energy to rise from an energy level and leave the atom entirely.

2.14.2 Uses to Astrobiology

The implications of the above equations and ideas to astrobiology can now be explained.

When a hot gas absorbs light the electrons in its different gases will jump energy levels and drop back down again, emitting light at discrete wavelengths. These give rise to emission spectra. Each gas has a very characteristic emission spectrum that is a finger print, if you will, of its atomic structure, with lines arranged across the spectrum corresponding to the different energy levels of its electronic orbitals (Figure 2.24b). This **emission spectrum** can be used to identify gases in astronomical objects. These effects are pressure dependent and at very high pressures, hot gases tend to produce continuous spectra without characteristic emission lines.

The detection of gases in the atmospheres of distant stars, beginning in the late nineteenth century, finally confirmed that distant stars are other Suns. By characterising these gases in stars of different colours and luminosities it became possible to systematically categorise stars into different spectral types.

By contrast, if light travels through a cold gas, the gas will tend to absorb the light at the particular wavelengths corresponding to the energies needed to make electrons jump energy levels, itself dependent on the atomic structure of the different gases. These create a characteristic **absorption spectrum**. Absorption spectra from stars are characterised by very distinct lines in the spectrum called Fraunhofer lines, named for German optician Joseph von Fraunhofer (1787–1826; Figure 2.24b). The temperature and density within a star affect the intensity of the lines and so they can reveal information about the characteristics of a given star.

As we shall see later, absorption spectroscopy allows us to investigate the gaseous composition not just of stars, but of planetary atmospheres, such as those of extrasolar planets. By using light from a star that has travelled through a planetary atmosphere we can determine the gaseous composition of the atmosphere and seek gases that are signatures of life (Chapter 19).

We leave the summary of this last section to German physicist Gustav Kirchhoff (1824–1887) who listed, before the structure of the atom was understood, what are sometimes called the three laws of spectroscopy:

1. A hot solid object gives off a continuous spectrum (called a blackbody – we will return to this in Chapter 8).
2. A hot gas, under low pressure, produces a bright-line or emission spectrum, which depends on the energy levels of the atoms in the gas.
3. A hot solid object surrounded by a cooler gas (e.g. light from a star passing through an exoplanet atmosphere) produces light with a spectrum (an absorption spectrum) that has gaps at discrete wavelengths depending on the energy levels of the atoms in the gas.

2.15   Conclusions

In this chapter we have explored the basic structure of matter relevant to astrobiology. We have investigated the five major bonding types in matter. In each of these cases we assessed the role of these bonding types in life with some examples. We saw how matter can be described in different states and by using simple phase diagrams we can explain a variety of features of the physical characteristics of extraterrestrial environments and the

behaviour of life. We looked at some unusual states of matter, which although not part of living things, certainly are distributed across the Universe and have consequences for the astronomical environment in which life resides. Finally, we investigated the principles of spectroscopy and how light interacts with matter in ways that are useful for scientists seeking to investigate the properties of gases in distant stars or planets. Equipped with this knowledge, we can now investigate how atoms and molecules are put together to make the basic molecules of life.

Further Reading

Books

Pavia, D.L., Lampman, G.M., Kriz, G.S. (2000) *Introduction to Spectroscopy*. Brooks/Cole, Kentucky.

Tabor, D. (1991) *Gases, Liquids and Solids: And Other States of Matter*. Cambridge University Press, Cambridge.

Papers

Attard, P. 1996. Patterns of hydrogen bonding in water and ice. *Physics A* **233**, 742–753.

Auffinger, P., Hays, F.A., Westhof, E., Ho, P.S. (2004) Halogen bonds in biological molecules. *Proceedings of the National Academy of Sciences* **101**, 16789–16794.

Autumn, K., Sitti, M., Liang, Y.A., Peattie, A.M., Hansen, W.R., Sponberg, S., Kenny, T.W., Fearing, R., Israelachvili, J.N., Full, R.J. (2002) Evidence for van der Waals adhesion in gecko setae. *Proceedings of the National Academy of Sciences* **99**, 12252–12256.

Burrows, A.S. (2014) Spectra as windows into exoplanet atmospheres. *Proceedings of the National Academy of Sciences* **111**, 12601–12609.

Drake, F.D. (1973) Life on a neutron star. *Astronomy* December, p. 5.

Fidler, A.L., Vanacore, R.M., Chetyrkin, S.V., Pedchenko, V.K., Bhave, G., Yin, V.P., Stothers, C.L., Rose, K.L., McDonald, W.H., Clark, T.A., Borza, D.B., Steele, R.E., Ivy, M.T., Aspirnauts, T., Hudson, J.K., Hudson, B.G. (2014) A unique covalent bond in basement membrane is a primordial innovation for tissue evolution. *Proceedings of the National Academy of Sciences* **111**, 331–336.

Gobre, V.V., Tkatchenko, A. (2013) Scaling laws for van der Waals' interactions in nanostructured materials. *Nature Communications* **4**, DOI 10.1038/ncomms3341.

Holm, R.H., Kennepohl, P., Solomon, E.I. (1996) Structural and functional aspects of metal sites in biology. *Chemical Reviews* **96**, 2239–2314.

Horowitz, S., Trievel, R.C. (2012) Carbon–oxygen hydrogen bonding in biological structure and function. *Journal of Biological Chemistry* **287**, 41576–41582.

Jones, E.G., Lineweaver, C.H. (2012) Using the phase diagram of liquid water to search for life. *Australian Journal of Earth Science* **59**, 253–262.

Jones, E.G., Lineweaver, C.H., Clarke, J.D. (2011) An extensive phase space for the potential Martian biosphere. *Astrobiology* **11**, 1017–1033.

Lattimer, J.M., Prakesh, M. (2004) The physics of neutron stars. *Science* **304**, 536–542.

Leckband, D., Israelachvili, J. (2001) Intermolecular forces in biology. *Quarterly Reviews of Biophysics* **34**, 105–267.

Raines, R.T. (1997) Nature's transitory covalent bond. *Nature Structural Biology* **4**, 424–427.

Tadmore, R. (2001) The London–van der Waals interaction energy between objects of various geometries. *Journal of Physics: Condensed Matter* **13**, L195–L202.

3

Life's Structure: Building the Molecules

Learning Outcomes

➤ Understand that the basic elements required by life are carbon, hydrogen, nitrogen, oxygen, phosphorus and sulfur (CHNOPS). Other elements can also be used for a range of functions specific to particular types of life.

➤ Understand why carbon is the most versatile backbone of molecules – life is 'carbon-based'.

➤ Know the characteristics of the major classes of biological macromolecules: proteins, carbohydrates, lipids and nucleic acids.

➤ Understand that some classes of molecules such as proteins and carbohydrates are chiral.

➤ Understand why water is the most versatile and plausible solvent for life.

➤ Understand some of the arguments for, and some of the limitations of, proposed alternative building block elements such as silicon and alternative solvents such as ammonia.

3.1 Building Life

In the previous chapter, we saw how different bonding types allow for the association of atoms, ions and molecules. Life assembles these units, a bit like a construction kit, into **macromolecules** (very large molecules) which themselves form the architecture of cells and in multicellular organisms like you and me, organs and whole organisms.

Building from the previous chapter we could start with a very simple question: what is the minimum set of elements that life needs to assemble molecules? The matter from which life is constructed is unsurprisingly restricted to what is in the Periodic Table and therefore its options are not unlimited.

Furthermore, we can assume that the Periodic Table is universal. This seems a reasonable working assumption as no astrophysicist has yet observed a place in the Universe that seems to have an entirely alien chemistry (in fact quite the contrary – as we shall see later, many elements of importance to life, such as carbon, seem to be surprisingly abundant throughout the Universe). Therefore, the observations we make about the chemistry of terrestrial life are likely to reveal important insights into the potential chemistry of life anywhere in the Universe.

3.2 The Essential Elements: CHNOPS

If we survey the molecules from which life is constructed, we find six that are found in all living things and have often been described as the 'building block' elements of life (Figure 3.1). They are C (carbon), H (hydrogen), N (nitrogen), O (oxygen), P (phosphorus) and S (sulfur). They are sometimes remembered as the odd word, CHNOPS or SPONCH – take your pick.

Carbon is used as the backbone of most complex molecules in life as we shall shortly see. Carbon comes in both inorganic and **organic carbon** forms. Inorganic carbon is carbon that lacks C–H bonds and includes

Astrobiology: Understanding Life in the Universe, First Edition. Charles S. Cockell.
© 2015 John Wiley & Sons, Ltd. Published 2015 by John Wiley & Sons, Ltd.
Companion Website : www.wiley.com/go/cockell/astrobiology.

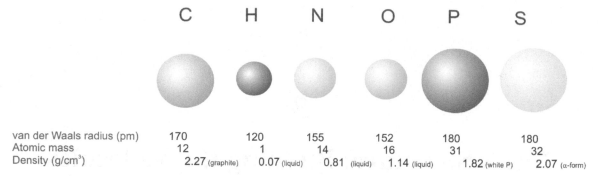

		C	H	N	O	P	S
van der Waals radius (pm)		170	120	155	152	180	180
Atomic mass		12	1	14	16	31	32
Density (g/cm³)		2.27 (graphite)	0.07 (liquid)	0.81 (liquid)	1.14 (liquid)	1.82 (white P)	2.07 (α-form)

Figure 3.1   *The six ubiquitous elements of life, CHNOPS. The van der Waals radius in picometres (pm) is the effective radius of the atom when modelled as a hard sphere. Also shown is the atomic mass and density for certain forms of the elements. The coloured spheres show the relative sizes of the atoms* (Source: Charles Cockell).

minerals such as calcium carbonate ($CaCO_3$). The form we are interested in, at least in terms of the assembly of life, is organic carbon, or carbon combined with elements such as H, N, and O. The name is rather unfortunate. It was originally derived from the view that such carbon could be found only in living things, although we now know that organic carbon compounds can be formed by non-biological processes.

What about the other five elements? Hydrogen is bound to many atoms in –CH, –NH_2, –SH groups and it is common in all organic compounds found in life (as well as being in life's solvent, water, in addition to oxygen). Nitrogen is found in many carbon-based ring structures and **amino acids** and confers a greater degree of complexity on the range of organic molecules possible. Oxygen is an essential atom in many organic compounds. It is found in ring structures, in '–OH' groups (for example in sugars and alcohols) and other bonds – again expanding the range of organic complexity. Phosphorus is found in the energy molecules of life and the **nucleic acids**, such as DNA. Sulfur is found in some amino acids, such as cysteine.

Of course, life gathers together many other elements, such as some metals. These include iron and copper which form part of proteins involved in energy acquisition in life (discussed in Chapter 5). In humans, iron is an essential component in haemoglobin, the protein that carries oxygen around in your blood. Calcium is used in bones. These elements are more specific and can be commandeered by life in different situations and for different biochemical purposes. So we can think of life as using CHNOPS to build its basic structures and then rummaging around in the Periodic Table to find other elements that could be useful in putting together particular molecules that perform specialised functions.

3.3   Carbon is Versatile

The elements of life come together in combinations to form molecules. The most common backbone of these molecules is carbon, for example in the simple amino acid, glycine in Figure 3.2, in which two carbon atoms form the core of this molecule. Most of these atoms, as shown for glycine, are bound together covalently. The abundance of carbon in biological molecules means that we refer to life on Earth as 'carbon-based'.

There are a variety of reasons why carbon is the basic building block of life. It forms stable bonds with many compounds, such as H, N, O, P and S, but these bonds are not sufficiently stable to make it necessary to use large quantities of energy to break them (if that were the case, it would be difficult for life to break down and metabolise compounds for food). The energy it takes to make (or break) bonds with H (413 kJ mol^{-1}), N (308 kJ mol^{-1}), O (360 kJ mol^{-1}), S (272 kJ mol^{-1}), P (264 kJ mol^{-1}) and other carbon atoms (347 kJ mol^{-1}) is quite similar, which means that carbon can interchange between these atoms without much energy being required or released. This gives carbon versatility in breaking down and forming new complex molecules. Furthermore, carbon forms stable carbon–carbon double and triple bonds, which again increases the diversity of possible compounds.

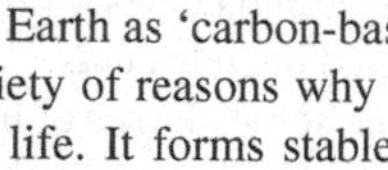

Figure 3.2   *The organic molecule, glycine, the simplest amino acid* (Source: Charles Cockell).

Molecules containing carbon range in two dimensional structure from chains to rings. The simplest carbon molecules are alkanes [with the formula $C_nH_{(2n+2)}$]. If $n = 1$ then the molecule is CH_4 or methane, which is very common on Saturn's moon, Titan, and underground on Earth where it is produced by microbes called **methanogens**. If $n = 2$ the molecule is C_2H_6, which is ethane, another common organic molecule in the Universe. The substitution of hydrogen with other atoms results in functional groups of wide use in different biochemical functions. For example, esters have the general formula –COO–R (where R is an alkyl group – an alkyl group is any group with the general formula C_nH_{2n+1}). These turn up in the membranes of cells. Amino (–NH$_2$) and carboxyl (–COOH) groups attach to carbon to form amino acids in proteins. The phosphate group, –PO$_4$, makes a whole variety of phosphate-containing molecules which turn up in the genetic information, DNA and energy storage molecules. Alcohols are carbon compounds with an –OH group, used by microorganisms in energy yielding reactions. And so the list goes on, a vast array of compounds made possible by the covalent bonding of carbon to the CHNOPS elements, including other carbon atoms.

3.4 The Chains of Life

If we had to identify just one feature of life that stands out when we are discussing the formation of molecules, we would probably say that it has a propensity to form chains. Perhaps this isn't surprising. Life is complex and if we want to build complex molecules we would intuitively suggest that the best way to do this is to take simple molecules and string them together into more complex chains. All of the major classes of molecules in life result from this process of putting single molecules (monomers) together into chains (polymers) in the process of polymerisation. Here we shall look at the four major classes of molecules from which life is made.

3.5 Proteins

The first example is proteins. Proteins are composed of chains of **amino acids**. These chains are referred to as **polypeptides**. Amino acids in proteins are carbon molecules made from an amine (–NH$_2$) and a carboxylic acid (–COOH) attached to a central carbon called an alpha-carbon. The alpha carbon also has a side group attached to it. It is this side group that is altered in the wide variety of amino acids found in biology (Figure 3.3). Although there are a vast variety of amino acids in nature, with over 500 known, only 20 of these compounds are commonly used in life (Figure 3.3), with two others more rarely used (selenocysteine and pyrrolysine). Life therefore uses a very select number. We don't really know why this is the case. It might be like asking why someone building a house doesn't use all of the wonderful variety of bricks that are available from their local garden centre. It makes no sense to use all of them because that would result in incompatibility between brick types and too much complexity to get the job done. If 20 amino acids allows for a life form to come into existence and reproduce, then there is no evolutionary selection pressure to use more amino acids.

In Figure 3.3 you can see the variety of amino acids created by changing the side group. Amino acids, as well as being designated by their full names, are also given a three-letter code (Figure 3.3) and a one-letter code (Appendix). The side groups confer upon amino acids different properties. Some are polar, which means they dissolve in water on account of their ability to take part in hydrogen bonding. They include serine, asparagine and histidine. Some are **hydrophobic** (they repel water), such as alanine, valine and leucine. Some are charged amino acids and play an important role in ionic bonds. They include lysine, aspartic acid and arginine. These properties and, by consequence, their different interactions with other molecules are essential for generating the vast number of ways in which the amino acid chains can be folded together.

Amino acids are assembled together in chains through **peptide bonds**, whereby the amino and carboxyl group of the amino acids react to form a bond with the release of a water molecule in a dehydration reaction (Figure 3.4). This process of polymerisation can go on until proteins are built that have many hundreds of amino acids.

The exact sequence of the amino acids determines what the protein will do in the cell. An obvious question to ask is how this long chain of amino acids (the **primary sequence**) is turned into something useful?

Some of the charged amino acids will bind with one another from different places on the chain to form ionic bonds (e.g. the positively charged aspartic acid binds ionically to the negatively charged lysine), as we saw in Chapter 2. Some amino acids will covalently bind to each other to form bonds, for example two cysteine amino acids that contain sulfur form a disulfide bridge, as also discussed in Chapter 2. Thus, the primary sequence comes together to form hairpins, helices and other

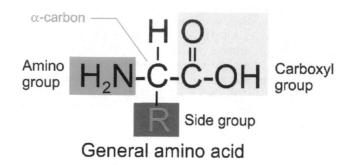

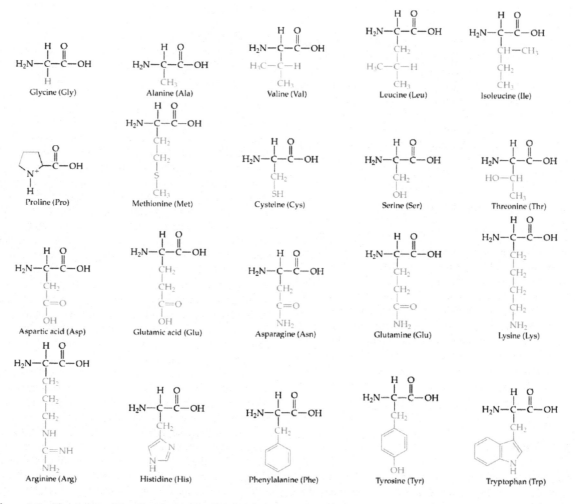

Figure 3.3   *The variety of the 20 main amino acids in life showing the different chemical structure of the side groups (in blue). Also shown in brackets is the three-letter designation of each amino acid* (Source: wikicommons).

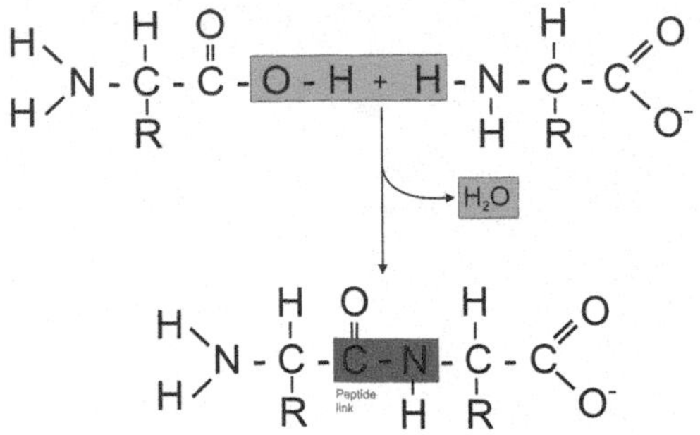

Figure 3.4 *The formation of a peptide bond between two amino acids. This dehydration reaction (involving the release of a water molecule) allows for the assembly of polypeptide protein chains* (Source: Charles Cockell).

three-dimensional structures which are referred to as the secondary structure. The complete atomic arrangement within a whole protein is called the tertiary structure. It is this three-dimensional structure that can now do useful biological work. The three-dimensional arrangement of amino acids in the protein can be capable of carrying out a chemical reaction that makes something useful like a sugar. The site within a protein where the amino acids are configured in such a way that their side groups can bind reactants and catalyse a chemical reaction is called the **active site**. Sometimes individual proteins come together to make an even larger protein. These protein subunits form a **multimeric** structure and we refer to this arrangement as the quaternary structure.

Proteins are involved in many functions. As intimated in the previous paragraph, they can act as catalysts, carrying out chemical reactions. These proteins are called **enzymes**. Proteins are also used in cell membranes to transport molecules, as structural materials and energy transfer molecules.

3.6   Chirality

An important feature of amino acids is that all of the biological amino acids, apart from glycine, are **chiral**. The best, and iconic, way to explain this is with your hands. Place your left hand on a table palm down. Now place your right hand on top of it, palm down. They don't exactly overlap. Your left hand's thumb sticks out to the right, your right hand's thumb to the left. That is because they are mirror images. They are non-superimposable (Figure 3.5).

Given four different side chains or more, molecules too can be assembled into left- and right-handed forms. They are said to be **isomers**, which are chemical compounds with the same chemical formula, but different structures. The conventional way to classify chiral molecules is based on the direction that polarised light, when shone at the molecules, is rotated (Figure 3.6). The different mirror images tend to rotate it one way or the other. If it is rotated to the left we call it a levorotatory molecule or the 'L' form. If it is to the right, or dextrorotatory, we call it the 'D' form. When we have an equal mixture of both L and D molecules we say that the mixture is **racemic**.

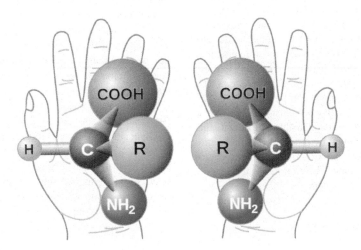

Figure 3.5 *Chirality illustrated with hands and the generic structure for amino acids* (Source: wikicommons).

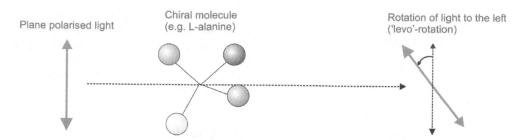

Plane polarised light   Chiral molecule (e.g. L-alanine)   Rotation of light to the left ('levo'-rotation)

Figure 3.6   *Chiral molecules have a tendency to rotate polarised light in particular directions. Schematically, the rotation of light to the left (levorotatory) is shown here with the amino acid, L-alanine* (Source: Charles Cockell).

Debate Point: Life with a different chirality

You might like to contemplate whether you agree with the statement in the last paragraph of this section. Is it possible to construct life forms in which D-amino acids are the dominant amino acids in protein-like structures? Or would life elsewhere also be biased towards L-amino acids? We will return to this when we discuss astrochemistry (Chapter 9) and after that chapter you might like to revisit this question. What information can you gather to shed light on why life on Earth chose L-amino acids? What range of problems and even advantages could you think of for a lifeform that used a racemic mix of D- and L-amino acids in its biochemical architecture?

Bonner, W.A. (1991) The origin and amplification of biomolecular chirality. *Origins of Life and Evolution of the Biosphere* **21**, 59–111.

Breslow, R., Cheng, Z.-L. (2009) On the origin of terrestrial homochirality for nucleosides and amino acids. *Proceedings of the National Academy of Sciences* **106**, 9144–9146.

Specifically, we use the term **enantiomer** to refer to two molecules that are chiral or optically active isomers.

Now a strange thing about life on the Earth is that almost all the amino acids used in life are of the 'L' form. The other form, 'D' (dextrorotatory), is very rare, although D-amino acids are found in the cell membranes of bacteria and their function is not entirely understood. They will reappear in the next chapter. The reason why life predominantly uses one form of amino acids is not fully understood, but it might have something to do with molecular recognition. As many biochemical interactions involve one molecule slotting into another to carry out chemical reactions, a system of life that was a mix of L and D forms would cause great complexity because we would need proteins with active sites that could recognise either the L or D form. It seems plausible to speculate that once a biochemical architecture emerged that had a preponderance to use either the L or D form, then that choice would have been perpetuated. Life on Earth developed a biochemistry based on the L form.

3.7   Carbohydrates (Sugars)

Chains are an important component of many other crucial characteristics of life. For example **carbohydrates** are used as structural support and as energy molecules (which is why people on diets are interested in how much carbohydrate they eat). Carbohydrates are hydrated carbon atoms with the generic formula CH_2O and multiples thereof. They are made up of chains of individual sugars such as glucose, $C_6H_{12}O_6$ or fructose, which has the same chemical formula as glucose, but a different structure (it is an **isomer**; Figure 3.7).

A sugar with six carbon atoms is called a hexose (hence despite their different structures, glucose and fructose are hexose sugars as they both contain six carbon atoms). A sugar with five carbon atoms is called a pentose. Ribose is such an example (Figure 3.7).

Sugars join together through a glycosidic bond to form chains, analogous to the peptide bonds in proteins. O-Glycosidic bonds are oxygen-bridged links between sugar molecules (Figure 3.8). Figure 3.8 shows the example of maltose formed by a link between the –OH bond of one glucose (the 4 carbon position) and the hemiacetal group of another glucose (that's the carbon with an –OH group and a link to the oxygen atom in the sugar ring – the 1 carbon position in glucose). This explains why they are sometimes called 1–4 glycosidic linkages. Alternatively, a link between the 1 carbon of a glucose molecule and the 2 carbon of a fructose molecule

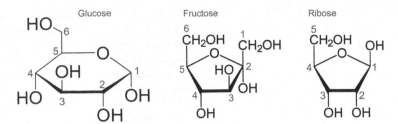

Figure 3.7 *The molecular structure of the sugars, glucose, fructose and ribose, showing the numbering convention on the carbon atoms. Note that the carbon atoms within the rings are not indicated with a letter, but occur where the numbers are shown* (Source: Charles Cockell).

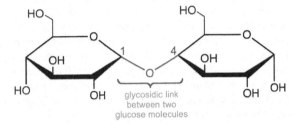

Figure 3.8 *Glycosidic bonds allow sugar molecules to be linked together. In this case, two glucose molecules have linked together to form the two-sugar molecule, maltose. By adding further units we can produce polysaccharides (carbohydrates)* (Source: Charles Cockell).

(the hemiacetal carbon) produces the sugar sucrose (a 1–2 glycosidic link).

Carbohydrate polymers can be produced by linking many sugar molecules together with glycosidic bonds. These carbohydrates are sometimes also called **polysaccharides**.

Sugars do not just have to bind to other sugars. One of the remarkable versatilities of the sugar rings is their ability to link to molecules through nitrogen (N-glycosidic bonds) and sulfur (S-glycosidic) bonds, thus generating a great variety of molecules with a complex structure. The N-glycosidic bond turns out to be essential for the binding of the sugar backbone of DNA to the base pairs that make up the genetic code as we shall see in a later section.

Like amino acids, sugars are chiral, coming in L and D forms. All life on Earth uses primarily D-sugars.

3.8 Lipids

Another class of compounds are the lipids, which encompass a wide diversity of chained and ring-containing

molecules. They include long-chained carboxylic acids (also called **fatty acids**) which are chains of carbon compounds joined together, mainly through single bonds (e.g. saturated fatty acids) and some containing double bonds (unsaturated fatty acids; Figure 3.9). One important group of lipids are the fats or triglycerides (Figure 3.9), which are a combination of glycerol and three fatty acids and are found widely in animal fats and plant oils. The energy stored in the many bonds of fatty acids and triglycerides has made them useful as energy storage molecules in life. Many lipids have side groups or other functional groups at their ends, which allow for important molecular interactions.

Many lipids have the important characteristic of having one end that is charged. This end tends to be attracted to water and the charge helps it to dissolve. We call it the **hydrophilic** end from the Greek *hydro* (water) and *philos* (love; it loves water). The other end, which is non-polar, does not dissolve so readily as it is uncharged. It is called the **hydrophobic** end because it dislikes water (from the Greek *phobos*, or fear). This property of having an uncharged and charged end means that these molecules are also referred to as **amphiphilic**. The phospholipids are one important class of lipids with a hydrophilic phosphate at one end. They are involved in cell membrane assembly. Like sugars and proteins, lipids too come in a vast variety, allowing for their complex and diverse functional roles in life. Some contain ring structures such as cholesterol (Figure 3.9), a component of animal cell membranes that is involved in maintaining membrane fluidity and integrity. It contributes to the ability of animal cells to change shape and move.

3.9 The Nucleic Acids

Last, and certainly not least, we come to the nucleic acids, the information storage molecules of life.

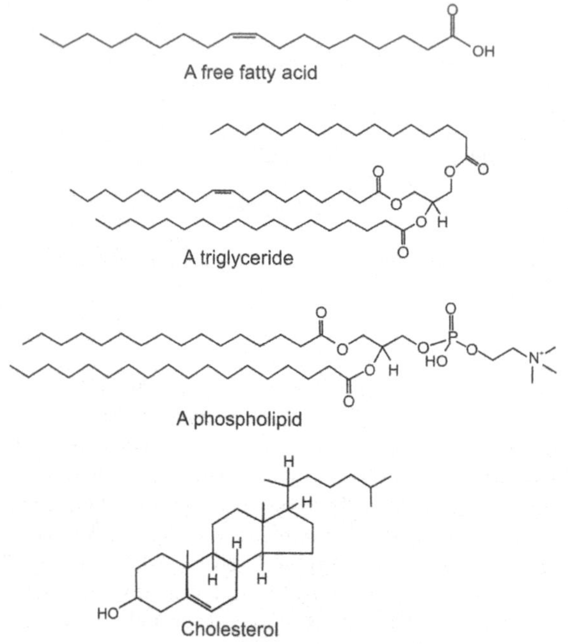

Figure 3.9 *The molecular structure of some lipids. Free fatty acids are found as energy storage molecules. Triglycerides are found in vegetable oils and animal fats. Phospholipids are involved in cell membrane formation in microbes and other organisms. Cholesterol is a structural component of animal cell membranes* (Source: Charles Cockell).

First, consider the most well-known of these, DNA or **deoxyribonucleic acid**. This molecule is a complex assembly, but when stripped down, it is beautifully simple. It takes up the three-dimensional structure of a double helix, but let's untwist this and flatten it down on the table and take a close look at it. First, we have the two backbones that run down the length of the molecule. They are made up of a pentose (ribose) sugar attached to a phosphate group that repeats over and over again along the length of the molecule as a sugar–phosphate backbone. You can think of this as scaffolding. It has an important role to play in holding the DNA molecule together and mediating the interactions between DNA and other molecules. However, the real nitty-gritty of this molecule is what is inside. The phosphate and pentose sugar are attached to a base (or nucleobase) through a glycosidic bond.

There are four bases in DNA – thymine (T), adenine (A), guanine (G) and cytosine (C) (Figure 3.10a). Let us call these bases the 'letters' of the DNA. These letters are the genetic code and we can attach them to the sugar–phosphate replicating backbone in any sequence we want to generate long sequences of them. It is this sequence of letters that codes the instructions for all biological functions, as shall become apparent later. Thymine and cytosine are also referred to as pyrimidines, which are a class of molecules possessing a single aromatic ring (Figure 3.10a) and adenine and guanine are purines, a class of molecule with both a pyrimidine ring and an imidazole ring (or heterocyclic ring because it is a ring that contains different elements, C and N in this case; Figure 3.10a).

The components of the base and its attached phosphate groups are defined as *nucleosides* (Figure 3.10b). The term **nucleotide** is a generic term generally used to describe the base with a sugar and at least one phosphate group. For example, in the case of adenine, the base attached to a ribose sugar and one phosphate group is AMP or Adenosine monophosphate. The molecule is a building block of DNA (containing the base and one unit of the backbone) and it is a nucleotide.

The DNA molecule has two backbones. How is this explained? Through the centre of the molecule each base is hydrogen bonded to another base in a very specific bonding pattern. The A binds to a T and vice versa and the G binds to a C and vice versa. We end up with two strands bound through their centre with hydrogen bonds across two bases. You can probably see why this is important. It certainly occurred to Francis Crick (1916–2004) and James Watson in 1953 when they discovered and first published this structure. If we unzip the molecule along the hydrogen bonds each strand can be used to re-assemble the complementary strand because, as we already know, each base can only bind very specifically to another particular base. In this way, DNA can be replicated. The two separate strands can be used to recreate two individual double helices.

It is also important to know that a DNA strand has a direction. It runs from the 3′ (pronounced 'three prime') to the 5′, where these numbers are defined by the numbering convention on the ribose sugar on the DNA backbone. The 3′ carbon has the hydroxyl group, the 5′ carbon has the phosphate group attached to it. You can see this numbering convention in Figure 3.10b and the labelling of the strands according to this convention in Figure 3.10a.

3.9.1 Ribonucleic Acid

DNA is not the only type of nucleic acid. Another important type is **ribonucleic acid (RNA)**. It shares the same

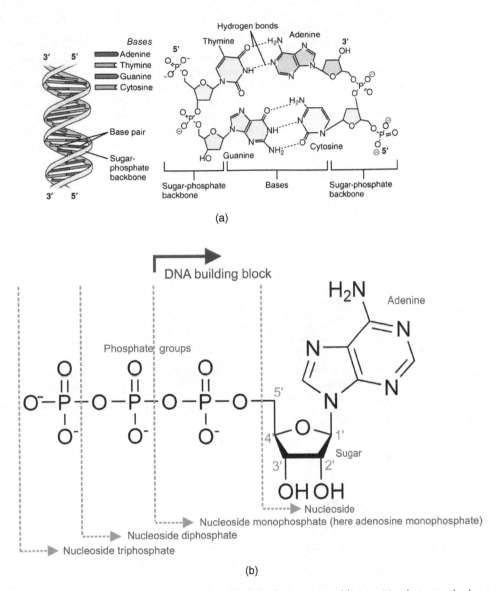

(a)

(b)

Figure 3.10 *(a) The structure of DNA showing the double helical structure and base pairing between the bases along the centre of the molecule (Source: OpenStax College). (b) The base pair components of DNA (shown for adenine) and the names given to the segments of these components. The nucleobase (here adenine), ribose sugar and one phosphate constitutes a basic building block of DNA (AMP or adenosine monophosphate). The numbering convention on the ribose sugar is also shown that accounts for the directionality in DNA expressed as the 3′ and 5′ ends (Source: Charles Cockell).*

Debate Point: How different could the genetic code be?

Research has shown that 'unnatural' base pairs can be incorporated into cells. The base pair d5SICS and dNaM is a synthetic base pair successfully incorporated into the DNA of the microorganism *Escherichia coli*. This research raises the question of whether the genetic code needs to have the four base pairs that we find in the natural terrestrial genetic code. Are they the serendipitous result of one evolutionary path? Is it possible that we could have an entirely different set of bases? A second, and related, question is whether the fundamental structure of DNA need be the same, even if the base pairs were different? You might like to discuss what sort of molecular structure an alternative information storage system could have, bearing in mind that it must replicate and be used to code information to construct other molecules (such as proteins). What required general characteristics does this impose upon an information storage system? Does such an information storage system have to be organic – could you imagine such a system using minerals or regular crystal structures to store the quantity of information needed to build life?

Malyshev, D.A., Dhami, K., Lavergne, T., Chen, T., Dai, N., Foster, J.M., Corrêa, I.R., Romesberg, F.E. (2014) A semi-synthetic organism with an expanded genetic alphabet. *Nature* **509**, 385–388.

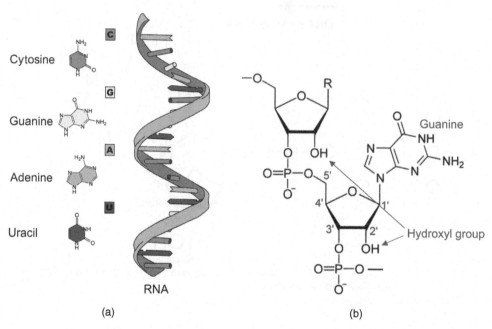

Figure 3.11   *(a) The schematic structure and bases of a single strand of RNA. (b) The molecular structure of the RNA molecule. The key difference with DNA is the presence of the –OH group (an extra oxygen atom) at the 2′ position of the ribose sugar in the sugar–phosphate backbone, hence the name, ribonucleic acid, in comparison to deoxyribonucleic acid (DNA).*

fundamental structure as DNA, but has a few crucial differences (Figure 3.11). First, the ribose sugar has an oxygen atom in the hydroxyl (–OH) group on the 2′ position on the sugar unlike the sugar in DNA (deoxyribose), hence its name, ribonucleic acid, as opposed to DNA – *deoxyribo*nucleic acid. Second, instead of thymine, the RNA has a uracil (U) base.

RNA is an important part of the architecture of reading the cell's information and we will return to it when we look at how the genetic code works in the next chapter. RNA is also found in a range of viruses. The RNA molecule, on account of its different chemistry and orientation of its bases, is generally more reactive than DNA. We will see later that on account of these properties, RNA plays a central role in some theories about the origin of life.

3.10 The Solvent of Life

We now turn to another essential requirement for building life. All of the molecules that have been discussed must be assembled in a liquid. Life requires a solvent for chemical reactions to occur. Chemical reactions cannot occur efficiently when a system is completely desiccated as the reactants that we need for chemical reactions cannot easily move and interact.

This solvent must have a number of characteristics which include a viscosity and density that allows molecules essential to biological function to be maintained at sufficient concentrations in the cells and to be moved around rapidly enough to allow chemical reactions to occur.

3.10.1 Water as the Best Solvent

One solvent that meets these needs, as life on Earth will testify, is water, or H_2O. Of the characteristics that make water a particularly suitable solvent for life, its **dipole moment** or **polarity** is a very important one (measured as permittivity or the dielectric constant).

The dipole moment of water (6.2×10^{-30} C.m or 1.85 Debye) is such that the molecule readily dissolves both salts and small organic molecules (Figure 3.12). Salts are important to life as a source of cations (positively charged ions such as K^+, Na^+, Fe^{3+}) and anions (negatively charged ions such as Cl^-), all of which play a role in a variety of functions such as stabilising membranes or as sources of energy. The dissolution of salts in water as charged ions contributes to the ability of water to act as a medium for chemical reactions that require charged species. The polarity of water also allows for the dissolution of small organic compounds such as amino acids. This property allows water to act as a mediator of the organic polymerisation reactions discussed in previous sections.

The polarity of water allows water molecules to bond together through hydrogen bonding. This property accounts for the wide temperature range of water, temperatures which overlap with environmental conditions on the surface and in the sub-surface of the Earth and physical conditions to be found on other planetary bodies such as the interior of the Jovian moon, Europa or the Saturnian moon, Enceladus. Without these characteristics,

Figure 3.12 Water dissolves a range of substances, including sugars (Source: Charles Cockell).

the small molecular weight of water would allow for a much smaller range of temperature conditions in the liquid state.

Other properties of water account for many of its beneficial uses as a biological solvent. Water has a high heat of vapourisation (in other words it takes quite a lot of energy to get it to vapourise), which promotes a stable liquid phase inside organisms and stabilises temperatures, enhancing the ability of organisms to cope with fluctuating environmental temperature regimens. By contrast, a high heat of vaporisation also implies a high energy loss during evaporation, which is used by multicellular organisms to achieve evaporative cooling against high temperatures in the environment.

Perhaps one of the most discussed properties of water that has been implicated in its biological usefulness is the lower density of ice than water, which we discussed in the previous chapter. The property that ice has of floating on water when frozen provides protection for organisms that can remain in liquid water beneath the ice layer. There is little doubt in saying that this property is beneficial for fish and many other aquatic organisms. However, many

Figure 3.13 *The wood frog can tolerate freezing temperatures by using glucose as an anti-freeze protectant in its tissues* (Source: wikicommons, Brian Gratwicke).

microorganisms can resist freezing, and the wood frog (*Lithobates sylvatica*; Figure 3.13), in a similar way to other North American frogs that hibernate close to the surface in soil or leaf litter, can tolerate freezing temperatures. The frog transforms glycogen in its blood to the sugar, glucose, in response to internal ice formation at the beginning of winter. These molecules act as protectants against damaging ice crystal formation. Frogs can survive freezing during winter if not more than about 65% of the total body water freezes. Although the wood frog is an unusual example, it is clear that evolutionary strategies do exist to tolerate freezing and that although the physical attribute that ice has to float on water may appear to favour life, it may not, in itself, be a fundamental requirement for life. However, the property has other important implications. Ice formed on the surface of a water body tends to trap energy underneath and thus maintain a liquid state over long time periods. If water ice sank, lakes, rivers and other water bodies would tend to freeze completely from the bottom up. So from a physical point of view the property is important for understanding aspects of planetary habitability.

It is also worth pointing out that water does have some properties that are not very conducive to life. In certain situations it can be a reactive solvent, causing the dissociation of many molecules, for example it can be disruptive to hydrogen bonding in proteins, making it sometimes unconducive to protein folding. Furthermore, as we have seen, the formation of some biological molecules, such as proteins, involves condensation reactions where water is removed from a bond, eliminating it from the chemistry. So we should be careful not to view

it as a perfect solvent. Life has to make compromises. However, of all possibilities, and despite some deleterious properties in certain situations, water might be the best solvent for carrying out a large variety of chemical reactions.

3.11   Alternative Chemistries

We have looked at the key requirements for building life on the Earth, but what about other possible chemistries? We have no empirical evidence to suggest that alternative chemistries can be used by life or that other life forms on some distant planet are using these chemistries, but to put the previous discussions into context it is worthwhile to briefly consider what alternatives have been discussed.

As the book progresses, you can consider for yourself whether our view that carbon as the ideal backbone element for life and water as the ideal solvent is a sensibly considered scientific conclusion or merely an Earth-centred prejudice.

3.11.1   Alternative Core Elements

What about alternative core elements? One popular suggestion is silicon. As a p-block element of group 14, below carbon in the Periodic Table, it shares many common chemical characteristics with carbon.

Silicon can be more reactive than carbon, which is attributed to three characteristics. First, silicon, like carbon, typically forms four bonds, but unlike carbon it can accept additional electrons and form five or six bonds (it can have a higher number of near neighbours than carbon) and this allows some reactions to occur at lower energies. Second, many silicon bonds with other elements are weaker than in carbon, requiring less energy to break them. The Si–Si bond strength ($222\,kJ\,mol^{-1}$) is lower than the C–C bond strength ($347\,kJ\,mol^{-1}$). For example, silane (SiH_4) combusts spontaneously in air even at $0\,°C$, whereas its carbon analogue, methane (CH_4), remains completely stable, even in pure oxygen, unless energy is imparted to it. Finally, silicon is more electropositive (a greater tendency to donate electrons) than carbon, leading to strongly polarised bonds with other non-metals which are much more susceptible to chemical reactions. Although the more reactive nature of silicon (at least in the absence of oxygen) may at first appear to be a disadvantage in any potential biochemistry, this high reactivity might make it more conducive to biochemical reactions at low temperatures.

Silicon has some impressive properties. It forms stable tetra-, penta- and hexa-coordinated compounds with N, C and O bonds and can form stable covalent bonds with N, P, S and many other elements that are associated with the generation of molecular diversity in carbon biochemistry. Although silicon cannot easily form a six-ring structure on its own like benzene, it can form a six-ring structure (siloxene) in which oxygen atoms hold together the silicon atoms. Cage-like molecular systems such as silsesquioxanes can be reacted with a wide diversity of side groups to allow for a remarkable diversity of molecules (Figure 3.14) that have industrial uses in chemical catalysis and making light emitting diodes (LEDs).

Although there are similarities, carbon and silicon have some significant differences that affect compound formation. The larger radius of silicon (233 pm) compared to carbon (154 pm) accounts for its weaker bond strengths, which means that, despite some of the variety of complex compounds it can form (Figure 3.14), in general it less readily forms stable complex compounds. In particular, bond angles of silicon compounds are generally larger because of its larger size, meaning that silicon cannot form very stable molecules analogous to ring compounds in carbon biochemistry. Stable **aromatic** (ring) compounds are found throughout carbon biochemistry and give huge versatility to the complexity of compounds that can be assembled and their bonding interactions; for example the base pairs within DNA. Few silicon compounds contain double and triple bonds, which are common in carbon compounds and produce many important compounds such as lipids.

Perhaps one of the most significant limitations of silicon is the tendency it has to form very inert structures with oxygen. Fully oxidised silicon is silica, a highly unreactive compound which makes up quartz and a wide variety of minerals. By contrast, carbon forms a double bond with oxygen to produce carbon dioxide, a gas which has a diversity of uses in biochemistry, not least as an easily accessible form of carbon for life, but also in energy yielding reactions such as biological methane production (**methanogenesis**). Indeed, in most settings under standard temperatures and pressures silicon forms unreactive silicates, which on Earth are found in a wide diversity of different rock types (Figure 3.15).

The silicates are all formed from the silica tetrahedron as the unit building block. When on their own and gathered together into an assemblage with cations (such as iron or magnesium) between them they form minerals such as the **olivine** class of minerals. When attached into a long chain they form the **pyroxene** class. When these chains are themselves linked together to form double chain silicates they form the **amphibole** class of minerals. When

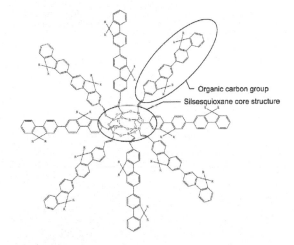

Silsesquioxane cage structure

Polyphenylsilsesquioxane ladder repeating unit.

Organic carbon group
Silsesquioxane core structure

Silsesquioxane structure with attached organic side groups

Figure 3.14 *Silicon can form extraordinarily complex structures, such as these silsesquioxane structures* (Source: Charles Cockell).

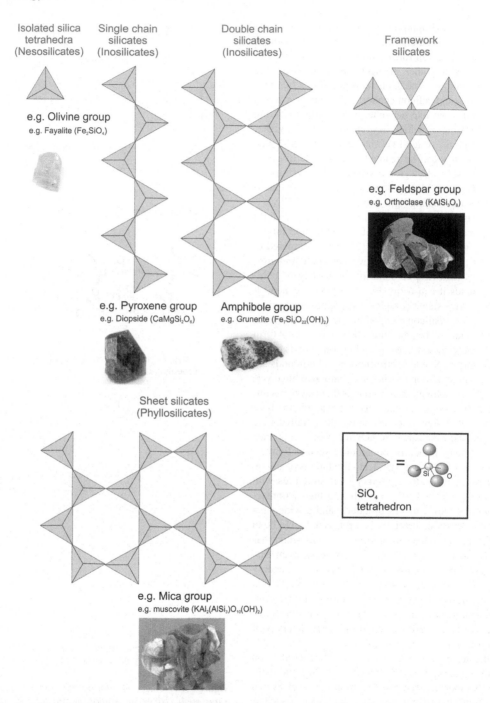

Figure 3.15   *A variety of biologically inert silicate structures formed when silicon binds to oxygen. These are the structures that make what we generally refer to as rocks and minerals. They are comprised of the core building block, a silica tetrahedron (SiO_4)* (Source: Charles Cockell).

the chains are assembled into a large layer, they form the sheet silicates. They include the **phyllosilicates**, a class of clays found widely on the Martian surface and many other environments where volcanic rock silicates have reacted with water. Three dimensional mineral structures form the framework silicates such as the **feldpsars**. You can see examples of all of these in Figure 3.15. We will return to minerals in Chapter 13. For now it is sufficient to summarise by saying that the presence of oxygen, which is common in the Universe, tends to drive silicon chemistry towards silicates and away from a tendency to form the sorts of diverse silicon compounds that might be of interest to a biochemist.

A more plausible chemistry involving silicon is a hybrid system with carbon. Silanes (Figure 3.16) are molecules made up of complex arrangements of silicon and hydrogen, analogous to the alkanes in carbon. They have the ability to form branched chains of molecules. By replacing the hydrogen atom in silanes with organic groups, greater stability is achieved because the silicon–hydrogen bond is weak. These organosilicon compounds are thermally stable and are chemically inert. Although laboratory experiments can be performed to generate these compounds, rarely are they found in the natural world. However, if life was based on silanes, chemical reactions would probably have to occur at low temperatures or in an essentially oxygen and water-free environment to prevent reactions of silanes, ultimately leading to reactions with oxygen and the production of inert silicon materials.

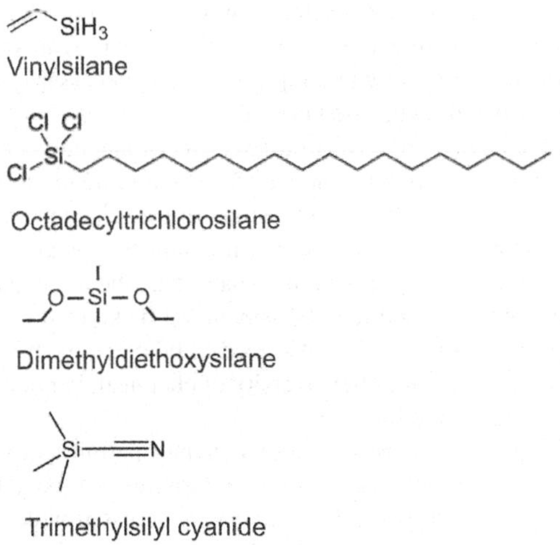

Vinylsilane

Octadecyltrichlorosilane

Dimethyldiethoxysilane

Trimethylsilyl cyanide

Figure 3.16 *Silanes include hybrid molecules with organic groups* (Source: Charles Cockell).

Other speculations have considered silicate-based life forms at high temperatures when the melting of rocks increases their reactivity. Some people have discussed the possibility of 'lavobes' and *'magmobes'*, organisms that inhabit molten rocks. Genetic information would be encoded within structural defects within the minerals. Although these ideas are intriguing, no evidence exists for them in the terrestrial rock record despite the fact that the Earth has had abundant environments containing molten rocks throughout its history.

An examination of the Periodic Table doesn't suggest many other suitable elements that even come close to silicon as plausible alternatives as a core building element for a diverse range of molecules.

We might look below silicon and carbon to germanium, the next element down in group 14 of the Periodic Table. It can form a set of three dimensional molecules called germanates that have many similarities to the silicates, forming tetrahedral GeO_4 arrangements. The element has a mass of 72 and is extremely rare in the Universe compared to less massive elements (the terrestrial abundance is about 1 ppm). It, like silicon, also lacks the capacity of molecular diversity seen in carbon.

The halogens (chlorine and fluorine) are too unreactive. The metals, including magnesium, iron, nickel, potassium, sodium, calcium and so on, do not form strong covalent bonds but instead form ionic bonds (Chapter 2), which although strong, do not generate a rich diversity of complex compounds associated with the variety of chains possible with covalent bonds. Oxygen and boron are also implausible candidates in most planetary environments because, although they form covalent bonds and are involved in carbon chemistry, they do not, in themselves, forms chains and other complex molecular arrangements.

3.11.2 Alternative Solvents

The possibilities for alternative solvents to liquid water have for a long time been discussed by the astrobiology community. Although water is the most abundant solvent in the universe, other solvents might be plausible in planetary environments with different physical and chemical environments. Unfortunately, few of these alternatives have been empirically investigated in any depth. Without a detection of a life form using them it is difficult to conceive of convincing biological experiments. Additionally, our knowledge of how the origin of life occurred is not sufficiently complete to be able to test alternative solvents in the laboratory as possible solvents for the origin of life. Nevertheless, some physical properties of alternative solvents might make them candidates.

It has been shown that many enzymes can be active in non-polar solvents such as benzene and that about 20% of the human DNA encodes membrane proteins that require the non-polar environments inside cell membranes to operate. Although these observations do not provide any direct evidence for the possibility of non-aqueous solvents being potentially successful media for a complete biochemistry, they show that even some terrestrial biochemistry can operate in non-aqueous solvents and in the case of membrane proteins, actually require these environments.

Ammonia has been one of the most discussed alternatives to water. Early discussions on the topic focused on the possibility of peptide bonds mediated by a –$CONH_2$ group as opposed to the –COOH group in amino acids. In Figure 3.17 the terrestrial peptide bond formation reaction is shown (Figure 3.17a) and the corresponding reaction that has been hypothesised for a liquid ammonia biochemistry (Figure 3.17b).

Comparing the physical properties of ammonia to those of water yields insights into their possible comparative advantages and disadvantages. Ammonia is less viscous than water [compare 1 centiPoise (cP) for water at room temperature to 0.265 cP for ammonia] and so molecules diffuse through it more quickly. However, ammonia has

a lower heat of vaporisation (1369 kJ kg^{-1}) than water (2257 kJ kg^{-1}) and so is less able to accommodate temperature fluctuations. The greatest differences from water are that ammonia is liquid at lower temperatures and has a smaller liquid temperature range at atmospheric pressure (−78 to −34 °C). For a cold planetary environment, its low-temperature liquid state could be advantageous. Its liquid range can be increased by increasing the pressure, such that about 20 atm, the boiling point is increased to around 50 °C.

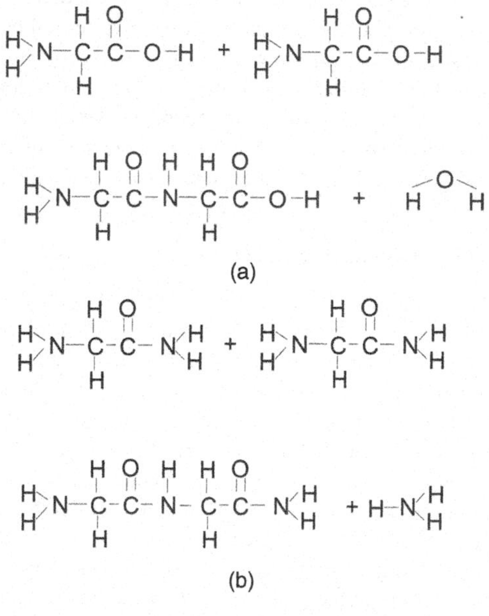

Ammonia presents other challenges for life, most notably the extremely high pH at which ammonia solutions form. Ammonia solutions of 1% or greater have pH values greater than 11.0 and biochemistries would require adaptation to these conditions, although organisms with adaptations to high pH are known.

Hydrofluoric acid has been discussed as another alternative solvent. The wide temperature range at which it remains liquid make it a possible candidate, with fluorine replacing oxygen in many molecular structures. However, the low cosmic abundance of fluorine and high reactivity with organic carbon molecules make its use limited. A range of other substances has been discussed (Table 3.1).

3.12   The Structure of Life and Habitability

Astrobiologists are very interested in looking for habitable conditions on other planets, a theme that will crop up throughout this book. Habitability can simply be explained as conditions suitable for the activity of at least one known organism, where activity is the ability of an organism to maintain itself, grow and reproduce. It is necessarily circumscribed by what we know about life. We do not have a choice since we have to assess other planetary bodies based on what we know about life we have already found. This explains why the discussions in this chapter are so important, because the minimum conditions required by life will ultimately determine what we define to be habitable conditions and the assessment we make of the habitability of other worlds. We need to be clear about what we think is the plausible diversity of chemical compounds required to make life.

So far, the discussions in this chapter permit us to list at least two requirements for an environment to be habitable. We need to have: (i) an environment that provides the basic elements for life (CHNOPS) and (ii) an environment where physical and chemical conditions will support the existence of a suitable solvent, which at the current time

Figure 3.17   *A scheme for the formation of a peptide-like bond in terrestrial proteins (a) compared to a hypothetical linkage involving ammonia (b).*

Table 3.1 *A range of possible solvents for life, their temperature ranges, heat of vapourisation, viscosity and dipole moments.*

Solvent	Molecular weight	Liquid range (K at atmospheric pressure)	Heat of vapourisation (kJ kg^{-1})	Viscosity (cP)	Dipole moment (Debye)
H_2O	18	273.1–373.1	2257	1.00	1.85
NH_3	17	195.4–239.8	1369	0.265	1.47
HF	20	190.0–292.7	374.1	0.256	1.91
H_2SO_4	98	283.5–611.1	56.0	48.4	2.72
CH_4	16	90.7–111.7	480.6	0.184	None

we assume is liquid water. We will see in Chapter 5 that we also require an energy supply. In Chapter 7 we will discuss in more detail the physical and chemical limits to life on the Earth, which also define the limits of habitable conditions.

3.13   Conclusion

Despite the huge variety of elements in the Periodic Table, the selection used by all life is very small. Six of them (CHNOPS) are used to construct the major molecules of life, with other elements commandeered by organisms for specific purposes. The chemical versatility of carbon manifested in its ability to form a wide diversity of covalent bonds with elements that have comparable bond strengths accounts for the millions of compounds it can form and the capacity to generate the molecular complexity required by life. Although the diversity of carbon molecules is vast in life, they broadly sit into some major groups including the proteins, carbohydrates, nucleic acids and lipids. Proteins and sugars exhibit chiral excesses. For a long time there have been many speculations about alternative chemistries: silicon as the most favoured alternative to carbon, ammonia as a possible alternative to water as the solvent in which biochemistry occurs on the Earth. Although we should not rule out life based on these different chemistries, we have seen that there are good chemical reasons to suspect that a carbon and water-based life would exhibit the greatest possibility of the chemical diversity and environmental tolerances required by living things. With this knowledge under our belt, we are now in a position to see how these molecules come together to build living systems. This is the focus of the next chapter.

Further Reading

Books

Grinspoon, D. (2004) *Lonely Planets*. Harper Collins, New York.

Schulze-Makuch, D., Irwin, U. (2008) *Life in the Universe*, Springer, Heidelberg.

Ward, P. (2005) *Life As We Do Not Know It*. Penguin, New York.

Papers

Bains, W. (2004) Many chemistries could be used to build living things. *Astrobiology* **4**, 137–167.

Benner, S.A., Ricardo, A., Carrigan, M.A. (2004) Is there a common chemical model for life in the Universe? *Current Opinions in Chemical Biology* **8**, 672–689.

Conrad, P.G., Nealson, K.H. (2001) A non-Earth-centric approach to life detection. *Astrobiology* **1**, 15–24.

Firsoff, V.A. (1965) Possible alternative chemistries of life. *Spaceflight* **7**, 132–136.

Kamerlin, S.C., Sharma. P.K., Prasad, R.B., Warshel, A. (2013) Why nature really chose phosphate. *Quarterly Reviews in Biophysics* **46**, 1–132.

McKay, C.P., Smith, H.D. (2005) Possibilities for methanogenic life in liquid methane on the surface of Titan. *Icarus* **178**, 274–276.

Mottl, M., Glazer, B., Kaiser, R., Meech, K. (2007) Water and astrobiology. *Chemie der Erde* **67**, 253–282.

Pace, N.R. (2001) The universal nature of biochemistry. *Proceedings of the National Academy of Sciences* **98**, 805–808.

Watson, J.D., Crick, F.H.C. (1953) A structure for deoxyribose nucleic acid. *Nature* **171**, 737–738.

Westheimer, F.H. (1987) Why nature chose phosphates. *Science* **235**, 1173–1178.

4

Life's Structure: Building Cells from Molecules

Learning Outcomes

➤ Understand the different types of cells found in life on the Earth.
➤ Understand the importance of compartmentalisation in life.
➤ Understand the structure of cell membranes and the different variations in cell membrane structure.
➤ Understand the processes of genetic transcription and translation and the principal molecules involved in these processes.
➤ Know how the genetic code is used to assemble proteins and understand the meaning of the degeneracy of the genetic code.
➤ Describe the characteristics of viruses and prions.

4.1 From Molecules to Cells

We have a good idea of how elements are assembled into atoms and how atoms form molecules and then the major classes of macromolecules from which life is constructed. The next stage of this journey must be to consider how these macromolecules are assembled into self-replicating living cells. From this point we can go on to consider how these cells diversified into the array of life that we witness on the Earth today, how this has changed over time and whether this might have occurred or be occurring on other planetary bodies in the Universe. This chapter is concerned with understanding how macromolecules are assembled into the cells of life.

4.2 Types of Cells

Cells are the packages that hold life together. The term 'cell' is derived from the Latin word, *cella*, which means 'small room'. In biology, the cell refers to the structure that encloses the apparatus that allows for the growth and reproduction of organisms. Indeed, the compartmentalisation of life's major biomolecules within cellular packages is so fundamental that some definitions of life have explicitly included compartmentalisation or cellular structure as a feature of living things. Cells were first observed and described by polymath Robert Hooke (1636–1703) in his book, *Micrographia*, in which he documented the structure of plant cells by observing thin slices of cork. He found them to look similar to honeycomb cells, an indication of compartmentalisation.

Broadly we can recognise two types of cells – prokaryotic and eukaryotic cells. Prokaryotic cells do not have a nucleus, hence their name, which derives from the two Greek words, *pro* (before) and *karyon* (nut or kernel – a reference to the nucleus in the biological context). They are single-celled organisms and include two major

Astrobiology: Understanding Life in the Universe, First Edition. Charles S. Cockell.
© 2015 John Wiley & Sons, Ltd. Published 2015 by John Wiley & Sons, Ltd.
Companion Website : www.wiley.com/go/cockell/astrobiology.

domains of life: **bacteria** and **archaea**. Domains are the highest hierarchy of life. The bacteria include many species of microorganisms that live in soils, the oceans, in your gut and in many other environments. Archaea is the domain that includes many of the species that live in extreme environments, for example microbes in hydrothermal vents. However, the archaea also include non-extremophile species that have important roles to play in natural environments, including soils, such as species that cycle nitrogen. We will encounter both of these groups more as we proceed through the book. In prokaryotic cells, the DNA is free-floating in the cell fluid or **cytoplasm**.

By contrast, eukaryotic cells are cells with a nucleus that contains the DNA. They make up the structure of multicellular organisms, including us, and some single-celled organisms like algae. Eukaryotic cells are usually larger (about 10–100 μm) than prokaryotic cells (1–10 μm).

Most of life on Earth is made up of prokaryotic cells. There are estimated to be about 10^{29} prokaryotes on the Earth. Prokaryotes were discovered in the seventeenth century by Dutch fabric maker, Anthony van Leeuwenhoek (1632–1723).

van Leeuwenhoek was keen to improve the quality of the cloth that he was selling, which enticed him to develop small microscopes to observe the fibres in his fabrics (Figure 4.1). As he was an inquisitive man, he used his new devices to examine samples of pond water. What he found were tiny creatures, which at the time he called 'animalcules', or little animals.

The creatures he observed had different shapes and he published numerous papers through The Royal Society in which he described these organisms and their appearances. Remarkably, he managed, even with his primitive microscopes, to observe some of the major shapes (morphologies) of microorganisms, including coccoids (spheres), rods, spiral organisms and microbes with a bent cell shape. He even observed bacterial movement or **motility**, which you can see in Figure 4.2.

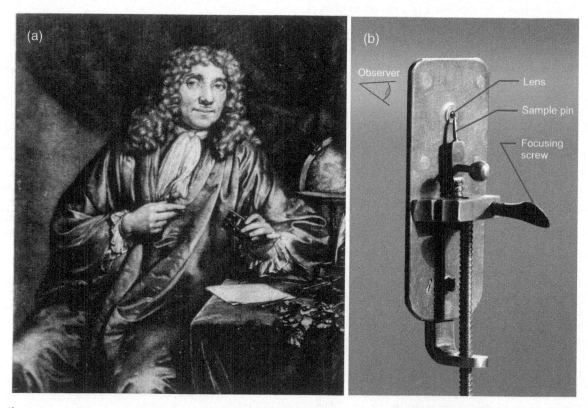

Figure 4.1   *Anthony van Leeuwenhoek (a) and one of his first microscopes (b). The microscope (~10 cm long) is held up to the eye and objects are observed through the tiny glass lens (Source: wikicommons, Jeroen Rouwkema).*

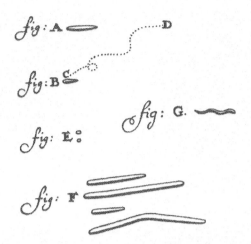

Figure 4.2 *van Leeuwenhoek's figures in the 1670s show the first drawing of prokaryotes and their diverse morphologies. In the part labelled 'fig: B (C–D)' you can see a depiction of bacterial motility.*

Leeuwenhoek even showed that he could kill bacteria. In a letter of 1684, he wrote: 'I took a very little wine-Vinegar and mixt it with the water in which the scurf [tooth plaque he had scrapped from his teeth] was dissolved, whereupon the Animals dyed presently'.

For the moment we will leave historical accounts. However, I have no compunction here in recommending the book '*The Microbe Hunters*' by Paul de Kruif, an amusing and classic account of the history of microbiology and its primary characters written in 1926.

4.3 Shapes of Cells

Cells take on a diversity of morphologies depending on the species, their local environment and their function. Eukaryotic cells in multicellular organisms can vary from the rectangular cells of plants whose dimensions range from 10 to 100 μm to the long thin cells of neurons that transmit nerve impulses through organs such as your brain, which although thin (from about 4 to 100 μm in width) can be up to 1 m long.

The most consistent patterns of morphology are found in the prokaryotes. Prokaryotic cells fall into a number of major classes of shapes (Figure 4.3).

Many microbes have a coccoid (spherical) morphology and are referred to as *coccus*. An example is *Staphylococcus* (Figure 4.3a), which is responsible for some infections in humans. In some species coccoids aggregate to form pairs (a diplococcus) or even collections of cells that look like bunches of grapes, such as with *Staphylococcus*. Many organisms are rod-shaped or **bacilli**, which are round-ended cylinders. An example of these organisms is bacteria in the group *Bacillus* or the gut bacterium, *Escherichia coli* (Figure 4.3b). To complicate matters there are even prokaryotes that are something of a mix between a coccus and a rod (coccobacillus) and can be considered to be a slightly elongated coccus.

Microorganisms can also be found that are filamentous, such as many cyanobacteria (Figure 4.3c), fungi and some species of soil bacteria such as certain microorganisms in the group *Actinobacteria*. Filamentous organisms can have branched morphologies, most commonly observed in fungi and cyanobacteria. Yet other organisms have a spiral shape, such as **Spirochaetes** (Figure 4.3d), which are responsible for some diseases such as Lyme disease (*Borrelia burgdorferi*). There are organisms that are disc-shaped, star-shaped and tapered.

Perhaps the most unusual shape is to be found in a member of the archaea, *Haloquadratum*, which adopts a square shape (Figure 4.3e). The organisms divide into sheets, looking somewhat like a sheet of postage stamps, achieving layers of microbes up to 40 μm in length. The reason for this shape is likely to be linked to its growth in briny pools and the resulting balance of osmotic stresses inside and outside the cells.

Some microorganisms can change their shape under stress. When exposed to nutrient stress or low temperatures certain species can become filamentous in growth. It is thought that the filamentous shape enhances the surface area for nutrient acquisition. Filamentous shapes may also improve attachment to surfaces and enhance the formation of layers of organisms or **biofilms**. It is clear that prokaryote shape is not just a serendipitous and ephemeral evolutionary feature, but that there is a whole diversity of environmental pressures that might influence microbial shapes, which are only just beginning to be understood.

4.4 The Structure of Cells

Despite the diversity of cell shapes and types, we can say that all cells have three basic components that are essential for an organism to function. They are: (i) a membrane to hold the cell contents in, (ii) an information storage system to direct molecular synthesis and ultimately reproduction of the cell itself and (iii) an energy

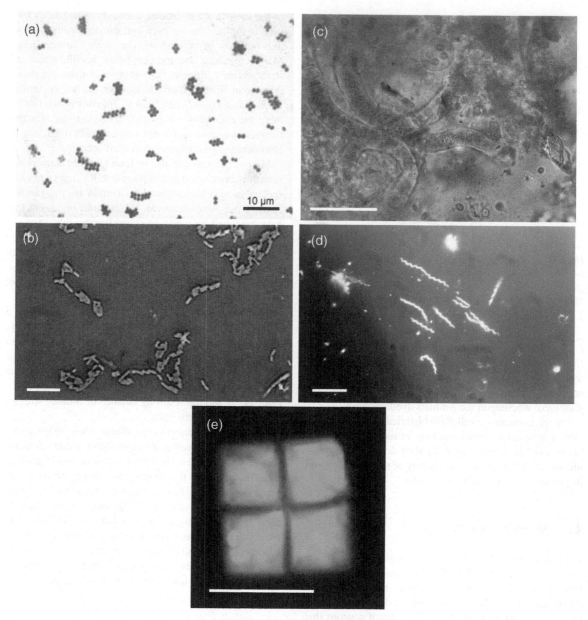

Figure 4.3   *The wide variety of prokaryote shapes. (a)* Staphylococcus *(coccoid bacteria forming grape-like clusters). Scale bar: 10 μm (Source wikicommons, Y Tambe). (b) Rod-shaped* Escherichia coli. *Scale bar: 5 μm (Source: Josef Reischig). (c) Filamentous cyanobacteria from the Arctic. Scale bar: 10 μm. (d) Spiral-shaped spirochaete bacteria. Scale bar: 10 μm. (e) The square-shaped bacterium,* Haloquadratum walsbyi, *here shown stained and observed under fluorescence. Scale bar: 5 μm.*

system for gathering energy from the outside environment to drive chemical reactions, growth, reproduction and so on. In this chapter we will look at the structure of cells, focusing on the membranes that enclose cells and the information storage system. In the next chapter we will focus on how life gets energy to maintain itself, grow and reproduce.

4.5   Membranes

To create a self-replicating cell we require some type of vessel in which to concentrate molecules. One might imagine a science fiction scenario of a garden pond, in which chemicals gain sufficient concentration for complex reactions and interactions with the outside environment to occur. The pond could essentially become a 'cell' defined by the physical boundaries of the pond with wind acting to mix its constituents, keeping them in suspension. However, in general, at least on Earth, the hydrological cycle has a tendency to dilute things, so that most water bodies are far too dilute for useful biochemistry to occur in – cells therefore require a membrane structure (and besides, the cellular garden pond would have no way of replicating, so its future is quite limited).

A cell enclosure, or membrane, provides a way to keep molecules together at relatively high concentrations. A membrane will also retain water, allowing for chemical reactions to proceed even when outside conditions are quite desiccating.

In the previous chapter we discussed the lipids, some of which contain a hydrophilic head (which is attracted to water) and a hydrophobic tail made from a fatty acid (which is repelled from water). One important class of lipids is the phospholipids in which the charged polar end is a phosphate group (Figure 4.4).

A remarkable characteristic of these so-called **amphiphilic** molecules is that when added to water they have a tendency to spontaneously assemble in such a way that the hydrophilic head is oriented into the water and the hydrophobic tails are attracted towards each other to expel water. The result is a lipid bilayer membrane (Figure 4.5). These bilayers themselves tend to assemble into vesicles, small spherical structures with fluid in the inside. They are in essence cellular membranes. This property is by no means rare. Even **fatty acids** on their own, such as the long-chained **carboxylic acid**, decanoic acid, extracted from meteorites (Chapter 11) have a tendency to form membranous vesicles, suggesting that this is a fundamental property of this class of molecules and that the assembly of the first cell membranes on the Earth may not have been so extraordinary.

In cells, in addition to the lipid membrane itself, other proteins are incorporated into the bilayer, resulting in a complex system that regulates the interaction of the internal cell environment with the outside environment. Many of these proteins are trans-membrane proteins, acting for example as thin channels both for molecules such as nutrients to move into the cell and for wastes to be expelled. Some of these proteins act as a line of communication,

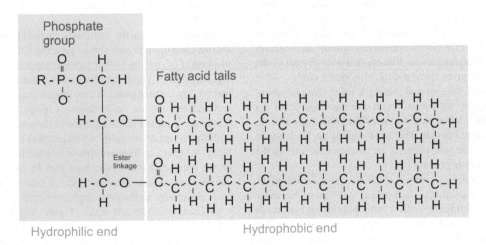

Figure 4.4 *Amphiphilic molecules such as phospholipids that make up some cell membranes. They have a hydrophilic end (attracted to water) and a hydrophobic end (repelled from water)* (Source: Charles Cockell).

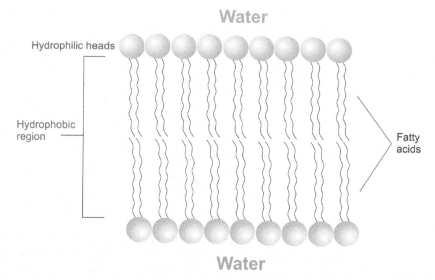

Figure 4.5   *A simplified diagram showing the structure of a lipid bilayer that makes up cell membranes* (Source: Charles Cockell).

allowing the cell to sense changes in physical and chemical conditions in the outside environment and respond to them. The membrane and its proteins are highly fluid.

4.5.1   Gram-negative and Positive Prokaryotic Membranes

The lipid bilayers found in life are assembled into more complex structures. In bacteria, for instance, they can be broadly separated into Gram-positive and Gram-negative membrane structures, a separation defined by how organisms stain in the **Gram stain**, a method of dyeing bacteria developed by microbiologist Hans Christian Gram (1853–1938) in the nineteenth century. In the Gram stain, microorganisms are first stained with crystal violet stain, which gives them a dark blue/violet colour. After washing with ethanol and adding a paler dye (safranin), Gram-positive organisms retain the dye and show up as dark blue, whereas Gram-negative organisms lose it, becoming a red colour. These differences are understood to be due to the quite different membrane structures in the two groups of organisms.

Gram-negative membranes (found, e.g., in the ubiquitous gut bacterium, *Escherichia coli*) have an inner and outer membrane with a thin **peptidoglycan** mesh sandwiched between them (Figure 4.6), containing just a few layers of the peptidoglycan molecules. The inner membrane is the plasma membrane. The space between the inner and outer membrane is termed the **periplasm**. It is somewhat more viscous than the **cytoplasm** and contains a wide variety of proteins involved in sugar and amino acid transport. The periplasm also acts to sequester harmful enzymes, preventing them from causing cell damage.

Peptidoglycan is a complex matrix made up of sugars cross-linked to amino acids. It is one of the few structures in life that contains D-amino acids (Figure 4.7). The peptidoglycan helps the cell maintain its shape and it is involved in cell division.

Other structures in the membrane include **porins**, which are one class of outer-membrane proteins. They control the movement of ions and other small compounds across the cell membrane. There are other trans-membrane proteins involved in communication and interactions between the cell and the outside environment. **Lipopolysaccharides** (LPS), long sugar chains attached to the outer membrane with lipids, are involved in the attachment of cells to surfaces. A variety of **lipoproteins** within the inner envelope of the outer cell membrane take part in mediating biochemical reactions. They also play a role in linking the outer membrane to the peptidoglycan layer.

In Gram-positive cell walls (Figure 4.6) there is one cell membrane, a periplasmic space above it and the cell is surrounded by a thick peptidoglycan mesh made up of

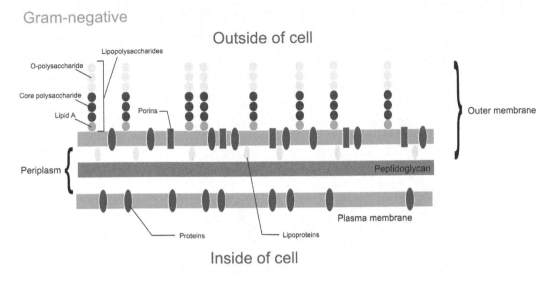

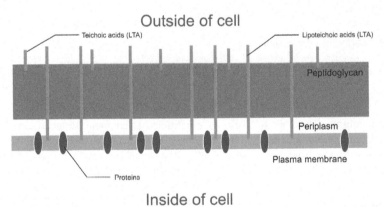

Figure 4.6 *Gram-negative and Gram-positive cell membranes* (Source: Charles Cockell).

around 40 peptidoglycan layers (between 30 and 100 nm thick, compared to the Gram-negative peptidoglycan layer which is just a few nanometres thick). **Teichoic** acids and **lipoteichoic** acids are long-chain sugars present on the surface and threading through the peptidoglycan layers. They are involved in the integrity of the peptidoglycan layers, but some also play a role in cell attachment, a bit like the lipopolysaccharides in the Gram-negative cell wall. An example of a Gram-positive organism is the very versatile bacterium, *Bacillus subtilis*, found in diverse places from soils to spacecraft assembly rooms.

The structures outside of the plasma membrane in both Gram-positive and negative bacteria are generally referred to as the **cell wall**.

The bacterial cell wall is often surrounded by a further layer called the **glycocalyx**. The glycocalyx is a network of polysaccharides. A distinct, gelatinous glycocalyx is called a capsule, whereas an irregular, diffuse layer is called a slime layer. Sugars attached to the outside of cells are sometimes called Extracellular Polysaccharide or **EPS**. The EPS acts as a protective layer of sugars that mediates interactions with surfaces. They allow cells to

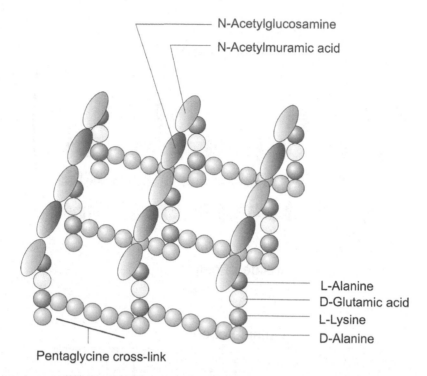

N-Acetylglucosamine

N-Acetylmuramic acid

L-Alanine
D-Glutamic acid
L-Lysine
D-Alanine

Pentaglycine cross-link

Figure 4.7 *A typical structure of peptidoglycan. The pentaglycine cross-link is made from five glycine amino acids. N-acetylglucosamine and N-acetylmuramic acid are sugars and are derivatives of glucose* (Source: Charles Cockell).

bind minerals and form biofilms. They are the substance that bond minerals and bacteria together into macroscopic **stromatolites** or domes of microorganisms. Cyanobacteria are particularly prolific producers of EPS.

4.5.2 Archeal Membranes

A fascinating difference in cell membranes is to be found in the archaea (Figure 4.8). They will be discussed more in the chapters on phylogenetic trees and life in extreme environments. In the archaea, the lipids that make up the membranes are different from those of the bacteria and eukaryotes showing that they are only distantly related. Instead of ester groups within the hydrophilic head of their lipids they have ether groups, which are more resistant to a variety of chemicals and could play a role in their resistance to extreme environments. The long tails of the lipids are also very different. They are called terpenoids (or isoprenoids) and have complex side chains, in contrast to bacteria where the lipid tails are long and simple (fatty acids). These side chains might make the membranes less leaky and more resistant to extreme

conditions. Even more strangely, in some archaea the bilayer is replaced by a monolayer where the tails from two lipids are fused together (Figure 4.8). This is the case in *Ferroplasma*, an archaeon (member of the archaea) that lives in very acidic environments. It is thought that this adaptation might make the membranes more resistant to extreme conditions, preventing the membranes from falling apart.

4.6 The Information Storage System of Life

We have a membrane containing the molecule of life. But an obvious question arises – how is the biochemistry of life controlled? What is the way in which instructions are read and directed to make molecules required for the cell to function and eventually to reproduce itself? The information storage system of life is **deoxyribonucleic acid, DNA**. We briefly explored the structure of this molecule in the previous chapter. We learned about the four bases, G, T, A and C that comprise the information code of DNA,

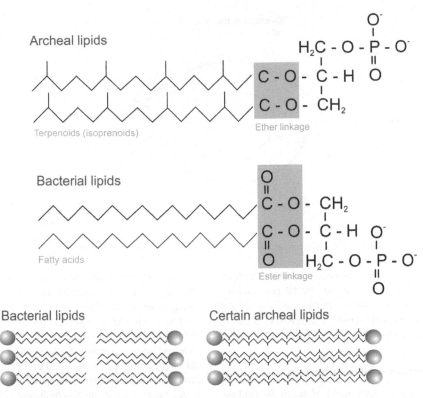

Figure 4.8   *The structure of archeal cell membrane lipids compared to bacterial lipids. The lower diagram also shows how archeal lipids can be joined in the middle* (Source: Charles Cockell).

strung together in a long sequence of different combinations of these four letters. Here we investigate how the code is turned into useful proteins.

4.6.1 Transcription – DNA to RNA

The question arises: how do we go from the DNA information storage system to making proteins and other molecules needed by the cell for growth and reproduction?

This extraordinary transformation is accomplished in two stages. First, DNA is used to make a similar molecule called **ribonucleic acid, RNA**, that was also described in the last chapter. RNA is similar to DNA except that thymine is replaced by uracil (U), it has a ribose sugar with an extra oxygen atom in a hydroxyl group (DNA is *deoxy*-ribose) and it is generally (but not always) single-stranded. These are three important differences to DNA. DNA and RNA are similar in that they both contain a base, sugar and phosphate in each structural unit.

RNA is made from DNA as the DNA unwinds. The DNA unwinds along a small part of its length (about 15 base pairs long) called the **transcription bubble** (Figure 4.9). At this point on the DNA, an RNA polymerase, which is a protein with five subunits, binds together with a **sigma factor**, a small specialised protein that recognises particular **promoters** on the DNA. Promoters are sequences of DNA that correspond to the beginning of particular sequences that are to be decoded and in colloquial language they tell the RNA polymerase: 'start decoding the DNA into an RNA strand here'. The RNA polymerase reads along the strand, generating an RNA molecule, called **messenger RNA** or **mRNA**, as it goes. This mRNA molecule is sometimes called the mRNA transcript or primary transcript. It is essentially a copy of the DNA code with each base in the RNA corresponding to its partner base in the DNA. It is sometimes referred to as a complementary copy of DNA. The production of the mRNA strand occurs from the 3' to the 5' end of the DNA strand.

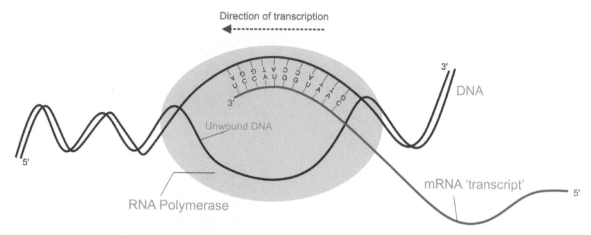

Figure 4.9   *The transcription of DNA into mRNA* (Source: Charles Cockell).

In prokaryotes, this process occurs in the cytoplasm of the cell. In eukaryotes, it occurs in the cell nucleus. In eukaryotic cells it must be transported out of the cell nucleus into the cytoplasm. The mRNA strand so produced can then act as the template for protein synthesis.

4.6.2   Translation – RNA to Protein

Another question now raises itself. How do we read the mRNA into protein? This is achieved in the next process called **translation**.

In translation, another type of RNA that is folded into a **ribosome** (sometimes called *ribosomal* **RNA** or **rRNA**) binds to the mRNA strand. It provides a scaffold on which other pieces of RNA called **transfer RNA** (**tRNA**) can bind. tRNAs are the molecules that bring in the amino acids to allow for the assembly of proteins. They can be considered adaptor molecules that bind to the mRNA and bring amino acids into alignment to add to a growing polypeptide chain.

Let's go through this process in a simplified way and consider what is happening. We have a mRNA strand, just transcribed from DNA. Around this strand the ribosome forms, made up of subunits which will provide the assembly point around which protein synthesis will occur. The ribosome is an impressive structure. In bacteria it is made up of a large subunit of RNA, consisting of two folded strands of RNA and no less than 31 specialised proteins. A small subunit, made up of one folded piece of RNA, called the 16S rRNA, together with another 21 proteins additionally binds to the mRNA (remember the **16S rRNA**, it will

become important later in the book when we explore the diversity of life on the Earth). This whole apparatus is now ready to make protein (Figure 4.10).

The ribosome provides a protected environment in which the tRNA can bind. The tRNA has an amino acid at one end and each amino acid has its own tRNA molecule associated with it. At the other end is an **anticodon**. The anticodon is a three-letter sequence of bases that matches three bases on the mRNA molecule (called the **codon**). Thus each codon or triplet code corresponds to a specific amino acid (Figure 4.11).

If we think about this quantitatively, we can immediately see that each codon can code for 64 combinations. There are four possible bases in the codons, G, C, A and U, and three codon positions, which gives us $4 \times 4 \times 4 = 64$ possible combinations. But we already saw in the previous chapter that there are generally 20 amino acids used by life. As a result, there is redundancy. Each amino acid can be coded for by more than one codon. We call this the 'degeneracy of the genetic code'. Figure 4.12 shows the mRNA codons and their corresponding amino acids.

Some codons code for the instruction to Stop reading and one of them (AUG – a methionine) to Start reading the mRNA strand.

Each amino acid brought to the mRNA in this way forms a peptide bond with the existing chain and so as new tRNAs bind to the mRNA a polypeptide or protein is synthesised, with the ribosomes moving along the mRNA strand. Thus the mRNA sequence has been translated into the primary protein sequence. This primary sequence will

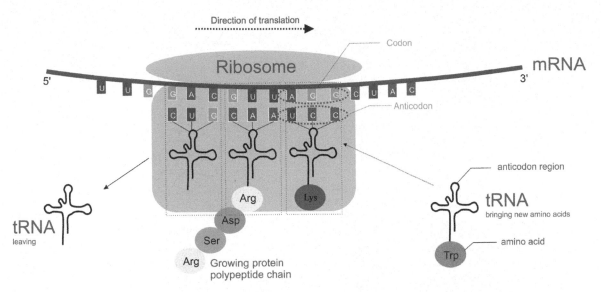

Figure 4.10   *The protein synthesis apparatus around the mRNA* (Source: Charles Cockell).

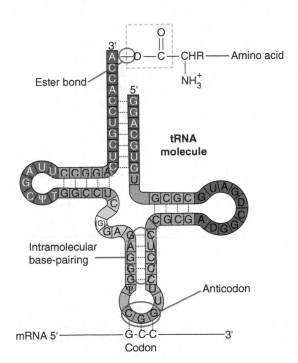

Figure 4.11   *The binding of a tRNA to mRNA with its amino acid. Notice the codon-anticodon binding* (Source: John Wiley & Sons, Ltd).

fold together to make the three-dimensional structure of a useful functioning protein.

The sequence of bases that codes for a single protein is called a **gene** and we call the entire complement of genes within an organism its **genome**. The genome size varies enormously between organisms. The human genome contains about 3240 megabases of DNA, bacteria have up to about 13 megabases of DNA depending on the species and they typically have about 4000 genes. The smallest genome of a free-living self-replicating organism belongs to *Carsonella ruddii*, which lives off sap-feeding insects. It has a genome of just 160 kilobases of DNA and 182 protein-coding genes. The smallest flu virus (which cannot replicate on its own) has only 11 genes. Although genome size is very loosely linked to complexity (bacteria tend to have smaller genomes than animals), this relationship is by no means general. Some **protozoa** (single-celled eukaryotes) have larger genomes than humans. This great difference between the genome sizes of organisms is called the **C-value paradox**. The 'C' refers the quantity of DNA between species, which early researchers thought was constant.

Some of the DNA in an organism is referred to as **non-coding DNA** as it has no known translation into protein. The amount of this non-coding DNA varies between species. In bacteria it can be around 2%, and it is up to 98% in humans. Sometimes called **junk DNA**, this is a

First position	Second position				Third position
	U	**C**	**A**	**G**	
U	UUU Phe	UCU Ser	UAU Tyr	UGU Cys	**U**
	UUC Phe	UCC Ser	UAC Tyr	UGC Cys	**C**
	UUA Leu	UCA Ser	UAA Stop	UGA Stop	**A**
	UUG Leu	UCG Ser	UAG Stop	UGG Trp	**G**
C	CUU Leu	CCU Pro	CAU His	CGU Arg	**U**
	CUC Leu	CCC Pro	CAC His	CGC Arg	**C**
	CUA Leu	CCA Pro	CAA Gln	CGA Arg	**A**
	CUG Leu	CCG Pro	CAG Gln	CGG Arg	**G**
A	AUU Ile	ACU Thr	AAU Asn	AGU Ser	**U**
	AUC Ile	ACC Thr	AAC Asn	AGC Ser	**C**
	AUA Ile	ACA Thr	AAA Lys	AGA Arg	**A**
	AUG Start (Met)	ACG Thr	AAG Lys	AGG Arg	**G**
G	GUU Val	GCU Ala	GAU Asp	GGU Gly	**U**
	GUC Val	GCC Ala	GAC Asp	GGC Gly	**C**
	GUA Val	GCA Ala	GAA Glu	GGA Gly	**A**
	GUG Val	GCG Ala	GAG Glu	GGG Gly	**G**

Figure 4.12 *The universal genetic code. Codons of mRNAs and their corresponding amino acids. The amino acids are shown with their three-letter designation (see Appendix). Note the degeneracy of the code.*

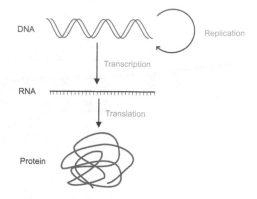

Figure 4.13 *A summary of the two main steps in going from DNA to proteins* (Source: Charles Cockell).

misnomer since it is becoming increasingly understood that a proportion of this DNA has biochemical functions, for example producing RNA molecules including ribosomes, or encoding viral DNA. Some of the sequences are pseudogenes. These are sequences that code for proteins that are not produced by the cell or are replicas of other genes that are not functional. Much of the C-value paradox is explained by non-coding DNA.

The process of going from DNA to RNA to protein is sometimes called the 'central dogma of molecular biology' (Figure 4.13). The word 'dogma' is always a troubling word in science, but the overall scheme broadly shows the two fundamental steps of reading the genetic code.

4.6.3 A Remarkable Code

There are a few remarkable things about this process worth mentioning and which should stun any self-respecting human being. First, the table of codons shown in Figure 4.12, bar some minor modifications in some organisms, is essentially universal to all life forms. This not only shows the great antiquity of the genetic code, but also strongly suggests that all life on Earth was derived from a single common ancestor in which this code

first emerged, presumably from more simple codes and structures that preceded it.

Second, the table is astounding because it is a code that allows a one-dimensional piece of information – the strand of a DNA molecule – to be transformed into the three-dimensional structure of a chemically active molecule. How did the structures emerge and what was the evolutionary process that linked one-dimensional information storage to three-dimensional biological function? This is one of the most fascinating questions in astrobiology, linked explicitly into our attempts to understand the origin of life and we will come back to this when we consider the origin of life.

Yet the information storage system is crucial to the cell since from this information all cell instructions and coordinated biological patterns emerge. Instructions for the cell to replicate itself and interact with its environment are provided. This information coordinates cells within complex multicellular organisms, allowing them to communicate with each other and create differentiated cell structures, with each cell playing a dedicated role in the whole.

4.6.4 DNA Replication

One of the most crucial events in the cell is DNA replication during cell division. This event underpins what many consider to be at least one characteristic of life – reproduction. DNA replication proceeds in three coordinated steps: initiation, elongation and termination.

In a cell, DNA replication begins at specific locations on the DNA, or **origins of replication**. In bacteria, there is a single origin of replication on their circular genome or **chromosome**, whereas in eukaryotes, that have longer linear chromosomes, they initiate replication at multiple origins. The unwinding of DNA at these locations and the synthesis of new strands results in a **replication fork** (Figure 4.14). Let's examine this process in more detail.

An enzyme called a helicase is used to separate the two strands of DNA, essentially 'unzipping' the hydrogen bonds. As the helicase separates the DNA at the replication fork, the DNA ahead of it is forced to rotate. This process results in a build-up of twists in the DNA and a resistance becomes established, which if not dealt with, would eventually halt the progress of the replication fork. A topoisomerase is an **enzyme** that temporarily breaks the strands of DNA, relieving the tension caused by unwinding the two strands of the DNA helix.

The replication fork generates two single strands of DNA. On one of the strands (Figure 4.14), DNA can be synthesised by a DNA polymerase that runs along the single strand following the replication fork as it moves forward. This is called the 'leading' strand. The DNA polymerase is an ancient **multimeric** enzyme that is responsible for assembling the nucleotides into the newly forming DNA strand.

However, on the other strand of DNA, the 'lagging' strand, synthesis must always restart near the replication fork as the DNA unzips (Figure 4.14). The reason for this is that DNA replication needs a source of energy to proceed. This energy is acquired by cleaving the 5'-triphosphate of the nucleotide that is added to the existing DNA chain, which means that DNA synthesis

can only proceed from the 5' to 3' direction. Thus, DNA replication has directionality. This results in the complication that the lagging strand must be synthesised in fragments. To accomplish this task an RNA primer is synthesised and attached to the DNA by a primase enzyme (which is actually a type of RNA polymerase since it makes RNA). DNA synthesis then proceeds from the RNA primer, accomplished by DNA polymerase. The strands of DNA produced in this way are called 'Okazaki' fragments. These fragments are stitched together with an enzyme called DNA ligase, generating the complete double strand.

There is one other problem during replication. Single-stranded DNA, produced after the helicase has separated the two DNA strands, tends to fold back on itself forming secondary structures. These structures would interfere with the movement of DNA polymerase along the strand. To prevent this, single-strand binding proteins bind to the DNA until the second strand is synthesised, preventing secondary structure formation.

Termination of DNA replication requires that the progress of the DNA replication fork be stopped. Termination involves the interaction between two components: a specific 'termination site sequence' in the DNA, and a protein which binds to this sequence to physically stop DNA replication. In various bacterial species, this protein is named the DNA replication terminus site-binding protein, or Ter protein.

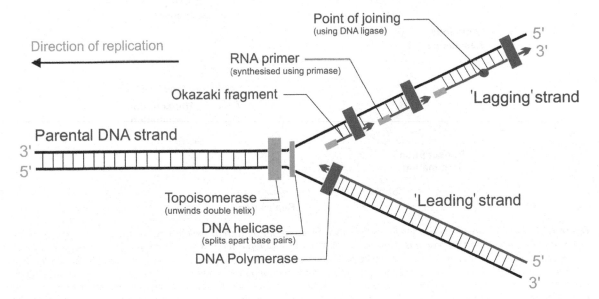

Figure 4.14 *The replication of DNA showing some of the diversity of machinery involved in the process* (Source: Charles Cockell).

Because bacteria have circular chromosomes, termination of replication occurs when the two replication forks that started out from the origin meet each other on the opposite end of the chromosome.

4.6.5 Plasmids

Organisms carry DNA in forms other than their primary chromosomal DNA. A plasmid is a small DNA molecule that is separate and can replicate independently of the chromosomal DNA in a cell, although it is considered to be part of the total genome of an organism (Figure 4.15). They are most commonly found as small circular, double-stranded DNA molecules in bacteria, although they have also been identified in archaea and eukaryotic organisms. They carry genes that can benefit the survival of the organism in the natural environment. Examples include genes for antibiotic resistance (which makes them of central significance in the transfer of antibiotic resistance between bacteria in hospitals) and resistance to heavy metals (which allows organisms to live in diverse environments with heavy metals from volcanic hot springs to human-polluted industrial sites). Plasmids can also provide bacteria with the ability to fix nitrogen gas from the atmosphere into useful nitrogen compounds or to degrade certain organic compounds that provide an advantage when nutrients are scarce. Plasmids vary in size from 1000 to over a million bases and the number of identical plasmids in a single cell can range from one to several thousand depending upon the environment and the species.

Artificial plasmids are widely used in molecular cloning, where they serve as 'carriers' of foreign genes that the researcher might wish to incorporate and have read inside a particular organism. For this reason, they are one of the most important tools in genetic engineering.

Plasmids are sometimes also called replicons, capable of replicating autonomously within a suitable host. However, plasmids, like viruses, are not considered to be a form of life. You might like to consider whether you agree with this view and why.

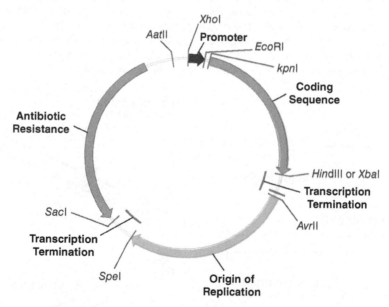

Figure 4.15 *Plasmids are most common in bacteria and can also be introduced by genetic engineers to make microorganisms produce industrially important products. This plasmid has an origin of replication, which is the sequence that allows it to be copied. It contains a 'coding sequence', which could be a DNA sequence producing an important drug. The 'promoter' is a DNA sequence that allows the coding region to be read. The transcription termination sequence tells the genetic machinery when to stop transcribing. It also contains a gene for antibiotic resistance which allows for bacteria successfully carrying the plasmid to be selected by researchers if they want to make sure that only the drug producing bacteria are growing. The other sites marked on this diagram (e.g. SpeI, KpnI) are locations where specific enzymes (restriction enzymes) will cut the DNA and allow the researcher to insert or remove bits of DNA.*

Debate Point: What is the minimum size of a cell?
The minimum size of a cell is an important measure, since it defines the size below which we would question whether a putative fossil or living cell was a self-replicating living organism. For a cell to grow and reproduce, it must have some minimal machinery to produce proteins needed for division and have other molecules and proteins to gather energy and carry out basic functions such as replicating genetic information. Consider the following statement from a United States National Research Council report on the 'Size Limits of Very Small Microorganisms': 'Free-living organisms require a minimum of 250 to 450 proteins along with the genes and ribosomes necessary for their synthesis. A sphere capable of holding this minimal molecular complement would be 250 to 300 nm in diameter, including its bounding membrane. Given the uncertainties inherent in this estimate, the panel agreed that 250 ± 50 nm constitutes a reasonable lower size limit for life as we know it'. Discuss this statement and whether you agree with the assessment of the minimum number of proteins required for life. How would the minimum size of a cell be different if it required ten times this number of proteins or 50 times this number?

Size Limits of Very Small Microorganisms: Proceedings of a Workshop, 1999. National Academies Press, Washington, DC (available online: http://www.nap.edu/books/0309066344/html/).

4.6.6 eDNA

One of the more enigmatic occurrences of DNA is extracellular DNA (eDNA), which is found outside the cell. Once thought to be DNA that had simply been released from dead cells, it is now understood to be involved in a variety of processes, including the formation of **biofilms** in bacteria. It may also provide a rich source of genetic information that can be sequestered by other organisms in the process of **transformation**, whereby naked DNA is taken up and incorporated into the genome of organisms. This will be discussed in more detail later in the book.

4.7 Cell Reproduction

In cellular organisms (the prokaryotes and in most eukaryotic cell division) cells divide by **mitosis** (called **binary fission** in the prokaryotes; Figure 4.16). During mitosis, DNA is replicated, generating two cells with exactly the same genetic composition at the original cell. All prokaryotes divide in this way. In prokaryotes, the rate of cell division can be sufficiently high that a cycle of DNA replication can be begun before the previous one is completed. Mitosis is used in multicellular eukaryotes for replicating cells such as hair and skin cells.

In many eukaryotes, including animals, plants and fungi, an additional form of replication is achieved, referred to as **meiosis** (Figure 4.17). Meiosis, put simply, is the process of making sex cells that can come together to make new individuals – this defines sexual reproduction. It seems logical then, that the central process of meiosis is to make cells with half the genetic complement,

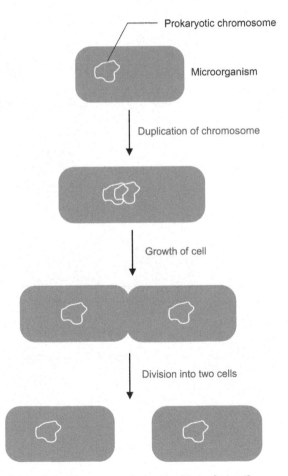

Figure 4.16 *The process of mitosis or binary fission* (Source: Charles Cockell).

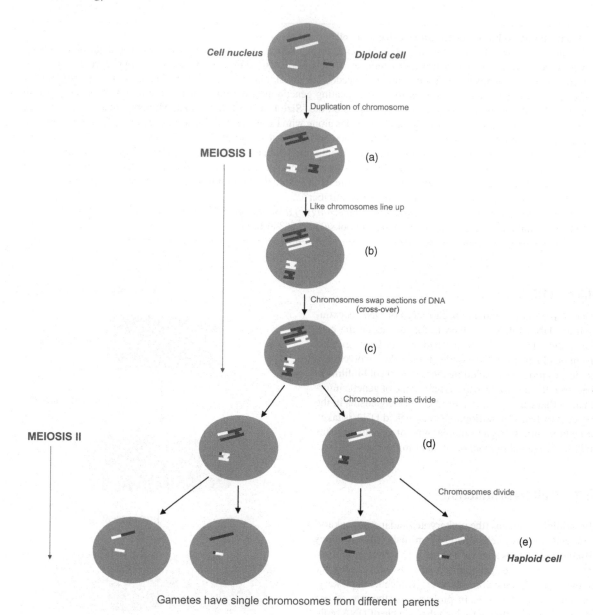

Cell nucleus   *Diploid cell*

Duplication of chromosome

MEIOSIS I   (a)

Like chromosomes line up

(b)

Chromosomes swap sections of DNA
(cross-over)

(c)

Chromosome pairs divide

MEIOSIS II   (d)

Chromosomes divide

(e)
Haploid cell

Gametes have single chromosomes from different parents

Figure 4.17   *The process of meiosis* (Source: Charles Cockell).

so that when the mother's and father's sex cells come together they produce a full complement again. This form of genetic replication is essential to generate the variety associated with sexual reproduction compared to mitosis, where variation is primarily dependent on errors introduced during DNA replication or the introduction of new genetic elements in what is otherwise the replication of one cell into two identical cells. We do not know exactly why sex evolved, but the production of variation by mixing genes and thus greatly increasing the potential number of variations from one generation to another may have provided some evolutionary advantage.

Meiosis can be explained by reference to typical animal cells involved in this process (Figure 4.17). Animal cells contain two sets of **chromosomes**. They are called **diploid** cells. One set has come from the mother and one set from the father. In the first stage, these sets of chromosomes are replicated in a diploid cell (Figure 4.17a). This process occurs before mitosis or meiosis. In mitosis these chromosomes would just separate into daughter cells, creating identical cells to the parent cell. However, in doubling up the chromosomes in meiosis, the cell has entered into the stage called Meiosis I. In the next step, the sets of doubled up chromosomes line up (Figure 4.17b) and then exchange genetic information (Figure 4.17c), generating chromosomes that are no longer identical to one another. This event is unique to meiosis and is called **cross-over**. By crossing over, segments of the chromosomes are mixed, generating variation. These chromosomes can now be divided into two new cells (Figure 4.17d). In

Meiosis II, the next stage of meiosis, these cells are again divided in an identical way to mitosis, generating four cells (Figure 4.17e). These are the sex cells. These sperm or eggs (gametes) contain half the genetic complement and are called **haploid** cells. They can join together in sexual reproduction in which chromosomes from the mother and father come together to generate new adult diploid cells, which begin the process again in the new individuals.

4.8   The Growth of Life

The modes of reproduction that we have just examined allow populations of cells to grow. In multicellular eukaryotes this process is controlled by a network of genes that lead to *cell* **differentiation**, whereby initially unspecialised cells change into the whole panoply of cell types in a given organism, such as skin cells, liver cells, heart cells. These cells are under the control of genes that produce **hormones** influencing cell differentiation and development. These pathways encompass an entire field of science that deals with cell developmental biology. We won't delve into this field any more here, but you are encouraged to find out more. Cell developmental biology provides many insights into the origin and evolution of plants and animals and particularly the evolution of cell differentiation.

Perhaps the best characterised growth patterns are to be found in the prokaryotic cells (Figure 4.18). They

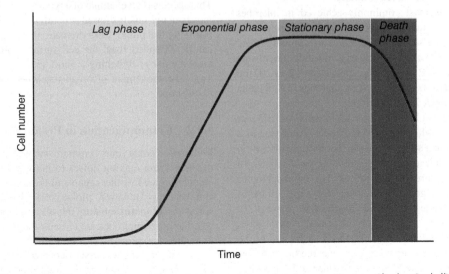

Figure 4.18   *The major phases of growth in a population of prokaryotes* (Source: Charles Cockell).

generally do not differentiate, although some fungi and slime moulds do exhibit primitive differentiation and even bacteria have specialised cell structures such as the **heterocysts** in cyanobacteria, which are specialised cells for fixing atmospheric nitrogen gas from the atmosphere into biologically available nitrogen compounds.

The pattern of growth follows a well-defined set of stages which is almost universal. In the first stage, the **lag phase**, the cells grow slowly. This represents the phase during which the cells are beginning to reproduce in the presence of newly available energy and nutrients. Following this stage, when the cells have adjusted to the new environment, they enter the logarithmic phase or **exponential phase** when growth is rapid. The organisms are not always growing according to an exact mathematical logarithmic function. The true growth rate will depend on the energy available. In the next stage, the cells run out of some essential nutrient or energy supply and they enter the **stationary phase**. Following the stationary phase, the cells will begin to die as they enter the **death phase**.

4.9 Moving and Communicating

4.9.1 Movement in Prokaryotes

Cells are not necessarily sessile. For many microorganisms, moving around is essential. By changing location they can get access to new energy resources, nutrients or move away from toxins (**chemotaxis**).

Movement is most commonly achieved in microbes with the use of a **flagellum** (plural flagella), a remarkable and exquisite cell structure that works by spinning round (Figure 4.19), driven by the energy molecule, **adenosine triphosphate**, ATP. Swimming speeds from 2 to 200 μm s^{-1} in water have been reported in microbes. The flagella are typically about 15–20 μm long.

A flagellum can be rotated at 200 revolutions per second. The main 'whip' of the flagellum is composed of a hollow cylindrical structure, about 20 nm (nanometres) in diameter, which is anchored to a rotating mechanism by a specialised structure called the hook. The chemical motor that rotates the flagellum is situated across the cell membrane and is made up of over 20 different proteins. To build the flagellum, new proteins are passed from the base, up through the hollow core, and added to the end of the filament, just underneath a cap protein at the tip. The flagellum is probably one of the most remarkable structures that have evolved in microorganisms.

In the simplest situation, there is one flagellum, or a group of flagella, at one end of the microbe, propelling the microbe forwards. To change direction, the flagella briefly rotate in the opposite direction, which causes the microbe to 'tumble' and so change the direction in which it is facing (Figure 4.20). When the original direction of rotation resumes, the microbe moves off in a new direction. In this way a microbe can move away from noxious substances or towards nutrients.

Not all microbes have flagella just at one end of the cell; some have groups of them at both ends. Some members of the family Vibrionaceae, which are widely found in the environment and include the organism responsible for causing the disease cholera, can either have flagella at one end of the cell or they can be all over the outside surface of the microbe, depending on whether they are attached to a surface or are free-swimming. Some of these different arrangements are shown in Figure 4.21.

Although flagella on the outside of cells are by far the most common means of achieving locomotion, it is by no means the only way microbes are able to move. Bacteria belonging to the group, the Spirochaetes, have numerous flagella located between the cell membrane and the cell wall. The microbes are helically shaped and can move using these internal flagella. However, to be able to move forwards, the flagella at both ends of the cell must move in opposite directions; if they moved in the same direction, the actions would cancel each other out and the cells would be stationary.

Pseudomonas aeruginosa is a common and well-studied microbe, found in soil, water and other moist locations. This species is an example of a type of microorganism that moves using much shorter structures than flagella, called **pili** (singular pilus). Less dramatic than the flagella, they can be extended from the cell surface and they achieve movement by 'twitching'. Short jerks can be observed as a slow movement of organisms when viewed down a microscope.

4.9.2 Communication in Prokaryotes

Multicelled eukaryotes communicate with a large array of methods from making noises to body movements. They are interpreted by other organisms through touch, eyesight and hearing. However, prokaryotes also communicate. Known as **quorum sensing**, this allows microbes to sense whether other microbes are close to them, and to regulate their chemical processes accordingly.

Quorum sensing was first discovered in the microbe *Vibrio fischeri* that colonises the light-producing organs of some fish and squid, where it is responsible for producing

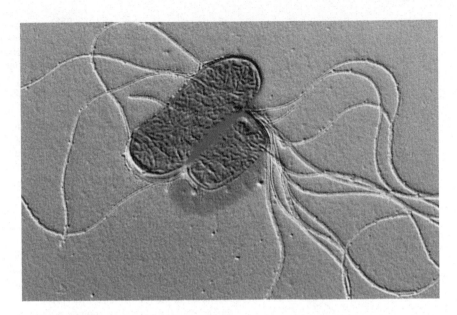

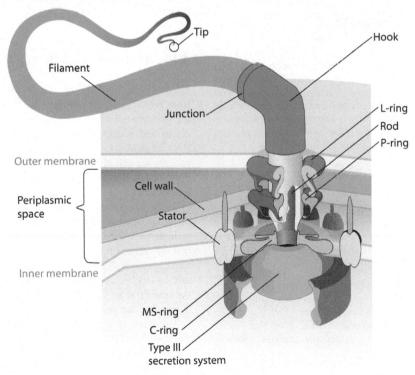

Figure 4.19 *The microbe,* Salmonella, *stained to show the flagella. Each organism is about 1 μm long* (Source: Centers for Disease Control). *Below is shown the structure of a bacterial flagellum in a Gram-negative bacterium. The structure is a motor embedded in the cell membranes and driven by ATP. It is comprised a number of protein subunits which are labelled* (Source: wikicommons, public domain).

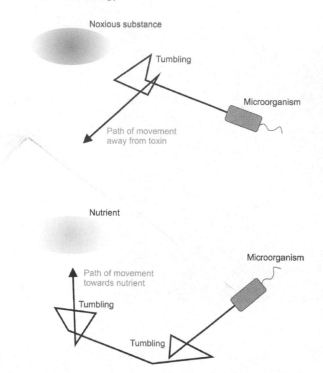

Figure 4.20 *A possible track of a microorganism that encounters a noxious substance or nutrients. Tumbling allows for the organism to randomise direction and, depending on whether it detects a greater concentration of noxious substance or nutrients after changing direction, it may initiate further tumbling events until it is either moving away or towards the concentration gradient (Source: Charles Cockell).*

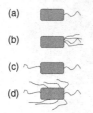

Figure 4.21 *Sketch showing different arrangements of flagella: (a) a single flagellum (monotrichous), (b) many flagella bundled at one end of the cell (lophotrichous), (c) one or more flagella at both ends (amphitrichous), (d) flagella distributed randomly over the surface (peritrichous) (Source: Charles Cockell).*

the light. When the *V. fischeri* microbes are in normal seawater, where they rarely come into contact with another member of the same species, they do not produce any light. However, once they reach a high concentration, such as inside the squid's light-producing organ, they sense the presence of others and the light-producing chemical reactions are triggered. Why would microbes sense other microbes? Quorum sensing provides microbes with a means to sense how much competition there might be for resources and to regulate their activity accordingly. In some ways it is a type of cooperation, but it also benefits each individual microbe to sense whether it is alone or not. For example, if there is intense competition for resources it might be better to use less energy and become more inactive. An example is to be found in the microbe, *Pseudomonas aeruginosa*, which can live as individual cells, but at a certain concentration they form a structured **biofilm** which can allow them to be more resistant to environmental extremes and even overcome host immune systems in the medical context. Quorum sensing is mediated by organic molecules such as very small chains of amino acids (peptides). The prokaryotic ability to engage in rudimentary forms of communication shows us that interactions between cells and complex social dynamics are not just the preserve of multicelled eukaryotes, but can also modify and shape prokaryotic populations.

4.10   Eukaryotic Cells

Eukaryotic cells have a number of crucial differences from prokaryotic cells (Figure 4.22), the most notable is the presence of a cell nucleus within which the DNA is stored and within which transcription occurs before the RNA is moved into the cytoplasm for processing. In the prokaryotic cell, the nuclear material is free floating in the fluid, or cytoplasm, of the cell and is often referred to as the nucleoid.

In the eukaryotic cell, the nucleus is surrounded by a membrane containing pores allowing for the movement of material in and out. Eukaryotic DNA also has important differences to prokaryotic DNA. The DNA contains sequences called **introns**. These are removed once the mRNA has been synthesised in the nucleus to leave only the **exons** (the parts of the mature mRNA that will be used to translate the genes).

A variety of other adaptations are found in specific eukaryotic cells (Figure 4.22). For example, a cell wall made of the polysaccharide, cellulose, which is constructed from repeating glucose molecules is found in plants. In fungi, a cell wall that contains chitin, a polysaccharide made from *N*-acetylglucosamine, a derivative of glucose, is found.

Many eukaryotic cells have a cytoskeleton, made up of a variety of structures including tubes and filaments

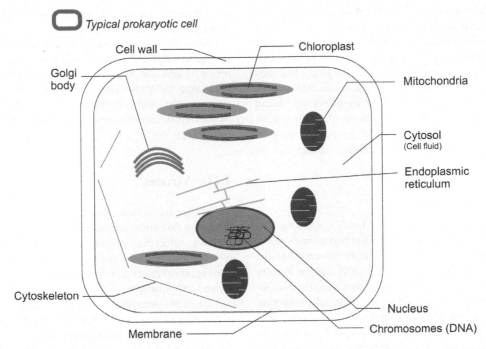

Figure 4.22 *A typical plant eukaryotic cell with some of its components. A typical size of a prokaryotic cell is shown for comparison* (Source: Charles Cockell).

of the proteins tubulin (microtubules) and actin (microfilaments), which provide a rigid structure to the cell and can allow it to change shape, such as when it is involved in tissue formation or movement towards nutrients and resources. The cytoskeleton plays a particularly important role in cell division, providing a scaffold during the separation of chromosomes. It is also used as a network for moving cell components around within the cell. Components travel along the microtubules like a railway track, attached to molecules such as kinesin. This is an ATP-requiring process.

Keratins are another type of filament made up of hard fibrous proteins. These are excreted in skin cells and account for the tough exterior of skin. When excreted outside the cell, keratins make hair and nails.

The cytoskeleton may not be completely unique to eukaryotes. Prokaryotes have been shown to have simple cytoskeletal structures made up of proteins such as crescentin, which seems to form a ring structure, providing shape to some bacterial cells.

Eukaryotic cells can contain a number of other organelles in which separate tasks are undertaken. For example, plant cells contain chloroplasts, the site of photosynthetic reactions, which we will explore more

in Chapter 5. Eukaryotic cells contain the endoplasmic reticulum, which is split into two types. 'Rough' endoplasmic reticulum is the site of protein synthesis, 'smooth' endoplasmic reticulum, the site of lipid synthesis. The Golgi body (or Golgi apparatus) is involved in packaging proteins from the endoplasmic reticulum, particularly proteins that are to be excreted from the cells. **Mitochondria** (singular mitochondrion), found in most eukaryotic cells, are ATP-producing organelles – the site of aerobic respiration, although they also have roles to play in cell signalling and differentiation. Although plants trap energy in chloroplasts, they still use mitochondria to break down the glucose they produce in photosynthesis as a source of energy for the cell. Animal cells have lysosomes, which are organelles containing enzymes that allow the cell to break down engulfed molecules as a source of food. Secretory vesicles are involved in the excretion of hormones and other chemical messengers.

The diversity of cellular structures made possible by the plethora of component parts allows for a wide variety of cell types. For example, the human body contains over 200 types of cells specialised for functions as diverse as passing electrical signals (neurons) or providing an external barrier (skin cells). Together these cells comprise the ~100 trillion cells that make up the human organism. Some of these cells last for a lifetime (some neurons) and some, such as white blood cells, can last for just a day. It is now recognised that this is not just a random case of cells dying when they get damaged or otherwise compromised. Many cells have a programmed cell death or **apoptosis**, which can be triggered by cell damage or when the cells are no longer required. These pathways presumably evolved to prevent rogue cells from causing damage to the organism.

4.10.1   Endosymbiosis

Where did these organelles come from? It was long suspected that maybe some of them had begun their origins as independent cells. For example in the mitochondria, the physically circular nature of their DNA and the content of the genetic code suggested that they were once bacteria that were acquired by eukaryotes, eventually becoming dependent on the host cell (Figure 4.23). This process of **endosymbiosis**, championed by biologist Lynn Margulis (1938–2011) is also thought to explain the acquisition of chloroplasts by eukaryotic photosynthetic organisms such

as plants and algae. The chloroplasts were once independent photosynthetic cyanobacteria.

Endosymbiosis has even been invoked to explain the emergence of the eukaryotic nucleus. Another explanation is that the nucleus was a part of the original eukaryotic membrane that split off to form a separate structure (the exomembrane hypothesis). The origins of the eukaryotic nucleus remain an intriguing problem in cell biology.

4.11   Viruses

Viruses are not cellular structures, but we must discuss them here for completeness, particularly since viruses are known to infect bacteria and archaea (as well as eukaryotes, such as humans) and they are known to be infectious agents in extreme environments such as volcanic hot springs and polar lakes.

Viruses are microscopic infectious particles that have a diversity of shapes and forms, from helical to bottle shaped. They are constructed of either single- or double-stranded DNA or RNA surrounded by a protein coat called a **capsid**. Some of them are surrounded by

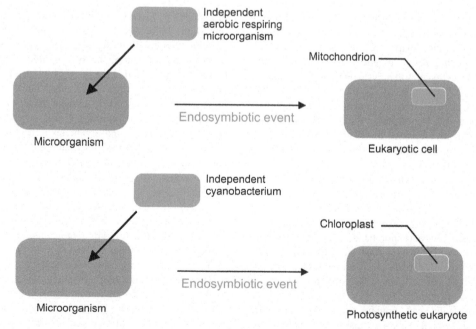

Figure 4.23   *A schematic illustration of the concept of endosymbiosis, whereby chloroplasts and mitochondria were acquired initially as independent prokaryotic cells* (Source: Charles Cockell).

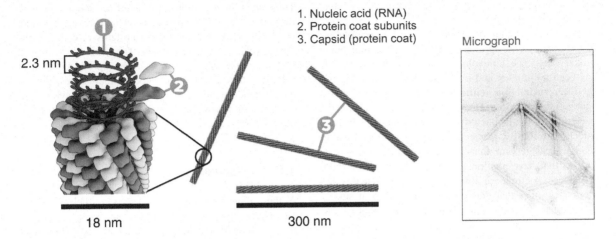

1. Nucleic acid (RNA)
2. Protein coat subunits
3. Capsid (protein coat)

Micrograph

2.3 nm

18 nm

300 nm

Figure 4.24 *The structure of the Tobacco Mosaic Virus (TMV) showing its nucleic acid and protein coat. Also shown is a micrograph of the virus* (Source: schematic from wikicommons, Thomas Splettstoesser).

a structure called the envelope, which is made of lipid and often has **glycoproteins** within it. The envelope plays a role in facilitating access to the host cell. In some sense, although viruses are not cellular, they still exhibit compartmentalisation of their structures. In the case of the Tobacco Mosaic Virus (TMV), the capsid forms a long, thin tube that surrounds the RNA (Figure 4.24). Viruses are typically 20–350 nm in size, much smaller than even prokaryotic cells. Viruses require a cell within which to replicate. This inability to replicate on their own has caused controversy about whether viruses can be considered 'life'.

The differences in the genetic material of viruses are important. RNA viruses carry the enzymes needed to copy and translate the RNA into proteins, meaning that they can often complete their life cycle in the cytoplasm of eukaryotes. Some RNA viruses such as the retroviruses (which include Human Immunodeficiency Virus, HIV, or AIDS) encode for the enzyme reverse transcriptase, which allows the RNA to be converted to DNA and then subsequently incorporated into the host genome as a **provirus**.

Viruses mediate important processes in the biosphere. They play an immensely influential role in the cycling of carbon in the biosphere by breaking apart or causing the *lysis* of bacterial cells in which they are reproducing, thus recirculating carbon in the oceans. They can mediate the transfer of genetic information from one cell to another as they infect and reproduce inside cells. Although they may not fit within a classical definition of life that includes the ability to reproduce, they are

certainly an important part of life on the Earth. There are about ten times as many virus particles on the Earth as there are bacteria – often about 10 billion of them per litre of seawater.

The origin of viruses is not certain. There have been diverse hypotheses. They include the ideas that viruses are independent entities that co-evolved with other cells, that they represent ancient degenerate cells that lost lots of genes and became dependent on other cells, or that they have evolved from the genes of larger organisms. RNA viruses have encouraged suggestions that they are the remnants of the **RNA World** (Chapter 10), prior to the evolution of the first cells. None of these hypotheses is adequate to explain the characteristics of all viruses and all of them have exceptions. It seems likely that viruses are very ancient and that they may have evolved several times in different ways.

The diverse patterns of nucleic acid structure in viruses and their non-cellular nature mean that they cannot readily be fitted into a traditional Tree of Life (Chapter 6), but rather form their own group of entities separate from cellular life. Nevertheless, their ancestry can be traced and many of them have ancient lineages. The Herpesviruses, enveloped DNA viruses, form a group of at least 150 viruses and are separated into three groups, the alpha, beta and gamma subfamilies. In humans they are responsible for diseases as diverse as chickenpox, cold sores and genital herpes. They are thought to have evolved about 400 million years ago in the Devonian and would have infected the first animals that emerged onto land. Their ancient lineage also accounts for the fact that they infect all human populations.

A prominent group of viruses are the **bacteriophages** or phages, a term applied to viral particles that infect prokaryotes. The co-evolution of bacteria and viruses has been a complex and prolonged interaction. Bacteria produce **restriction endonucleases**, enzymes which splice (cut up) virus DNA after its injection into the bacterial cell by the bacteriophage, preventing successful infection. **CRISPR** sequences are codes of DNA within the bacterial genome that correspond to sequences from viruses that have previously infected them. Using these sequences, bacteria are able to synthesise **nucleases** and RNA sequences that destroy the nucleic acid of similar phages that infect them at a later stage. This is a type of acquired immunity.

One reason why astrobiologists are interested in viruses is because of this intimate association between them and the prokaryotic world. Their role in information transfer between extant prokaryotes (Horizontal Gene Transfer; Chapter 6), complicates our efforts to build evolutionary trees depicting life on Earth and to unravel the origin of particular metabolic and biochemical pathways. An example of these complications is illustrated by the cyanophages that infect cyanobacteria. These viruses are capable of transferring photosynthetic genes and it has been estimated that 10% of the world's cyanobacterial photosynthesis is carried out by genes that have been transferred by cyanophages. There are many other examples of this phenomenon that suggest a need to understand the extent to which the activity of gene transfer by viruses influences the accuracy of inferred evolutionary relationships between organisms based on genetic information.

4.12 Prions

It would be incomplete not to mention **prions**. These entities are made of misfolded proteins that have disease-causing characteristics. One of the best characterised of these is the agent responsible for Scrapie, one of several **transmissible spongiform encephalopathies**, which affect brain and neural tissue. Prions can induce normal proteins in cells to misfold into a stable configuration, which can then cause other proteins to misfold, thereby generating a chain reaction of misfolded proteins.

These misfolded proteins are folded in such a way as to make them resistant to **proteases**, which are enzymes that normally break down defective proteins in cells. Prions are not yet known in prokaryotes. They have been reported in fungi. Like viruses, they can be reproduced in host cells, but they cannot reproduce by themselves, putting them outside most definitions of life. They invite us to advance our discussion of what operational definition we use for 'life'.

4.13 Conclusions

At the cellular level life is complex. Nevertheless, we can identify cell types and we can broadly recognise certain features about them that they all share. The basic biochemical structure of cell membranes, made of amphiphilic lipids, is common to all cellular life, although there are different types of membranes in different organisms. In prokaryotes, Gram-positive and Gram-negative membranes are the two main groups found in bacteria, with archaea showing unusual adaptations that might reflect their origins in extreme environments. Similarly, the reading of the genetic code, from DNA to RNA to protein, is common to life on Earth. The genetic code is exquisitely evolved for its role in storing information, yet we also have seen how it contains redundancy in the amino acids for which it codes (the degeneracy of the genetic code). Eukaryotic cells are more complex than prokaryotic cells, but exhibit some of the same fundamental structures. Despite our view of them as 'simple', prokaryotes exhibit some remarkably complex behaviours such as movement and quorum sensing. Life on Earth exhibits a vast variety of cellular structures and shapes and it interacts with non-cellular structures such as viruses. We must now explore where all these cellular structures acquire their energy to operate.

Further Reading

Books

Allen, T., Cowling, G. (2011) *The Cell: A Very Short Introduction*. Oxford University Press, Oxford.

Pross, A. (2014) *What is Life? How Chemistry Becomes Biology*. Oxford University Press, Oxford.

Papers

Bell, S.P., Dutta, A. (2002) DNA replication in eukaryotic cells. *Annual Reviews of Biochemistry* **71**, 333–374.

Bray, D. (1995) Protein molecules as computational elements in living cells. *Nature* **376**, 307–312.

Jacob, E.B., Becker, I., Shapira, Y., Levine, H. (2004) Bacterial linguistic communication and social intelligence. *Trends in Microbiology* **12**, 366–372.

Koonin, E.V., Novozhilov, A.S. (2009) Origin and evolution of the genetic code: the universal enigma. *Life* **61**, 99–111.

Kovacs, G.G., Budka, H. (2008) Prion diseases: from protein to cell pathology. *American Journal of Pathology* **172**, 555–565.

Lurland, C.G., Collins, L.J., Penny, D. (2006) Genomics and the irreducible nature of eukaryotic cells. *Science* **312**, 1011–1014.

Le Romancer, M., Gaillard, M., Geslia, C., Prieur, D. (2007) Viruses in extreme environments. *Reviews in Environmental Science and Biotechnology* **6**, 17–31.

Lombard, J., López-Garciá, P., Moreira, D. (2012) The early evolution of lipid membranes and the three domains of life. *Nature Reviews Microbiology* **10**, 507–515.

Martin, W., Roettger, M., Kloesges, T., Thiergart, T., Woehle, C., Gould, S., Dagan, T., Rogers, K.L., Amend, J.P., Gurrieri, S. (2012) Modern endosymbiotic theory: getting lateral gene transfer into the equation. *Journal of Endocytobiosis and Cell Research* **23**, 1–5.

Satkudo, A., Ano, Y., Onodera, T., Nitta, K., Shintani, H., Ikuta, K., Tanaka, Y. (2011) Fundamentals of prions and their inactivation. *International Journal of Molecular Medicine* **27**, 483–489.

Young, K.D. (2006) The selective value of bacterial shape. *Microbiology and Molecular Biology Reviews* **70**, 660–703.

5

Energy for Life

Learning Outcomes

➤ Understand the different types of energy acquisition in organisms.

➤ Describe the differences between chemo- and phototrophic forms of energy acquisition.

➤ Understand chemiosmosis, electron transport and some of the biochemical pathways of energy acquisition.

➤ Be able to calculate the proton motive force.

➤ Understand the fundamental role of ATP in energy acquisition.

➤ Describe the importance of different modes of energy acquisition in biogeochemical cycles.

➤ Understand the structure of a microbial mat and how it is formed.

➤ Be able to calculate the available Gibbs free energy for different microbial metabolisms using known redox reactions.

5.1 Energy and Astrobiology

Life requires energy. It needs it to do work, such as to grow, to reproduce, to repair and to synthesise all of the molecules from which cells are constructed. Indeed, the presence of an energy supply is known to be one of the crucial ingredients for an environment to be habitable (together with a source of liquid water, CHNOPS elements and a variety of other elements and compounds specific to particular organisms). So if we are to understand how life has persisted on the Earth for over 3.5 billion years and how it might exist elsewhere, if it does, then we need to know something about the energy sources that it might use. This part of astrobiology is so important we'll look at the variety of energy acquisition processes in life in some detail.

The behaviour of energy is governed by the laws of thermodynamics. Energy cannot be created or destroyed. This is the First Law of Thermodynamics and it is a formulation of the law of the conservation of energy. However, energy can be converted to mass and vice versa according to Albert Einstein's (1879–1955) famous relationship, $E = mc^2$, where E is the energy, m is the mass and c is the speed of light. The Second Law of Thermodynamics states that an isolated system will tend towards greater thermodynamic equilibrium or **entropy**. Systems will tend towards equilibrium either through the dissipation of mass or energy. A good analogy is the way that a cup of hot water added to cold water will lead to dissipation and homogenisation of the energy until the combined water is warm. Sometimes entropy is referred to as a measure of disorder. Colloquially this is acceptable, but it is not necessarily accurate in the sense that the important concept is the homogenisation of energy. A messy disordered bedroom, for example, may not have higher entropy than a well-ordered bedroom since the energy and mass in the two situations may be equally inhomogeneous. The Second Law is consistent with a number of observations about life. An organism cannot merely extract energy from the environment when it first comes into existence and then circulate this energy within the organism indefinitely. Reaction pathways are

Astrobiology: Understanding Life in the Universe, First Edition. Charles S. Cockell.
© 2015 John Wiley & Sons, Ltd. Published 2015 by John Wiley & Sons, Ltd.
Companion Website : www.wiley.com/go/cockell/astrobiology.

never 100% efficient and energy will be lost, for example in heat. Thus, life must continually obtain energy from the environment. Life manages to maintain this order against the tendency for entropic homogenisation of energy because it is not a closed system, but receiving energy from the external environment. A particularly important example is plants, which use light from the Sun as a source of energy.

5.2   Life and Energy

To order our discussion of how life gets energy from the Universe around it, it is first worth clarifying some definitions. These definitions are not only important for classifying life, but they also provide some structure to a discussion on what is a quite complex and involved area of biology and biochemistry.

We refer to the process by which an organism obtains its energy, carbon and other raw materials from the environment as catabolism. The term 'other raw materials' used here is a very loose one that includes a wide variety of items such as different elements (for example, calcium needed for bones and iron needed in blood in some animals) and in many organisms, vitamins, which are small organic compounds. When you eat your lunch you are engaged in catabolism. The process by which the products of this feeding are used to build up new molecules and cells is referred to as anabolism and we might also refer to this as **biosynthesis**. We can put both of these processes together into the generic term **metabolism**. So metabolism is the process that involves breaking down molecules to obtain energy and other essential molecules or elements and the process of building up new molecules. People have a tendency to use the word, metabolism, to describe either catabolism and/or anabolism and this is the word we'll use from now on.

We can classify metabolisms based on some basic needs. Metabolism is classified into nutritional groups on the basis of two major criteria:

1. **The source of energy**. Used for maintenance (such as repair), growth, reproduction and all other cellular functions.
2. **The source of carbon**. Carbon is the major backbone element of life. Although life also needs H, N, O, P and S and a variety of other organism-specific elements, the central importance of carbon means that it has taken its place as one of the defining requirements in the classification of different metabolisms.

Considering energy first, we can broadly split life into two groups. One group is the phototrophs that use light as an energy source. More precisely they use photons from the Sun, but in principle phototrophs could use photons from any star. The other group is the chemotrophs that obtain their energy from all other chemical sources, which we'll come to in a while.

We can also split organisms that gain their carbon from different sources broadly into two major groups. Autotrophs get their carbon from carbon dioxide (or the bicarbonate ion, HCO_3^- if the CO_2 is dissolved in water). These organisms are particularly important because they are the primary producers, organisms that turn carbon from the atmosphere into organic carbon compounds (sometimes called 'fixing' carbon) that can be used by the rest of the biosphere including the heterotrophs. Heterotrophs get their carbon from organic compounds. The heterotrophs include you and me. There is also a third group which we should mention here. The mixotrophs can get their carbon from both organic compounds and by fixing carbon dioxide. This is a versatile mode of existence because when carbon compounds are in short supply, they can switch instead to using atmospheric CO_2. Some varieties of pond algae are mixotrophs.

Now we can combine these two concepts into the following groups:

Photoautotrophs. Obtain energy from light and carbon from carbon dioxide.
Photoheterotrophs. Obtain energy from light and carbon from organic compounds.
Chemoautotrophs (sometimes called chemolithotrophs, or chemolithoautotrophs). Obtain energy from chemical compounds and carbon from CO_2.
Chemoheterotrophs (sometimes called chemoorganotrophs). Obtain energy from chemical compounds (primarily organic compounds) and carbon from organic compounds.

The flow diagram in Figure 5.1 summarises this complex, but quite logical set of names. The examples shown in Figure 5.1 will be discussed in this chapter.

5.3   The Central Role of Adenosine Triphosphate

In terrestrial life, the essential molecule used to transfer energy within the cell is adenosine triphosphate (ATP; Figure 5.2). Although other molecules such as lipids are

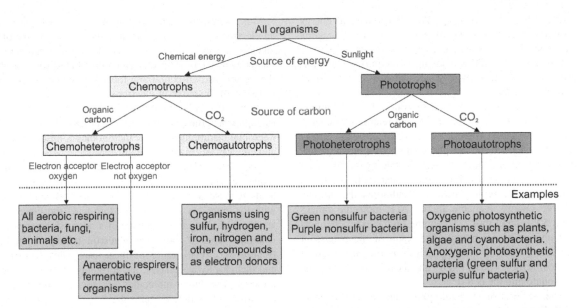

Figure 5.1 The variety of different metabolisms in life and some example groups of organisms that use them.

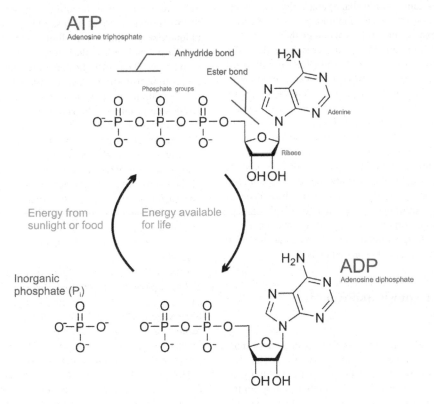

Figure 5.2 Adenosine triphosphate (ATP), its structure and cycle in the cell (Source: Charles Cockell).

also used to store energy, ATP is the universal energy currency for life. It is a small molecule that can be readily transported to sites where energy is required to carry out metabolism. ATP is made up of the purine molecule, adenine, linked to a ribose sugar, which is linked to three phosphate groups.

ATP is generally unstable in water. This instability, makes it ideal as a molecule that can release free energy quickly in places where it's needed. Most of the energy in ATP resides in the last phosphate bond (the anhydride bond) which has about twice the energy of the ester bond (the bond at which the triphosphate is linked to the ribose). Under typical cellular conditions, the amount of free energy available in the ATP molecule ranges from 50 to 65 kJ mol^{-1} depending on the exact chemical and physical environment. The result of the release of this energy during ATP hydrolysis is the formation of adenosine diphosphate (ADP) and the release of inorganic phosphate into the cell (P_i).

The addition of inorganic phosphate to ADP to form ATP (Figure 5.2) is referred to as **phosphorylation**. This is accomplished with an ATP synthase, a remarkable apparatus of protein subunits that operates like a rotor and is positioned across the cell membrane. The following sections will explore how this is accomplished.

It is also worth reminding ourselves here that ATP is used in other functions. For example, we have seen that it is the precursor to AMP (adenosine monophosphate), a nucleotide in DNA providing an adenine base. This versatility reflects its important role in biology and probably its very ancient origins. It is not necessarily the case, however, that ATP was the first and only energy molecule available to early life. Two phosphate groups can form an anhydride bond with each other to produce the energy-rich molecule, pyrophosphate. The addition of more phosphate groups results in the formation of polyphosphates. Both pyrophosphate and polyphosphates are found in microorganisms, suggesting that other permutations of phosphate-containing molecules are also ways to store energy in phosphate bonds.

5.4   Chemiosmosis and Energy Acquisition

To make molecules, to grow and reproduce, cells must gather energy. Although the overarching system is very complex, the principle behind it is quite simple. All organisms, whatever their metabolisms, ultimately rely on the movement of electrons to generate energy. How does this work?

Most organisms gather energy by first making use of an electron donor, or a compound that wants to give off an electron. An example would be electrons from organic matter, which acts as the electron donor. In humans, this is the organic carbon-containing food that you and I eat – it is essentially a source of electrons.

Within the cell membrane are a series of proteins through which the electrons move (a process that occurs in microbial membranes and the membranes of mitochondria and chloroplasts in eukaryotes).

The chemical properties of each molecule in the sequence (shown in Figure 5.3) are such that each one tends to have a greater affinity for electrons compared with the previous molecule. In this way, the electrons are shunted along the sequence in a successive series of reactions. This is called the **electron transport chain** (Figure 5.3). The nature of these molecules is varied and includes hydrogenases, flavoproteins and iron–sulfur proteins. The latter molecules are thought to be very primitive and include proteins that have Fe_2S_2 and Fe_4S_4 clusters, which are known as ferrodoxins. We'll see later that these iron–sulfur clusters have been suggested to be relics of an iron–sulfur mineral origin of life at deep-sea hydrothermal vents. Other molecules include ubiquinones, which are highly soluble non-protein **aromatic** molecules that readily diffuse within the membrane. The cytochromes are another ubiquitous electron transporting class of proteins. They have an iron-containing porphyrin ring (a cyclic structure). Different side groups on the ring generate a variety of these molecules (cytochrome a, b, c, d, and o) all with slightly different electron potentials and therefore all occupying slightly different places in different electron transport chains. Cytochromes are ancient and very diverse. Some of them serve the purpose of transporting electrons from one carrier to the next in the chain, some gather electrons from electron donors in the external environment and some are responsible for transferring the electron to the final player in the chain (the terminal electron acceptor).

Eventually the electron gets to the end of the chain and gets passed to a terminal electron acceptor. A good example of an electron acceptor is oxygen. In you and me this oxygen is the terminal electron acceptor in aerobic respiration – organic matter used as the electron donor, oxygen as the electron acceptor. We eat organic matter and breathe in oxygen to make energy.

How does this chain of electron movement from a donor to an acceptor harness energy? As the electrons are moved along the membrane they generate free energy (they become more stable and move to successively

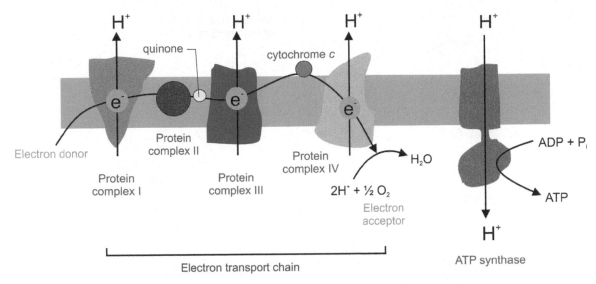

Figure 5.3 *A schematic example of an electron transport chain linked to ATP formation. Here the electron transport chain in mitochondria is shown. Electron transport involves protein complexes that pump protons (H⁺) out and an ATP synthase, which makes ATP, driven by protons moving back in* (Source: Charles Cockell).

lower energy states). This free energy is used to actively pump protons across the cell membrane from the inside of the cell to the outside. More accurately, in bacteria they pump protons from the cytoplasm to the periplasm in Gram-negative cells and to the outside of the cell in Gram-positive organisms.

The movement of protons to outside the cell results in the storage of energy in a proton gradient. It is stored in two forms. Chemical energy exists in the pH difference that is generated (ΔpH). Protons are the basis of acidity and so the gathering of protons outside the cell also implies a reduction in the pH outside the cell membrane. Energy also exists in an electrical potential ($\Delta\Psi$), caused by a net positive charge in the protons now outside the cell membrane.

These two components, a pH and charge difference give a tendency for protons to move back across the membrane to equalise the charge and pH. This is called the **proton motive force (pmf)**.

The pmf can be determined by using the following equation:

$$\text{pmf} = \Delta\Psi - (2.3RT/F)(\Delta\text{pH}) \qquad (5.1)$$

where R is the universal gas constant (8.314 J mol⁻¹ K⁻¹), T is the temperature, and F is the Faraday constant (96.48 kJ V⁻¹). A typical value for the pmf ranges from 150 to 200 mV.

The pmf can be used to generate ATP. As the membrane is generally impermeable to protons, the protons are forced to preferentially move back through an ATP synthase (sometimes also called the F-type ATPase) lodged in the membrane. This complex multi-subunit (**multimeric**) protein generates ATP from ADP and P_i. The ATP synthase has specific active sites that align the ADP with P_i in just the right way to cause them to form a covalent bond. The protons act to rotate the enzyme round like a ratchet – each ratchet changes the shape of the enzyme to bring an ADP and P_i together, generating a new ATP molecule (Figure 5.4).

As well as generating energy, the pmf can be used to carry out a variety of other useful biochemical functions. For example, a proton can be exchanged with a cation such as Na⁺ which is exuded from the cell in an electroneutral exchange involving a trans-membrane transporter. The pmf can also be used to drive the uptake of organic molecules, the electroneutral uptake of anions such as sulfate ions and other trans-membrane processes.

What we have been looking at here – the electron transport chain causing the pumping of protons resulting in a proton motive force which is then employed to make ATP, is collectively termed the **chemiosmosis theory** (Figure 5.5). It was first proposed by Peter Mitchell (1920–1992) in 1961. He won the Nobel Prize for this discovery in 1978. It is an extraordinary process because

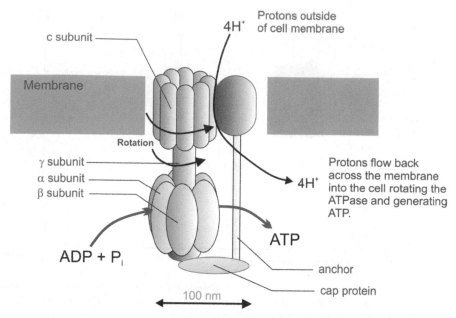

Figure 5.4 *The ATP synthase (or F-type ATPase) complex. Schematic showing ATP formation through the ATP synthase with some of the complex variety of subunits involved in the assembly of its structure. Typically three to four protons (four shown here) are thought to be required to generate one ATP molecule* (Source: Charles Cockell).

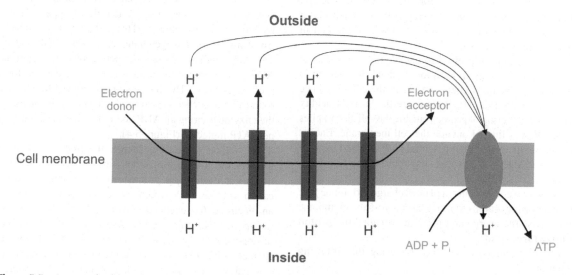

Figure 5.5 *A general schematic providing an overall summary of the principles of chemiosmosis showing electron transport, proton pumping to the outside of the cell and ATP formation. This is one of the most important means of harnessing free energy in life on Earth and uses the movement of subatomic particles (electrons and protons) to drive ATP production* (Source: Charles Cockell).

it essentially involves the shunting of subatomic particles (electrons and protons) to generate energy for life. Here we see an indivisible link between particle physics and biology.

5.5   What Types of Electron Donors and Acceptors can be Used?

In the previous section, organic compounds were discussed as examples of electron donors and oxygen as an electron acceptor. When organic compounds and oxygen are combined in this way to release energy we refer to this as aerobic respiration, but there are many other electron donors and acceptors that do the job. We can quantify just how good any element or compound is at being an electron donor or acceptor by specifying its electrode potential. The **electrode potential** (or oxidation/reduction potential; ORP) is a measure of the tendency of a chemical species to acquire electrons and thereby be reduced. It is measured in volts (V), or millivolts (mV). Each chemical species has its own intrinsic reduction potential. The more negative an electrode potential, the more the substance wants to give up an electron (it makes a good electron donor). The less negative, or positive, the potential, the more it behaves as an electron acceptor. The electrode potential is relative to a standard (usually H_2). Standard electrode potentials, E^o, are given for 25 °C and 1 atm, but in biological systems we also specify the standard electrode potential at pH 7 (written as $E^{o\prime}$).

We can rank electron donors and acceptors from the most negative to the most positive in a table (for examples, see Figure 5.6) which is called the electromotive series or **electron tower**. You can also see immediately that if we want an electron transfer to make energy we need electrons to move from a substance that has a more negative electrode potential to one that is more positive. Provided the electron donor is above the electron acceptor in the tower we can use a whole variety of electron donors and acceptors to transfer electrons and make energy.

It's also unsurprising that we call these electron transfer reactions, **reduction–oxidation reactions** or **redox reactions**.

Oxidation is the removal of electrons from a substrate. This is the reaction of the electron donor:

$$S_d \text{(the electron donor, here indicated with a}$$
$$\text{subscript } d) \rightarrow P_d \text{ (product)} + e^-.$$

The addition of electrons to the electron acceptor is reduction:

$$S_a \text{ (the electron acceptor, } a) + e^- \rightarrow P_a \text{ (product)}.$$

We sometimes refer to each reduction and oxidation reaction as a half reaction. Two half reactions or *redox couples* come together to make a complete electron transfer from one substance to another.

We'll return to how we quantify the thermodynamics of these reactions later.

5.6   Aerobic Respiration

The electron transfer chain we have just seen is part of a much larger apparatus in many organisms. We need to take a step back now and see how it fits into the larger scheme of metabolism. To begin with, the food must be broken down into smaller pieces to get at those electrons. You can't feed a sandwich directly into an electron transport chain.

Let's first return to aerobic respiration and explore the basic steps of how carbon molecules and oxygen are ultimately used to make energy.

First, the organism must break down the complex carbon compounds in food. There are a large variety of pathways for doing this depending on the substrate.

Lipids (fats), for example, are broken down by the process of beta oxidation, whereby the fatty acids are sequentially broken down by enzymes into the molecule, acetyl coenzyme A (acetyl-CoA). This molecule is a thioester between the compound, coenzyme A (a thiol) and acetic acid (an acyl group carrier). The structure is shown in Figure 5.7a. It is sometimes called the 'Hub of Metabolism' because of its central importance in metabolic pathways. We'll see in a moment where this product goes. Proteins are broken down by proteolytic enzymes that deaminate them, removing their amino acids. Long-chain sugars are broken down by enzymes that split the glycosidic links and produce the constituent sugar units, such as glucose.

These products of organic molecule metabolism enter into new pathways that break the products down further. One of the most important of these is the glycolytic pathway (glycolysis) or Embden–Meyehof pathway, which breaks down glucose into pyruvate (Figure 5.7b). This is accomplished by a series of reactions mediated by enzymes. Sugars other than glucose are modified by enzymes so that they too can enter into the glycolytic pathway. Pyruvate, the end product of glycolysis, as

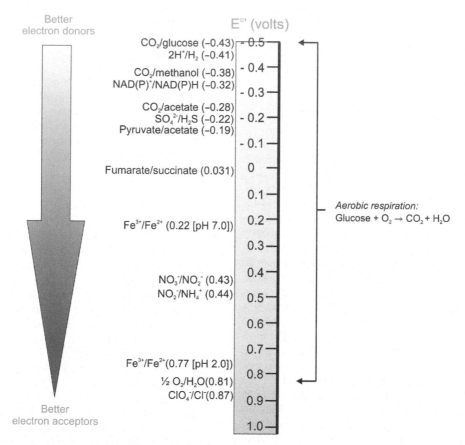

Figure 5.6   *An electron tower showing some examples of half reactions (redox couples) and their electrode potentials at standard conditions and at pH 7.0. Each half reaction is shown as a pair. If it is acting as an electron donor, the substance giving up the electron (e.g. glucose) is shown on the right and on the left the product that results (e.g. CO_2). If it is acting as an electron acceptor, the substance accepting the electron (e.g. O_2) is on the left and on the right, the product (e.g. H_2O). By coupling two of these reactions, energy can be generated from the flow of electrons from an electron donor to an acceptor. The example of aerobic respiration is shown. In this reaction, glucose is the electron donor and oxygen the electron acceptor. Later in the chapter the free energies from these reactions are quantified (Source: Charles Cockell).*

with lipids and proteins, is ultimately converted into acetyl-CoA.

Although the glycolytic pathway requires some ATP energy to get it going (steps 1 and 3; Figure 5.7b), it involves a series of steps during which phosphate molecules can be transferred to ADP to make ATP (steps 7 and 10). This process of making energy, by using a high-energy substrate (in this case the metabolic products of breaking down glucose) to generate ATP is called substrate-level phosphorylation. In breaking down one glucose molecule, the cell can generate two ATP molecules by this process.

Other pathways exist to break down sugars. The pentose phosphate pathway, which like glycolysis can operate with or without oxygen, is found in the cytoplasm of prokaryotes and eukaryotes and is used to break down sugars.

The result of the breakdown of sugars, proteins and fatty acids is a lot of acetyl-CoA. This molecule now enters the **Krebs cycle**, sometimes known as the tricarboxylic acid (**TCA**) cycle or just the citric acid cycle. The cycle is an ancient one and parts of it, or even reversals of its pathways, are found in a variety of organisms. The details of each step of the cycle and the enzymes involved are shown in Figure 5.7c. One essential function of the cycle is to convert the acetyl-CoA into CO_2 molecules (the ultimate fate of the organic molecules eaten by an aerobic respirer). It also plays a role in producing intermediate carbon compounds used, for example, in the *de novo* synthesis of amino acids in the cell.

Now in the process of breaking down the acetyl-CoA into carbon dioxide we can gather a lot of energy. As one acetyl-CoA molecule is broken down it produces two

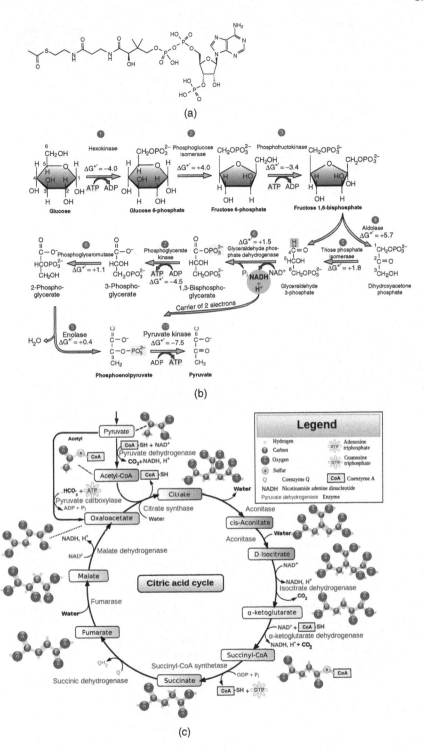

Figure 5.7 *The structure of acetyl Co-A and two ancient and important biochemical pathways. (a) The structure of one of the most important intermediates in metabolism, acetyl-CoA. (b) The glycolytic pathway and the structures of the intermediates* (Source: John Wiley & Sons Ltd). *(c) The Krebs or citric acid cycle* (Source: wikicommons, public domain).

molecules of carbon dioxide and releases a lot of electrons. Four pairs of electrons are transferred to a molecule called nicotinamide adenine dinucleotide (NAD^+) and one pair to flavine adenine dinucleotide (FAD^+). These two molecules are referred to as co-enzymes and their function in biochemistry is generally to accept electrons and then pass them on to other useful biosynthesis pathways. In biology electrons are often transferred in hydrogen atoms (which, remember, contain one proton and one electron). In the case of NAD^+ or FAD^+, two hydrogen atoms are transferred from a donor molecule to them, resulting in the transfer of two electrons to the molecule. However, the molecule can only accept one proton from the two hydrogen atoms, so one proton is left free and is liberated into solution. The overall reaction for adding electrons to NAD^+ for example is:

$$RH_2 + NAD^+ \rightarrow NADH + H^+ + R,$$

where R is the electron donor molecule.

These electron carriers can now enter the electron transport chain, whereupon the electrons travel through the membrane, generating a proton motive force and produce ATP as those protons move back across the membrane through ATP synthase. Eventually the electrons that move through the membrane will be taken up by oxygen in aerobic respiration as the terminal electron acceptor.

The energy gained from these various steps is impressive. For one molecule of glucose, a final total of 38 ATP molecules can potentially be produced in prokaryotes for energy-requiring reactions in the cells. The break-down of the ATP budget is as follows: 2 from glycolysis, 2 from the Krebs cycle and about 34 from the electron transport system. However, this yield is never in reality reached because of losses, such as leaky membranes. More realistic yields are 29–30 ATP molecules per glucose molecule.

Now we understand how the breakdown of carbon molecules is linked into the Krebs cycle which is itself linked into the electron transport chain and ultimately the electron acceptor (oxygen in aerobic respiration), yielding, in the process, ATP with which cells can go about their business. This overall process, which can be considered to be a giant metabolic funnel, is summarised in Figure 5.8.

5.7   Anaerobic Respiration

Oxygen is not the only electron acceptor that will mop up the electrons that have travelled through the process we

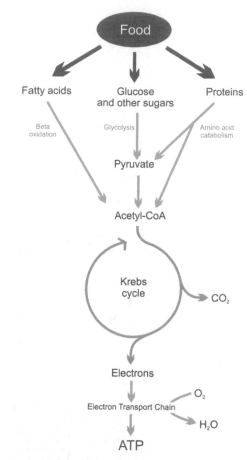

Figure 5.8   *A summary of the pathways for the break down and generation of energy from a diversity of carbon compounds in respiring animals* (Source: Charles Cockell).

have just described. There are plenty of other molecules and elements that will do the job, albeit with less positive electrode potentials than oxygen (i.e. further up the electron tower).

These **anaerobic** modes of growth produce less energy than aerobic respiration, typically about 10 times or even less, but they produce sufficient energy for microbial growth.

One of the most similar pathways to aerobic respiration is anaerobic respiration using nitrate (NO_3^-) as the electron acceptor. The nitrate is reduced to nitrite (NO_2^-) or ammonia (NH_3) or it is reduced to nitrogen gas (called **denitrification**) using a series of reductase enzymes. In some organisms, the nitrate is reduced to nitric oxide (NO) and nitrous oxide (N_2O) gases. The latter, in high

Focus: Astrobiologists. Tori Hoehler

What do you study? I study how life – particularly microbial life – distributes itself against the varying backdrop of physical and chemical parameters that comprise the diverse environments of the Earth, the Solar System and beyond. For Earthly ecosystems, which are my bread and butter, this is basically just microbial ecology. But I seek to understand the relationship between environment and life at a sufficiently fundamental level that the organising principles I find can be extended to thinking about the habitability of environments beyond Earth.

Where do you work? I've been fortunate to work in some amazing places. I've done quite a lot of work in Baja California (Mexico), as well as Yellowstone. I've been to the bottom of the ocean to work on hydrothermal vents and way above the Arctic Circle, to Svalbard. Currently, a lot of my work is closer to home, as I study subsurface microbial communities in the *ophiolites* of Northern California. I am based at the NASA Ames Research Center.

What was your first degree subject? I got my BS in chemistry and PhD in oceanography.

What science questions do you address? I am particularly interested in how the flow of energy – the supply from the environment, and the demand from life – influences ecology and habitability. This has led to a deep interest in understanding how life copes with extremely limited energy supply, as is likely the case in much of the habitable volume of our Solar System.

What excites you about astrobiology? What excites me about science in general is the creativity it requires. I suspect that many people do not think of science as a creative endeavour, but the best science is exactly that. I love that astrobiology requires a lot of creativity, a lot of 'mental gymnastics'. In particular, there is a careful line to walk between learning as much as possible from the specific example of Earthly life, but not being bound by it – in thinking as broadly as possible. The questions are big and fundamental; the thinking must be similarly so.

concentrations, is a potentially distinctive biosignature of life since it is difficult to achieve high levels of the gas by chemical processes alone. It has been suggested as a potential way to seek the gaseous signatures of life in exoplanet atmospheres, which we'll return to in Chapter 19. A large variety of bacteria including the genera *Pseudomonas*, *Bacillus* and *Staphylococcus* are capable of nitrate reduction. Many organisms are capable of both nitrate reduction and aerobic respiration and flip between the two, depending on the availability of oxygen and nitrate.

Other organisms are capable of using oxidised iron as an electron acceptor using organics as the electron donor. Oxidised iron is available in a huge variety of forms from dissolved Fe^{3+} to minerals such as hematite (Fe_2O_3) or non-crystalline iron oxides such as ferrihydride, a widespread form of iron with the formula $(Fe^{3+})_2O_3 \cdot 0.5H_2O$. *Geobacter* and *Shewenella* are common groups capable of carrying out this conversion. Iron-reducing bacteria are to be found in the Bacteria and the Archaea. Iron reduction results in the releases of reduced iron (Fe^{2+}) into the environment. The process is

immensely important in the oxidation of organic matter in the biosphere and the reduction of iron into soluble, and more accessible, dissolved Fe^{2+}.

A challenge faced by the iron-reducing bacteria is that the electron acceptor is quite often in the solid form of iron oxide minerals. A variety of evolutionary innovations have been identified that facilitate the ability of the organism to move electrons from the electron transport chain to solid oxidised iron compounds as the terminal electron acceptors. They include pili, which are small outer appendages that improve attachment of microbes to iron oxides, the production of quinones which act as electron shuttles from the cell to the iron oxides, and even the excretion of nanowires, long conductive appendages that feed the electrons from the organism to the metal surface. Iron reductase enzymes on the outer surface of cells enhance the transfer of electrons to the mineral surface.

Yet other bacteria are capable of using sulfate as the electron acceptor in sulfate (SO_4^{2-}) reduction. They include both Bacteria and Archaea (e.g. *Desulfomaculum* and *Desulfosporosinus*) and many ancient lineages from hot spring environments (e.g. *Thermodesulfovibrio*).

As sulfate is a stable molecule, cells must first activate it, using up ATP. The enzyme, ATP-sulfurylase, adds the sulfate ion to the phosphate group of an ATP. The activated product, Adenosine phosphosulfate (APS) can then be reduced to sulfite which itself can be reduced to sulfide.

The end product of sulfate reduction is sulfide, which is released as H_2S gas or binds with minerals to produce metal sulfides. The reduction of sulfate in this way can occur through a series of intermediates including trithionate ($S_3O_6^{2-}$) and thiosulfate ($S_2O_3^{2-}$) which if released into solution may be used by other organisms in sulfur reduction reactions, leading to a consortium of different sulfur-metabolising organisms.

The overall reaction for sulfate reduction using organic carbon as the electron donor is:

$$2CH_2O + SO_4^{2-} + 2H^+ \rightarrow H_2S + CO_2 + 2H_2O$$

Some of these anaerobic organisms display considerable versatility and are capable of switching between nitrate, iron and sulfate reduction depending on the availability of these different electron acceptors. They are broadly split into two groups, one group that can use a range of different organic molecules as the electron donor and one group that exclusively use the simple organic acid, **acetate**.

The anaerobic metabolisms just described are the tip of the iceberg. Uranium, arsenic and a variety of trace metals can act as electron acceptors, ultimately allowing for energy acquisition. Indeed, almost every theoretical redox couple that could generate energy has been found to be used by some type of organism. Even perchlorate (ClO_4^-) can be used as an electron acceptor for anaerobic growth. This anion has been found in the surface of Mars at ~0.4–0.6 wt% (Chapter 17).

Anaerobic respirations have great practical use. For example, iron-reducing microorganisms that can reduce uranium can be used in nuclear contaminated sites to carry out **bioremediation**. By reducing the uranium from its very soluble oxidised hexavalent state (U^{6+}) to its reduced U^{4+} state, which is less soluble, they can be used to immobilise uranium that might otherwise get into groundwater and cause serious contamination and a threat to human health.

These anaerobic metabolisms are of astrobiological significance because they represent the capacity to gather energy in anaerobic environments using locally available elements. The presence of iron oxides, abundant sulfate salts and perchlorates on the anaerobic surface of Mars raises obvious questions about whether that environment could have supported anaerobic respiration, perhaps using organic matter from carbon-rich meteorites if, of course, there was ever life there to take advantage of these redox couples.

5.8 Fermentation

In aerobic respiration we saw how glucose can be broken down to pyruvate (or pyruvic acid – pyruvate is just the ion formed when pyruvic acid is dissolved in water) in the glycolytic cycle. In the absence of oxygen we have a potential problem. The NAD^+ that we saw earlier, which has now gathered some electrons from carbon compounds (to become NADH) must get rid of them somehow. If there is no oxygen or other electron acceptors available, the NADH will build up and cell metabolism will grind to a halt. Many microbes get round this by using pyruvate itself as the electron acceptor for the reoxidation of NADH. The process is so effective that many organisms use it even when oxygen is available. This process is **fermentation** and it is sometimes called a disproportionation reaction in which a chemical compound is simultaneously reduced and oxidised to form different products.

The variety of fermentations that are carried out by microbes are extraordinary (Figure 5.9). Alcohol fermentation results in alcohols of obvious use to the wine, beer and drinks industry. Lactic acid fermentation occurs in animal skeletal muscles and gives you unpleasant cramps after sudden exercise. It is also important in cheese production. Bacteria turn the lactose sugar in milk into lactic acid, which causes the casein protein in the milk to denature and entangle, forming the whey. The addition of rennin, a **protease** which breaks down the whey, results in the curd, the raw material of cheese. A diversity of mixed acid and other acid fermentations have industrial and medical importance. The astrobiological relevance of fermentation is the observation that organisms on a planetary surface can obtain energy from organic materials without requiring a complete externally available redox couple.

5.9 Chemoautotrophs

So far, our discussion has been primarily limited to different electron acceptors and we've assumed that the electron donor is organic material. However, this is not mandatory and many organisms, the so called chemoautotrophs, use inorganic electron donors (see below).

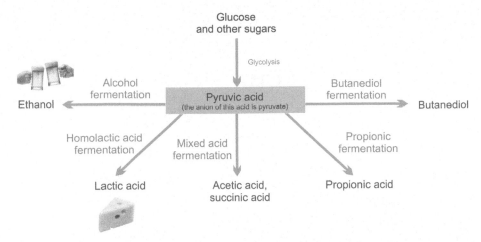

Figure 5.9 *Pathways for fermentation and just a few of the variety of possible fermentations. Some of them have obvious use in producing food and drink. However all of them are pathways for important base chemicals used in industrial chemical syntheses.*

It is worth noting here that the word **chemolithotroph** is used interchangeably with chemoautotroph. Almost all organisms that use inorganic electron donors are autotrophs (use CO_2 as the source of carbon) since they do not have access to readily formed organic compounds as a source of carbon (otherwise they could potentially use them to make energy).

The variety of electron donors is huge and includes hydrogen, reduced iron, ammonia, sulfur and other compounds. Some of these compounds, unlike many organic compounds that chemoheterotrophs use, are not readily dissolvable and so chemoautotrophs often have oxidase enzymes positioned in their membranes that facilitate the uptake of electrons from the external substances.

Some examples of chemoautotrophic redox couples found in microorganisms are shown in Table 5.1.

Autotrophy means that chemoautotrophs must use energy expensive alternative pathways such as the **Calvin cycle** (which will be discussed later when we consider the phototrophs) or the reverse Krebs or reverse tricarboxylic acid (**TCA**) cycle to fix CO_2 into carbon molecules. The reverse TCA cycle is essentially the TCA cycle run backwards to take up CO_2, instead of releasing it (Section 5.6). However, it is not a simple reversal. It has some additional specific enzymes, many of which contain iron–sulfur clusters (ferrodoxins) that act as electron storage and transport molecules.

The acetyl-CoA pathway is yet another means of fixing CO_2 used by many anaerobic autotrophs such as the **methanogens**. It makes substantial use of iron–sulfur proteins and enzymes that use transition metals such as nickel and cobalt. It is thought to be ancient.

Chemoautotrophy is a challenging means of gathering energy. The organisms must endure the relatively low energy yield of using compounds less suitable as electron donors (less negative electrode potentials) compared to organic molecules. Furthermore, many of the electron donors used by chemoautotrophs (apart from, e.g., hydrogen) have lower electron potentials than the half reactions required to produce **NAD(P)H** from $NAD(P)^+$

Table 5.1 *Some examples of chemoautotrophic redox reactions.*

Electron donor	Electron acceptor	Products	Microbial process
$4H_2$	$+ CO_2$	$\rightarrow CH_4 + 2H_2O$	Methanogenesis (anaerobic)
CH_4	$+ 2O_2$	$\rightarrow CO_2 + 2H_2O$	Methanotrophy
$10Fe^{2+}$	$+ 2NO_3^- + 12H^+$	$\rightarrow 10Fe^{3+} + N_2 + 6H_2O$	Iron oxidation (anaerobic)
$4H_2$	$+ SO_4^{2-} + H^+$	$\rightarrow HS^- + 4H_2O$	Sulfate reduction (anaerobic)
NH_4^+	$+ 4O_2$	$\rightarrow 2HNO_3 + 2H_2O$	Nitrification (aerobic)
$2S$	$+ 3O_2 + 2H_2O$	$\rightarrow 2H_2SO_4$	Sulfur oxidation (aerobic)

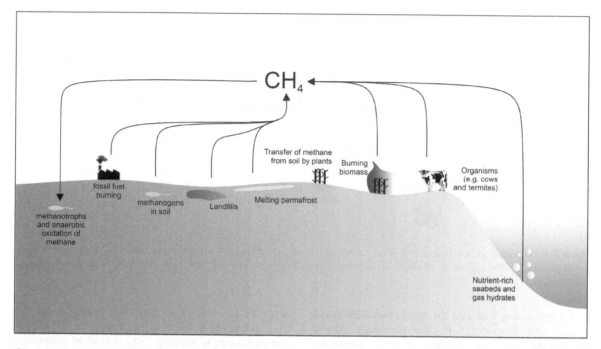

Figure 5.10 *The biological methane cycle showing some of the major sources of the gas (both natural and anthropogenic) and its microbial sink* (Source: Charles Cockell).

(Figure 5.6) used as electron donors in biosynthetic pathways such as the Calvin cycle. The consequence is that many chemoautotrophs, as well as gaining ATP from any given chemoautotrophic redox couple, must also use some ATP to push electrons 'uphill' to $NAD(P)^+$ so that they can produce the $NAD(P)H$ required to transport and donate electrons to their biosynthesis pathways.

5.9.1 Methanogens and Methanotrophs

The use of hydrogen as an electron donor is common and it can be coupled to the reduction of CO_2 in **methanogens**, which use the CO_2 both as an electron acceptor and their source of carbon. Methanogens are found in many anaerobic environments and most of them are Archaea. Their ability to use CO_2 from the atmosphere and H_2 (produced by geological processes such as **serpentinisation**), has made them of special interest to astrobiologists since they may represent the sorts of organisms that could persist independently of photosynthesis and exist in the deep subsurface of other planetary bodies.

An important group of chemoautotrophs are the **methylotrophs** that use various one-carbon compounds

as electron donors, including methane (these are the **methanotrophs**), methanol, methylamine and formaldehyde (Figure 5.10). The methanotrophs are particularly significant since they are responsible for the natural metabolism of methane and probably play a role in metabolising methane from deep sea sediments, methane that would otherwise accumulate in the atmosphere and contribute to the greenhouse effect. Therefore, understanding their contribution to methane breakdown has become a high priority in the light of the need to understand what might happen if frozen reserves of methane were eventually released through climate change.

Methane oxidation can also occur when methanogens take up residence with sulfate reducers in the oceans. The methanogens reverse their metabolism and instead of producing methane start to consume it. However, this only works if the hydrogen they produce is removed to prevent it building up and thus encouraging the reverse reaction (methanogenesis). Sulfate reducers that can use hydrogen as the electron donor instead of organic carbon (this reaction is shown in Table 5.1) use up the hydrogen produced by these 'reverse methanogenesis' organisms and so allow anaerobic methane oxidation to occur.

5.9.2 Sulfur Cycling

A prominent and geographically widely distributed group of chemoautotrophs are the sulfur oxidisers. These are organisms that can use reduced sulfur compounds such as elemental sulfur or sulfides as the electron donors. They broadly fall into two groups. One group includes *Thiobacillus*, *Sulfolobus* and other organisms that live in acid environments and can oxidise sulfide they find in the environment, coupling it to oxygen as the electron acceptor to generate energy. The other group, such as *Beggiatoa*, live at interfaces between reduced sulfide regions and low oxygen regions when oxygen is typically about 5% of atmospheric concentration (Figure 5.11). At these interfaces they have access to sulfide or other reduced sulfur species as well as oxygen as the electron acceptor. They are a very good example of how life takes up residence in places where there is a geochemical disequilibrium – where oxidised and reduced compounds co-exist. In the case of *Beggiatoa*, they can oxidise the sulfide to elemental sulfur, with the sulfur accumulating within the cells. The sulfur itself may then be used as an energy source – being oxidised to sulfate.

5.9.3 Iron Oxidisers

Metals are substrates for some organisms. One group of chemoautotrophs is the iron oxidisers. Fe^{2+} (ferrous iron) is generally unstable at neutral pH and rapidly oxidises to Fe^{3+} (ferric iron) forming rust and other compounds. However at acid pH it is relatively stable, accounting for the widespread presence of iron-oxidising organisms like *Acidithiobacillus ferrooxidans* found in acidic environments. These acidophiles live on the thermodynamic edge. They are only about 20% efficient in retrieving the energy from the redox couple and the energy available from iron oxidation with oxygen is very low, leading to energy yields on the order of 35 kJ mol^{-1} (compared to 2879 kJ mol^{-1} for aerobic respiration). They need over 100 moles of Fe^{2+} to ultimately produce 1 mole of glucose. Some organisms may be able to accomplish iron oxidation at neutral pH, an example being *Gallionella*, which form remarkable stalks that are thought to be involved in preventing iron toxicity (Figure 5.12). Although the energy yield for iron oxidation is greater at circumneutral pH compared to acidic pH, there is still some controversy about the extent to which reduced iron is actually used as an energy source under these conditions.

Iron oxidation can also be accomplished under anaerobic conditions. Species have been reported that can couple iron oxidation to the reduction of nitrate as the electron acceptor (anaerobic iron oxidisers). This reaction is shown in Table 5.1.

Often in the environment these organisms co-exist with the anaerobic iron-reducing bacteria (e.g. *Geobacter*), which reduce oxidised iron to ferrous iron, thus generating an iron cycle in which iron is both biologically oxidised and reduced.

5.9.4 Nitrogen Cycling and the Chemoautotrophs

When we looked at anaerobic respiration, we saw how denitrifiers turn oxidised compounds such as nitrates into ammonium or nitrogen gas and other intermediates. Nitrifiers are the groups of chemoautotrophic organisms that take ammonium (NH_4^+; at neutral pH) and turn it into nitrite and then nitrate, essentially the reverse set of reactions (with nitrogen gas first being fixed into ammonium by nitrogen-fixing bacteria).

Nitrifying chemoautotrophic organisms are abundant in any environment where there are fixed nitrogen compounds, for example notably in places where fertiliser run-off accumulates or in coastal regions where there is upwelling of nutrients near the shore. Two groups of organisms dominate; those that oxidise ammonium to nitrite (including *Nitrospira* and *Nitrosomas*) and those that convert nitrite to nitrate (such as *Nitrobacter* and *Nitrococcus*).

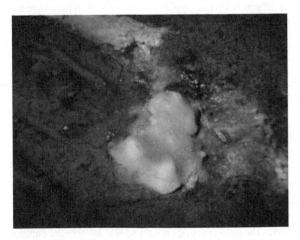

Figure 5.11 *White mats of* Beggiatoa *(centre) in the deep oceans that play a widespread role in sulfur metabolism. The red marks are laser markers from the deep-sea submersible taking the image* (Source: NOAA).

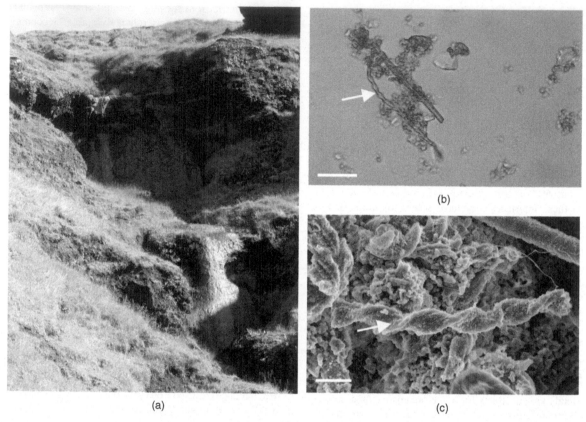

(a)

(b)

(c)

Figure 5.12 *(a) An environment in Iceland where iron oxidation and iron cycling are evident in the bright red compounds that form on a stream bed. The stream is about 1 m across. (b) An image taken in a light microscope showing the characteristic spiral structures that* Gallionella *produce (white arrow). The scale bar is 10 μm. (c) These structures can be seen in much greater detail at the higher magnifications possible in an electron microscope (white arrow). The scale bar is 2 μm. The stalks have been proposed to play some role in reducing iron toxicity.*

5.10 Energy from Light: Oxygenic Photosynthesis

We now turn to a ubiquitous form of energy acquisition – photosynthesis – using starlight to gather energy. There are two types of photosynthesis. Anoxygenic photosynthesis does not produce oxygen. Oxygenic photosynthesis does produce oxygen. Organisms that carry out oxygenic photosynthesis include cyanobacteria, algae and plants (Figure 5.13).

Oxygenic photosynthesis can probably be described, without over-exaggeration, as the most important metabolism on the planet. It generates most of the biomass on the Earth and fixes a vast quantity of CO_2 into carbon compounds each year – a net amount of about 260

billion tonnes of CO_2 is fixed. It also supplies the oxygen which acts as a terminal electron acceptor in a wide variety of metabolisms, including aerobic respiration and chemoautotrophic reactions. As we'll see later, it is also probably responsible for allowing the rise of multicellular animals, which depend on aerobic respiration. Animals require about 10% oxygen in the atmosphere to carry out aerobic respiration.

How does this remarkable mode of metabolism work? The process of oxygenic photosynthesis can be split into two parts, the light and dark reactions. As the name suggests, the light reactions are those that involve the capture of photons from a star, the dark reactions require no light and are the reactions that fix CO_2 into carbon compounds.

Let's first consider the light reactions. Photosynthesis works by the capture of light using light sensitive

Oxygenic photosynthesis
$$6CO_2 + 12H_2O^* + 48 \text{ photons} \rightarrow C_6H_{12}O_6 + 6H_2O + 6O^*_2$$

Figure 5.13 *The reaction for photosynthesis. The star indicates that the oxygen comes from water molecules. Below are some examples of oxygenic photosynthesisers. From top left: Cyanobacteria (Lyngbya), algae (Botryococcus braunii; Source: wikicommons, Neon ja), cactus (Parodia microsperma; Source: wikicommons, Dake) and tulip tree (Liriodendron tulipifera).*

pigments, the **chlorophylls** (Figure 5.14). These are large planar molecules belonging to a class of molecules called the chlorins. The chlorophylls are evolved for particular light environments, the most common in oxygenic photosynthesis being chlorophyll *a*, which absorbs light at 665 nm in the red region of the spectrum, but also blue light at 430 nm. The absorption of this red and blue light and reflection of green light is what gives oxygenic photosynthetic organisms their green colour.

In addition to chlorophyll, many organisms have accessory pigments such as beta-carotene, a carotenoid involved in trapping blue and green light; phycoerythrin, a red pigment which absorbs at about 550 nm; and phycocyanin which maximally absorbs at about 620–640 nm. All of these pigments are organised very specifically. The chlorophylls and accessory pigments are structured into antennas, containing up to 300 molecules, which trap the photons within reaction centres. There are two kinds of reaction centres in oxygenic photosynthesisers. Photosystem I (PS I) optimally absorbs light at 700 nm and photosystem II (PS II) at 680 nm.

This light-driven system, like all energy-acquiring systems we have already discussed, needs a source of electrons. They come from water (Figure 5.15). Water is split in the water oxidising complex. Every water molecule split generates two electrons. The electrons are transferred to reaction centre P680 in PSII. Photons captured by the reaction centres energise the electrons in P680 and allow them to reduce a pheophytin molecule, a chlorophyll molecule with the magnesium replaced by hydrogen. The electrons then shuttle through plastoquinone (sometimes called the 'Q cycle' because it involves quinones) to a cytochrome and eventually to a copper-containing protein, plastocyanin or the functionally equivalent protein, cytochrome c_6. In this process they generate a proton motive force which generates ATP. These electrons are then donated to the next photosystem in the chain, PSI.

In PSI, light is again used to excite electrons in the reaction centre P700, which then donates an excited electron to iron–sulfur proteins. The electrons are transferred eventually, via ferrodoxin, through the electron transport chain

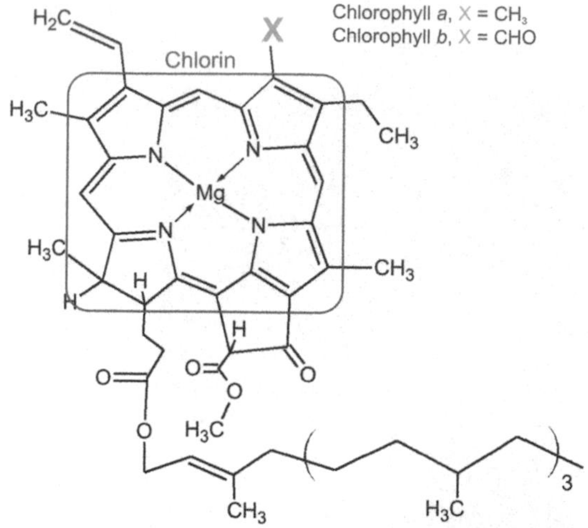

Chlorophyll *a*, X = CH₃
Chlorophyll *b*, X = CHO

Chlorin

Figure 5.14 *The structure of chlorophyll. The side group that gives rise to two types, chlorophyll* a *and* b, *is shown. The brackets at the bottom of the structure denote a repeating unit.*

again generating a proton motive force for ATP formation (Figure 5.15). These electrons can move back to P700 and begin another round of ATP generation. This is referred to as cyclic phosphorylation. However, there is another pathway: non-cyclic phosphorylation. It does not involve the production of a proton motive force. Instead, the electrons are taken up by the electron acceptor **NADP⁺** to form NADPH (nicotinamide adenine dinucleotide phosphate). This molecule will go on to take part in turning the CO_2 taken up by the plant into sugar.

The two-step process we have just discussed is described as the 'Z' scheme of photosynthesis and it

is shown in Figure 5.15. You will notice that the electrons pass into PSII *before* PSI. The naming is a quirk of history. Photosystem I was discovered before photosystem II.

The final part of photosynthesis are the dark reactions. Once ATP has been generated in the light reactions it can contribute to the fixation of CO_2 into organic molecules. In oxygenic photosynthesisers this pathway is the **Calvin cycle**. CO_2 is taken up by an enzyme called **Rubisco** (ribulose bisphosphate carboxylase/oxygenase) and shunted through a series of enzymatic transformations whereby ATP is used to generate fructose-6-phosphate, a molecule used to synthesise glucose or other organic carbon compounds. Other pathways include the reductive citric acid cycle, the ribulose monophosphate pathway and the serine pathway, but they all achieve the same end – the use of the products of the light reactions to drive the fixation of CO_2 gas (or bicarbonate ions, HCO_3^-, in aqueous environments) into carbon compounds.

5.11   Anoxygenic Photosynthesis

There are photosynthetic organisms that do not use water as a source of electrons to drive energy acquisition from starlight. The process they used is called **anoxygenic photosynthesis**. This process is probably much more ancient than oxygenic photosynthesis, but we looked at oxygen photosynthesis first to get to grips with what is the most globally important and pervasive type of photosynthesis.

Anoxygenic photosynthesis is found in a variety of organisms including: green sulfur bacteria (e.g. *Chlorobium*), green non-sulfur bacteria (e.g. *Chloroflexus*), purple sulfur bacteria (e.g. *Chromatium*), purple non-sulfur bacteria (e.g. *Rhodobacter*).

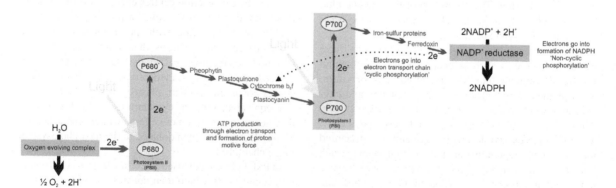

Figure 5.15 *The 'Z' scheme of oxygenic photosynthesis* (Source: Charles Cockell).

There is a diverse list of features that distinguish oxygenic photosynthesisers from anoxygenic photosynthesisers, but some key differences will be highlighted here.

First, they lack photosystem II (PS II). Instead they just have a photosystem like photosystem I (PS I) and they perform cyclic phosphorylation using this photosystem. The photosystem is like PSI in that it is not itself capable of extracting electrons from water. Instead of receiving electrons from the cleaving of water, the electrons come from hydrogen, hydrogen sulfide, elemental sulfur, reduced iron or organic molecules (accounting for the fact that they do not produce oxygen, hence anoxygenic photosynthesis).

Second, they have very different pigments called the **bacteriochlorophylls** that absorb at longer wavelengths than chlorophylls. Bacteriochlorophyll *a*, for instance, typically has absorption maxima at 364 and 770 nm when extracted in polar organic solvents (Figure 5.16), although in living cells, such as purple bacteria and green sulfur bacteria, the absorption maximum is typically 830–890 nm. Bacteriochlorophyll *b* has absorption maxima of 835–850 and 1020–1040 nm in living purple bacteria. These wavelengths suit the habitats of these organisms in sediments and in 'mats' of microbes where other wavelengths have either been absorbed by sediments or even by oxygenic photosynthesisers living above them.

A good example of an anoxygenic photosynthesis cycle can be seen in Figure 5.17, illustrated for green sulfur bacteria. The photosystem P840 bacteriochlorophyll is excited by light to a higher state, with the electrons moving through iron–sulfur proteins and cytochromes back to the reaction centre in cyclic phosphorylation as additional electrons are provided by sulfide reduction (Figure 5.17). The additional electrons are required because cyclic phosphorylation is not a perpetual motion machine. Leakage causes a loss of electrons, which must be replenished.

The green sulfur bacteria, such as *Chlorobium*, are strict anaerobic phototrophs and require sulfide to do photosynthesis. They are usually to be found in deep sediments or at the bottom of microbial mats. The organisms produce elemental sulfur from sulfide which they accumulate inside their cells. If the sulfide is limited they can oxidise the elemental sulfur to sulfate and generate energy.

The green non-sulfur bacteria are populated by organisms such as *Chloroflexus*, which are thermophiles that inhabit hot springs (Figure 5.18). They form distinctive mats around the edges of hot springs which are held together by their entangled filamentous growth. They are metabolically rather capable, being able to grow photoheterotrophically on simple organic compounds like amino acids or they can use sulfur, sulfide and reduced iron as electron donors for photosynthesis. They possess

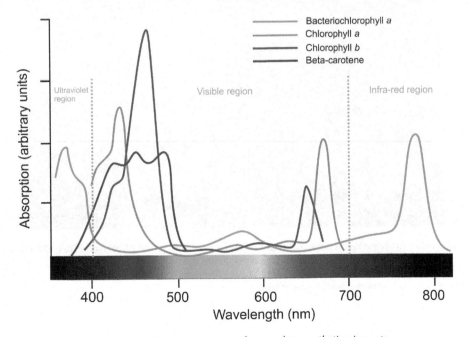

Figure 5.16 *Absorption spectra of some photosynthetic pigments.*

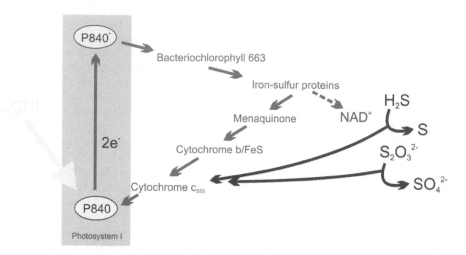

Figure 5.17 *Cyclic phosphorylation in green sulfur bacteria. The electron donors, unlike water in oxygenic photosynthesis (Figure 5.15), are sulfur compounds (black reaction schemes)* (Source: Charles Cockell).

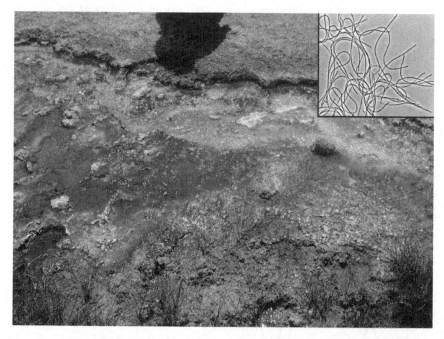

Figure 5.18 *Microbial mats in the Yellowstone National Park include the anoxygenic photosynthesiser,* Chloroflexus, *seen here forming a brown microbial mat. The image is about 1 m across. In the inset a micrograph of the filamentous organism (which causes the formation of the mats) can be seen. The width of the filaments is about 1–2 μm.* (Source: main image, Lunar and Planetary Institute; inset, Sylvia Herter, Joint Genome Institute).

bacteriochlorophyll *c* which has an absorption maximum at 740 nm.

The purple sulfur bacteria, such as *Chromatium*, possess the long wavelength varieties of bacteriochlorophylls (e.g. *a* and *b*) and a variety of red and purple pigments that absorb short wavelength radiation (between 450 and 550 nm) which allow them to live in deep water and sediments where infrared light may be filtered out, but the carotenoids can capture the short wavelength light. The organisms are again very versatile and can use hydrogen, thiosulfate or reduced iron as electron donors. Some of them can grow chemoautotrophically and others (such as *Thiocapsa*) can grow heterotrophically, feeding off available organics. Like the green sulfur bacteria, they can accumulate sulfur within the cell. When sulfide is limited they can use this as an electron donor, oxidising it to sulfate.

Finally, the purple non-sulfur bacteria include organisms like *Rhodobacter*. Unlike the purple sulfur bacteria, they cannot live in high concentrations of sulfide, requiring them to live in different habitats from the purple sulfur bacteria. The *Rhodobacter* can use reduced iron and hydrogen as electron donors. Some species can produce elemental sulfur outside the cell as a result of anoxygenic photosynthesis. They can also metabolise organics and grow photoheterotrophically using simple organic compounds which must either be provided abiotically in the environment or produced by the breakdown of complex macromolecular

organics by other heterotrophs. We call a situation where the waste products of one microbe are the food for another a **syntrophy**.

Finally, we should mention for completeness the *Heliobacteria*, which are genetically distinct from the previous groups of organisms and are capable of photoheterotrophy.

5.12 Global Biogeochemical Cycles

We have examined a variety of different metabolisms, but as was mentioned when we discussed the iron oxidisers and reducers, the combination of these metabolisms results in cycles. Sulfur is oxidised by sulfur bacteria to sulfate, sulfate is reduced back to sulfur compounds by anaerobic respirers (or chemoautotrophs using hydrogen as the electron donor to reduce sulfate). Carbon is broken down by chemoheterotrophs to carbon dioxide, carbon dioxide is fixed into carbon compounds by autotrophs and so on.

These transformations, driven by the need for energy and carbon, have profoundly altered our planet. By changing oxidation states and solubility, microbes mobilise elements, cause the formation of new minerals and drastically change the atmosphere. We'll learn more about some of these changes later in the book, but for the time being let us consider nitrogen (Figure 5.19).

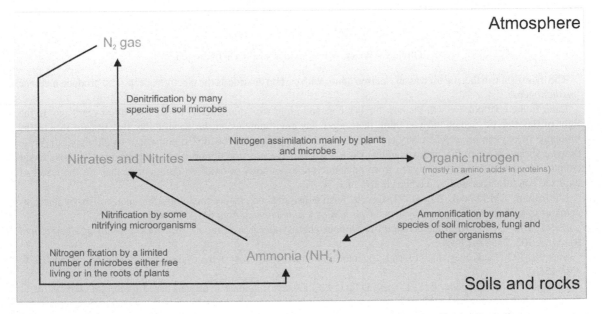

Figure 5.19 *The planetary scale biological nitrogen cycle on Earth* (Source: Charles Cockell).

Each year about 160 million tonnes of nitrogen gas is fixed by bacteria directly from the atmosphere into ammonia. This is done by both free-living nitrogen-fixing bacteria (e.g. *Azotobacter* and *Azospirillum*) and **symbiotic** nitrogen fixers found in roots of legumes (e.g. *Rhizobium* and *Bradyrhizobium*). Cyanobacteria also contribute significantly to nitrogen fixation, such as *Trichodesmium* in nutrient-poor tropical and subtropical waters and *Crocosphaera* in tropical waters. Nitrogen fixation is carried out using **nitrogenase**, an enzyme containing iron and either molybdenum or vanadium as a co-factor. It is a very energy intensive process. About 16 ATP molecules are required to split one N_2 molecule.

Once nitrogen is fixed in this way it is then transformed into other nitrogen compounds in nitrification as discussed earlier. For example, ammonium (NH_4^+) is fixed to nitrite (NO_2^-) by organisms such as *Nitrosomonas* and then transformed to nitrate (NO_3^{2-})

by organisms such as *Nitrobacter* in chemoautotrophic energy and carbon acquisition.

The nitrate produced in this way can be reduced into organic nitrogen and incorporated into biomass in assimilatory nitrate reduction, done by many microbial species including certain *Pseudomonas* and *Bacillus* species.

Eventually, the nitrogen is returned to the atmosphere. This occurs in denitrification reactions performed by anaerobic respirers (chemoheterotrophs) as we saw earlier. Thus, in nitrogen fixation we find chemoautotrophy and chemoheterotrophy working as part of a huge planetary scale cycle (Figure 5.19). Some of the products of this cycle (e.g. the gas N_2O), produced during denitrification, are considered biosignature gases with promise as a means to attempt to detect life on extrasolar planets using spectroscopy.

No less important is the carbon cycle. The *lithosphere* is only 0.032% carbon by weight, yet carbon is cycled through it and the atmosphere on an immense scale. A net

---

Debate Point: Life independent of photosynthesis

Many of the energy-gathering systems we have looked at ultimately depend upon oxygenic photosynthesis. Organisms that use oxygen as the electron acceptor, including aerobic respiring and some chemoautotrophic organisms, depend upon oxygenic photosynthesis to produce oxygen at sufficient concentrations. Organic matter produced by photosynthesis powers most chemoheterotrophic metabolisms that use organics as the electron donor (although abiotic organics could potentially provide another source as we shall see in Chapter 9). There have been very few descriptions of ecosystems that can run entirely independently of photosynthesis. One such system could be in deep rocks where organisms would use hydrogen produced from serpentinisation, in which minerals (such as olivine) react with water to make hydrogen in the following reaction:

$$3Fe_2SiO_4 + 2H_2O \rightarrow 2Fe_3O_4 + 3SiO_2 + 2H_2$$

Olivine + Water → iron oxides + silica + hydrogen

The hydrogen can then be used as an electron donor with carbon dioxide as the electron acceptor to produce methane (*methanogens*).

Later in the textbook we will see how oxygen concentrations rose dramatically about 2.4 billion years ago. In the light of this rise of oxygen, do you think there are many environments on the Earth today that remain free of the influence of oxygen on geochemistry and therefore on energy supplies for life? If so, where would they be? Write down a variety of microbial redox couples. Decide whether the electron donors and acceptors used in these redox couples ultimately depend on oxygen from oxygenic photosynthesis or oxidised compounds that are the result of oxidised conditions caused by oxygen photosynthesis.

McCollom, T.M., Shock, E.L. (1997) Geochemical constraints on chemolithoautotrophic metabolism by microorganisms in seafloor hydrothermal systems. *Geochimica Cosmochimica Acta* **61**, 4375–4391.

Shock, E.L. (1997) High temperature life without photosynthesis as a model for Mars. *Journal of Geophysical Research* **102**, 23687–23694.

Stevens, T.O., McKinley, J.P. (1995) Lithoautotrophic microbial ecosystems in deep basalt aquifers. *Science* **270**, 450–455.

Anderson, R.T., Chapelle, F.H., Lovley, D.R. (1998) Evidence against hydrogen-based microbial ecosystems in basalt aquifers. *Science* **281**, 976–977.

total of over 260 billion tonnes of atmospheric CO_2 is fixed by photosynthesis annually. The majority of the fixation occurs in marine environments. The scale of this process accounts for the fact that **Rubisco**, involved in the Calvin cycle, is the most abundant protein on the Earth. The distribution of carbon on Earth is instructive. The oceans contain about 38 500 billion tonnes, but the lithosphere contains much more than 70 000 trillion tonnes, locked up in carbonates and **kerogens**. Although fossil fuels (including coal and oil) account for a relatively small 4100 billion tonnes, the atmospheric total of 720 billion tonnes of carbon is small compared to this and other pools. Thus changes in the balance of atmospheric sources and sinks in exchange with the geosphere and biosphere (including human-induced changes) can significantly change the atmospheric content of carbon, for example in CO_2 and CH_4, ultimately influencing climate.

5.13 Microbial Mats – Energy-driven Zonation in Life

You may have noticed that when muddy puddles dry out, sometimes there is a green film on the surface of the mud. These layers of microbes also occur around hot springs in places such as Yellowstone National Park, United States. The films of microbes tend to be associated with surfaces such as rock or soil where the microbes can attach and form a stable community, in some sense using the surface as an anchor. These layers of microbes are called **biofilms** and the thicker ones are **microbial mats**. In both cases, they usually contain more than one species of microbe and often there is a clear internal structure with different microbes being found at different levels within the structure. These layers are sometimes mutually beneficial with the waste products of some microbes providing food for others in **syntrophy**. These biofilms show that most microbes do not live on their own in isolation, but they often live with other microbes in communities where they can gain benefits from each other.

Microbes in biofilms tend to have different properties from free-living individuals of the same species, partly because of the protection provided by the biofilm, and partly because the outer layers of cells protect those beneath. These properties include the production of polysaccharides that allow them to more effectively attach to a surface. They even have significance in the medical context since biofilm growth can increase the resistance of microorganisms to antibiotics, making them more difficult to remove from the human body.

Figure 5.20 *A cross-section through a microbial mat showing typical energy metabolisms found in well-ordered layers with dark photosynthetic layers at the top containing a variety of pigments, and the other metabolisms, showing up as whitish layers, underneath. The thickness shown here is about 1 cm* (Source: wikicommons Botanical Garden, Delft University of Technology).

The microbes within mats are often ordered in vertical layers according to the availability of their essential requirements, particularly energy. On the surface there might be photosynthetic microbes, such as the cyanobacteria (Figure 5.20) carrying out oxygenic photosynthesis. Below them are anoxygenic photosynthesising bacteria and below them other non-photosynthetic metabolisms. The surface layers are often packed with pigments including UV radiation-screening pigments.

Moving down into the mat, oxygen levels fall as oxygen is used up by those microbes that respire aerobically. Below them we move into a zone where chemoautotrophy or anaerobic respiration occurs. The use of raw materials and energy sources by microbes at one depth will influence their availability for the microbes that grow below or above. Only certain species will thrive at particular depths in the mat and thus a series of layers of microbes is formed.

Microbial mats are significant for astrobiologists because they are a macroscopic manifestation of microscopic life and so relatively easy to find in ancient rocks. Evidence for mats (**stromatolites**) has been suggested in the fossil record of the early Earth (Chapter 12). Furthermore, the biogeochemical cycles within mats can be regarded as a microcosm of the biogeochemical cycling that occurs on the planetary scale. By studying what goes on in a microbial mat, we learn how energy and nutrients are cycled on a planetary scale.

5.14   The Thermodynamics of Energy and Life

We've looked at the various metabolisms that life uses. We now need to think about how we can quantify the energy yields from such reactions so that we can grasp how life might be able to use energy to make a living and how we might be able to compare different energy yields from metabolisms in extreme environments. Beyond the Earth, we might want to assess an environment as a plausible location for life. One way we could do this is to work out how much energy it might offer using different theoretical electron donors and acceptors and therefore whether it might be a good location for life.

5.14.1   Gibbs Free Energy: The Energy in Reactants and Products

We define the energy available in any given redox reaction as the **Gibbs free energy**.

One way to calculate the Gibbs free energy of a reaction is to look at how much energy there is in its constitute reactants and products. If one takes any compound, it is possible to work out the energy required to make that compound. This is the standard free energy of formation, ΔG°_f. You can look this up in a table and it tells you the difference between the compound's free energy and the free energy from the elements of which it's formed. In other words, it tells you how much energy is 'available' in that compound if you were to break it down.

So we could work out the energy given up or taken in by a reaction simply as the difference between the sum of the free energies of formation in the products and the free energies of formation of the reactants that went into it. That would tell us the net energy difference that exists in a reaction.

That allows us to formulate an equation for Gibbs free energy, simply given as:

$$\Delta G = \Sigma G_f^\circ (\text{products}) - \Sigma G_f^\circ (\text{reactants}) \qquad (5.2)$$

We sum the energy of formation or Gibbs free energy of the products $[\Sigma G_f^\circ (\text{products})]$ and subtract the sum of the free energy of the reactants $[\Sigma G_f^\circ (\text{reactants})]$, taking care to keep the signs consistent, to get the overall energy given off in the reaction.

You can see from this relationship that we can now make some basic statements:

- Products and reactants are in equilibrium when their free energies of formation are equal.

- If the overall Gibbs free energy of the reaction is negative then the redox couple is exergonic – it gives off energy since there is more energy in the reactants than the products. In biological terms, it means that the reaction is favourable for life to make energy. If the Gibbs free energy is positive then the products contain more energy than the reactants. This means that the reaction is endergonic – it requires energy to make the products and it is generally not favourable for life unless organisms use some ATP to make the reaction happen.

5.14.2   Gibbs Free Energy: The Concentration of Compounds

There is another way to calculate the Gibbs free energy and that is to use the concentrations of the products and reactants.

The standard Gibbs free energy of the reaction: $aA + bB \leftrightarrow cC + dD$ (where the capital letters are the compounds and the lower case letters are the relative quantities of each) can be given by:

$$\Delta G^\circ = -RT \ \ln([C]^c [D]^d / [A]^a [B]^b) \qquad (5.3)$$

where the square brackets denote the concentrations of each compound, or:

$$\Delta G^\circ = -RT \ln K_{eq} \qquad (5.4)$$

where K_{eq} is the reaction equilibrium constant, here given as $[C]^c[D]^d/[A]^a[B]^b$, T is the temperature and R is the universal gas constant. The equilibrium constant is just the ratio of the speed of the reaction going forwards to the speed of the reverse reaction.

As an example, in the case of aerobic respiration the Gibbs free energy could be determined using the reaction for this process:

$$C_6H_{12}O_6 + 6O_2 \rightarrow 6CO_2 + 6H_2O$$

which gives a Gibbs free energy of:

$$\Delta G^\circ = -RT \ \ln([CO_2]^6 [H_2O]^6 / [C_6H_{12}O_6] [O_2]^6)$$

5.14.3   Gibbs Free Energy: Using Redox Reactions

These two previous approaches do have several drawbacks. Most notably it is laborious to calculate the Gibbs free energy of formation of every compound involved in a reaction. It is also the case that we do not always know the exact concentration of the different reactants and products in any given reaction, particularly for biological reactions. So we need to consider an alternative way of calculating the energy in reactions.

We can think of reactions in a different way: as redox reactions. You'll recall that reactions can be written as the half reactions:

Oxidation is the removal of electrons ($S_d \rightarrow P_d + e^-$), that is the reaction of the electron donor (shown here as the substrate, S_d, which gives up an electron to become the product, P_d).

Reduction is the addition of electrons ($S_a + e^- \rightarrow P_a$), that is the reaction of the electron acceptor (shown here as the substrate, S_a, which accepts an electron to become the product, P_a).

A redox reaction used by life to make energy needs a reducing and oxidising half reaction. The energetics of redox reactions can be considered as combined electrical potentials of these two 'half reactions'.

You will also recall that we can conceptualise this on an 'electron tower' that shows the various half reactions of interest with their potentials. Reactions with stronger tendency to give up electrons are higher (more negative) on the tower.

If we want to work out the potential for a whole reaction, we could simply calculate the total difference between the electrode potential of the two half reactions.

But note that the total potential difference must take into account that the scale goes from negative values to positive ones! So if one half reaction has a potential of -1.2 V and the other is -0.5 V, then the total reaction potential will be the difference between these two half reactions, that is 0.7 V. However, if one reaction is -1.2 V and the other is $+0.3$ V, then the total reaction potential is 1.5 V (-1.2 V on one side of zero and $+0.3$ V on the other side; look at Figure 5.6 to ensure you are clear about this point).

This difference is denoted E° and it is the difference in electrode potential of the half reactions.

We can then state that Gibbs free energy is given by:

$$\Delta G^\circ = -nFE^\circ \qquad (5.5)$$

where ΔG° is the standard Gibbs free energy change, n is the number of electrons transferred in moles (usually given as one mole in tables). F is Faraday's constant [96.48 J $(V \cdot mol)^{-1}$] and E° is the redox potential difference under standard conditions ($25\,^\circ$C).

We can also express E° as the electron activity, or pE°. This is what is often given in tables in standard texts and is sometimes shown as $p(\varepsilon)^\circ$.

The electron activity is analogous to pH. It can also be expressed in terms of the reaction constant as follows:

$$\log K_{eq} = n(pE^\circ) \qquad (5.6)$$

But remember that we already said that Gibbs free energy can be expressed as:

$$\Delta G^\circ = -RT \ln K_{eq} \qquad (5.7)$$

So we can also express Gibbs free energy as:

$$\mathbf{\Delta G^\circ = -2.303nRT(pE^\circ)} \qquad (5.8)$$

where the factor 2.303 comes from the conversion of a natural logarithm to a log.

This particular equation is important because it allows us to simply calculate a Gibbs free energy by using electron activities for half reactions given in tables and assuming standard conditions.

Consider the example of aerobic respiration of glucose at $25\,^\circ$C. The pE° values for the two relevant half reactions (glucose oxidation and oxygen reduction) extracted from a published table are given below (note that these are not the same as the $E^{\circ-}$ values shown in Figure 5.6):

Glucose oxidation $0.25\ CH_2O + 0.25\ H_2O$

$$= 0.25\ CO_2 + H^+ + e^- \qquad -8.20 \qquad (5.9)$$

Oxygen reduction $0.5\ O_2 + H^+ + e^-$

$$= 0.5\ H_2O \qquad +13.75 \qquad (5.10)$$

The reactions are each shown for a mole of electrons (e^-). The pE° value for the total reaction of glucose with oxygen is therefore $13.75 + 8.20 = 21.95$ (the values straddle the zero value in the electron tower and so we add them to get the total pE° between the two reactions).

The ΔG° for the reaction can therefore be calculated using Equation (5.8) as -125 kJ mol^{-1} electrons. As a glucose molecule provides 24 electrons for each molecule oxidised we could also express this as a Gibbs free energy for the consumption of one mole of glucose which would give us a value of 3000 kJ mol^{-1} of glucose.

This approach gives us quantitative estimates. For example, we could compare the Gibbs free energy associated with aerobic respiration to sulfate reduction with glucose (about -25 kJ mol^{-1} electrons) and iron reduction with glucose (about -40 kJ mol^{-1} electrons). These numbers quantitatively show how aerobic respiration produces more energy than anaerobic modes of metabolism.

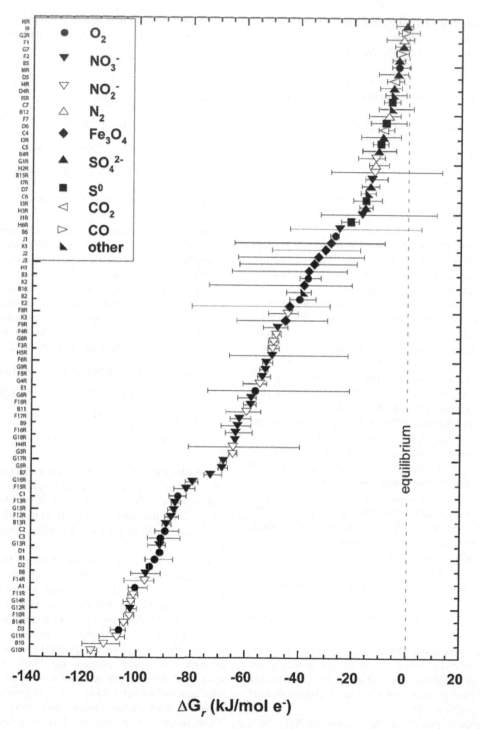

Figure 5.21   *Some examples of predicted Gibbs free energy yields for the Vulcano hydrothermal system in Italy. Values of Gibbs free energy (kJ mol⁻¹ e⁻) are shown for 90 chemolithotrophic reactions averaged across nine sites in the Vulcano hydrothermal system. Horizontal bars indicate the range of values for each reaction across the sites. Reactions are arranged in order of decreasing energy yield (lower left to upper right) and organised according to terminal electron acceptor (in the legend). Reaction numbers on the y-axis correspond to those given in Rogers et al. (2007).*

5.15   Life in Extremes

These approaches can be used to predict energy availability in extreme environments. An excellent example is the use of thermodynamics to predict what metabolisms might be favourable in volcanic systems. This was done for the Vulcano hydrothermal system in Italy (Figure 5.21) in which a whole variety of theoretical metabolisms was examined and their Gibbs free energy calculated under the geochemical conditions directly measured at the site. It was then possible for the researchers to search for these metabolisms in DNA sequences from these sites and confirm the presence of organisms theoretically capable of carrying out these reactions. This sort of study illustrates how Gibbs free energy calculations can be used to generate hypotheses about energy acquisition in different environments which can then be tested by looking for microbes with these metabolisms.

One complication in calculating Gibbs free energies is that they only consider thermodynamics. Reactions can be kinetically limited. For example, an iron-reducing microbe might be able to get a certain quantity of energy from a defined iron oxide, but if this iron is locked up in a mineral form within a rock it may be kinetically difficult for this energy to be released or for the organism to access it. So it is not always the case that thermodynamic calculations provide a reliable prediction of whether a particular metabolic pathway will occur in any given natural environment. At least as important is whether the theoretical concentrations of available electron donors or acceptors are in an accessible state.

5.16   Conclusions

In this chapter, we have investigated some of the range of energy-yielding reactions to be found in life. Many varieties of possible energy have been exploited by life and we looked at the primary phototrophic and chemotrophic modes that have been well characterised on the Earth. Despite the huge variety, they all share some common core characteristics, not least the use of electron transfer to generate proton gradients which are in turn used to manufacture ATP. We have seen how these transformations, most of them mediated by microorganisms, are not trivial in scale – they are responsible for planetary-scale biogeochemical cycles. Finally, we have seen just some of the approaches we can take to calculating theoretical energy yields from different reactions, which is one tool by which we can assess the habitability of environments on the Earth and other planetary bodies, at least with respect to energy availability. As this chapter has adequately emphasised, life, particularly microbial life, can be categorised into a vast number of groups based on their species and metabolic capabilities. In the next chapter we'll see how astrobiologists can make sense of this variety and order the information we have about the menagerie of life on Earth.

Further Reading

Books

Gorman, H.S. (2013) *The Story of N: The Social History of the Nitrogen Cycle and the Challenge of Sustainability.* Rutgers University Press, Chapel Hill.

Morton, O. (2009) *Eating the Sun: How Plants Power the Planet.* Fourth Estate, London.

Papers

Coates, J.D., Achenbach, L.A. (2004) Microbial perchlorate reduction: rocket-fuelled metabolism. *Nature* **2**, 569–580.

Cotner, J.B., Biddanda, B.A. (2002) Small players, large role: microbial influence on biogeochemical processes in pelagic aquatic ecosystems. *Ecosystems* **5**, 105–121.

Hoehler, T.M. (2004) Biological energy requirements as quantitative boundary conditions for life in the subsurface. *Geobiology* **2**, 205–215.

Hoehler, T.M., Jørgensen, B.B. (2013) Microbial life under extreme energy limitation. *Nature Reviews Microbiology* **11**, 83–94.

Mitchell, P. (1961) The chemiosmotic hypothesis. *Nature* **191**, 144–148.

Nitschke, W., McGlynn, S.E., Milner-White, E.J., Russell, M.J. (2013) On the antiquity of metalloenzymes and their substrates in bioenergetics. *Biochimica et Biophysica Acta* **1827**, 871–881.

Petersen, J.M., Zielinski, F.U., Pape, T., Seifert, R., Moraru, C., Amann, R., Hourdez, S., Girguis, P.R., Wankel, S.D., Barbe, V., Pelletier, E., Fink, D., Borowski, C., Bach, W., Dubilier, N. (2011) Hydrogen is an energy source for hydrothermal vent symbioses. *Nature* **476**, 176–180.

Rogers, K.L., Amend, J.P., Gurrieri, S. (2007) Temporal changes in fluid chemistry and energy profiles in the Vulcano Island Hydrothermal System. *Astrobiology* **7**, 905–932.

Seager, S., Schrenk, M., Bains, W. (2012) An astrophysical view of Earth-based metabolic biosignature gases. *Astrobiology*, **12**: 61–82.

Shively, J.M., van Keulen, G., Meijer, W.G. (1998) Something from almost nothing: carbon dioxide fixation in chemoautotrophs. *Annual Reviews in Microbiology* **52**, 191–230.

Spear, J.R., Walker, J.J., McCollom, T.M., Pace, N.R. (2005) Hydrogen and bioenergetics in the Yellowstone geothermal ecosystem. *Proceedings of the National Academy of Sciences* **102**, 2555–2560.

Suzuki, Y., Sasaki, T., Suzuki, M., Nogi, Y., Miwa, T., Takai, K., Nealson, K.H., Horikoshi, K. (2005) Novel chemoautotrophic endosymbiosis between a member of the *Epsilonproteobacteria* and the hydrothermal-vent gastropod *Alviniconcha* aff. *hessleri* (Gastropoda: Provannidae) from the Indian Ocean. *Applied and Environmental Microbiology* **71**, 5440–5450.

6

The Tree of Life

Learning Outcomes

➤ Understand the concept of phylogenetic trees.
➤ Be able to describe and explain some of the problems in interpreting phylogenetic trees.
➤ Understand how ribosomal RNA has been used to construct phylogenetic trees and how we can use phylogenetic trees to quantify evolutionary distance between organisms.
➤ Understand the concept of phylogenetic bracketing.
➤ Describe how phylogenetic trees can be used to test scientific hypotheses.
➤ Explain the concept of the Last Universal Common Ancestor (LUCA).

6.1 A Vast Diversity of Life

In Chapter 5, we explored the enormous variety of metabolisms and energy-acquiring pathways in life on Earth. This diversity is similarly reflected in the sheer number of species on the Earth. We don't know exactly what the number is, but it could be about 10 million for the eukaryotes alone. If you had limited knowledge about life on Earth, you would probably find yourself asking some pretty basic questions about this giant menagerie of creatures. For example: How are they related? How closely related is one organism to another? Are they actually related or could life on Earth have come from different origins of life? What was the nature of the ancestors of all life on Earth? We can only answer these questions by developing a systematic way of categorising organisms and studying their relationship to one another.

We already have an idea of some of the answers. In Chapter 4 we learned how the genetic code, as manifested in the table of codons that encode amino acids, is found to be similar in all life we know on Earth, suggesting that the biochemistry of all life on Earth came from a common origin. But how can we unravel the evolutionary relationships between living things since these early events?

Underpinning the relationships between organisms on the Earth is evolution, specifically Darwinian evolution. The idea that organisms evolve from earlier organisms was not first conceived by Charles Darwin (1809–1882). The Greek philosopher Anaximander (circa 611–546 BC) proposed that humans arose from fish, although it is difficult to know to what extent these early ideas arose from fables or observations. Darwin provided the first explanation for the mechanism based on the collection of a large number of observations that he had made primarily during his global voyage on HMS *Beagle*. Small variations within a population of organisms make them better or less well equipped to deal with their environmental conditions. Those that have more suitable characteristics have a greater chance of surviving, reproducing and passing on those characteristics to the next generation. As environmental conditions change, so different adaptations are required, imposing changing selection pressures on the different variations within populations. They evolve. This idea was in contrast to that espoused by Jean-Baptiste Lamarck (1744–1829), who proposed that organisms acquire characteristics in response to their environment

Astrobiology: Understanding Life in the Universe, First Edition. Charles S. Cockell.
© 2015 John Wiley & Sons, Ltd. Published 2015 by John Wiley & Sons, Ltd.
Companion Website : www.wiley.com/go/cockell/astrobiology.

during their lifetimes that they pass on to their offspring. The Darwinian mechanism of natural selection, elaborated in Darwin's book, *On the Origin of Species by Means of Natural Selection, or the Preservation of Favoured Races in the Struggle for Life*, usually just called *The Origin of Species*, published in 1859, is viewed, if not in detail, as broadly the correct mechanism for evolutionary processes. Indeed, evolutionary change in the Darwinian sense has often been alluded to as a defining characteristic of life (Chapter 1).

Darwin, of course, did not know about DNA and so he had no mechanism for explaining how this process would work at the biochemical level. In the pre-DNA discovery era, the basic principles by which variation can be generated and passed on through generations was experimentally investigated by Gregor Mendel (1822–1884), a friar who famously carried out a large number of fertilisation experiments between pea plants. By observing the characteristics of pea plants, such as their height, flower colour and seed shape, he was able to determine laws of heredity. These observations provided empirical evidence for the vertical transmission of information from one generation to the next, the dominant and recessive nature of some characteristics and insights into the way in which this information was 'mixed' between parents to generate the characteristics of offspring. Unfortunately, he and Darwin did not know of each other's work.

Darwin's and Mendel's insights can broadly be said to set the theoretical and empirical stage for understanding how and why organisms on Earth are related and what processes drive the variability that leads to new evolutionary lines of descent and hence an interconnected 'tree of life'. The era of molecular genetics and biochemistry would drive forward this enterprise by allowing us to read the genetic code itself and thus order, with much greater accuracy, the relationships between living things on Earth.

6.2   The Tree of Life

Although we will examine the details of how phylogenetic trees are constructed as we go through this chapter, it is useful to begin with some basic ideas.

Broadly, life on Earth can be split into three 'domains'. Domains can be thought of as the largest scale division possible of life on Earth. These three domains are the **bacteria**, which include many of the organisms that live in your gut, the soil and almost any other inhabited environment on the Earth; the **archaea**, which include a large variety of microorganisms in different environments,

but are famously known for their inclusion of many extreme-loving or tolerant organisms; and finally, the **eukaryotes**, organisms with a cellular nucleus, which include plants and animals. It is important to understand that although eukaryotes tend to be associated with complex multicellular organisms, they also include many single-celled organisms such as algae and some fungi. Bacteria and archaea are sometimes put together as the group, 'prokaryotes' (Chapter 4).

Figure 6.1 shows a basic phylogenetic tree with some putative relationships between these domains. Each line denotes an evolutionary 'distance' between the groups shown, which reflects the time since they split from a common ancestor. There is no special significance to the groups shown. They merely illustrate some examples. You will notice that the tree is 'rooted' in a single ancestor. This is generally called the **Last Universal Common Ancestor** or LUCA.

6.3   Some Definitions

At this point we need to understand some definitions important in constructing trees. **Phylogeny** concerns the evolutionary history of a species or groups of related species. The tree shown in Figure 6.1 is an example of a phylogeny. The discipline of **systematics** classifies organisms, determines their names, describes them and seeks to understand their evolutionary relationships. Systematists use fossil, molecular, and genetic data to infer evolutionary relationships between organisms. **Taxonomy** can be regarded as a subset of systematics and is specifically that part of the science that deals with the ordered division and naming of organisms and is not focused on the evolutionary relationships.

6.4   Classifying Organisms

In the eighteenth century, Swedish natural historian, Carolus Linnaeus (1707–1778) published a system of taxonomy based on resemblances. This was the first attempt to order organisms in a rational way. Resemblances recognise that many organisms share common characteristics. Before the birth of genetics this was a useful way of inferring relationships. In Figure 6.2 you can see an example of four different animals that all look quite similar. Their similarities do indeed represent a shared descent. They are all members of the family Felidae (the cats). By systematically comparing traits such as number

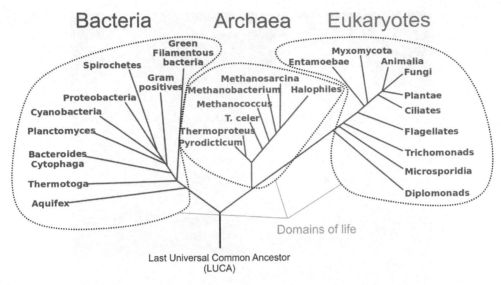

Figure 6.1   *A phylogenetic tree of life, showing a variety of groups of organisms associated with each of the three domains of life. This tree is based on comparing sequences of ribosomal RNA.*

and sizes of limbs, the presence of features such as fur, or whether an organism has eggs or fully grown young, early natural historians were able to start to build up a systematic study of the relationship between organisms.

Two key features of Linnaeus's system remain useful today: two-part names for species and hierarchical classification.

The Linnean system of classification involves generating a two-part scientific name of a species, called a **binomial**. The first part of the name is the **genus** (the level of hierarchy above species). The second part, called the **specific epithet**, is unique for each species within the genus. The first letter of the genus is capitalised, and the entire species name is italicised or underlined. Both parts together name the species. For example, the wolf is called *Canis lupus. Canis* is the genus and *lupus* the specific epithet. Sometimes, if there is a strain or variant of a species a third name may be introduced. A particularly good example is *Canis lupus familiaris*, which is the domestic dog, derived from wolves.

In addition to ordering species in this way, Linnaeus introduced a system for grouping species in increasingly narrow categories. The taxonomic groups from broad to narrow are domain, kingdom, phylum, class, order, family, genus and species. A taxonomic unit at any level of hierarchy is called a **taxon** (plural **taxa**). This hierarchical classification of species has stood the test of time and is today the basis of our way of classifying species.

Systematists depict the true evolutionary relationships in branching phylogenetic trees such as that shown in Figure 6.3.

What do phylogenetic trees show us? Phylogenetic trees show patterns of descent. The horizontal lines might correspond to the time since two groups or species diverged or they could correspond to the amount of genetic change between different groups or species. It depends on what data were used to construct the tree.

In Figure 6.3a you can see an example of a phylogenetic tree of large mammals. The tree also includes a salmon at the bottom. This is the **outgroup** that is a species distantly related to the ones we are interested in and helps us root the tree. As it is distantly related it causes the organisms we are interested in to tend to cluster into a single group. Starting from the left we see that all the organisms shown in the tree (apart from the fish) share a common ancestor in the Order Carnivora (the carnivores). Then they split at a **node**, which defines the point (the time) at which a cohesive population divided into two genetically distinct populations. These separations are caused by effects such as the geographical separation of two populations that then evolve independently, or two populations taking up two different life styles in the same geographical location that eventually leads to them becoming distinct **taxa** by evolution.

In Figure 6.3 we can see three groups that diverged at the family level, the Felidae (the cats), the Mustelidae (a group

Figure 6.2 *For some large organisms resemblances can be used to infer common descent, such as the features of these four cats. From top left to bottom right: leopard (*Panthera pardus pardus*; Source: wikicommons, JanErkamp), lion (*Panthera leo bleyenberghi*; Source: wikicommons, Appaloosa), domestic cat (*Felis silvestris catus*; Source: wikicommons, Panther), jaguar (*Panthera onca*; Source: wikicommons, 663highland).*

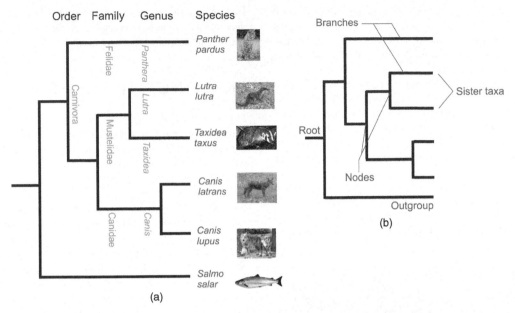

Figure 6.3 (a) A phylogenetic tree of some large mammals. (b) Some terminology with respect to phylogenetic trees (Source: Charles Cockell).

that includes the badgers and weasels) and the Canidae (the dogs). The vertical lines show where divergences have occurred between different groups. Moving further to the right, at the top of the tree we see a representative of the Felidae, *Panther pardus* (the leopard). Beneath this we see an example of divergence at the level of the genus where within the family Mustelidae there is divergence into the *Lutra* (otters) and the *Taxidea* (badgers), represented by *Lutra lutra* (the European otter) and *Taxidea taxus* (the American badger), respectively. In the lower part of the tree we see an example of a divergence at the level of the species. In the genus *Canis* there is divergence into *Canis latrans* (the coyote) and *Canis lupus* (the wolf).

In Figure 6.3b the tree has been stripped of the names and some definitions of different parts of a phylogenetic tree are shown: the root, the **nodes**, the 'branches' and **sister taxa**. The overall pattern or shape of any tree is called the **topology** and trees themselves are sometimes referred to as **dendrograms**.

6.5 Homology and Analogy

One factor we have to be very careful about whilst building a tree and determining similarities between organisms is the problem of homology and analogy. Homology is similarity due to shared ancestry like between a wolf and

a dog. The observed similarities between these two taxa would have led early taxonomists to conclude that they were related. They would have been correct, but only by luck. Analogy is similarity caused by evolutionary pressures resulting from exposure to similar conditions or adaptation pressures, but in two organisms that otherwise do not share a recent common ancestor. Analogies are also known as **homoplasies** (from Greek, 'to mould the same way').

Therefore, Linnaean classification and true phylogeny can sometimes differ from each other because although we might name things based on resemblances, this does not necessarily mean we have got it right! Two organisms might look the same, but be genetically different. Even Linnaeus fell afoul of analogies. In the first edition of his *Systema Naturae*, which described the links between living things, he grouped whales with fishes. In the 10th edition in 1758, 23 years later, it was corrected – whales were grouped with mammals.

A good example of analogy is the Australian and North American moles. Both of these organisms have attributes typical for small subterranean animals. They are cylindrical, have small eyes, reduced hindlimbs and short powerful forelimbs for digging. North American moles belong to the family Talpidae (the 'true moles'). Australian moles belong to a family of marsupial mammals that are more closely related to kangaroos and koalas than they are to the Talpidae moles. The similarities that result from being exposed to the same environmental pressures (pressures associated with being optimally adapted for burrowing underground) are known as **convergent evolution** (Figure 6.4).

Homology can be distinguished from analogy by comparing the fossil evidence. For example, in the case of the moles, it might be shown from fossil evidence that there are intermediate stages between Australian marsupial mammals and their extant underground descendants, demonstrating a different evolutionary ancestry than the Talpidae moles.

Another approach is to compare complexity. The more complexities that are shared between two organisms, the more likely it is that they are homologous. This method is based on a statistical consideration. The chances that two

similar-looking organisms are descended from a common ancestor will generally increase with the greater the number of characteristics that are shared.

6.6 Building a Phylogenetic Tree

Phylogenetic trees can be constructed by using resemblances. With large organisms, such as cats, this approach can work well. Large organisms with many characteristics, structures and surface features have a greater probability of being correctly assigned to their respective closest relatives. However, using resemblances fails when we start to assign evolutionary relationships to much simpler organisms. The obvious example of this problem is the building of phylogenetic trees for microorganisms, which as we saw in Chapter 4, adopt a very limited range of shapes. Few of them have surface features that can be distinguished, although **flagella** do take up different configurations and could be used to categorise some groups of microbes. Categorising thousands of species of microbes by resemblances would be impossible.

A much better way of approaching the problem of the construction of phylogenetic trees is to look at genetic information. All known cellular organisms on the Earth share certain genetic characteristics such as DNA as the information storage system and ribosomes (themselves containing RNA) as the molecules responsible for translating RNA into protein (Chapter 4). One approach to building a tree would therefore be to find a part of this genetic apparatus that is highly conserved, in other words DNA that doesn't change much over time. However, we want a part of this apparatus to change just a little, so that we can use those changes to build a tree and relate those changes to evolutionary distance. The closer two organisms are as relatives, the smaller the number of changes that would have occurred by the process of **mutation**. The greater the evolutionary distance between them (i.e. the greater the time when they diverged from their common ancestor) the more changes in the molecules we would expect to see.

Ribosomal RNA (rRNA), which was introduced in Chapter 4, is one such type of nucleic acid that changes only rarely over long time periods. As it plays a role in a most fundamental core process (translating RNA into protein), any radical mutations that modified ribosomal RNA in any substantial way would cause serious disruption to the whole cell, which would then cease to replicate. There are very strong evolutionary pressures for changes to be minimised. The DNA sequences that

Figure 6.4 *Analogy results from structures that result from organisms being exposed to the same environmental pressures. Here this is illustrated with moles. At the top is a 'true' mole (*Talpa europaea; *Source: wikicommons, Hundehalter*). Beneath is a marsupial mole from Australia (*Notoryctes typhlops; *Source: wikicommons, Bartus.malec*). The moles are a few centimetres in length.*

encode for ribosomal RNA allow scientists to construct phylogenetic trees that reflect evolutionary relationships over time periods up to billions of years.

The genes for ribosomal RNA have been used to construct some of the most important phylogenetic trees of life on Earth (such as the one at the beginning of this chapter). An example of such a gene is the 16S ribosomal RNA gene (16S rRNA), which encodes the 16S ribosome in prokaryotes. The 16S ribosome is itself part of the 30S 'small subunit' ribosome of prokaryotes. Its function is to act as a scaffold to determine the position of the ribosomal proteins and stabilise the interaction of the tRNA, which brings in the amino acid for protein synthesis to the mRNA (review Chapter 4 if this is not familiar). Trees of bacteria and archaea are constructed by sequencing 16S rRNA genes from different organisms.

Yet another ribosomal gene, 18S rRNA, a ribosomal subunit of eukaryotes, is used to construct trees of many eukaryotes such as algae or fungi.

Trees have been constructed from a large number of different genes. For example, if we are interested in the evolutionary relationships of particular functions important in the biosphere, we could sequence functional genes, such as *nif* genes, which are involved in the biological fixation of atmospheric nitrogen gas into biologically available nitrogen compounds such as ammonia and nitrate (nitrogen fixation). Genes for sulfur and carbon cycling and a whole host of other important processes in the biosphere can be sequenced and trees made to understand their evolutionary relationships (Table 6.1).

The sequences of genes are obtained by extracting the DNA from given organisms and then applying modern sequencing methods to get the full sequence of nucleotides from the gene. In the case of the 16S rRNA gene, the total length is 1500–1600 nucleotides.

The process of using DNA or other molecular data to construct phylogenetic trees is called **molecular systematics**. Systematists use computer programs and mathematical tools when analysing comparable DNA segments from different organisms. Clearly the longer the gene being sequenced and compared, the more likely a better tree can be constructed as more bases provide a greater confidence in seeing small differences, particularly when comparing species that are very closely related. The best phylogenetic trees are assembled by sequencing the largest segments of a gene, so that in the case of 16S rRNA trees for instance, scientists will attempt to sequence the whole gene.

How do the mutations that generate these changes in the genetic code occur? Mutations that generate variation in DNA sequences are the result of a number of natural processes. Errors in DNA replication (no process is 100% accurate) can introduce changes in the code. Exposure to agents that damage DNA can also cause mutations, for example intense radiation or noxious, specifically mutagenic or mutation-causing, chemicals in the environment. The mutations that result can take a number of forms. If a purine (A or G) mutates into a pyrimidine (C or T), this is referred to as a transversion. The change of a purine into another purine (an A changing to a G or vice versa) or pyrimidine changing to a pyrimidine (a C changing to a T or vice versa) is called a transition (Figure 6.5). Other types of mutations that cause differences between the DNA in different organisms include insertions and deletions (Figure 6.6). Insertions are caused by single bases or even sequences of bases being inserted into the gene sequence. Deletions are the loss of one or more bases.

Although it is generally true that the more genetic differences that exist, the further is the evolutionary distance between two organisms (Figure 6.6a), systematists have to distinguish homology from analogy. DNA might mutate, for example a C to T transition, resulting in a T in the DNA sequence at an identical location to another unrelated species that also has a T at that position. Analysis of the data could fool a scientist into concluding that two DNA sequences were homologous, when in fact they are analogous (homoplasy). Mathematical tools help to identify molecular homoplasies, or coincidences, particularly since mutation rates are not the same between all nucleobases.

Figure 6.6b shows that when trees are constructed they usually have a scale bar that allows a researcher to quantify the differences between species. For trees constructed using genetic information this scale bar will usually correspond to the average number of nucleotide changes at any position. Trees often have numbers at the nodes (Figure 6.6b) that give the confidence in that node being correct. This is calculated by bootstrapping,

Table 6.1 *Some examples of genes that can be sequenced and used to build phylogenetic trees of metabolisms or other biochemical processes in organisms.*

Gene	Gene encoded	Function
*dsr*AB	Dissimilatory sulfite reductase	Sulfate reduction
msr	Methyl-coenzyme M reductase	Methanogenesis
nif	Nitrogen fixation genes	Nitrogen fixation
amo	Ammonia monooxygenase	Ammonia oxidation

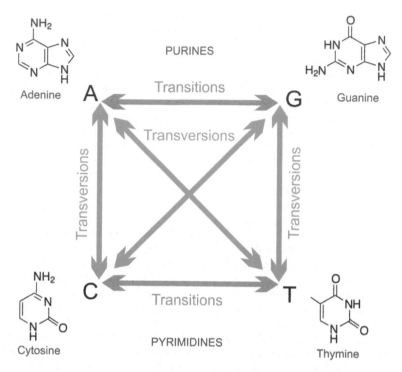

Figure 6.5 *Transversions and transitions are just two ways in which mutations cause DNA sequences to diverge* (Source: Charles Cockell).

a process whereby different trees are constructed using different arrangements of the nodes and this is repeated many times. A 'consensus' tree is built from all of these trees that reflects the most likely correct interpretation of the data. The number of times that the node is found in all the iterations compared to the consensus tree is then calculated (for example as a percentage). Thus, it provides a confidence level in that particular node. As a general rule, if the bootstrap value for a given node is 95% or higher, then the topology at that branch is considered 'correct'.

6.7   Some Definitions and Phylogenetic Trees

It is time to review some further definitions. Once we have a phylogenetic tree, we can categorise different parts of it and define some features of the overall tree.

Often a tree needs some sort of 'root'. What we mean by this is that we need an organism very different from the ones we are examining, which will act as a common ancestor for all the organisms we are studying.

For example, if we drew a phylogenetic tree based on the genetic information from mice and we added into the tree the same genetic information from a fish, we would find that the DNA sequences from the mice would tend to cluster together, but the fish would be the odd one out because it is much more distantly related to all the mice sequences than the mice are to each other. It would root the tree. As described earlier, we refer to such an organism as the **outgroup** and you can see this illustrated in Figure 6.3. All the other organisms that are of interest to the researcher are called the ingroup. Systematists compare each ingroup species with the outgroup to differentiate between characteristics. Homologies shared by the outgroup and ingroup are ancestral characters that predate the divergence of both groups from a common ancestor.

Cladistics is the part of taxonomy that groups organisms by common descent.

A **clade** is a group of species that includes an ancestral species and all its descendants. Another term for a clade is **monophyletic**, signifying that it consists of the ancestor species and *all* its descendants. This was most simply summarised by evolutionary biologist Robert O'Hara who explained a clade (a monophyletic group) like this:

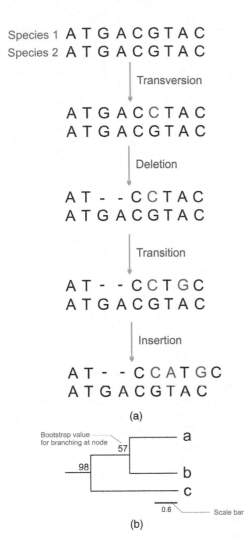

(a)

(b)

Figure 6.6 *Using DNA sequences to build phylogenetic trees. (a) A variety of mutations (red) result in divergence in genetic sequences which allow for evolutionary distance to be inferred and phylogenetic trees to be constructed. Here just some changes to one species (species 1), compared to the unchanged sequence of species 2 from the same ancestor, are shown for illustrative purposes. As is apparent from this figure, the longer the genetic sequence obtained, the more likely it is that statistically robust inferences about the time of divergence can be obtained. (b) Phylogenetic trees usually have a scale bar which allows for the horizontal branches to be quantified. For trees constructed using genetic information, it usually corresponds to the average number of nucleotide changes at a given position over that distance. The numbers at the nodes are the confidence levels for that branching. They are calculated using bootstrapping (Source: Charles Cockell).*

'If you were to grab hold of a tree at any point, and cut immediately below your grip – the chunk of tree in your hand would by definition be a clade'.

If we examine a tree we can also see that we can identify other types of groups within it. A polyphyletic grouping consists of various species that have similar traits, but the group lacks a common ancestor (Figure 6.7). A paraphyletic grouping consists of an ancestral species and some, but not all, of the descendants (Figure 6.7).

6.8   Types of Phylogenetic Trees

There are different types of phylogenetic trees. In some trees, the length of a branch can reflect the number of genetic changes that have taken place in a particular DNA sequence in that lineage. In other trees, branch length can represent chronological time (Figure 6.8), and branching points can be determined from the fossil record. These trees are designed to convey different information or combined information. The choice of tree will depend upon the information that a researcher is interested in conveying. You will also find trees drawn in different ways, some use diagonal branches (Figure 6.1) or curved branches. Many of these convey exactly the same information as trees drawn with horizontal branches and vertical lines at the nodes (principally used in this chapter), they are just different methods of presentation.

Systematists can never be sure of finding the best tree in a large data set, so they narrow possibilities by applying the principles of maximum parsimony and maximum likelihood.

Maximum parsimony assumes that the tree that requires the fewest evolutionary events (appearances of shared characters) is the most likely tree to be correct.

The principle of **maximum likelihood** states that, given certain rules about how DNA changes over time, a tree can be found that reflects the most likely sequence of evolutionary events. For example, the rate of mutations, such as transversions or transitions, can be accurately defined from DNA sequence and chemical knowledge. These rules can be incorporated into models to predict the likelihood that certain changes in the sequences have occurred.

6.9   Using Phylogenetic Trees to Test Hypotheses

Phylogenetic trees are useful for classification and ordering, but they are not just a means to make a 'stamp album'

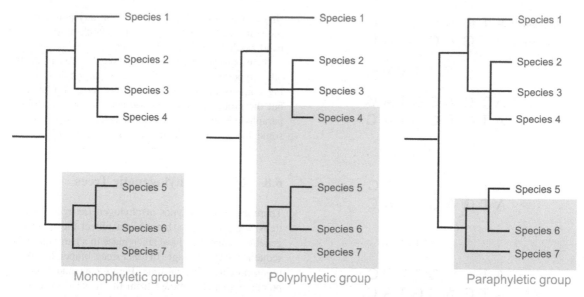

Species 1
Species 2
Species 3
Species 4
Species 5
Species 6
Species 7
Monophyletic group

Species 1
Species 2
Species 3
Species 4
Species 5
Species 6
Species 7
Polyphyletic group

Species 1
Species 2
Species 3
Species 4
Species 5
Species 6
Species 7
Paraphyletic group

Figure 6.7 *Some groups in phylogenetic trees recognised by cladistics* (Source: Charles Cockell).

of life and order information about relationships between organisms. They can also be used to test scientific hypotheses. It is in this realm that they become useful for attempting to understand how life on Earth is related and how it might develop in the future. The best hypotheses for phylogenetic trees fit the most data: morphological, molecular and fossil.

One example of hypothesis testing using phylogenetic trees is **phylogenetic bracketing**. This is an approach that allows us to infer features of an ancestor from features of its descendants. A good example is dinosaurs. There are no dinosaurs alive today, but other descendants that share a common ancestor do exist. Birds and crocodiles 'bracket' dinosaurs (Figure 6.9) and there are many examples of these alive on the present-day Earth. Both birds and crocodiles lack hair, so we might infer that many dinosaurs did as well. Birds and crocodiles lay eggs and so we might infer that dinosaurs did as well. Both of these inferences turn out to be supported by the fossil record. Fossil dinosaur eggs have been found on numerous occasions and well-preserved dinosaur fossils do not exhibit evidence for hair. In this way, we can use phylogenetic trees to understand fossil evidence and re-affirm fossil observations. In addition, we might even be able to infer things about physiology that are not so easily preserved in fossils.

There are three types of inferences that can be developed based on phylogenetic bracketing. Concluding that

dinosaurs could lay eggs is called a type I inference because both the ancestors bracketing dinosaurs possess this characteristic. A type II inference is made when one of the extant relatives has the trait and the other does not. In the case of the dinosaurs, a type II inference would be to conclude that some dinosaurs could have had feathers like birds, although crocodiles do not. The accuracy of this inference would depend upon when feathers evolved. From phylogenetic bracketing alone one could equally conclude that dinosaurs had crocodile-like skin. Type III inferences are those that are made even when neither of the extant relatives possesses the trait. Concluding that dinosaurs had hair would be a type III inference – a characteristic not exhibited by extant birds or crocodiles. Type III inferences are only warranted if there is some strong positive evidence that the extinct creature possessed the trait, despite the fact that extant bracketing organisms did not (such as, e.,g. solid fossil evidence).

Phylogenetic trees have been used very powerfully to look at the evolution of species and even their geographical origins.

An example is investigating human evolution. Where did humans originate? Earlier we saw how 16S rRNA is very conserved and is excellent for building a phylogenetic tree of life on the Earth. However, if we want to examine humans, the changes in rRNA are generally too few during the last three million years of human evolution. Instead, we select genes that have changed more rapidly and can

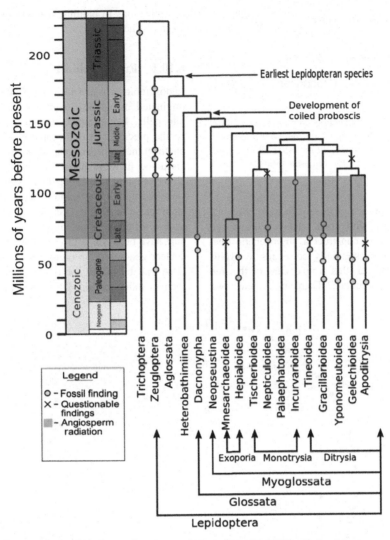

Figure 6.8 *An example phylogenetic tree that is aligned with geological time on the left. This tree shows major Lepidoptera (butterfly and moth) taxa aligned against geological time. The purple block shows when angiosperms (flowering plants) radiated. The tree is oriented vertically* (Source: wikicommons, Jerry A Powell, Conrad Labandeira).

be used to resolve changes in the human lineage over the last few million years. Mitochondrial DNA performs this function well.

Each tip of the phylogenetic tree shown in Figure 6.10 is one of 135 different mitochondrial DNA (mtDNA) types found among 189 individual humans. African mtDNA types are clearly basal on the tree, with the non-African types derived from them. This suggests that humans originated in Africa (Figure 6.10).

6.10   Complications in Building the Universal Tree of Life

This chapter so far has given the impression that by comparing characteristics, including DNA, and quantifying the differences, we can construct a phylogenetic tree. This is broadly true. However, underpinning this assumption is the idea that information is only transmitted vertically – in other words, DNA and its ultimate **phenotypic**

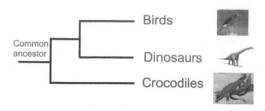

Figure 6.9 *Phylogenetic bracketing can be used to infer characteristics of long-extinct organisms, here illustrated with dinosaurs by studying extant organisms (such as birds and crocodiles) that share a common ancestor* (Source: Charles Cockell).

(physical) manifestations are merely passed from one generation to another with small changes introduced that we can then use to construct a phylogenetic tree. The extent of changes essentially correlates to evolutionary distance. However, this simple view of the tree of life is not entirely accurate. There are at least two

complications in building a phylogenetic tree which are discussed here.

6.10.1   Endosymbiosis

Endosymbiosis is the idea that some organelles in cells originated as whole organisms that were incorporated into cells. For example, cyanobacteria are thought to be the precursors to chloroplasts within the eukaryotic photosynthetic organisms, algae and plants. This was discussed in more detail in Chapter 4. Endosymbiosis clearly creates a problem for making phylogenetic trees since it implies that some genetic information (such as chloroplast and mitochondrial DNA) began as DNA from a different organism and once it was incorporated may have taken on a different evolutionary path than the nuclear genome of an organism. Although this organelle DNA can be useful for tracking evolutionary changes, such as mitochondrial DNA used to build phylogenetic

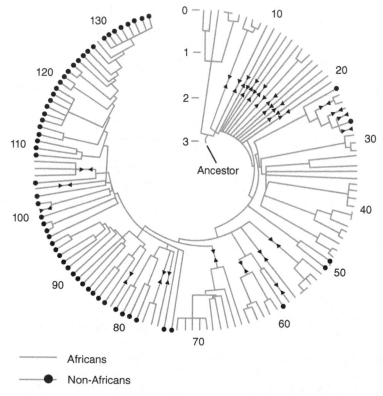

Africans
— ● — Non-Africans

Figure 6.10 *Using a phylogenetic tree of mitochondrial DNA to infer the evolutionary changes and thus the origins of humans. Non-Africans are a more recent addition to the tree of human evolution.* Source: Vigilant *et al.* 1991. Reproduced with permission of American Association for the Advancement of Science.

trees of human evolution, endosymbiosis cautions us to consider the origin of genetic information used to build phylogenetic trees.

6.10.2   Horizontal Gene Transfer

The movement of genetic material between organisms complicates the tree of life. It is very common in prokaryotes. Unlike 'vertical' transmission of information whereby information is passed sequentially from one generation to the next, 'horizontal' gene transfer refers to the direct transmission of nucleic acid between two extant organisms. **Horizontal gene transfer** (HGT) provides some important potential advantages to organisms. It allows for the relatively rapid acquisition of information that can be used for adaptation to the environment compared to the process of acquiring traits from former generations with small incremental changes. However, HGT makes life difficult for phylogeneticists since they cannot assume that changes observed in nucleic acid are always mutations that have occurred from previous generations of that particular organism or that the DNA and its encoded function evolved in the lineage in which it was found. The process is sometimes called lateral gene transfer (LGT). There are three primary ways by which prokaryotes achieve horizontal gene transfer.

1. *Transformation* Prokaryotes can take up naked DNA from the environment. This is a remarkable capacity that requires that the cells enter into a state of **competence**. When cells enter into the **stationary phase** of growth they are thought to become more susceptible to transformation. The evolutionary reasons for this process are still not fully understood. One hypothesis is that it developed as a means to repair DNA damage when cells are in stationary phase and may have fewer copies of their genome. The external DNA would patch up errors. The biochemical process is not a simple one. In the bacterium, *Bacillus subtilis*, over 40 genes are involved in making a cell competent to receive DNA. Transformation can occur between organisms of the same or different species. Cells can be made competent for transformation by chemical methods, such as by soaking them in $CaCl_2$, which makes the cells more 'leaky' and able to take up DNA from their surroundings. This has found use in genetic engineering as a way to incorporate DNA into cells and turn them into 'factories' for producing a compound (such as a drug) of interest.

2. *Transduction* DNA can be transferred into a microbe through an agent such as a bacteriophage, a virus that infects bacteria (Figure 6.11). The virus particle may take over the cell and use its genetic machinery to produce copies of itself, or it may go into a state of

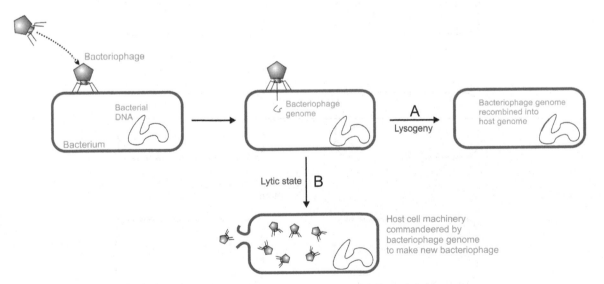

Figure 6.11   *Transduction is achieved when a bacteriophage injects DNA into a bacterium. The DNA can be incorporated into the host genome and be replicated as the organism divides (route A) known as the lysogenic state, or it can direct the cell apparatus to make new copies of itself in the lytic state (route B)* (Source: Charles Cockell).

lysogeny, whereby the nucleic acid is incorporated into the genome of the host and remains there for many hundreds or thousands of generations. Like transformation, transduction is used artificially to introduce genetic information into a cell.

3. *Conjugation* Genetic information can be transferred between microorganisms in the process of **conjugation** (Figure 6.12). A donor cell transfers information to a recipient cell through a specialist structure called a **pilus**. The information is typically transferred on **plasmids**, which are small circular pieces of DNA (Chapter 4) or transposons, which are linear mobile genetic elements. Plasmids are found in all three domains of life and are capable of replicating

autonomously within the host cell. However, the transfer of a plasmid from one cell to another through conjugation requires a plasmid equipped with *tra* genes, which encode the pilus and other factors that are required for DNA to be transferred. These plasmids are called Fertility plasmids or F plasmids.

Imagine if I told you that I could walk up to a person and, by touching them, their eye colour would change to mine. Conjugation essentially achieves this remarkable feat. Conjugation is notable for its ability to achieve inter-domain transmission of genetic information. For example, the nitrogen-fixing bacterium *Agrobacterium* is capable of transferring genes into plant cells and thus

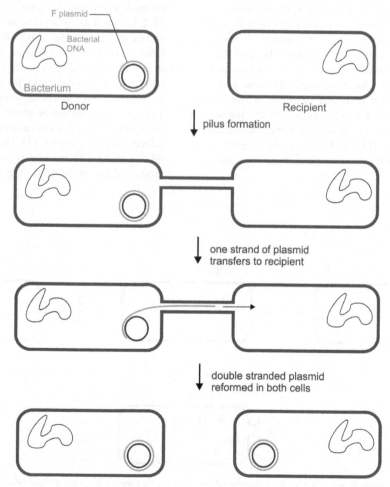

Figure 6.12 *Conjugation results in the transfer of genetic material from one microorganism to another. The recipient organism then becomes competent to transfer the new plasmid into other microorganisms* (Source: Charles Cockell).

altering the growth of plant cells in such a way as to generate habitats (galls) within plant cells in which the bacteria can reside, a type of **symbiosis**. The plants get the benefit of fixed nitrogen, the bacteria get the benefit of plant exudates and organic materials on which they can feed.

These forms of HGT are enormously important from two perspectives.

First, they are responsible for antibiotic, heavy metal and toxin resistance in bacteria making them important for medicine and environmental sciences. The introduction of heavy metal resistance into bacteria or the selection of bacteria with heavy metal resistance carried on plasmids is used in **bioremediation** – the cleaning up of sites polluted with heavy metals.

Second, they complicate the tree of life. If we find a set of genes in a large and unrelated group of organisms, but not in other branches of the tree, we have to wonder about several possibilities. Did the genes emerge in an ancient common ancestor? Were they originally in many organisms, but lost in some? Were they transferred between organisms by HGT? HGT invites us to consider a tree of life that might more properly be described as a web of life, whereby the simplistic phylogenetic relationships that depict vertical transmission of information must be augmented with the possibility of widespread transmission of information across the branches of the tree of life and even across domains.

6.11 The Last Universal Common Ancestor

It is clear that all organisms known on the tree of life share a common ancestor. One of the most important challenges for phylogeneticists and astrobiologists has been determining the nature of the last universal common ancestor (LUCA).

The first attempt to do this was based on an analysis of paralogue protein couples. These are proteins encoded by genes that have duplicated. If the gene duplication is found in all three domains, this implies that the duplication occurred before the three domains diverged. Thus, by examining the distribution of a variety of paralogous genes, one can infer when domains diverged. This led Carl Woese (1928–2012) and others to infer that the tree of life is rooted in the bacteria with a later divergence between the archaea and eukaryotes. A more recent examination using sequenced whole genomes to identify genes found in all three domains of life revealed approximately 100 common genes which were then used to construct a phylogenetic tree, with similar results.

In contrast to this view is one which places the LUCA nearer the eukaryotes with a later divergence between bacteria and archaea. This hypothesis is based on the idea that many complex processes involving RNA are found in eukaryotes, including the splicing of RNA, which has been hypothesised to be a relic of an ancient prebiological **RNA world**, which will be discussed in Chapter 11.

A concern with making these deep inferences is that biases can be introduced into tree reconstruction by varying rates of evolutionary change in different species. For example, the evolutionary rates of ribosomal RNA changes vary by up to 100 times in **foraminifera**. Artefacts in analysis are also possible. An artefact known as long branch attraction (LBA) results in the clustering of the longest branches, corresponding to the most rapidly changing sequences, regardless of their true phylogenetic affiliation. This phenomenon occurs because in rapidly evolving sequences some bases mutate into bases identical to other sequences. Thus, they artificially make the sequences look phylogenetically closer than they are. When a universal tree of life is constructed making special efforts to use only slowly evolving genes, then the root of life does not so clearly sit within the Bacteria and instead suggests that the rooting in the bacterial lineage could be an artefact.

Part of the problem in defining the true evolutionary root of the tree is the low number of genes being considered. As a greater number of organism genomes are sequenced, so the quantity of information that can be used to determine the root will increase and so the accuracy of the inferences.

A question of particular interest is what the functional capabilities of the LUCA were. Some biochemical pathways, such as the **TCA cycle** and components of electron transport chains, are very widely distributed and present in all three domains of life, suggesting that they are ancient and could have been present in the LUCA.

An early analysis of the tree of life suggested that the deepest branching organisms were heat-loving (the **thermophiles** and **hyperthermophiles**; Chapter 7). This led to numerous hypotheses to explain the observations. For example, intense impact bombardment in the early history of the Earth was suggested to have created an evolutionary bottleneck selecting for high temperature tolerant organisms. After bombardment, when impacts were more infrequent and the Earth cooler, life could have radiated into a greater diversity of microorganisms adapted for growing at lower temperatures.

However, the hot environment ancestry of life has been questioned. The heat-loving bacteria have acquired up to 24% of their genes from the archaea suggesting considerable horizontal gene transfer. The DNA reverse gyrase gene, which encodes for an enzyme that introduces a supercoil to DNA, making it more difficult to unwind and apparently conferring resistance to high temperatures, is an example of a gene that is likely to have been passed by horizontal gene transfer from archaea. It is found in bacteria such as *Thermatoga* and *Aquifex* that live in high temperature environments. If the root of life is bacterial, then the existing data might argue against a hot origin of life if they acquired these genes from elsewhere. There are several complexities in unravelling the functional capabilities of the LUCA. First, the open question of where the tree of life is rooted must be resolved as only then can it be determined whether high temperature adaptation is a primary inherited trait. Second, even if the tree of life does not root in organisms adapted to living at high temperatures, this does not rule out a hot origin of life, since a considerable period of time, and extinct organisms, might exist between the origin of life and the root of the tree as defined by extant organisms.

An intriguing approach to finding out the capabilities of early organisms is to recreate proteins that may have existed in them. One line of research investigated the elongation factors (EFs) involved in bringing tRNAs to the ribosome for protein synthesis. By using the extensive genome sequencing that is now available, it was possible to work out ancestral, ancient EF sequences that could have existed about two billion years ago. When these EFs were resurrected in the laboratory using synthesised DNA whose sequence was based on the presumed ancient sequences, they were found to work best at ~65 °C, suggesting a thermophilic ancestry for these proteins.

Finally, of course, we are assuming that the LUCA's capabilities and characteristics are defined by the three existing domains of life. We cannot be sure that there were not domains that went extinct early in the history of life and took with them crucial information.

6.12   Molecular Clocks

Phylogenetic trees are based on the idea of small evolutionary changes through time in given structures or molecules. Related to this is the notion of **molecular clocks**. In the early 1960s Linus Pauling (1901–1994) and Emile Zuckerkandl (1922–2013) noticed that the number of amino acid differences in haemoglobin changed roughly linearly with time. Later it was noticed that the differences in the amino acids of cytochrome *c*, a component of energy acquisition (electron transfer) in the mitochondria (Chapter 4), also correlated with the time elapsed since divergence from another species. The observations led to the idea that changes in amino acid compositions in proteins (and by inference the codons coding for them) were constant over time.

Since these observations it has become clear that the rates of change are not constant. Furthermore, the rates of molecular change vary greatly between taxa. Turtles have molecular clocks running slower than in many mammals by up to a tenth of the rate. Population sizes, differing metabolisms and other factors all influence the rate of molecular clocks. Nevertheless, when correlated to fossil evidence, the concept of the molecular clock has proven useful for molecular systematics and has facilitated the determination of the timing of molecular events and the construction of phylogenetic trees.

Debate Point: Shadow biospheres?

The phylogenetic tree represents the evolutionary relationship between all known organisms on the Earth. One intriguing question that has been raised is whether, if there was more than one origin of life, there could be other organisms on the Earth with entirely different genetic make-up that are not represented on the tree of life. These so-called shadow biospheres would constitute independent biochemistries (e.g. organisms that use a different chirality of proteins or sugars) that might be hiding out somewhere in an extreme environment undetected. What do you think about this proposal? What methods could you think of to detect organisms with a different biochemistry to the life that we already know on the Earth? How likely do you think it is that a shadow biosphere would have escaped our notice with the various microbial and molecular means we use to characterise life?

Davies, P.C.W., Benner, S.A., Cleland, C.E., Lineweaver, C.H., McKay, C.P., Wolfe-Simon, F. (2009) Signatures of a shadow biosphere. *Astrobiology* **9**, 241–249

6.13　Alien Life

If life with an independent origin from life on Earth was ever discovered in the Universe we would expect it to sit on an entirely different tree of life, one rooted in the evolutionary origins of that particular life and its particular genetic information. The construction of the Earth's phylogenetic tree therefore allows us to define uniquely, in a single diagram, the relatedness of life on Earth with which we might ultimately make comparisons to other trees of life if we ever find them.

Another hypothetical scenario would be the discovery of life elsewhere in our own Solar System, but which was related to life on Earth, for example if it was transferred to another planetary body in rocks launched from a planetary surface in asteroid and comet impacts, or indeed if life on Earth resulted from an origin elsewhere. This distantly related life would presumably fit on the tree of life with which we are familiar, but would be rooted very deeply depending on when the transfer of life occurred. If such life existed, it would provide a way in which additional information could be garnered to determine the nature of LUCA. At the very minimum, thinking about these theoretical alien scenarios gives us a perspective from which to understand better the terrestrial tree of life.

6.14　Conclusions

The diversity of life on the Earth is enormous. We do not know the total number of species but we do want to understand how these organisms can be named systematically, how we can describe their evolutionary relationships and from what common ancestor they were derived. In this chapter we have examined some of the approaches to building phylogenetic trees and some of the problems that arise in attempting to do this, most notably the different ways in which horizontal gene transfer can occur. Building phylogenetic trees is not just a means to classify organisms and build up a catalogue of life on the Earth. They also provide us with a means to test scientific hypotheses about the evolution of life and ultimately they may even give us a way to study alien life, if we ever find it.

Further Reading

Books

Gould, S.J. (2000) *Wonderful Life: Burgess Shale and the Nature of History*. Vintage, New York.

Wilson, E.O. (1999) *The Diversity of Life*. Norton and Co., New York.

Papers

Baldauf, S.L. (2003) Phylogeny for the faint of heart: a tutorial. *Trends in Genetics* **19**, 345–351.

Bousseau, B., Blanquart, S., Necsulea, A., Lartillot, N., Gouy, M. (2008) Parallel adaptations to high temperatures in the Archaean eon. *Nature* **456**, 942–945.

Ciccarelli, F.D., Doerks, T., Von Mering, C., Creevey, C.J., Snel, B., Bork, P. (2006) Toward automatic reconstruction of a highly resolved tree of life. *Science* **311**, 1283–1287.

Delsuc, F., Brinkmann, H., Philippe, H. (2005) Phylogenomics and the reconstruction of the tree of life. *Nature Reviews Genetics* **6**, 361–375.

Gregory, T.R. (2008) Understanding evolutionary trees. *Evolution, Education and Outreach* **1**, 121–137.

Hedges, S.B., Blair, J.E., Venturi, M.L., Shoe, J.L. (2004) A molecular timescale of eukaryote evolution and the rise of complex multicellular life. *BMC Evolutionary Biology* **4**, 2–12.

Mora, C., Tittensor, D.P., Adl, S., Simpson, A.G.B., Worm, B. (2011) How many species are there on Earth and in the Ocean? *PLoS Biol* **9**, e1001127; doi:10.1371/journal.pbio.1001127.

Pace, N.R. (1997) A molecular view of microbial diversity and the biosphere. *Science* **276**, 734–740.

Pace, N.R. (2009) Mapping the tree of life: progress and prospects. *Microbiology and Molecular Biology Reviews* **73**, 565–576.

Puigbò, P., Wolf, Y.I., Koonin, E.V. (2013) Seeing the Tree of Life behind the phylogenetic forest. *BMC Biology* **11**, 46–49.

Sanderson, M.J., Driskell, A.C. (2003) The challenge of constructing large phylogenetic trees. *Trends in Planet Sciences* **8**, 374–379.

Vigilant, L., Stoneking, M., Harpending, H., Hawkes, K., Wilson, A.C. (1991) African populations and the evolution of human mitochondrial DNA. *Science* **153**, 1503–1507.

Woese, C.R. (1998) The universal ancestor. *Proceedings of the National Academy of Sciences* **95**, 6854–6859.

Woese, C.R., Kandler, O., Wheelis, M.L. (1990) Towards a natural system of organisms: proposal for the domains Archaea, Bacteria, and Eucarya. *Proceedings of the National Academy of Sciences* **87**, 4576–4579.

7

The Limits of the Biospace

Learning Outcomes

➤ Understand the concept of the biospace and why it is important to astrobiologists.
➤ Be able to describe and explain the adaptations of different extremophiles.
➤ Understand the practical uses of research on extremophiles.
➤ Understand the role of habitats in protecting from extremes.
➤ Describe the importance of multiple extremes and polyextremophiles.
➤ Understand the importance of the deep subsurface as an extreme environment.

7.1 The Biospace

One way to think about life on the Earth is to think about how the vast number of species is evolutionarily related. This is the approach that we took in Chapter 6. Another way to conceptualise the biosphere is to consider its tolerances to physical and chemical extremes. One can view all organisms to be within an enormous biological zoo surrounded by a fence, where the fence is a set of physical and chemical extremes beyond which life cannot adapt – a fence that separates life from death. Figure 7.1 shows, in a conceptual way, what this looks like in two dimensions. We might call this the *biospace*. These boundaries can be defined differently depending on the limits of choice. They could be defined by the limits of reproduction, the limits

of metabolic activity or the limits of survival. We would expect the limits to generally be successively greater for each one of these cases (organisms can generally survive conditions that are more extreme than the conditions they require to successfully reproduce). Therefore it is necessary to define a biospace with respect to the particular biological behaviour being investigated.

7.2 The Importance of the Biospace for Astrobiology

Investigating the limits of life is important for astrobiology because it allows us to compare conditions that we measure on other planetary bodies to the biospace and to ask the question: 'Do these extraterrestrial conditions overlap with conditions on the Earth that we know are suitable for life?'. Of course, life needs many other things other than just appropriate physical and chemical conditions, as we have already seen. However, the biospace gives us a basis for a first-order assessment of the habitability of other planetary bodies. If physical and chemical conditions lie outside the biospace that we have defined, then the fact that other requirements for habitability are met may become irrelevant. Conversely, if we find a planetary environment that has physical and chemical conditions suitable for life, we are motivated to explore it further to see if it has liquid water, energy supplies and basic elements, CHNOPS.

The biospace is necessarily limited by what we understand about life on the Earth. It is a conservative

Astrobiology: Understanding Life in the Universe, First Edition. Charles S. Cockell.
© 2015 John Wiley & Sons, Ltd. Published 2015 by John Wiley & Sons, Ltd.
Companion Website : www.wiley.com/go/cockell/astrobiology.

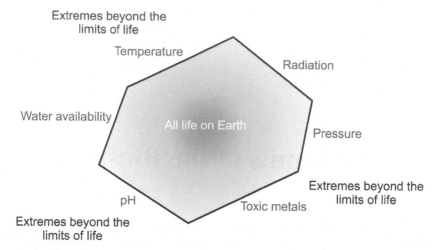

Figure 7.1   *A simplified depiction of the biospace, the space defined by physical and chemical extremes within which life can exist. There are many more extremes that define the limits of life than those shown here. These limits could be defined for survival, metabolic activity (growth) or reproduction* (Source: Charles Cockell).

boundary that does not take into account future advances in investigating life in extremes, which might expand the boundaries. Furthermore, it focuses only on carbon-based Earth-like life using water as a solvent. If alternative chemistries, such as those discussed in Chapter 3, exist elsewhere they would form a completely separate biospace which at the current time would fall outside our assessment of physical and chemical conditions suitable for life. We must keep an open mind and be ready to expand our ideas about the extent of the biospace. Nevertheless, without other empirical data, this is a limit we must accept and it provides us with a useful operational starting point.

7.3   The Edges of the Biospace are Dominated by Microbes

Within the confines of the biospace life can take on a huge diversity of wonderful and weird forms, from carnivorous kangaroos to dragonflies with one meter wingspans and all the other varieties of life alive today or revealed by the fossil record. But as astrobiologists, the edges of the biospace interest us most, because it is here that we come up against the limits of life.

We find that the edges of the biospace are dominated by microbes. As conditions become more extreme it becomes increasingly difficult to modify the entire biochemistry of complex multicellular organisms with their multiple cell types, giving single-celled microbes an advantage.

We observe that extreme conditions generally result in a reduction in the diversity of life. This is conceptually a significant point. Some people say that the phrase, 'life in extremes', is *anthropocentric* and that if microbes could express their views on us they would call us extreme. This view isn't supported by the evidence. We really can define physical and chemical 'extremes' where the abundance and diversity of life is significantly diminished by the demands of these environments.

The organisms that inhabit extremes generally fall into two categories. There are the *extremophiles* (from the Greek word *philos* or love). These are organisms that not merely tolerate extremes but have biochemical adaptations that require them to grow under given extremes. An example would be many microbes that require high temperatures to grow. Then there are *extremotolerant* organisms. These are organisms not optimally adapted to growth in extremes, but having the capacity to grow in them. An example would be many cyanobacteria in the Arctic that have optimal growth temperatures well above typical Arctic temperatures, but they succeed there by slow persistent growth.

We categorise different extremophiles and extremotolerant organisms according to the different physical or chemical extremes they require or can tolerate. Some examples are shown in Table 7.1.

Many of the record holders for extremophiles are in the domain Archaea, but extremophiles are found in the domain Bacteria and there are some remarkable survival

Table 7.1 *Examples of microbial extremophiles (italics) and extremotolerant (bold) organisms and some of their habitats. In each case an example of the challenge they face and an adaptation is shown.*

Type of extreme and name of extremophile/ extremotolerant organism	Conditions that allow growth (green) or that can be tolerated (red)	Example microorganism	Example habitat	Example challenge	Example adaptation
Temperature extremes					
Hyperthermophile	>80 °C	*Pyrolobus fumarii* (Archaea)	Hydrothermal vents, hot springs	Thermal breakdown of molecules	Thermostable enzymes
Thermophile	45–80 °C	*Thermus aquaticus* (Bacteria)	Hot springs	Thermal breakdown of molecules	Thermostable enzymes
Psychrophile	<15 °C	*Dioszegia cryoxerica* (Eukaryotic yeast)	Ice sheets, glaciers	Loss of membrane fluidity	More unsaturated bonds in membranes
pH extremes					
Acidophile	Low pH (< pH3)	*Acidithiobacillus ferrooxidans* (Bacteria)	Sulfur-rich acid rivers and springs	Acidification of cell interior	Effective proton pumps
Alkaliphile	pH > 9	*Natronobacterium magadii* (Archaea)	Carbonate-rich lakes	Hydroxide ions inside cell	Negatively charged outer cell wall/membrane
Pressure extremes					
Barophile (piezophile) or **barotolerant**	Pressures higher than atmospheric	*Halomonas salaria* (Bacteria)	Deep oceans, deep crust	Reduction in capacity to pump materials in and out of cell	Production of more effective membrane transporters
Extremes of salt					
Halophile	2–5 M NaCl	*Haloferax lucentensis* (Archaea)	Evaporative ponds	Extreme osmotic stress	Equalisation of osmotic stress using ions or **compatible solutes**
Extremes of radiation					
Radiotolerant (Radiation tolerant)	Tolerant of ~ > 10 Gy	*Deinococcus radiodurans* (Bacteria)	Desiccated soils and rocks (see text)	Free radical damage to molecules	Molecules to quench **oxygen free radicals**, repair molecules such as DNA
Extreme metal concentrations					
Metallotolerant	Tolerant of high concentrations of metals	*Cupriavidus metallidurans* (Bacteria)	Volcanic environments, polluted areas	Toxic effects of high metal ion concentrations	Effective pumps to remove metal ions
Extremes of desiccation					
Xerotolerant	Tolerant of Desiccation	*Sphingomonas desiccabilis* (Bacteria)	Deserts	Breakdown of molecules	Synthesis of cell-protecting sugars

strategies in the Eukaryotes as well – extremophiles are found in all three domains of life.

Embarking on a vast shopping list of extremes and describing the biochemical adptations to them is not a very useful learning exercise. However, in the following sections we look at some particular examples to explore the principles by which organisms can adapt to extreme environments and provide some key understanding important for astrobiology.

7.4 Life at High Temperatures

Many organisms live in high-temperature environments that include hot springs on land such as Yellowstone National Park, United States, deep sea hydrothermal vents (Figure 7.2a) and regions deep in the Earth's crust. Organisms that need to live at these extremes to grow are called *thermophiles* if their optimal growth range is 45–80 °C and *hyperthermophiles* if they have an optimal growth temperature above 80 °C. The range of their growth optima is schematically illustrated in Figure 7.2b in comparison to the *mesophiles*, which grow at temperatures between about 15 and 45 °C.

The first studies of thermophiles were carried out by Thomas Brock in Yellowstone National Park in the 1970s, where he examined organisms growing at temperatures around 70 °C. Since that time a large variety of organisms

have been identified and cultured that grow between this and much higher temperatures.

Deep ocean hydrothermal vents are one important habitat for hyperthermophiles (Figure 7.2). The high pressures in the deep oceans increase the boiling point of water to above 100 °C (see the discussion in Chapter 2 on phase diagrams) and thereby allow for organisms with optimum growth temperatures above this threshold. *Methanopyrus kandleri*, which grows and reproduces at 122 °C, is an archaeon (a member of the domain archaea) that inhabits black smokers and defines the current upper temperature limit for life. *Pyrolobus fumarii* is yet another hyperthermophile, an archaeon with a growth range of 90–113 °C, isolated from a black smoker.

Microorganisms are too small to regulate their internal temperature against the outside temperature and so having a biochemistry that can operate at the high environmental temperature is a requirement. The challenge for these organisms includes overcoming the problems caused by the thermal breakdown of biomolecules and thermally induced membrane disruption. Organisms must also gather enough energy to repair damaged molecules and synthesise new ones.

Organisms at high temperatures have evolved a number of adaptations, including the evolution of thermostable proteins and enzymes. Thermostability can be achieved by introducing new ionic or covalent bonds into proteins which essentially pin the amino acid chains together more

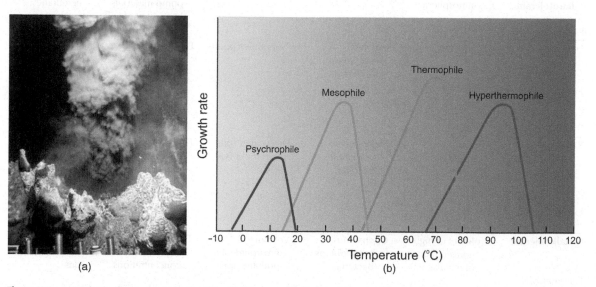

(a)   (b)

Figure 7.2 (a) A hydrothermal vent in the Atlantic Ocean, called a 'black smoker' because of the smoky black sulfide minerals produced as the fluids come into contact with cold sea water (Source: NOAA). (b) Schematic showing the typical temperature growth range of example organisms adapted to high temperatures compared to mesophiles and the cold-adapted psychrophiles.

Table 7.2 *Some examples of products (mainly enzymes) and processes involving extremophiles or extremotolerant organisms.*

Extremophile	Enzyme or whole organisms	Application
Thermophiles	Amylases, xylanases, proteases, DNA polymerases	Paper bleaching, food sweeteners, PCR
Psychrophiles	Proteases, dehydrogenases	Dairy manufacture, biosensors for detecting pollutants
Acidophiles	Whole organisms	Desulfurisation of coal, biomining (bioleaching) of economic metals
Alkaliphiles	Cellulases	Polymer degradation in detergents
Halophiles	Whole organisms	Polymer production
Metallotolerant	Whole organisms	Bioremediation of polluted industrial sites
Radiation-tolerant organisms	Whole organisms	Bioremediation of radiation-contaminated sites.

effectively and makes them less liable to denature when exposed to high temperatures.

Inevitably these adaptations are not perfect and protein damage does occur. One way to mitigate this problem is simply to synthesise more replacement protein. Another way is to prevent or reverse unfolding of proteins under high temperatures. Organisms adapted to life at high temperatures have a variety of *heat shock proteins* and *chaperonin* proteins which attach to other proteins, stabilising them or assisting in re-folding denatured proteins. Thus a complex and mutually reinforcing set of biochemical adaptations all work together to allow the cell to operate at high temperatures.

7.4.1 Uses for Thermostable Molecules

Many thermostable enzymes have found enormous use in industrial and everyday applications. There are a multitude of processes and products that require high temperatures, for example paper making that makes use of thermostable enzymes that can break down *cellulose*.

Perhaps best known is the enzyme used in the *polymerase chain reaction (PCR)*. Polymerase enzymes are involved in replicating DNA (Chapter 4). PCR is the laboratory process whereby small amounts of a target strand of DNA can be replicated to produce many thousands to millions of copies to be analysed in molecular biology (to make phylogenetic trees), forensics (to find a criminal) or paternal analysis (to find out who your real father is).

However, in order to 'melt' or break apart the double-stranded DNA to be able to replicate it, it must

first be heated. In the old days of molecular biology, the polymerase would have to be added at room temperature because of its sensitivity to high temperatures at which it would denature. This necessitated adding new enzyme at each separate replication step, which made it very laborious. However, the isolation of *Taq polymerase* from the thermophile bacterium, *Thermus aquaticus*, originally isolated from a volcanic hot spring in Yellowstone National Park by Thomas Brock and Hudson Freeze, allowed DNA replication to be achieved at high temperatures, considerably simplifying PCR and allowing for much greater automation of the process.

These examples illustrate how by studying extreme environments astrobiologists make discoveries of practical importance for society. Table 7.2 shows some examples of classes of useful compounds from different extremophiles.

7.5 Life at Low Temperatures

About 20% of the Earth hosts permafrost – permanently frozen ground. The temperature of much of the deep oceans is about 4 °C. As oceans cover 71% of the Earth, the habitat for low-temperature organisms is enormous. Aside from permanently frozen land, cold temperatures are generally common on much of planet Earth and so it is not surprising that many organisms have adapted to live in these conditions (Figure 7.3). These are the cold-loving microorganisms, known as *psychrophilic* or sometimes 'cryophilic' organisms, with an optimum

Figure 7.3 *Habitats for psychrophiles – the ice sheet between Qikiqtarjuaq and Cape Dyer on Baffin Island, Nunavut, Canada* (Source: wikicommons, Mike Beauregard).

growth temperature of less than 15 °C. Some organisms are cold tolerant (psychrotolerant) organisms that grow below this threshold, but do not require low temperatures to grow. The habitats of these organisms include polar soils, snowfields and glaciers.

These organisms have always been of great interest to astrobiologists because many of the planetary bodies that might host habitable conditions, at least in our Solar System (such as Mars, Jupiter's moon, Europa and Saturn's moon, Enceladus), are cold.

An example of the psychrophiles are some members of the bacterial genus, *Arthrobacter*, which inhabit many polar environments. Perhaps better known are the snow algae, a colourful group of eukaryotes that inhabit snowfields and often confer upon them yellow, red and orange colouration because of the *carotenoids* that they produce, pigments that screen ultraviolet radiation and quench the

oxygen free radicals produced in this environment when they are exposed to UV radiation.

A major challenge to life at low temperatures is loss of membrane fluidity. Membrane fluidity is important for ensuring the effective transfer of molecules such as nutrients from the outside to the inside of the cell and vice versa for waste, and for ensuring the movement of proteins and other molecules in the membrane. As membranes are made of lipids that have long fatty acid chains, the best analogy to understand this is to think of a material that contains fatty acids – butter. Butter becomes soft when you leave it on your kitchen table, but place it into a refrigerator and it becomes hard. In exactly the same way, microbes in cold environments lose fluidity in their membranes – they start to solidify.

One of the remarkable ways in which organisms can adapt to low temperatures is to introduce *unsaturated fatty acids* into their membranes. The introduction of double bonds creates kinks in the fatty acids (Figure 7.4) and as a result the membrane is less tightly packed. An analogous material is olive oil, which contains about 60% unsaturated fatty acids. It is a liquid at room temperature. If you place it into your refrigerator, unlike butter which has a higher proportion of saturated fatty acids, it remains a liquid, showing how unsaturated fatty acids can enhance fluidity.

I call this adaptation remarkable because it shows how a very simple chemical change, the introduction of a single additional bond between two carbon atoms, causes a profound change in membrane behaviour that helps organisms to adapt to an entire planetary environment – cold habitats. It illustrates how simple chemical physics is at the foundations of many of life's adaptations to its planetary environment.

Other adaptations used by life at low temperatures include the production of sugars such as trehalose which allows cells to mitigate the effects of damaging ice crystals that disrupt membranes and destroy macromolecules.

Focus: Astrobiologists. Christopher McKay

What do you study? I study the physical ecology of extreme environments.

Where do you work? NASA Ames Research Center.

What was your first degree subject? Physics.

What science questions do you address? I conduct field research in extreme environments with a focus on how the physical factors in the environment limit and shape microbial life. My particular focus is on dry and cold environments. From this data on Earth we extrapolate to dry and cold environment on other worlds.

What excites you about astrobiology? The prospect of finding second genesis of life somewhere in the Solar System.

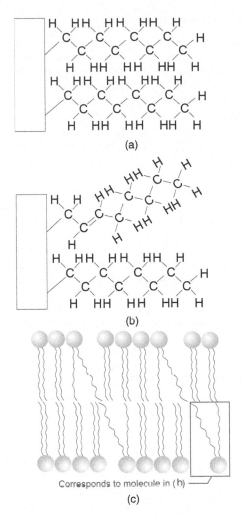

Figure 7.4 *An example of adaptation to low temperatures. The introduction of unsaturated fatty acids into the membrane phospholipids [from (a) to (b)] causes a kink in the fatty acid (b), the consequence of which is that the membrane is less closely packed and more fluid (c)* (Source: Charles Cockell).

able to cause disorder in molecules, such as magnesium chloride. By slightly disordering molecules, such as membrane lipids, they reduce the extent to which the molecules are made immobile or inactive by low temperatures.

One particularly interesting habitat in which cold adapted organisms can live is between the boundaries of ice grains in ice sheets. As ice freezes in ice sheets and glaciers, salt water is excluded from the ice and forms channels at the grain boundaries. As the *eutectic point* of NaCl is $-21.2\,°C$ (the lowest possible temperature at which it will freeze in solution with water), the presence of salt in inter-grain spaces can increase the availability of liquid water habitats, even below $0\,°C$. These brine pores and channels have been shown to be interconnected at temperatures down to $-15\,°C$, demonstrating that ice is a potentially vast habitat. Living within this environment is not without its challenges. Although the cold, restricted environment can exclude larger predators, some sea ice habitats have been shown to have very large numbers of viruses ($\sim 10^8$ viruses ml^{-1}), as they become concentrated in the small enclosed pore spaces. Virus attack is a serious challenge for many microorganisms in these environments.

The ice sheet habitat illustrates that to grow at low temperatures in these environments, microbes must also have adaptations to high salt concentrations. Rarely are environments subject to only one extreme.

The precise lower temperature limit of life is unknown. A large number of claims and counter-claims have been made for microbial reproduction and metabolic activity at sub-zero temperatures. The current limit for microbial reproduction lies at $-15\,°C$, with reports of metabolic activity at $-25\,°C$. Furthermore, it is important to define the criteria for the lower limit of life. The absolute lower limit of metabolic activity (e.g. the activity of some enzymes) is lower than the temperature limit of cell division that requires the interaction and activity of many biochemical processes.

7.6   Salt-loving Organisms

In a variety of environments found on Earth there are high salt concentrations. At ice grain boundaries that we met in the last section salt concentrations build up. In warmer environments, at the edges of salt lakes and seas, where water evaporation rates are high, microbes must adapt to high salt concentrations, particularly halite (NaCl). Salt-loving (*halophilic*) organisms need to grow in salt concentrations of 15–37% NaCl. Some of them live in solid salt crusts known as evaporites that build up

By synthesising sugars, the freezing point can be lowered and ice crystal formation prevented.

These types of biochemical adaptations allow life to adapt to a range of cold environments. Their numbers vary hugely from the quite sparse ~ 200 cells ml^{-1} in the ice at the South Pole to up to 10^8 cells ml^{-1} in permafrost. Glacial ice can be inhabited to values of about 10^7 cells ml^{-1}.

The biological challenge of low temperature can also be partly offset by compounds that are *chaotropic*, that is

Figure 7.5 *Cyanobacteria forming a green layer within an evaporite from California. The sample is about 3 cm in thickness* (Source: Lynn Rothschild).

at the edges of these natural habitats. Examples of natural evaporite habitats are The Dead Sea on the border with Israel and Jordan, the salt flats of Baja California and Great Salt Lake, Utah (Figure 7.5). Not all salt crusts are made from halite. Microbes also inhabit gypsum ($CaSO_4$) evaporites and other crusts that form from precipitating salts under evaporation.

The challenge of living in these environments is to deal with the osmotic stress. Water will tend to move from a region of low salt to a region of high salt, so when a microbe is immersed in very salty water, there will be a tendency for water to flow from the inside of the microbe to the outside, depriving the microbe of the essential solvent required for it to operate. Microbes have adapted ways to deal with this challenge, which generally fall into two categories, the salt-in and salt-out approaches.

7.6.1 Salt-in Strategy

Microbes using the salt-in strategy allow for a high concentration of ions to collect inside the cell, thus generating a solution inside the cell which is osmotically in equilibrium with the outside environment. This is usually achieved by pumping K^+ ions into the cell. However, the high concentration of ions inside the cell requires that the proteins and other biochemical machinery are adapted to high salt concentrations. This includes proteins with a higher number of ionic bridges. *Haloferax volcanii*, a bacterium that was originally isolated from the Dead Sea, is an example of an organism that uses the salt-in strategy. Many of these organisms are *obligate* halophiles.

7.6.2 Salt-out Strategy

Microbes using this strategy pump out the salts to prevent them from damaging the cell interior. The question is then: how do organisms deal with the huge osmotic stress that would be caused by the salt difference inside and outside the cell? This is dealt with by the production of *compatible solutes* that prevent osmotic pressure from developing across the cell membrane by ensuring that osmotic pressure is equalised. Many of these compounds, although they equalise the osmotic pressure, are less damaging than small ions.

Compatible solutes fall into two classes: (i) Amino acids, such as glycine betaine, glutamine, glutamate, proline, ectoine and (ii) Sugars and polyols (sugar alcohols), such as trehalose, glycerol, di-*myo*-inositol-1,1'-phosphate (DIP) or sulfotrehalose. These solutes are synthesised until they reach sufficient concentration to counter the osmotic potential of the outside environment.

Organisms that live in moderate or fluctuating salt concentrations, including many bacteria, use the salt-out strategy. This group is occupied by a large number of *facultative* halophiles.

7.6.3 Low Water Activity

Related to the stress of high salt (and desiccation) is low *water activity* (referred to as a_w). Water activity is a measure of the availability of water. It is formally defined as the ratio of the vapour pressure of a substance to the vapour pressure of pure water. Distilled water therefore has a water activity of 1. The lower the water activity, the less the availability of water. The lower water activity limit of life is around 0.6, set by a variety of desiccation resistant fungi. However, most organisms require a water activity great than $\sim$0.9 to grow. A saturated solution of NaCl has a water activity of 0.75, illustrating that it poses a challenge to microbes, although a variety of microbes are known that inhabit saturated salt. Many foods have low water activity, for example honey typically has water activities of between 0.5 and 0.7, explaining why it rarely becomes contaminated with microorganisms.

7.7   pH Extremes

pH is a measure of the acidity or alkalinity of a solution. Mathematically, pH is the negative logarithm of the activity of the hydronium ion (H_3O^+), more often expressed as

Debate Point: Salts in the Universe

One challenge in astrobiology is to consider how studying life in extremes on the Earth might tell us about the possible conditions for life elsewhere. Salts are ubiquitous in the Universe. They have been detected on Mars and on the surface of Jupiter's moon, Europa. Most of the studies that have been conducted on the Earth have focused on sodium chloride-rich environments and so our understanding of microbial adaptations has focused on these salts. However, on other planetary bodies, salts such as sulfate compounds dominate. Explore the literature about the salts that have been found on other planetary bodies. How might our knowledge of life in extreme salty environments on the Earth inform our understanding of the habitability of these extraterrestrial salts? What might be the limitations of our knowledge?

Martínez, G.M., Rennó, N.O. (2013) Water and brines on Mars: Current evidence and implications for MSL. *Space Science Reviews* **175**, 29–51.

McKinnon, W.B., Zolensky, M.E. (2003) Sulfate content of Europa's ocean and shell: evolutionary considerations and some geological and astrobiological implications. *Astrobiology* **3**, 879–897.

the measure of the hydronium ion concentration. It may be thought of as the concentration of protons (H^+) since hydronium is protonated water. Acidic solutions have higher proton concentrations and thus lower pH.

Organisms that inhabit extremes of pH are called the *acidophiles* (usually below pH 3) and *alkaliphiles* (usually above pH 9) A wide variety of natural environments exhibit extremes of pH. Where pyrite (iron sulfide) reacts in water, the sulfide within it is oxidised to sulfuric acid and the result is an extremely acidic river, such as the RioTinto in Spain (Figure 7.6), which has been a natural low pH environment for thousands of years, with a pH value around 2. Iron Mountain in northern California, United States, has water with a pH of 0 to 1. Other naturally acidic environments include volcanic hot springs where water comes into contact with sulfur from volcanic activity. High pH environments are found where water comes into contact with certain volcanic rocks.

Mono Lake in California, United States (Figure 7.6), is an enclosed lake formed about 760 000 years ago. Salts accumulate within the enclosed lake and the result is a shallow water environment with a pH of 10.

Although organisms are adapted to pH extremes, most of them maintain a nearly neutral pH inside the cell.

In the case of the acidophiles, the high concentration of protons in the environment is kept out of the cell by actively pumping them out. This is sometimes achieved by the production of large numbers of proton pumps and other membrane transporters that increase the efficiency of proton pumping across the cell membrane. Highly acidic conditions within the cytoplasm result in protein denaturation (unfolding of the protein), providing a selection pressure for acidophiles to evolve low proton permeabilities across cell membranes to prevent internal cell damage. It is not always possible to completely restrict all protons from entering the cell.

Figure 7.6 *Life at pH extremes. Left: The acidic Rio Tinto River, Spain* (Source: wikicommons, Carol Stoker). *Right: Alkaline Mono Lake, United States* (Source: wikicommons, Brocken Inaglory).

Some acidophiles have adaptations to tolerate a low pH including the evolution of acid-stable proteins. In many acid stable proteins there is higher abundance of the two acidic amino acids (aspartic and glutamic acid) which have a carboxyl group (–COOH) and tend to lose protons at the pH in the cell, becoming negatively charged (–COO⁻). This minimises protein destabilisation at low pH caused by a build-up of positive charge on the protein. In a specialised case of acid stability, the NAPase protein (which breaks down the long-chain protein compound *keratin*) from the bacterium *Nocardiopsis alba* was shown to have relocated acid-sensitive ionic bonds away from exposed regions to less exposed regions to prevent protein denaturation.

Organisms that tolerate high pH values are less well understood, but suffer from the problem of low proton concentrations outside the cell which make it challenging to maintain an efficient proton gradient across the cell membrane, necessitating the evolution of efficient proton transport mechanisms across the cell membrane. The hydroxide ion (OH⁻), generated at high pH, is damaging to molecules and many of these organisms have negatively charged cell walls that prevent these ions from entering the cell. Some of the bacteria belonging to the *Bacillus* genus are alkaliphiles, such as *Bacillus alcalophilus*, which can grow at pH 12.

Figure 7.7 *The high-pressure habitat of the Challenger Deep, Mariana Trench (10 900 m depth), Pacific Ocean being explored by the one-person Deepsea Challenger submarine* (Source: Deep Sea Challenger project).

7.8 Life Under High Pressure

By far the greatest physical extent of habitat is under the surface of the Earth and in the depths of the oceans. In the oceans, pressure increases by approximately one atmosphere for every 10 m depth so that, at the Mariana Trench (10 970 m in the Pacific Ocean; Figure 7.7), pressures exceed 1000 atm. The *piezophilic* or *barophilic* ('pressure-loving') microorganisms inhabit high-pressure environments. For organisms underground, high pressure is ubiquitous. It still remains a point of contention the extent to which organisms in these environments actually need high pressure or whether most of them are merely pressure tolerant. An example of a pressure tolerant organism is the bacterium, *Halomonas salaria*, which can grow at a pressure of 1000 atm (100 MPa) and a temperature of 3 °C.

One of the challenges faced by life at high pressure is the pressure-induced packing of molecules and loss of fluidity in the membrane. Intriguingly, one way in which cells adapt to high pressures is to increase membrane fluidity with unsaturated fatty acids, much like the adaptation to cold temperatures that was discussed earlier. This illustrates a general principle that adaptation to some extremes can be synergistic. Instead of requiring adaptations to each individual extreme that life might face in any given environment, in some cases one biochemical adaptation will help provide the required alterations for another extreme. As high-pressure deep ocean environments are often cold, these two stresses are found in the same place.

Other adaptations to high-pressure environments are the production of trans-membrane proteins that span the cell membrane and aid in the uptake of small molecules, such as nutrients, and the removal of waste from the cells.

7.9 Tolerance to High Radiation

Some organisms have the capacity to deal with very high levels of radiation, including *ionising radiation* such as *gamma radiation*. For example, doses of 10 000 *Gray* (Gy)

Facilities Focus: Underground Laboratories

One way that scientists go about studying extremes is to set up permanent facilities in environments in which they can study life, unravelling which microorganisms grow there and how they adapt. On the surface of the Earth, this is mainly a matter of finding a place to put a building, getting the right permits and building your laboratory. But what happens when the life you want to study is 1 km underground? Scientists can build permanent underground laboratories. One such facility is the Boulby Underground Science Facility north of the seaside town of Whitby in England run by the United Kingdom Science and Technology Facilities Council (STFC). The Boulby Mine is an active mine that extracts potash for fertiliser and halite for salting roads. For over a decade, it has been home to a *Dark Matter* laboratory that is using the 1 km of salt above it to cut down cosmic radiation and look for signs of Dark Matter in the Universe. The laboratory is internet connected and air conditioned. The facility is also home to the Boulby International Subsurface Astrobiology Laboratory (BISAL), that is used by astrobiologists to study life in deep, dark hypersaline brines (Figure 7.8). How does life make a living here without photosynthesis? What sort of organisms live underground? Laboratories can provide a way for scientists to set up a permanent presence in extreme environments and thereby establish investigations over many years. As well as being a base for studying extreme life using the over 500 km of roadways in the mine to access the deep dark biosphere, BISAL is also a base for scientists to test technology for the exploration of Mars and other planetary bodies. The isolated environment and the ancient salts provide an ideal way for teams of researchers to get together and test their technologies.

Figure 7.8 *Scientists use a deep underground astrobiology laboratory (left and centre) to sample the microorganisms that live in deep, dark brine seeps (right).*

can be tolerated by the bacterium *Deinococcus radiodurans* (Figure 7.9). *Chroococcidiopsis*, a photosynthetic cyanobacterium that lives in rocks in cold and hot deserts as well as other environments (Figure 7.10) can tolerate 15 000 Gy of gamma rays. To get some perspective, 5 Gy of equivalent ionising radiation is sufficient to kill a human. These organisms are tolerant to radiation, but do not require it to grow. They are *radiotolerant*.

Adaptations to high radiation include having multiple copies of the genome so that damage to one genome might be overcome by another copy. Effective DNA repair mechanisms, pigments such as *carotenoids* and high manganese concentrations that quench *oxygen free radicals* are other adaptations.

High radiation resistance might seem to be a mystery in cells since very few natural environments have high radiation doses. Where did this resistance arise? This conundrum has led some people to outlandish speculations that some of these organisms might have come from space. However, we do not have to resort to these hypotheses to provide an explanation.

Any environment that causes direct and indirect damage to cells and their constituent molecules will result in an evolutionary selection pressure for organisms capable of withstanding or repairing that damage. One such environment is desiccated soils and rocks, where damage caused by the desiccation of cells includes DNA damage and the formation of reactive oxygen species.

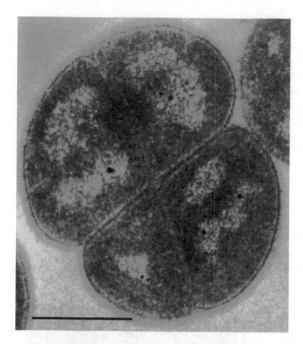

Figure 7.9 *The radiation resistant* Deinococcus radiodurans *scale bar 2 μm across.*

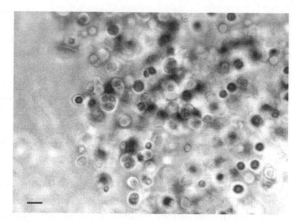

Figure 7.10 Chroococcidiopsis, *an extreme radiation-tolerant cyanobacterium that lives on, in and under rocks in hot and cold deserts of the world. Scale bar 2 μm.*

Many of the radiation-resistant organisms discovered are also organisms that inhabit deserts and soils and so it seems reasonable to suppose that high radiation resistance can be a serendipitous result of adaptation to desiccated environments that cause damage to biological macromolecules, including DNA. Radiation resistant organisms have also been isolated from high temperature environments, where organisms also have to cope with high temperature-induced DNA damage.

7.10   Life in Toxic Brews

In 1976 a discovery was made of an extraordinary bacterium, *Cupriavidus metallidurans* (Figure 7.11), inhabiting a toxic heavy metal dump that had high concentrations of zinc and other heavy metals toxic to humans. This microbe was subsequently found to contain small circular pieces of DNA, or plasmids (Chapter 4). One of the plasmids was found to encode genes that allow the organism to export cobalt, zinc and cadmium, another cobalt and nickel. These plasmids allow the cell to maintain low concentrations of heavy metals inside the cells that would otherwise disrupt protein function.

These metallotolerant organisms are adapted to living in a range of natural environments, such as volcanic hot springs, where the circulation of hot fluids through rocks leads to a build-up of toxic metals. Particularly important is the encoding of these genes on plasmids, providing an example of how adaptations to extremes can be encoded on DNA that could potentially be passed from one microbe to another in conjugation (Chapter 6), a mechanism by which horizontal gene transfer can allow for adaptation to extreme environments. These organisms generally do not require heavy metals to grow, they tolerate these conditions.

7.11   Life on the Rocks

In the previous sections we have looked at some biochemical adaptations to particular extremes and how different adaptations can allow microbes to grow and reproduce in extreme environments. Another way in which life can deal with extreme conditions on a planetary surface is to occupy habitats which provide shelter or amelioration of extreme conditions.

An example of such habitats is rocks. This habitat is of special interest to astrobiologists because, of course, Earth-like planets are made of rock and so if we want to understand how life can persist on and in a planetary crust for several billion years, it is useful to investigate the relationship between life and the crust.

There is an entire nomenclature to describe habitats on and in rocks (Figure 7.12) and although we should not

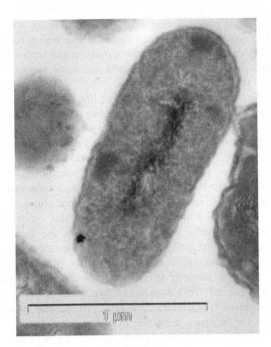

Figure 7.11 Cupriavidus metallidurans, *a heavy metal tolerant organism.*

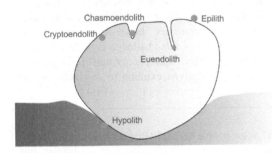

Figure 7.12 *Types of habitats in and on rocks* (Source: Charles Cockell).

of interconnected pore spaces within the rocks rather than merely the existence of a crack from the surface of the rock. The prefix, crypto (Greek, *kruptos* – hidden), reflects the fact that because they grow inside the rock you cannot see them from the surface unless you break the rock open. Euendoliths are organisms that actively burrow into rocks and they include cyanobacteria that live on carbonates in coastal regions and dissolve their way into the rocks, creating a long thin tunnel in which they live (from the Greek, *eu*, which means well or happy; this description is a little more mysterious, although it can be taken to mean that the organisms are good at living in rocks). Epiliths live on the surface of rocks (Greek, *epi* – upon) and hypoliths live at the soil-rock interface – in other words under a rock (Greek, *hypo* – under).

An example of cryptoendoliths is organisms that live in the interstices of rocks shocked by an asteroid or comet impact (Figure 7.13). Under the intense heat and pressure of an asteroid or comet impact a rock can be transformed into a very porous material. The porosity of these rocks is up to 25 times greater than unaltered unshocked rocks. Within the fractures and vesicles produced by the impact the microbes can make a home. The example shown in Figure 7.13 is from the Haughton impact crater in the Canadian High Arctic, which was formed by an asteroid or comet impact 39 million years ago. The material is shocked gneiss, a type of *metamorphic* rock, which will be discussed in Chapter 13. The rock is inhabited by *Chroococcidiopsis* and other microbes. The photosynthetic microbes form a distinctive green band inside the rock. The surface of the rock is dry and wind-swept and the microbes cannot grow on it. The lowest depth of the

get too diverted by definitions, it is perhaps useful here to briefly summarise them.

Chasmoendoliths are microbes that live in rock fractures, such as fractures caused by freezing and thawing of the rock, or by other physical disruption of a rock. This long word is constructed from Greek (*khasma* – void; *endo* – inside; *lithos* – rock). Cryptoendoliths are microbes that live in the rock interstices – inside the pore spaces of rocks. This is a little more specific than the chasmoendolithic habitat because it requires the existence

Figure 7.13 *Impact-shocked gneiss – a habitat for photosynthetic cryptoendolithic life in asteroid and comet craters. The rock shown here is about 6 cm across* (Source: Charles Cockell).

band within the rock is set by the depth at which light is reduced to below the minimum required for photosynthesis. This is an example of how even a catastrophic event such as an asteroid or comet impact can generate new habitat space for life.

The importance of these rock habitats for understanding life in extremes is that rocks can mitigate extreme conditions. Rock substrates rapidly attenuate ultraviolet (UV) radiation, which is absorbed and scattered by the substrate. Pore spaces in rock hold water, so that after rain or snow melting, water can be trapped for many days, providing an aqueous environment for microbial growth. Rocks can also provide a thermal advantage. Low albedo rocks (rocks that reflect very little solar radiation because they are dark or contain dark minerals) absorb solar radiation and become warmer than the air temperature, allowing for an improved thermal regimen for growth within the rock. Concomitantly with the availability of liquid water, this allows for growth and reproduction in the rock substrate.

7.12   Polyextremophiles – dealing with Multiple Extremes

In the natural environment, extremes rarely come on their own. The deep oceans are cold and at high pressure, hot volcanic environments are sometimes acidic and the surfaces of rocks are often desiccated and exposed to UV radiation. In some environments two or more extremes are encountered. Organisms adapted to two or more extremes

are called *polyextremophiles*. *Chroococcidiopsis*, which we encountered under high radiation and within rocks, is an example of a polyextremophile. Some strains are desiccation and radiation tolerant and many strains are also found growing in salt crusts.

To truly understand the limits of life, it is the polyextremophiles that we must investigate. Although earlier in the chapter we showed the biospace in a simplified two-dimensional form, we could redraw this in many dimensions, corresponding to different extremes. In Figure 7.14 you can see it drawn in three dimensions depicting the biospace for pH, salinity and temperature for the growth range of a selection of extremophiles.

A feature of these sorts of plots is the 'missing' polyextremophiles at some extremes. Is the lack of acidophilic halophiles, for example, caused because it is very difficult to adapt to both low pH and high salt or is it because we haven't explored enough low pH, high salt environments to find these organisms? Ultimately life will reach a limit, so we can expect that the biospace does not extend indefinitely. However, it is almost certainly the case that we have much work to do in exploring more extreme environments. Studying microbes in the laboratory will allow scientists to define the limits of life under many extremes, especially since most studies investigate temperature, pH and salt, so we cannot complete three-dimensional plots for all the extremes shown in Figure 7.1. One little-known aspect of this challenge is the energetics of life in extremes. How much energy must be acquired by an organism for a given extreme to be dealt with and

---

Debate Point: The limits of life – are they universal?

What ultimately defines the limits of life – biochemistry or physics? Some people have said that the limits of life on the Earth are an accident of biochemistry and that if life exists elsewhere it would have entirely different capabilities. But is this the case? The carbon–carbon bond for example, has a defined energy that is universal. At a certain temperature the energy put into these bonds will surely exceed any biochemistry's ability to repair these molecules or produce new ones fast enough. Ultimately, therefore, we might confidently say that for high temperature extremes, physics and chemistry must define a limit of life, not merely accidents in evolutionary biology. What about other physical and chemical extremes? Are the limits of organisms on Earth at pH, metal toxicity, pressure or salinity, for example, defined by quirks of biochemistry fixed into cells long ago, or does life operate at the limits of physics? What might this tell us about how universal these extremes are?

Terrestrial life has had over 3.5 billion years to evolve methods to cope with physical and chemical extremes. One could hypothesise that the limits exhibited by terrestrial life represent universal boundaries of biophysical capabilities, although the boundaries may subtly vary depending on the exact architecture of key molecules that become incorporated into any particular carbon-based life. The discovery of life elsewhere would allow us to test this hypothesis.

Clarke, A. (2014) The thermal limits to life on Earth. *International Journal of Astrobiology* **13**, 141–154.

Corkrey, R., McMeekin, T.A., Bowman, J.P., Ratkowsky, D.A., Olley, J., Ross, T. (2014) Protein thermodynamics can be predicted directly from biological growth rates. *PLoS ONE* **9**, e96100. doi:10.1371/journal.pone.0096100.

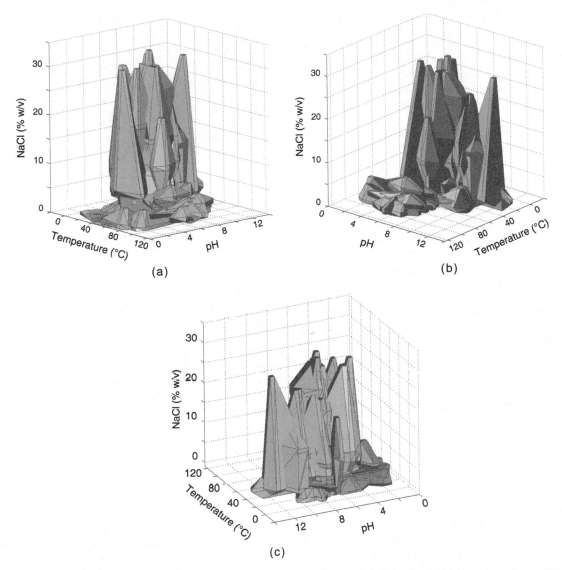

Figure 7.14 *The biospace of pH, temperature and salinity (a–c) calculated using the growth ranges of a variety of extremophiles, illustrating the multiple extremes to which life is normally exposed* (Source: Harrison et al. 2013). *The axes are shown with different orientations to display the 3D space.*

what are the energetic costs and limits to given multiple extremes?

7.13   Life Underground

The subsurface of the Earth is an extreme environment that in many ways exemplifies the multiple extremes to which

life is often exposed. Lack of nutrients, high pressures and temperature, salinity, desiccation and high concentrations of toxic elements are just some of the extremes experienced separately or in combination in different locations in the subsurface. The depth of life underground is ultimately set by the depth at which temperatures exceed the upper temperature limit for life. This is probably between about 5 and 10 km depth (assuming an upper temperature limit of

122 °C) depending on whether you are in oceanic crust or continental (land-based) crust. In the latter the temperature increase (about 25 °C km^{-1}) is about 10 °C km^{-1} higher than oceanic crust.

Microbes have been found in deep oil reservoirs, hydrothermal systems, deep granites and other rocks. In the deep subsurface, where photosynthesis is not possible, many of the extremophiles are chemoautotrophs and chemoheterotrophs, which were introduced in Chapter 5 when discussing microbial metabolism.

Microbes living in the deep subsurface are adapted to extremely low energy availabilities. They are capable of long periods of dormancy and their metabolic activity almost approaches time periods that are geological with doubling times exceeding centuries or even millennia.

The extent of the deep biosphere is not fully known, but it might constitute between 10 and 50% of the biomass of life on Earth. The presence of organisms that live on low energy availabilities in darkness and using chemoautotrophic pathways under multiple extremes has made the deep subsurface biosphere of great interest to astrobiologists seeking models to understand possible conditions for deep biospheres on other planets, if they exist.

7.14   Dormancy in Extreme Conditions

Some microorganisms have specialised responses to extreme conditions that allow them to go into a state of dormancy until better conditions arrive or until they are transported to a place with better conditions.

When environmental conditions become extreme some bacteria, including many *Bacillus* and *Clostridium* species, have the capacity to form *endospores* (or simply spores) a specialised type of cell (Figure 7.15). The onset of environmental extremes, such as extreme nutrient limitation, triggers a series of changes in the cell. DNA is replicated and whilst one copy of the DNA is destroyed as part of the cell withers away, the other copy becomes enclosed in a series of spore coatings. From the outside in, the spore is surrounded by the exosporium (not all spores have this layer), a *glycoprotein* layer. The protein-rich spore coat beneath it acts as a molecular sieve, preventing destructive enzymes and other molecules from reaching the interior. Beneath it is the *peptidoglycan*-containing cortex. Beneath the cell wall and plasma membrane, the next layer down is the core. The DNA within the core is stabilised by calcium dipicolinate, which can make up between 5 and 20% of the dry weight of the spore. The

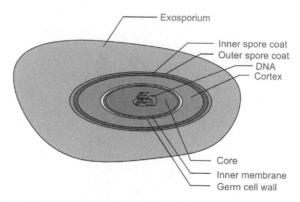

Figure 7.15   *Schematic diagram of a typical bacterial spore. The exact nature and presence of the different layers varies for different spores. A spore is typically about 1 µm long (Source: Charles Cockell).*

DNA within the core is also coated in small acid-soluble proteins (SASPs), which protect it from radiation and chemical damage.

Spores have remarkable environmental resistance, tolerating vacuum desiccation and large temperature variations. They survived for over six years on the Long Duration Exposure Facility (LDEF) satellite in *low earth orbit* and are found in spacecraft clean rooms. The maximum longevity of spores is not known. There have been controversial suggestions of their survival in salts many tens of millions of years old. They have been reported from 500 000 year old permafrost. They have a clear interest to astrobiologists as they are a mechanism for life to withstand environmental assaults over long time periods.

Although they are resistant to many environmental extremes such as high temperatures, spores, like vegetative cells, are sensitive to UV radiation and are rapidly killed by the short wavelength radiation since it penetrates to the DNA in the spore. For instance, they would be quickly killed on the surface of Mars by the intense UV radiation on that planet. However, it does not seem unreasonable to suggest that protected within the interior of unsterilised spacecraft or under layers of dust, viable spores might persist on the surface of Mars. The presence of contaminating endospores in spacecraft raises questions about the extent to which we are likely to contaminate other planetary surfaces during planetary exploration. This concern is called planetary protection and we will return to it later in the book.

Even vegetative cells can survive great extremes. Some polyextremophiles have been shown to survive the ultimate extreme – outer space. In space, organisms are exposed to the combined extremities of UV radiation, extreme vacuum desiccation and low temperatures. In the European Space Agency's EXPOSE facility, attached to the outside of the International Space Station, organisms were exposed to the extremes of space conditions in low earth orbit. Yet a variety of organisms have been able to survive these conditions, including cyanobacteria such as *Chroococcidiopsis* (Figure 7.10) and lichens. It should be stressed that these organisms were not growing in the environment, but they were exposed as desiccated, inactive cells.

7.15 Eukaryotic Extremophiles

Although bacteria and archaea dominate extreme environments, it would be wrong to paint a picture that excludes the eukaryotes. Many fungi have remarkable extreme tolerance, such as the fungus *Trichosporonoides nigrescens*, which grows on jam and can grow at a water activity of 0.75. Indeed, the lower limit of growth at low water activity ($a_w = 0.67$) is set by the fungus *Xeromyces bisporus*. Many fungi are also resistant to high radiation, acid, desiccation and a wide range of other stresses.

Some multicellular eukaryotes are tolerant to extremes. Apart from the microbes, Mono Lake, the pH 10 lake we met earlier as a home for alkaliphiles, is home to alkali flies (*Ephydra hians*) and brine shrimp (*Artemia monica*).

One of the adaptations of the alkali fly larvae is the precipitation of calcium carbonate inside the larval body (the 'lime gland'). The carbonate ions, which contribute to the high pH values of the lake, are sequestered into *biominerals*. In this way they prevent the extremely high carbonate concentrations of the lake interfering with cell functions.

The adaptations of the Mono Lake brine shrimp are still not fully understood. High concentrations of the simple sugar compound, glycerol, occur in their cysts (dormant resting states), which is thought to allow them to survive in the high ion waters. Brine shrimp are remarkably resilient, not just to salt, but to desiccation, making them polyextremophiles. Cysts from another species, *Artemia salina*, were taken to lunar orbit and back during the Apollo 16 and 17 missions. *Cosmic rays* were found to prevent full development of the larvae, although some developed normally.

Another spectacular example of a multi-cellular extremophile is *Lithobates sylvaticus*, the tree frog, which can survive freezing during the winter (Chapter 3). It inhabits the forests of North America. During winter

glycogen in its blood is converted to glucose and acts as a cryoprotectant. The frog is capable of remaining inert during the winter and then returning to activity in the spring. These adaptations to freezing are shared by insects, fish and variety of *taxa* that live in polar environments.

Many multicellular polar organisms, including insects and fish produce *antifreeze proteins* which are *glycoproteins* that inhibit ice crystal formation within cells. These proteins have enormous commercial interest in food preservation and even potential use in the cold storage of organs for organ transplant.

The smallest animals that show an impressive resistance to extremes are the tardigrades (water bears or moss piglets; Figure 7.16). Tardigrades belong to the phylum Tardigrada, an ancient group with fossils dating from the Cambrian period, 530 million years ago. They are typically about 0.5 mm long when fully grown, containing about 40 000 cells. They have been found throughout the world, including in polar regions, equatorial deserts and high mountains.

Tardigrades can survive an impressive range of stresses. They can be heated for a few minutes to 151 °C or cooled to −200 °C. They can withstand vacuum and more than 1200 times atmospheric pressure. Some species can even withstand a pressure of 6000 atm, which is nearly six times the pressure of water in the deepest ocean trench, the Mariana Trench. They can survive in a dry state for nearly 10 years. When exposed to extremely low temperatures, their body composition goes from 85% water to only 3%. They can withstand a thousand times more ionising radiation than other animals and are highly resistant to shortwave UV radiation. They were the first known animal to survive

Figure 7.16 *A tardigrade: an extreme tolerant animal. The creatures are usually about 0.5 mm long* (Source: wikicommons, Darron Birgenheier).

in space. In 2007, dehydrated tardigrades were taken into low earth orbit on the FOTON-M3 mission carrying the BIOPAN astrobiology payload. For 10 days, groups of tardigrades were exposed to the vacuum in Low Earth orbit. After being rehydrated back on Earth, over 68% of the organisms protected from high-energy UV radiation revived, but subsequent mortality was high.

Despite these impressive eukaryotic adaptations, which reveal a great deal about the versatility and mechanisms of biochemical adaptation, the microorganisms provide us with our most comprehensive set of data on environmental extremes and provide us with the key to defining the limits of life under multiple extreme conditions.

7.16   Are there Other Biospaces?

Do other biospaces exist in the Universe that are defined by the physical and chemical extremes of entirely different biochemistries that we discussed in Chapter 3? The possibility of multiple biospaces which life can occupy with different biochemistries is conceptually illustrated in Figure 7.17. As yet, there is no chemical evidence for life that uses an alternative elemental building block than carbon and an alternative solvent to water. As yet, we know of only the biospace occupied by carbon- and water-based life exemplified by life on the Earth. Nevertheless, the alternative solvents and elemental building blocks that have been suggested invite astrobiologists to do experiments to find out whether our own biospace defines the real boundaries to life in the Universe.

A future challenge in astrobiology is to test the hypothesis that alternative biospaces exist on other planetary bodies and thereby to take us nearer to a complete description of the limits of life.

7.17   The Limits of Life: Habitability Revisited

In Chapter 3 we saw that for an environment to be habitable it must be an environment that provides the basic elements for life (CHNOPS) and where physical and chemical conditions will support the existence of a suitable solvent, which at the current time we assume is liquid water. In Chapter 5 we also saw that a habitable environment requires a plausible energy supply for life that is accessible and can be sustained over long time periods, so that life can grow and reproduce. In this chapter

we have also learned that a habitable environment must have chemical and physical conditions not too extreme for life, independent of the factors above. From the point of view of a physical space or particular environment, these requirements define the physical and chemical conditions that must exist in a place for it to be habitable for given species.

This observation is important because the presence of elements, a solvent and energy do not imply, on their own, that an environment is habitable. Consider, for example, a low pH, low water activity brine on Mars. These solutions have been predicted from geochemical studies. Some of these brines may have water activities as low as 0.5, below the currently known biological limit for water activity. Such a brine is aqueous (it contains liquid water), it could contain CHNOPS (C in the form of carbon dioxide from the Martian atmosphere; H in water; N perhaps in fixed nitrogen such as nitrate; O in water; P in P-containing minerals such as *apatite*, which are found in volcanic rocks and could be dissolved in the brine; and S in sulfate compounds which are known on Mars and are likely to be in brines). The brine could also contain an energy source, for example it could have organic material in it or if it was on the surface it might allow for anoxygenic photosynthesis. This hypothetical brine, from the point of view of CHNOPS, water and energy availability, is habitable. However, if its water activity is below that required for life, then the brine is uninhabitable. Thus, physical and chemical conditions extremes, on their own, are an important part of defining the habitability of an environment.

7.18   Conclusions

Life on Earth can be considered as a biospace surrounded by physical and chemical extremes. Understanding the limits of life has importance to astrobiology because it provides us with an empirical basis from which to assess the habitability of extraterrestrial environments, particularly with respect to their physical and chemical extremes. Few organisms are exposed to a single extreme and although we know much about how organisms adapt to specific extremes, most natural environments require adaptation to multiple stresses. Many environments on the Earth are of interest to astrobiologists; extreme pH, briny and deep subsurface environments, for example, provide astrobiologists with knowledge of how biochemistries evolve under a range of planetary conditions. In some cases, the organisms that live in these extremes provide novel products of industrial use. One important question

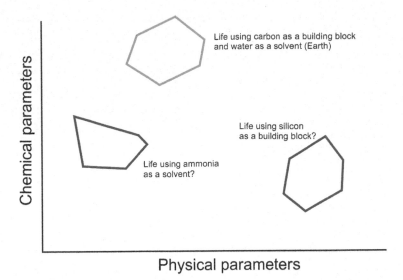

Figure 7.17 *Are there other biospaces in the Universe? A simple conceptualised 2-D diagram showing biospaces of life within a given set of physical and chemical extremes* (Source: Charles Cockell).

in astrobiology is to what extent life on Earth has explored the absolute physical and chemical limits possible and if there is life elsewhere, is it bounded by the same extremes?

Further Reading

Books

Gross, M. (2001) *Life on the Edge: Amazing Creatures Thriving in Extreme Environments*. Basic Books, New York.

Postgate, J.R. (1995) *The Outer Reaches of Life*. Cambridge University Press, Cambridge.

Papers

Amaral Zettler, L.A., Gómez, F., Zettler, E., Keenan, B.G., Amils, R., Sogin, M.L. (2002) Microbiology: eukaryotic diversity in Spain's River of Fire. *Nature* **417**, 137.

Billi, D., Friedmann, E.I., Hofer, K.G., Grilli Caiola, M., Ocampo-Friedmann, R. (2000) Ionizing-radiation resistance in the desiccation-tolerant cyanobacterium *Chroococcidiopsis*. *Applied and Environmental Microbiology* **66**, 1489–1492.

Bowers, K.J., Wiegel, J. (2011) Temperature and pH optima of extremely halophilic archaea: a mini-review. *Extremophiles* **15**, 119–128.

Brock, T.D., Hudson, F. (1969) *Thermus aquaticus* gen. n. and sp. n., a nonsporulating extreme thermophile. *Journal of Bacteriology* **98**, 289–297.

Clarke, A. (2014). The thermal limits to life. *International Journal of Astrobiology* **13**, 141–154.

Cox, M.M., Battista, J.R. (2005) *Deinococcus radiodurans* – the consummate survivor. *Nature Reviews Microbiology* **3**, 882–892.

Edwards, K.J., Wheat, C.G., Sylvan, J.B. (2011) Under the sea: microbial life in volcanic oceanic crust. *Nature Reviews Microbiology* **9**, 703–712.

Friedmann, E.I. (1982) Endolithic microorganisms in the Antarctic cold desert. *Science* **215**, 1045–1053.

Grant, W.D., Gemmell, R.T., McGenity, T.J. (1998) Halophiles. In: *Extremophiles: Microbial Life in Extreme Environments*, ed. K. Horikoshi and W.D. Grant, Wiley-Liss, New York, pp. 99–132.

Harrison, J.P., Gheeraert, N., Tsigelnitskiy, D., Cockell, C.S. (2013) The limits for life under multiple extremes. *Trends in Microbiology* **21**, 204–212.

Horneck, G., Bücker, H., Reitz, G., Requardt, H., Dose, K., Martens, K.D., Mennigmann, H.D., Weber, P. (1984) Microorganisms in the space environment. *Science* **225**, 226–228.

Mesbah, N.M., Wiegel, J. (2008) Life at extreme limits: the anaerobic halophilic alkalithermophiles. *Annals New York Academy Sciences* **1125**, 44–57.

Oger, P.M., Jebbar, M. (2010) The many ways of coping with pressure. *Research in Microbiology* **161**, 799–809.

Pedersen, K. (1997) Microbial life in deep granitic rock. *FEMS Microbiology Reviews* **20**, 399–414.

Reed, C.J., Lewis, H., Trejo, E., Winston, V., Evilia, C. (2013) Protein adaptations in archaeal extremophiles. *Archaea* **2013**, 373275.

Rothschild, L.J., Mancinelli, R.L. (2001) Life in extreme environments. *Nature* **409**, 1092–1101.

Wackett, L.P., Dodge, A.G., Ellis, L.B.M. (2004) Microbial genomics and the periodic table. *Applied and Environmental Microbiology* **70**, 647–655.

Whitman, W.B., Coleman, D.C., Wiebe, W.J. (1998) Prokaryotes: The unseen majority. *Proceedings of the National Academy of Sciences* **95**, 6578–6583.

Whittaker, R.H., Levin, S.A., Root, R.B. (1973) Niche, habitat and ecotope. *The American Naturalist* **107**, 321–338.

Wilson, Z.E., Brimble, M.A. (2009) Molecules derived from the extremes of life. *Natural Product Reports* **26**, 44–71.

Wynn-Williams, D.D., Edwards, H.G.M. (2000) Antarctic ecosystems as models for extraterrestrial surface habitats. *Planetary and Space Science* **48**, 1065–1075.

8

The Formation of the Elements of Life

Learning Outcomes

➤ Understand current theories for the beginning of the Universe.

➤ Understand how stars form.

➤ Describe the difference between low and high mass stars and the implications for the elements formed within them.

➤ Understand the concept of a blackbody and be able to calculate the wavelength of maximum energy emission, the intensity distribution and power of different blackbody objects.

➤ Understand how the heavy elements that make up life were formed.

➤ Explain how planets form in protoplanetary discs and be able to calculate the Jeans mass and the magnitude of gravitational focussing by planetesimals.

➤ Be able to calculate the lifetime of stars from their mass.

➤ Understand how rocky planets form and why.

➤ Understand the types of objects that make up the Solar System and the laws governing their motion.

➤ Have a basic understanding of major types of meteorites and what they might tell us about early materials in the Solar System.

8.1 In the Beginning

In the previous chapters we explored life on Earth. We examined the molecular structure of life, where it gets its energy and the interrelationships between organisms. Indeed, one might simply summarise by saying that the first part of this book was about understanding the one data point we have of life in the Universe. Equipped with knowledge about the structure of life we now set out to explore life in its cosmic context. The rest of this textbook is focused on that task. Our first objective is to understand where the elements from which life is assembled came from. To do this we must first expand to the astronomical scale.

As we transition to the cosmic scale you will notice that the distances of which we speak are vast. Metres and kilometres are inadequate to capture them. Even if we do use these measurements, we end up with unwieldy numbers. Throughout this chapter you'll see the term **Astronomical Unit** (AU) used, the equivalent distance of the Earth to the Sun. As the Earth's orbit is not perfectly circular, this distance used to be the average Sun–Earth distance. It is now defined to be exactly 149 597 870 700 m. A **light year** is another distance term used to describe the distance that light travels in a year, 9.46×10^{15} m. Yet another measure is the **parsec**. A parsec is the distance from the Sun to an astronomical object which has a parallax angle of one arcsecond (In other words, when comparing a star one parsec away to stars a very long way away from it, the nearby star would be seen to move, tracing an angle of one arcsecond (one-three thousand six hundredth of a degree) in relation to the more distant 'fixed' stars when the Earth has moved a straight line distance of 1 AU). A parcsec is equal to 3.26 light years.

So here we are, orbiting a star in the Milky Way Galaxy, which is about 100 000 light years across (the diameter is not exactly defined because the galaxy 'thins out' at the edges and so its diameter is rather arbitrary). Our Solar System is within the Orion Arm, a minor spiral arm of the galaxy about 3500 light years wide and 10 000 light years long. The Sun is about 26 000 light years from the galactic centre and currently about 21 light years above the galactic plane. We are travelling around a supermassive black hole in the centre of the galaxy that has about four million times the mass of the Sun. We orbit it once every 250 million years at about 250 km s^{-1}. Our galaxy, a fairly average-sized **barred spiral** *galaxy*, is close to the Andromeda Galaxy, a spiral galaxy about 2.5 million light years away which is moving rapidly towards us and will merge with our galaxy in about four billion years. Around us is a group of much smaller galaxies (more than 50). Together we form the Local Cluster of the Local Galactic Group. Our Local Group is itself part of a much bigger group called the Virgo Supercluster (Figure 8.1), itself part of the vast Laniakea Supercluster of galaxies. In the inner region of the supercluster is the Virgo Cluster, a giant collection of galaxies ~50 million light years away.

How did we get into this situation? What was the origin of the elements that eventually came together to produce all these galaxies, stars, the Earth and eventually, the biosphere in which we live?

Let's go back now to the very beginning of the Universe and consider where the elements for life came from and how the stars and planets formed. To return to the beginning of the Universe about 13.8 Ga ago is to embark on the study of cosmology. There are an enormous number of questions that are unanswered in cosmology. Some of these questions have obvious astrobiological corollaries. Are we the only Universe that exists? This question forces us to ask whether there are other life forms and civilisations in other universes, the so-called **Multiverse**. Was the Big Bang merely the end of a previous Universe, in which case did that previous Universe contain life forms and civilisations now long since destroyed? These questions sit at the intriguing interface between cosmology and astrobiology. The problem is that at the current time, however interesting they are, they are quite intractable questions. We won't detain ourselves any more with these questions, but instead we will focus on the one Universe we do know. We also take as our starting point the Big Bang and assume that this theory of the Universe is the correct one. There are lines of evidence that support it, so this seems a sensible point of departure.

At the time of the Big Bang (Figure 8.2), the entire Universe may have been compressed into a **singularity**. The

Figure 8.1 *Our Solar System in the cosmic context* (Source: wikicommons, Andrew Z. Colvin).

earliest stages of the Universe (the first millionth of a second) are poorly understood. Approximately 10^{-37} s into the history of the Universe, a cosmic inflation occurred during which the Universe grew exponentially. After inflation stopped, the Universe consisted of elementary particles and energy. The Universe continued to decrease in density and fall in temperature. At about 10^{-6} s, the temperature of the Universe was about 10^{13} K and particles were being created as matter-**antimatter** pairs, which then annihilated each other to produce energy consistent with Einstein's famous equation, $E = mc^2$, whereby mass (m) can be converted into energy (E) and vice versa.

There is an intriguing problem that immediately besets astronomers. We are all made of matter, not antimatter, and as far as we know all of the other galaxies around us are made of matter. Therefore, there must have been some asymmetry in the early Universe that ensured that when matter–antimatter collisions occurred, there was a small amount of matter left over or else all the material in the Universe would have been turned back into energy. The asymmetry has to be about one in a billion. How this occurred remains one of the mysteries of the steps in the early formation of the Universe.

After the first microsecond of the Universe, temperatures were low enough for continuous annihilation to calm down and allow protons and neutrons to form. At these remarkably high temperatures, neutrons cannot bind with protons to form nuclei. They are separated by thermal energy.

Soon after this and up to about 3 min into the history of the Universe (although the exact timings are uncertain) the temperature, now below about one billion K, allowed for the formation of the first atomic nuclei. The first to form was deuterium, an *isotope* of hydrogen with one neutron and one proton. Other nuclei formed, included tritium, a hydrogen isotope with a proton and two neutrons, and helium-3, containing two protons and one neutron. However, all of these nuclei are unstable and during collision they would have formed the much more stable helium-4. Helium-4 (^{4}He) has two neutrons and two protons, giving it an atomic mass number of 4. It is sometimes called an **alpha particle**. In addition, a very small amount of lithium-7, the most common isotope of lithium with three protons and four neutrons, would have formed in these early **nucleosynthesis** (nuclear fusion) reactions.

These nuclei were the stuff of the Universe for a long time. Temperatures were still too high for electrons to be captured by nuclei to form atoms. The Universe remained a plasma for about 380 000 years or until temperatures dropped below 3000 K when the first atoms could begin to form (such as atomic hydrogen). This is the period known as recombination. The formation of atoms had a crucial side effect. Until this point, the photons present interacted with free electrons making the universe opaque to electromagnetic radiation. However, once atoms formed it allowed those photons to travel freely.

This allowed matter and electromagnetic radiation to separate. The electromagnetic radiation began to cool down. Today, this energy is detectable in space as a radio hiss at an equivalent temperature of 2.7 K. This **cosmic background radiation** was discovered by Arno Penzias and Robert Wilson (1927–2002) in 1965 when they turned an old radio receiver in New Jersey to the sky. Initially they thought the hiss they had discovered was electronic noise caused by pigeon droppings in the receiver, but after removing the pigeons, the hiss persisted. The researchers had discovered the noise of the remnant radiation from the formation of the Universe.

At the same time that the photons were becoming separate from matter, the atoms were not perfectly distributed. Density fluctuations became established. More dense areas would have sucked in material by gravitation, less dense areas less so, in a feedback effect whereby the dense areas became denser. Density fluctuations caused matter to start to coalesce into clumps that would be the earliest galaxies and stars. These fluctuations would be expected to be reflected in fluctuations in the density of the cosmic background radiation as the photons and matter separated. Accurate measurements of the background radiation reveal fluctuations of ten parts in a million. Remarkably, the measurements allow cosmologists to estimate the density of the nuclei in the very earliest stages of the Universe and these estimates completely, and independently, agree with the density estimates that are made based on the abundance of hydrogen and helium isotopes today.

The evidence for the expansion of the Universe is to be found by observing the light transmitted by distant galaxies and stars. The light of distant objects is shifted towards the red end of the electromagnetic spectrum. This **redshift**, a cosmological red shift, is related to the traditional **Doppler Effect** (an effect caused by an object moving towards or away from you in analogy to the changing pitch of an ice cream van as it drives disappointingly towards and then away from you). It is somewhat more complicated and involves the 'stretching' of space between the objects.

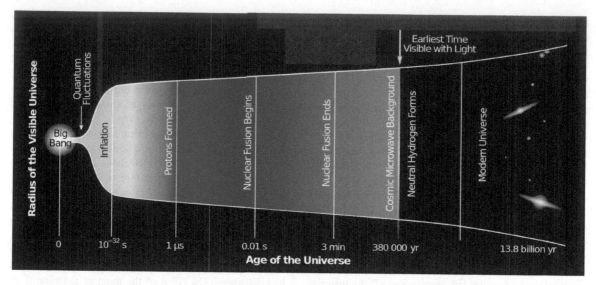

Figure 8.2 *The events following the Big Bang* (Source: Wikipedia, Yinweichen).

Further evidence for this expanding universe is provided by the linear relationship between the distance of other galaxies and the speed at which they are receding (Figure 8.3), sometimes known as Hubble's Law. The fact that objects further away appear to be receding faster suggests that if you rewind the clock the Universe was much smaller.

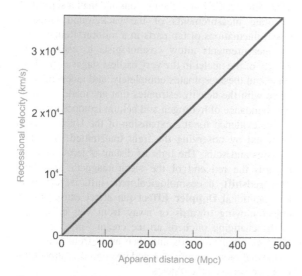

Figure 8.3 *Hubble's Law: A simple schematic graph showing the distance–velocity relationship of objects in the Universe* (Source: Charles Cockell).

The clumps of mass formed in the early Universe began to collapse and form more dense regions so that a few hundred million years into the history of the Universe the first protogalaxies, about a billion times more massive than our Sun, but smaller than present-day galaxies, had formed. In the next few billion years, these structures would themselves begin to coalesce into the billions of galaxies that we are familiar with in the Universe today.

During the time in which galaxies were forming, something important was happening. Here and there pockets of gas were collapsing under gravitational attraction. Within these balls of material a battle was being played out. Gravitational force was pulling material in; gas pressure was pushing outwards. When the mass exceeded a critical value allowing the gravitational pressure to exceed the gas pressure pushing outwards then the material collapsed into a star. This critical mass is called the **Jeans mass**, after astrophysicist, James Jeans (1877–1946), who calculated these conditions. The Jeans mass needs to be very large when the only material you have to work with is the early hydrogen and helium and temperatures were much greater than today. The earliest stars probably had a mass about a few tens to a few hundred times greater than our Sun. These stars were so large and so hot that they produced enormous quantities of radiation, including ultraviolet radiation, which stripped electrons off atoms and ionised matter back into the plasma state it was in at the earliest stages of the Universe. This period of the Universe is therefore called 're-ionisation'. The Universe, however, was transparent to photons because of the low density of ionised material at this stage.

These early stars are called Population III stars for historical reasons that need not detain us, but they constitute the first stars of the first galaxies of the Universe. However, until this point the Universe was limited in its chemistry. None of the heavy elements, including carbon, nitrogen, phosphorus, oxygen, iron and the numerous heavy elements used in living systems existed. This was not a Universe for life. It was a pre-astrobiological Universe. However, the material produced in the early Universe would allow for subsequent generations of stars, which would produce the material of life. In the following sections we will begin to understand where these elements came from.

To do this we will fast forward to the present day to look at the characteristics of different types of stars.

8.2 Low Mass Stars

To start with, we examine low mass stars. Our own Sun is such a star. Low mass stars can be broadly defined as stars with a mass ~2.5 times our Sun or less. They have a surface temperature below about 8500 K and a luminosity equal to or less than 20 times the luminosity of the Sun. As with the earliest stars discussed in the previous section, the formation of the Sun resulted from a trade-off between the collapse of a local gas cloud and the gas pressure pushing outwards.

Where does the energy produced by the Sun come from? If we calculate the energy that the Sun might have from gravitational energy alone we find that the energy is not nearly sufficient to keep the Sun burning for the time it has been around so far (4.56 billion years). Instead the Sun's energy comes from the conversion of part of its mass into energy in **nuclear fusion** reactions. Nuclear fusion is the process whereby smaller mass elements are fused together to make more massive elements and in the process release energy. This process is distinct from **nuclear fission** in which a large unstable element, such as uranium, breaks apart into smaller more stable elements with the release of energy. Our nuclear power plants depend on fission, but within stars nuclear fusion is the process generating energy.

The underlying process of nuclear fusion within the core of the low mass star is the **proton–proton chain** (Figure 8.4). Four hydrogen nuclei (four protons) fuse to form a helium nucleus (two protons and two neutrons) through a number of steps. Firstly, two hydrogen nuclei fuse to form deuterium, giving off a **positron** (a positively charged electron) and a **neutrino**, a low mass subatomic particle. The deuterium goes on to fuse with another proton to form helium-3, giving off a gamma ray. Two helium-3 nuclei can then fuse to form helium-4, releasing two protons. In this entire process energy has

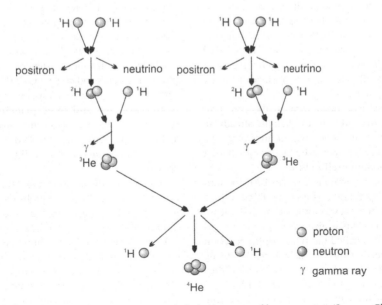

Figure 8.4 *Schematic showing the proton-proton chain in the cores of low mass stars* (Source: Charles Cockell).

been released in the form of gamma rays, neutrinos and positrons, the latter collide with electrons, annihilating each other to produce photons. The energy produced per helium nucleus is 4×10^{-12} J, but of course this small quantity is offset by the vast number of these reactions that are occurring. In a star like our Sun, the proton–proton reaction occurs about 10^{38} times per second.

The energy released from this process takes more than ~100 000 years to travel from the core to the surface of the star because of the many interactions that occur along the way. In the process they contribute to the gas pressure that prevents the star from collapsing under gravitational pressure. The size of the star is held in equilibrium in a feedback process. If the core contracts it heats up. More fusion reactions increase the gas pressure, forcing the star to expand. A core expansion causes cooling and reduces the efficiency of fusion reactions, lowering the gas pressure pushing outwards and allowing the core to collapse.

Eventually the hydrogen in the core will be exhausted as it is all transformed into helium, resulting in a helium core. For our own Sun this will occur in about five billion years. Once the fuel is exhausted, fusion in the core ceases. The gas pressure is thereby reduced and the star begins to collapse. The collapse will cause matter to contract and at this stage core collapse will cease. However, this collapse causes hydrogen around the helium core to compress to the point where it can initiate fusion. The fusion, because it is not contained within the core causes the material to expand and the star balloons outwards, achieving a radius about a hundred times greater than its 'normal' radius. It has entered its **red giant** phase. These types of giant stars, ones that contain an inert helium core and a shell of hydrogen-burning, are usually called **Red Giant Branch (RGB) stars**, a reference to the region of the Hertzsprung–Russell diagram, which we shall shortly discuss, that they populate.

The hydrogen fusion occurring around the core produces more helium that adds to the helium already in the core. Eventually after about one billion years the temperature of the core has increased to a point where it is now possible for helium itself to fuse.

The temperature of the core increases to 100–200 million K and at this temperature it is now possible for three helium nuclei to fuse together to form carbon and release energy. As a helium nucleus is also called an alpha particle, this reaction is also called the **triple alpha reaction** (Figure 8.5). It proceeds from two helium nuclei fusing together to form a beryllium-8 nucleus, which itself then fuses with a helium nucleus to produce carbon-12.

The probability of the triple alpha process occurring by the simultaneous collision of three alpha particles

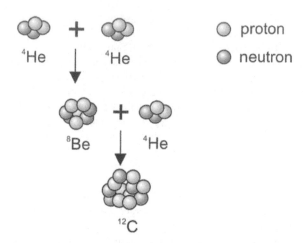

Figure 8.5 *The triple alpha reaction that produces carbon* (Source: Charles Cockell).

is extremely low. However, the beryllium-8 has almost exactly the energy of two alpha particles, favouring this interaction. In the second step when beryllium combines with another alpha particle, the process has almost exactly the energy of an energetically excited state of carbon, again favouring it. This so-called resonance increases the probability that an alpha particle will combine with beryllium-8 to form carbon. The existence of this resonance was predicted by astronomer, Fred Hoyle (1915–2001), before being observed.

Intriguingly, the resonance effect that enhances the probability of these reactions occurring was predicted by Fred Hoyle on the basis that they have to exist, because if they did not, we would not exist since heavy elements would fail to come into existence. This notion, that certain physical features and constants of the Universe must have certain values or we could not be around to observe them is known as the **anthropic principle**. The triple alpha reaction may be one of the first instances in which the anthropic principle was invoked to predict a physical process on the basis that if it didn't exist then we could not exist.

Another significant point about the triple alpha reaction is that it produces the first astrobiologically important element beyond hydrogen – carbon. The reaction is the source of most of the carbon in the Universe.

Back to our star. The helium fusion occurring in the core reaches a state where the energy accumulates sufficiently to cause runaway fusion in the star. It reaches about 100 billion times the star's normal energy production for a few seconds. The runaway fusion eventually causes a core helium flash which consumes about 60–80% of the helium in the star. This process occurs in the core and the star remains intact. It is therefore not directly observable from outside the star.

In the next stage of the life of the star we begin with a carbon core and a layer of helium outside of it which is undergoing fusion to produce yet more carbon. This helium layer is itself covered in a layer of hydrogen which is undergoing fusion to produce helium. This double-layered cake is undergoing 'double shell burning'. The outer layers of the star expand yet further, to even greater radii than the red giant stage. The double shell burning stage lasts just a few million years. This star has now become an **Asymptotic Giant Branch star** (**AGB star**). What distinguishes it from a red giant is the presence of a carbon core and double shell burning.

Low mass stars do not have sufficient mass to cause the carbon core to contract yet further and initiate carbon burning, which requires a temperature of about 600 million K. This is the end of the line for them. At this stage the outer layers of gas in the star are ejected into space in a series of thermal pulses. The radiation being emitted from the star ionises these layers causing them to glow brightly. The dissipating gas shells are called a planetary nebula, although this is an historical misnomer because they are not associated with planets. It dates back to William Herschel, who thought that such objects looked like planets and so bestowed this name upon them.

Meanwhile, in the remaining star, helium fusion begins to ramp down. The star becomes a **white dwarf** star, held up by electron degenerate matter (Chapter 2). In some stars the carbon from the interior is convectively cycled onto the surface of the star, forming a carbon star. The white dwarf finale is the fate of most of the stars in the Universe.

This long-term future, which has been ascertained from theory and by observing low mass stars at different stages of their lives elsewhere in the Universe, gives us some certainty about the fate of the Earth. In about five billion years, as the Sun turns into a red giant, the atmosphere of the Earth will be stripped away and although the orbit of the Earth might well expand as the Sun expands, at this point our planet will be transformed into an atmosphere-less ball of rock. However, as we shall see later (Chapter 16), the Earth will become uninhabitable long before this stage.

8.3   High Mass Stars

At the other end of the spectrum of star types are high mass stars. These stars have a mass about eight times or more

than that of the Sun. The consequence of their larger mass is that their cores contract to a much greater density and achieve higher temperatures – thus allowing fusion reactions to occur beyond the production of carbon.

In high mass stars, hydrogen fusion to helium is dominated by a more complicated cycle called the CNO cycle (Figure 8.6).

In the CNO cycle, carbon acts as a catalyst for the production of helium (alpha particles). One hydrogen nucleus (proton) fuses with carbon-12 (with the emission of a gamma ray) to produce nitrogen-12, which then decays to carbon-13 producing a positron and a neutrino. The carbon-13 so produced takes up yet another proton to form nitrogen-14 with a gamma ray given off. This nitrogen-14 takes up the next (third) proton to form oxygen-15 with the emission of a gamma ray and this oxygen decays to nitrogen-15 with the production of a positron and a neutrino. Finally, this nitrogen-15 fuses with the fourth proton with the release of a helium nucleus, producing carbon-12 which can begin the cycle again. The overall effect – the fusing of four hydrogen nuclei (protons) to form one helium nucleus – is the same as the proton–proton reaction in low mass stars.

The fusion rate within high mass stars, because of the high temperatures achieved, means that they are short-lived compared to low mass stars.

Like low mass stars, the interior of a high mass star arrives at a state where it has an internal carbon core, surrounded by a helium layer and a layer of hydrogen.

The interior of a high mas star reaches 600 million K, sufficient to initiate carbon fusion into neon which

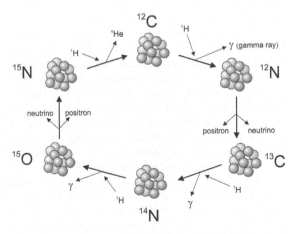

Figure 8.6 *Schematic of the CNO cycle in high mass (and intermediate) stars. The atomic structures in the diagram are only illustrative* (Source: Charles Cockell).

lasts just a few hundred years. As carbon runs out, so the core shrinks under gravitational pressure, achieving temperatures at the centre that then allow for neon to fuse and produce magnesium and oxygen. Neon fusion then ceases, yet again causing core collapse and triggering oxygen fusion, producing silicon. Each successive step takes a shorter time, such that silicon fusion ends in about a day. The final fusion reactions with silicon result in the formation of iron. We now have a star that in cross-section looks a little like an onion (Figure 8.7). The most common way in which these sequential elements are formed is by the fusion of helium nuclei, although heavy elements, at sufficient temperatures, can directly fuse together.

Iron is the final product because it is the most stable element (it has the highest binding energy per nuclear particle). Elements heavier than iron tend to break down to form iron by fission, elements lighter than iron tend to fuse to make iron (Figure 8.7). Once the core of the star is composed of iron, the star is doomed.

At this stage a critical and dramatic transformation occurs. The iron is no longer undergoing fusion. There is insufficient gas and radiation pressure to push out against the gravitational pressure pulling in. The gravity becomes sufficiently great that the electrons are driven into fusing with protons to form neutrons, thus producing neutron degenerate matter (Chapter 2). The star collapses to form a **neutron star** just a few kilometres in diameter. As this cataclysm occurs, an enormous amount of energy is released, blowing away the outer layers of the star in a **supernova** explosion. If the neutron star is sufficiently massive, it may collapse further and become a black hole.

Up until now you may have noticed a problem. If stellar nucleosynthesis can only go as far as iron, where does the rest of the periodic table come from? Most of the other elements are generated in the intense supernova explosion associated with the collapse of the star. During this event, iron is bombarded, very intensely, with neutrons. These neutrons are thought to be provided in an intense neutron wind from the young neutron star. When a single neutron is added to the nucleus it may decay into a proton and an electron. The electron is lost (**beta decay**) and we end up with a nucleus containing an additional proton. The rapid bombardment of neutrons, however, allows for elements to be formed that have both higher numbers of neutrons and protons. This process is called the rapid neutron capture process (or **r-process**), whereby rapid bombardment of seed nuclei such as iron with neutrons allows for heavier elements to form. There is an alternative process, the slow neutron capture (or *s-process*) that occurs in **AGB stars** and similar stars and requires pre-existing heavy nuclei. Here, instead of a rapid accumulation of neutrons in a nucleus, nuclei slowly accumulate one at a time within the star interior, generating heavier elements.

The evidence for these processes is to be found in the cosmic abundances of elements in the Periodic Table (Figure 8.8).

Hydrogen and helium are found to be the most abundant, as we would expect from their primordial abundance. Elements with atomic masses divisible by four are abundant, which is what we would predict if elements beyond carbon up to iron were produced in fusion reactions in which helium nuclei (with its two protons and two neutrons) were sequentially fused to heavier elements in high mass stars (C, O, Ne, Si). We would expect, as is observed, that increasingly heavy elements would become less abundant as they require successively more neutron bombardment in supernova explosions. You will also notice that elements with even numbers of protons and neutrons are more abundant than those with odd numbers. This is a function of nuclear architecture and the relative stability of nuclei with even numbered neutrons and protons. Finally, the abundance of iron is observed to be high, which is consistent with its role as the final element in nucleosynthesis in high mass stars.

Before we complete this section we should not forget the intermediate stars, between ~2.5- and eight times the mass of our own Sun. In these stars fusion occurs by the CNO cycle, but they do not go on to generate heavier elements. Their evolution follows a similar path to low mass stars.

8.4   The Elements of Life

Quite apart from gaining some understanding of the evolution of stars, these past sections also explain where the elements required to build life come from. Let's briefly review this story from an astrobiological perspective. Hydrogen and helium are the most common elements in the Universe and were the first elements to be formed. Helium has little relevance to life, but hydrogen, as well as being an essential element in carbon-based compounds, can also be used as an electron donor for growth (Chapter 5). Of course, the presence of this electron donor in great abundance was astrobiologically irrelevant in the early age of the Universe because life could not exist at this time. In the cores of low and high mass stars, the element carbon was formed from nuclear fusion, which provides the key 'backbone' element for the construction of the array of millions of organic compounds required by life (at least the life we know on Earth). Within the cores of high

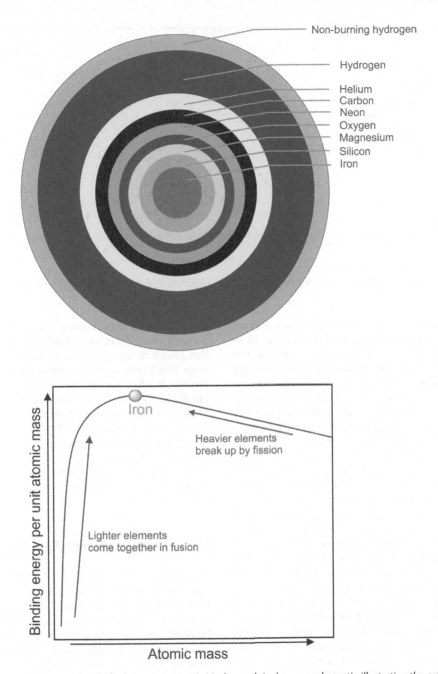

Figure 8.7 *The interior structure of a high mass star (top). Underneath is shown a schematic illustrating the principle by which elements heavier or lighter than iron converge on iron as the element with the highest binding energy per unit atomic mass, that is the most stable element, accounting for an end to fusion when an iron core has formed (Source: Charles Cockell).*

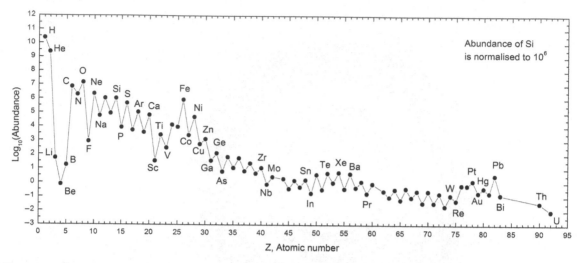

Figure 8.8 *The cosmic abundance of different elements. The abundance of the elements is shown relative to Si which is normalised to a value of 10^6.*

mass stars elements up to and including iron were formed, including the other four crucial elements for life, nitrogen, oxygen, phosphorus and sulfur. Thus, once the first high mass stars came into existence and exploded, the CHNOPS elements required for life were strewn into the Universe. These stars also produced elements used extensively in enzymes and as trace elements (including iron and magnesium) and in their supernova demise heavier elements were generated, including zinc, cobalt and many other trace elements that are used as enzyme cofactors and in other biological roles. Many of these elements and the compounds they form can be used as electron donors and acceptors in energy generation in life, such as the different oxidation states of iron. Thus, the presence of elements from high mass stars generated the required materials for redox couples to come into existence.

Astronomers often refer to any elements other than hydrogen and helium as 'metals', which is rather confusing given the usual use of the word. However, the nomenclature is often worth remembering because when astronomers talk of low **metallicity** objects, they mean objects with a low concentration of elements beyond helium – with the obvious consequences for life. High metallicity means an abundance of elements heavier than helium and by implication high metallicity planet-forming regions are generally better for life than low metallicity ones, depending on the absolute abundance of the elements.

It is often said, to the point of being a cliché, that we are all 'stardust'. We can see that this is correct. We can also see that several generations of stars might lead

to successively more heavy elements. For example, the presence of iron and heavier trace elements in our own Sun cannot be accounted for by fusion within the Sun, since it is a low mass star. These elements must have been there when the Sun first formed. It is thought that the Sun is at least a third generation star (a so-called Population I star, in contrast to the first stars in the Universe, Population III stars, and more recent, heavy-metal poor Population II stars, some of which still exist in the Universe). The Sun and the protoplanetary disc from which the planets formed incorporated heavy elements from previous stars and supernova remnants from which ultimately life on Earth emerged.

8.5 The Hertzsprung–Russell Diagram

We have touched upon some different star types. We can systematically characterise stars using the The Hertzsprung–Russell diagram. The diagram is a scatter graph of stars showing the relationship between the star luminosity versus its spectral type (essentially its temperature or colour). The temperature unconventionally runs from higher to lower along the *x* axis.

The luminosity of an astronomical object is the total amount of energy emitted per unit time (the power of a star).

The luminosity (L) is related to the temperature (T) of a star by the relationship known as the Stefan–Boltzmann law:

$$L = \sigma 4\pi R^2 T^4 \qquad (8.1)$$

Facilities focus: Space Telescopes and Observatories

The Earth's atmosphere causes all sorts of problems for astronomers and astrobiologists seeking to observe the Universe. Some wavelengths, such as X-rays, are absorbed by the atmosphere and in regions of the spectrum where light penetrates, it is subject to light pollution or 'twinkling' caused by atmospheric effects. Twinkling can be removed by adaptive optics that divide out the effects of atmospheric perturbations. However, some of these different limitations can be overcome by launching telescopes into space. Since the beginning of space exploration a range of telescopes have been launched into space to study the whole electromagnetic spectrum. To observe gamma rays, produced in supernovae, neutron stars and other objects, gamma ray observatories such as the Fermi Gamma Ray Space Telescope, which started operating in 2008, have been used. X-rays, also produced during violent astronomical events, have been studied using platforms such as the Chandra X-ray Observatory, which has operated in Earth orbit since 1999. Ultraviolet radiation is yet another type of radiation, like gamma-rays and X-rays, which is absorbed by the atmosphere (the wavelengths shorter than about 290 nm). The Galaxy Evolution Explorer, which operated from 2003 to 2013 is just one example of a telescope used to study UV emissions that are produced by stars and galaxies. Moving into the visible region of the spectrum, the most iconic (and the first) space telescope was the Hubble Space Telescope that returned stunning images of the cosmos. Galaxies, stars and other objects emit or reflect visible light, so that these telescopes can be used to observe a whole range of phenomena. The search for exoplanets in the visible region has been undertaken using instruments such as the Kepler Space Telescope (Figure 8.9), which searched for planets using the **transit method**. As we push into longer wavelengths of the infra-red we start to observe cooler objects in the

Figure 8.9   *The Kepler Space Telescope. The 1040 kg telescope was launched into space in 2009 to detect extrasolar planets in the direction of the constellations Cygnus, Lyra and Draco, seeking out planets about the same distance from the galactic centre as our Solar System. The telescope generated a vast amount of data about extrasolar planets including the detection of the first rocky terrestrial-type planets (Chapter 19; Source: NASA).*

(continued)

(continued)

Universe such as brown dwarfs, interstellar clouds, planetary bodies and organic materials. The Wide-field Infrared Survey Explorer (WISE), launched in 2009, discovered over 150 000 new Solar System objects, including near-Earth asteroids and comets. Continuing again into the longer wavelengths of the spectrum, we move into the microwave region where the cosmic background radiation can be observed and mapped (such as by the Cosmic Background Explorer). In the longer radio range, observations can be made of galaxies and supernova remnants. This is the region used in the Search for Extraterrestrial Intelligence (SETI). The RadioAstron mission, launched in 2011, was designed to map and study the structure of radio sources inside and outside our Galaxy. Yet other observatories are designed to look for cosmic rays and high energy particles such as the Solar Anomalous and Magnetospheric Particle Explorer (SAMPEX), launched in 1992, which investigated the radiation environment around the Earth.

where R is the radius of the star and σ is the Stefan–Boltzmann constant. This constant is an amalgam of other constants where k is the Boltzmann constant, c is the speed of light in a vacuum and h is Planck's constant:

$$\sigma = \frac{2\pi^5 k^4}{15c^2 h^3} = 5.670\,400 \times 10^{-8}\,\mathrm{Js^{-1}m^{-2}K^{-4}}$$

Therefore, given a temperature (x axis) and luminosity (y axis) for a star type, the Hertzsprung–Russell diagram is also telling us about stellar radius.

Hertzsprung–Russell diagrams are also referred to by the abbreviation H–R diagram (Figure 8.10). The diagram was devised by astrophysicists, Ejnar Hertzsprung (1873-1967) and Henry Norris Russell (1877–1957) and can be considered to be a summary of the evolution of stars or a map of their 'lives'.

Most stars occupy the region in the diagram along the line called the **main sequence**. The main sequence is the region of the diagram in which stars are carrying out stable hydrogen fusion in their cores. Although there are different forms of the graph, they all share the same general format: stars of greater luminosity are toward the top of the diagram, and stars with higher surface temperature are toward the left side of the diagram.

The H-R diagram also reflects the mass–luminosity relationship of stars, which is given by the general relationship:

$$L_{\mathrm{star}} / L_{\mathrm{Sun}} = (M_{\mathrm{star}} / M_{\mathrm{Sun}})^a \tag{8.2}$$

where a is some value between 1 and 6, and is generally taken to be 3.5 for main sequence stars. In other words, low mass stars are towards the bottom (lower luminosity) and high mass stars with high luminosity towards the top of the main sequence. Almost all stars begin their lives with the same composition (about 98% hydrogen and helium), which means that the luminosity of a star is determined by its mass.

Stars can also be classified by a **spectral class** that runs from hot stars to cool stars on the x axis of the diagram. Our own Sun is a G main sequence star. Hotter shorter-lived stars include F stars. M stars are longer-lived, cooler stars. The sequence of stars runs in an odd, historical series of letters: O, B, A, F, G, K, M from hotter to cooler. The mnemonic, in olden times, was, 'Oh Be A Fine Girl/Guy, Kiss Me'. The spectral types are based on absorption lines seen in their spectra (the principles of which were described in Chapter 2). Spectral types can be regarded crudely as just another way to define temperature. Spectral characteristics are a way to classify stars which gives information about their temperature in a different way – particular absorption lines are observed in their spectra only for a specified range of temperatures because only in that range are the relevant atomic energy levels involved in absorption active. Each spectral type is divided into 10 bands from 0 to 9 with a lower number being hotter. Our own Sun is a G2V star. The Roman numeral 'V' simply designates it as being on the main sequence.

Modern versions of the H-R diagram replace the spectral type by a colour index of the stars. This type of diagram is often called an observational Hertzsprung–Russell diagram, or a colour–magnitude diagram (CMD).

The H-R diagram also depicts the position of stars when they leave the main sequence, for example at the end of their lives when they start to fuse elements other than hydrogen. You can see on Figure 8.10 the giants, which are characteristic of low mass stars like our Sun when they have expanded and moved into the phase of helium fusion. Equation (8.1) shows why this size transition occurs. Luminosity is proportional to the radius of the star squared and the temperature to the power of four so if the luminosity increases, but the temperature decreases, then the radius must increase – and quite dramatically.

When low mass stars finally die down, they join the ranks of the white dwarfs towards the bottom left that

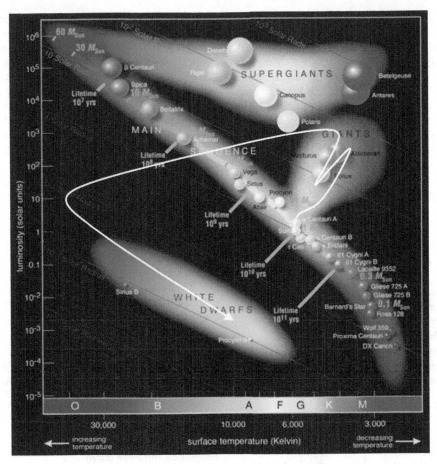

Figure 8.10 *A Hertzsprung–Russell diagram. The white line shows an approximated trajectory for the Sun in the future* (Source: ESO).

have low luminosity and high temperature (and therefore a small radius). Therefore, on any H-R diagram we could draw a line showing the trajectory of a star as it is born, joins the main sequence and eventually leaves as it dies, with the exact trajectory depending on the mass of the star. Such a general schematic trajectory for the Sun is shown in Figure 8.10.

The numbers of stars of different types is not uniformly distributed. G stars make up about 7% of stars in the Universe, whereas M stars constitute about 75%. Furthermore, and of profound consequences for habitability, stars also have different lifetimes. Hot stars, which are more massive, burn through their fuel more efficiently because of the greater temperatures achieved on account of the gravitational forces in the star. O stars (which make up about 0.001% of all stars) have lifetimes of less than a million

years. G stars, like our own, have lifetimes of about 10 billion years. M stars, which are the most numerous, have the longest lifetimes. An M star with a mass 0.1 times that of the Sun (i.e. 0.1 solar masses) could burn for 10 trillion years. Thus, the most numerous stars in the Universe are also the longest lived, suggesting that these may be the best places to look for life, although as we will see later, M stars have very specific challenges to habitability.

The life time of a star (t_{star}) is related to the mass (M_{star}). In comparison to our own Sun is can be stated with the following approximation:

$$t_{star}/t_{Sun} \sim (M_{star}/M_{Sun})^{-2.5} \qquad (8.3)$$

So a star about one-fifteenth the mass of the Sun (for example an M star) will have a life time about 650 times longer than the Sun – well over a trillion years.

8.6   The Sun is a Blackbody

The title of this section might seem curious. We use the term **blackbody radiation** to refer to the electromagnetic radiation emitted by an object at thermodynamic equilibrium, or in more basic terms, a body at a certain temperature. If we measure the wavelengths of this radiation we find that it produces a characteristic curve (Figure 8.11), where the peak of the curve is related to the temperature of the body. In fact this relationship can be exactly stated as Wien's displacement law. The brightness per unit wavelength shows a characteristic curve, where the peak is given by:

$$\lambda_{max} = b / T \qquad (8.4)$$

where b is Wien's displacement constant, given by 2.898×10^{-3} K·m and T is the temperature of the blackbody.

This law actually accords with our everyday experience. Very hot metal glows white/blue and as it cools it becomes orange and then red, a consequence of the peak wavelength shifting to higher wavelengths as it cools down. Exactly this colour change occurs as we go from hot F stars (bluish) to G stars like our own (yellow) to cooler M stars (red). It explains the x axis in Figure 8.10. As we go from hotter to cooler temperatures we go from spectral type stars that are bluish to redder M stars. Any object can emit blackbody radiation. The Earth, whose surface is ~300 K, has a blackbody emission in the infra-red (Figure 8.11).

We can calculate the intensity (given as the power per unit area per unit wavelength (W m^{-2} μm^{-1})) at any given wavelength (formally called the spectral radiant exitance or the radiant emittance). The equation that allows us to calculate this intensity value is Planck's Law of blackbody radiation:

$$M_\lambda(T) = 2\pi hc^2 / (\lambda^5(e^{hc/\lambda kT} - 1)) \qquad (8.5)$$

Where $M_\lambda(T)$ is the spectral radiant exitance at temperature T, h is the Planck constant, c is the speed of light in a vacuum, k is the Boltzmann constant, λ is the wavelength of the electromagnetic radiation and T is the absolute temperature of the body. An example of these calculations is shown in Figure 8.11.

The frequency distribution can be calculated by remembering that the relationship between wavelength and frequency (v) is given by:

$$v = c / \lambda \qquad (8.6)$$

We can also specify the power of a blackbody (the energy per unit time over all wavelengths per unit area,

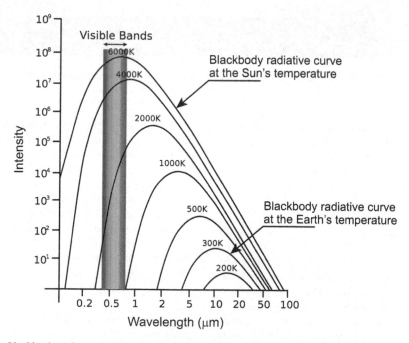

Figure 8.11   *Blackbody radiation curves for an object at different temperatures* (Source: wikicommons Ant Beck).

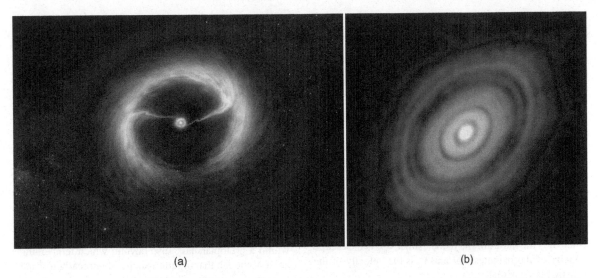

Figure 8.12   *(a) An artist's impression of the early protoplanetary disc around HD 142527, a young star in the constellation Lupus (HD refers to the Henry–Draper catalogue of astronomical objects [Source: wikicommons, ALMA (ESO/NAOJ/NRAO)/M. Kornmesser (ESO)/Nick Risinger (skysurvey.org)]. (b) A real protoplanetary disc formed around the star, HL Tau, less than a million years old and 450 light years from Earth in the constellation Taurus. Its radius is about 2000 AU (Source: Alma).*

A, emitted at a given temperature, *T*). This is given by the Stefan–Boltzmann law shown earlier applied to a discussion of stars (Equation (8.1)):

$$F/A = \sigma T^4 \qquad (8.7)$$

where *F* is the power of an object (for stars equivalent to the luminosity, *L*), A is the surface area ($4\pi r^2$ for a large planetary object or star) and σ is the Stefan–Boltzmann constant.

Blackbody curves are important because they allow us to calculate how much energy is being emitted by a star and therefore received by a planet. We can use them to work out what we might expect the temperature of a planet to be. We will explore this in greater detail when we discuss habitability in Chapter 16.

8.7   The Formation of Planets

The Solar System formed 4.56 billion years ago in a molecular cloud. Perturbations in the cloud generated local conditions where the gravity was sufficient to cause the gas to collapse. The law of the conservation of angular momentum (angular momentum refers to the momentum that an object has in moving around another object) meant that as the cloud collapsed the material spun into a disc.

A **protoplanetary disc** was formed. At the centre of this rotating disc (Figure 8.12) a large accumulation of matter occurred. This was the **protostar**. The rate of rotation of the disc meant that not all the matter was accumulated into the protostar. Some of it remained in the disc.

What defines the initial collapse of the cloud? For the cloud to initially collapse the gravitational force in the cloud must exceed the gas pressure and thermal or radiation energy pushing outwards. The minimum mass that allows for this collapse is the Jeans mass and it is given by:

$$M \text{ (Jeans mass)} = 5kTR/Gm \qquad (8.8)$$

Where *k* is the Boltzmann constant, G is the Gravitational constant, *T* is the temperature of the disc, *R* is the radius and *m* is the mass of the disc.

By observing in the infra-red region, it is possible to see disc-forming regions around new stars. These observations show that newly formed discs are short-lived, usually on the order of about five million years, suggesting that within this time, the planets are formed.

Within the disc, dust and gas begin to accumulate by electrostatic and gravitational interactions into larger grains and conglomerates including melted globules of silicate rock and iron/nickel called **chondrules**. These chondrules can be found in some **meteorites** and give us a great deal of information about the material conditions in the early Solar System. As the grains grow from small particles in the disc they become decoupled from the rotating fine dust and gas, essentially feeling a 'wind' against them which further enhances their growth. This process is by no means neat and tidy. Turbulence is established within the disc.

Eventually the material collects by **accretion** into large objects. These **planetary embryos** or **planetesimals** mop up material by gravitation. This process is known as **gravitational focussing**. We can consider a planetesimal to have an effective radius that is wider than its physical radius and allows for the accretion of material around the object. Its effective radius is given by:

$$\text{Effective radius} = R \ (1 + v_{esc}^2/v_o^2) \qquad (8.9)$$

where R is the radius of the planetesimal, v_{esc} is the **escape velocity** of the planetesimal and v_o is the velocity of the near-by planetesimals.

As the planetesimal grows larger so its escape velocity will also increase as a square, thus dramatically enhancing its ability to mop up yet more material around it. Once objects of 1000–5000 km range are formed they can be gravitationally attracted to form planet-sized objects. This period in which the objects are few and similar in size, colliding to form planets, is called the period of 'oligarchic growth'.

Encounters between planet-sized objects lead to scattering events during which giant planets can be forced into highly eccentric or inclined orbits, accounting for the wide diversity of eccentricities observed in exoplanet populations (Chapter 19). Planets can even be ejected from star systems to become lone planets.

If planets form early enough they can attract an outer shell of low molecular weight gases such as hydrogen and helium from the original disc. Gas giant planets such as Jupiter and Saturn are examples of these bodies. If the planets form later, for example if they are in the hot interior of the cloud near the protostar and all of the gas has dissipated by the time they are fully formed, they will form small rocky planets, such as the Earth and Mars. The sequence of materials that form from the disc is defined by what is termed the condensation sequence. At temperatures in the region of 1700 K and below, Ca, Ti and Al condense. Fe then reacts with condensing S to form sulfides (FeS). Early rocky silicates contained more Mg since this condenses before oxidised iron can form (before this time the Fe is present as a metal). Eventually, at sufficiently low temperatures, **volatiles** such as water condense. Thus, a well-defined sequence of condensed materials occurs.

The nature of planets formed will also depend critically on where they are in the disc. The differences in composition between the terrestrial planets in our Solar System, for example their varying ratios of Al and Mg to Si, show that there are complex variations and effects occurring within the disc. Furthermore, as the radiation from the protostar will reduce as the square of the distance, so temperatures become rapidly cooler as one moves from the interior to the outside of the disc. At a certain distance, about 2.7 AU for a disc surrounding a Sun-like star, the temperature is low enough that ice can form on the dust grains, increasing the inventory of volatiles such as water. This distance is known as the iceline or **snowline** and also accounts for the high abundance of volatiles and gases in gas giant planets formed beyond the snowline.

In addition to the snowline, we can also define a point within a protoplanetary disc beyond which temperatures are low enough that organics such as hydrocarbons can be formed, sometimes called the soot line.

During this early stage, other important processes are occurring. The observation that the Sun contains about 99.9% of the Solar System's mass, but only has about 2% of the angular momentum suggests that it lost a lot of mass during its formation in the early years of the Solar System. Stars in the early stage of development, called the **T-Tauri** stage, prior to joining the main sequence (and thus initiating stable fusion burning), have been observed to release energy in intense radiation and wind emissions which would have contributed to blowing away material in the disc and partly accounting for the loss of angular momentum in the star. This T-Tauri phase would have cleared small dust grains from the Solar System, ultimately contributing to the end of the planetary accretion process. Eddies and circulations within the disc would have caused the conversion of motion into heat and angular momentum, a process that would cause the shedding of angular momentum outwards into the disc.

The theory that the Earth and other planets formed from a giant gas cloud, a nebula, was first proposed by French mathematician Pierre-Simon Laplace (1749–1827). His nebula hypothesis gained increasing credence versus an alternative idea, proposed by Georges Buffon (1707–1788), that the planets formed as a result of a giant impact between the Sun and a comet. Yet another theory was that the planets had formed when two stars passed close to one another with the gravitational attraction of the passing star pulling material out, which condensed to form the planets.

The solar nebula hypothesis was successful at explaining many of the features that we have just discussed: why we tend to get gaseous planets in the outer regions of the Solar System, and small rocky planets in the inner regions (explained by the dynamics of the way in which a gas cloud condenses and collapses), why the planets tend to orbit in the same direction as the Sun's rotation (as a consequence of angular momentum) and why there are many fragments of **asteroids** and **comets** in the Solar System (they are the leftovers of the early process of **accretion**). Although the nebula hypothesis is largely accepted today, the details still need refining. The discovery of extrasolar planets such as **hot Jupiters**, immense Jupiter-sized planets orbiting near to their stars (Chapter 19), has required ideas of **planet migration** and other refinements to be incorporated into the model.

8.8   Types of Objects in our Solar System

It is well worth it, at this stage, to review some of the variety of bodies in our Solar System that emerged from the accretion disc. We will explore the features of some of the Solar System bodies, such as Mars and the icy moons, in more detail later in the book.

The size of the Solar System is controversial – it simply depends on your definition. If you include the comet-bearing Oort Cloud it could be 50 000 AU in radius. Most of the major bodies in the Solar System are embedded in the heliosphere, about 100–150 AU in diameter, which is the region where the **solar wind** pushes away the interstellar medium. One might also consider this to be the extent of the Solar System.

The protoplanetary disc origin of our Solar System seems confirmed by the observation that all the objects in our Solar System are almost in the same plane and they all orbit in the same direction around the Sun with orbits that are quite circular. There are some notable exceptions, including comets and asteroids, which can have very **eccentric** orbits. All of the planets, with the exception of Venus and Uranus, which may have been subject to early massive collisions, spin in the same sense as the direction of their orbits – in the **prograde** direction (from a point far above the Earth's North Pole, the prograde direction is anticlockwise).

Starting from inside and moving out, we begin with the Sun. We have already seen that it is a G2V star. It has a surface temperature of 5800 K and a radius about 110 times the Earth (695 500 km). It has a mass 330 000 times the Earth or 1.989×10^{30} kg. Its composition, by mass, is 71% hydrogen, 27.1% helium with about 2% heavier elements or 'metals' in the lexicon of astronomers.

The centre of mass of our Solar System is inside the Sun, so although technically other planetary bodies orbit around the common centre of mass with the Sun, colloquially we talk of planetary bodies 'orbiting the Sun'.

A planetary body has been defined by the International Astronomical Union as an object that orbits the Sun, is massive enough that its gravity causes it to be spherical and it has largely cleared its neighbourhood of other bodies. Dwarf planets include Ceres in the asteroid belt and Eris and Makemake in the Kuiper Belt. Pluto has been classified as a dwarf planet on the basis that it does not meet the third criterion. This decision has proven controversial in the planetary sciences community. That leaves eight planets in our Solar System (Mercury, Venus, Earth, Mars, Jupiter, Saturn, Uranus, Neptune).

Planets are generally further divided into the rocky (or 'terrestrial') planets, which include Mercury, Venus, Earth, Mars and the gaseous or icy planets, which encompass the remaining four. The separation between these two classes is not merely a matter of definition, but reflects the differences in planet formation beyond the snowline, as discussed earlier.

Many planets have moons, which orbit them in the same plane and direction. The four Galilean moons of Jupiter (Io, Europa, Ganymede and Callisto) were the first moons (other than our own) to be detected in space (Figure 8.13). All four of the gaseous planets have ring systems. Although Saturn's rings are well known and clearly imaged, similar, much thinner ring systems have been found around the other gas giants. They are comprised of ice and dust and may be only a few tens of metres thick.

From an astrobiological point of view the **asteroids** are a significant class of objects. Most of them orbit between Mars and Jupiter, but some of them have trajectories that bring them close to the Earth. These **Near-Earth Objects** (NEOs) are recognised as a hazard to life and may have contributed to the end-Cretaceous extinction that made the dinosaurs and many other organisms extinct (Chapter 15). The largest of them is called 1036 Ganymed, a 32-km diameter rock. Asteroids vary from small rocks to objects several hundred kilometres in diameter.

Beyond the orbits of the planets are **comets**, ice-rich objects containing organics and other materials. Short-period comets have orbital periods of about 200 years or less and primarily come from the Kuiper Belt, which extends from the orbit of Neptune (30 AU) to about 50 AU. There are over 100 000 objects greater than 100-km in diameter and billions smaller than this in the Kuiper Belt. Pluto is just one of these objects.

The long-lived comets come from the Oort Cloud, a spherical cloud of objects that surrounds the Sun between about 20 000 to 100 000 AU away, or nearly a light

Figure 8.13   *The Galilean moons of Jupiter, so named after Galileo Galilei who first observed them (Source: NASA).*

year. The Cloud has not been directly observed, but is hypothesised based on the orbital characteristics of comets and is thought to contain several trillion objects greater than 1 km in diameter. Its origins are thought to lie in the original protoplanetary disc, with gravitational influences driving the objects outwards on highly elliptical orbits. As the Sun undergoes its journey around the Milky Way Galaxy it is possible that passing stars gravitationally disrupt these objects and send some of them into the inner regions of the Solar System.

Comets are very clearly observed in the night sky, particularly when they get close enough to show off their tails and excite the media. Typically they have two tails (Figure 8.14), one of which is the dust tail, made of fragments of material leaving the object opposite to the direction of travel. The second tail, or plasma tail, is caused by charged particles in the comet's coma (its atmosphere of vapourising material) picked up in the **solar wind** and dragged away at rapid speed (about 400 km s^{-1}) forming a very different and distinctive tail that can be projected ahead of the direction of movement as the comet moves away from the Sun. Long after the comet has passed the Sun and left the inner regions of the Solar System, fragments of it can be swept up by the Earth in its orbit. These produce meteor showers (such as the Geminids and Leonids) as they burn up in the upper atmosphere.

8.9   Laws Governing the Motion of Planetary Bodies

It has been a tireless quest of astronomers and mathematicians to find the laws and rules that govern the

Figure 8.14   *Comet tails produced by Comet Hale–Bopp, imaged in April 2007 (Source: wikicommons, E. Kolmhofer, H. Raab; Johannes-Kepler-Observatory, Linz, Austria).*

planetary bodies and stars of which we have been speaking. Ptolemy (90–168 AD), a Greco-Roman writer and philosopher, imagined a Universe in which all bodies orbited the Earth. Earlier geocentric models suffered from a number of problems, not least the strange affinity of Mercury and Venus for the Sun and the apparent retrograde motion or backflips that some planets appear to make. Ptolemy's model has planets moving around small circles that themselves moved around the Earth. It was rather a desperate attempt to make a model fit the

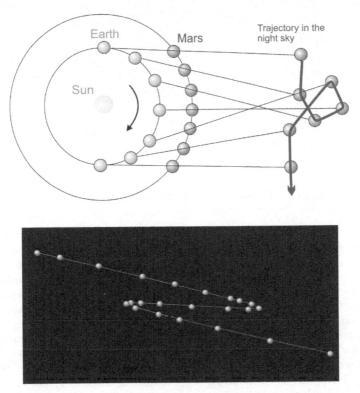

Figure 8.15 *Some planets appear to move in a retrograde motion or to do a backflip as elaborated by the diagram at the top (seen from the point below the Sun's south pole). The schematic at the bottom shows how Mars appears in the night sky over the period of time it does retrograde motion. These movements perplexed early astronomers and were difficult to explain with a geocentric model of the Universe* (Source: Charles Cockell and Tunc Tezel).

data. He wouldn't be the last one in science to suffer this affliction.

The early geocentric view was eventually replaced by the Copernican or heliocentric view, which placed the Sun at the centre of the Solar System, formulated by Nicolaus Copernicus (1473–1543), a Renaissance mathematician, in his book, *De Revolutionibus Orbium Coelestium* (On the Revolutions of the Celestial Spheres). This was a landmark text in astronomy, and therefore astrobiology. Copernicus was not the first to consider this possibility. Aristarchus (310–230 BC), a Greek philosopher, considered a heliocentric model of the heavens, but we do not know to what extent he developed this idea.

Copernicus's view provided explanatory power for the affinity of Mercury and Venus to the Sun (they were closer to the Sun) and it also posited that the distance from the Sun to the Earth was very small compared to the distance to the stars. His model also famously explained

retrograde motion (Figure 8.15), caused by differential rates of movement around the Sun.

Despite the new view of the universe, Copernicus remained wedded to the idea that planets moved in perfect circles, which stymied the accuracy of his model and its ability to reliably predict planetary motion.

The rules governing the movement of planetary bodies were first mathematically elaborated by Johannes Kepler (1571–1630), a German mathematician and astronomer, and protégé of Danish nobleman Tycho Brahe (1546–1601). Brahe was a scrupulous observer of the heavens and his abundant and very meticulous data provided the theoretically-minded Kepler with the information he needed to fathom that planets did not move in perfect circles, instead they moved in ellipses. The data also provided him with the information he needed to elaborate the more general rules that govern the motion of planets. The three rules he described were

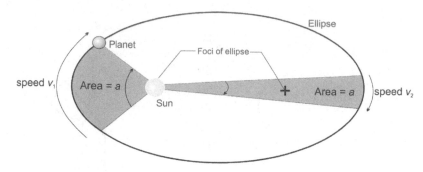

Figure 8.16   *Diagram illustrating Kepler's first two laws: that the Sun is at one of the foci of an ellipse (this applies to all planetary objects in a non-circular orbit around another object) and that the object sweeps out an equal area in an equal time (implying that speed v_1 is greater than v_2) (Source: Charles Cockell).*

themselves a direct consequence of Newton's laws of motion. They are:

Kepler's First Law: The orbit of each planet is an ellipse with the Sun at one focus (the point at which a planet is closest to the Sun in this orbit is called **perihelion**, in contrast to the most distant point, which is **aphelion**). This law applies to all objects in a non-circular orbit around another object. This law is illustrated in Figure 8.16.

Kepler's Second Law: An imaginary line from the Sun to the planet sweeps out equal areas in equal time. In other words, at perihelion a planetary body will be travelling faster than at aphelion since it must travel faster to sweep out an equal area in the same time. This law is illustrated in Figure 8.16.

Kepler's Third Law: The square of the **sidereal** period (the time it takes to complete a full orbit of its star) is proportional to the cube of the **semimajor axis** (one-half of the major axis of the ellipse).

Kepler's third law allows us to work out the sidereal period of a planet and is therefore enormously important for working out the orbital distance and mass of extrasolar planets. This will be discussed again in Chapter 19. As a consequence of its importance, it is worth deriving the mathematical relationship here.

Kepler's third law can be more exactly mathematically defined using Newton's laws by considering two bodies of masses m_1 and m_2, orbiting their (stationary) centre of mass at distances r_1 and r_2 and at speeds v_1 and v_2 (Figure 8.17).

As the gravitational force acts only along the line joining the centres of the bodies, both bodies must complete one orbit in the same period P, the sidereal period, although they move at different speeds v_1 and v_2. The forces on each body due to their centripetal accelerations are given by:

$$F_1 = m_1 v_1^2 / r_1 = 4\pi^2 m_1 r_1 / P^2 \qquad (8.10)$$

$$F_2 = m_2 v_2^2 / r_2 = 4\pi^2 m_2 r_2 / P^2 \qquad (8.11)$$

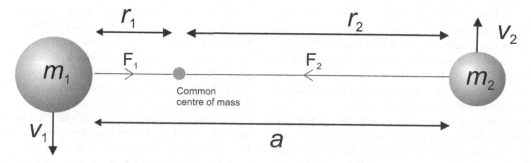

Figure 8.17   *Two bodies orbiting their common centre of mass (Source: Charles Cockell).*

The middle expression is derived from Newton's second law (F = ma). Centripetal acceleration, a, is given by the relationship, v^2/r. Thus, the force for each body is $F = mv^2/r$. The right hand expression is derived by recognising that v, the speed, is given by distance divided by time. This is given by $2\pi r$ (the length of the orbit, which is the circumference of a circle with radius, r) divided by the time it takes to go once round the orbit – the sidereal period, P.

Newton's third law (for every force there is an equal and opposite force) tells us that $F_1 = F_2$, and therefore:

$$4\pi^2 m_1 r_1 / P^2 = 4\pi^2 m_2 r_2 / P^2, \qquad (8.12)$$

thus:

$$r_1 / r_2 = m_2 / m_1 \qquad (8.13)$$

This relationship tells us that the more massive body orbits closer to the centre of mass than the less massive body (the larger m, the smaller the corresponding r to maintain the relationship).

However, the total separation of the two bodies, a (Figure 8.17), is given by:

$$a = r_1 + r_2 \qquad (8.14)$$

Note that here the term, a, is used for the distance between the planets not the acceleration!

This gives (substituting Equation (8.14) into (8.13)):

$$r_1 = m_2 a / (m_1 + m_2). \qquad (8.15)$$

Substituting this equation into the equation for F_1 derived above (Equation (8.10)) and then equating this to the force derived from Newton's third law, the Law of Gravitation ($F_{grav} = F_1 = F_2 = Gm_1 m_2 / a^2$), where G is the Gravitational Constant, gives Newton's form of Kepler's third law:

$$P^2 = 4\pi^2 a^3 / G(m_1 + m_2). \qquad (8.16)$$

If body 1 is a star and body 2 is a planet, then m_1 is much greater than m_2 and m_2 can be ignored. Hence the right side of the equation approximates to the equation which gives us the relationship between the sidereal period of a planet and the mass of the star:

$$P^2 = 4\pi^2 a^3 / GM_{star}. \qquad (8.17)$$

This equation is important for the study of exoplanets, because the mass of a star can be estimated from its colour according to well-known relationships in stellar evolution (graphically illustrated in the H-R diagram), P can also be measured using the **radial velocity method**, allowing us to estimate the distance of a planet from its star and to calculate its mass, as we shall see in Chapter 19.

8.10 Meteorites

Not all planetary objects remain firmly in fixed Keplerian orbits. They can take erratic orbits or become dislodged by collisions and gravitational influences, some of them eventually reaching the Earth and arriving at its surface. Some of these objects are very large and can cause local or global destruction to ecosystems. However, most are small objects that break up in the atmosphere and land intact on the Earth's surface.

These meteorites (not to be confused with the term, meteoroids, which applies to these objects as they travel through space and the term, meteors, applied to these objects when they burn up in the sky as shooting stars) provide a rich source of Solar System material for astrobiologists helping them to understand the abundance and distribution of the elements for life in the early Solar System, particularly carbon. They provide rocks to work on from distant bodies without having to send a spacecraft to collect them, although space missions are invaluable for collecting fresh unaltered material. There are tens of thousands of meteorite finds and if you are lucky enough to actually recover it immediately after it lands it is formally classified as a 'fall'. Most of them have come from the deserts of Australia, the Sahara, the Arabian peninsula, the ices of Antarctica and other locations where their characteristic shiny, and often black, **fusion crust**, caused by the intense heating of the outer layers of the rock as it ablates during entry into the Earth's atmosphere at over $10\,km\,s^{-1}$, stands out from the surrounding rocks, making them relatively easy to find (Figure 8.18).

The classification of meteorites is involved and complex, but we can recognise some major general types.

Around 5% or less of meteorites that arrive on the Earth are iron meteorites with intergrowths of iron–nickel alloys, such as kamacite and taenite and enriched in elements such as iridium. Most iron meteorites are thought to come from the cores of asteroids that were once molten. Like the Earth (Chapter 10), some asteroids experienced **differentiation**, with the denser metal, separated from silicates, eventually sinking to the centre of the asteroid to form an iron–nickel core. After the asteroid solidified, it was broken up in collisions with other asteroids, scattering these iron-rich fragments. The wonderful patterns that some of these iron meteorites contain, called **Widmanstätten patterns** (Figure 8.19), are long crystals of iron and nickel that are formed in the iron cores of these asteroids. They cool very slowly, allowing these crystals to form.

About 1% of meteorites are the stony-iron meteorites. They are a mixture of iron-nickel metal and silicate

Debate point: Old meteorites

Meteorites offer us material from the earliest Solar System and by studying them we gain glimpses into how the Solar System formed and what sorts of materials might have existed at that time. However, what problems might there be with meteoritic material, particularly material that has been sitting on the Earth for a long period before it was collected? Consider a meteorite that has just landed on the Earth and discuss what processes on Earth's surface might compromise the information it contains. How would those processes affect the pristine nature of a meteorite a few days after it landed, what about several centuries after? Even if a meteorite is collected as soon as it falls, how might the way it is collected and curated affect its usefulness as a scientific resource? Imagine you are given a meteorite that was collected as soon as it fell. It has been in a wooden display case in a museum for a hundred years. Would you have trust in the pristineness of the minerals it contains? What about if you wanted to study the original organic materials it contained? Discuss what this thought experiment tells you about how meteorites should be collected and stored.

Watson, J.S., Pearson, V.K., Gilmour, I., Sephton, M.A. (2003) Contamination by sesquiterpenoid derivatives in the Orgueil carbonaceous chondrite. *Organic Geochemistry* **34**: 37–47.

Toporski, J., Steele, A. (2007) Observations from a 4-Year contamination study of a sample depth profile through Martian meteorite Nakhla. *Astrobiology* **7**: 389–401.

Figure 8.18   *Recovery of a meteorite in Antarctica by members of the United States Antarctic Search for Meteorites (ANSMET) expedition. (Source: wikicommons, NASA).*

Figure 8.19   *The Casas Grandes iron–nickel meteorite with its Widmanstätten pattern. The image is about 3 cm across (Source: wikicommons, James St. John).*

minerals. One type, called pallasites, is thought to have originated in broken up asteroids at the boundary just above the core region where iron meteorites originated.

About 86% of the meteorites that fall on Earth come from a different and distinct class called the chondrites, which are named because they contain small, round particles. These **chondrules** are composed mostly of silicate minerals that melted while they were free-floating objects in space, generating their spherical shapes as the material minimised its surface tension in the gravity-free conditions in space. If you've ever seen an astronaut drinking water in the cabin of a space station you'll recall how it tends to forms 'balls' of water. Molten rock would have behaved similarly in the protoplanetary disc. Chondrites are about 4.55 billion years old and are thought to be material from the asteroid belt that never formed into large planetary bodies, making them some of the earliest and most primitive material in the Solar System. In other words, unlike the previous two classes of meteorites they come from undifferentiated bodies. Some types of chondrites, the **carbonaceous chondrites**, contain organic matter, including amino acids, suggesting that they may have formed at low temperatures. They are possible candidates for bringing prebiotic organic material to Earth.

About 8% of the meteorites that fall on Earth are achondrites as they do not contain chondrules. Most achondrites are ancient rocks, and are thought to represent crustal material of asteroids. They have some similarities to volcanic basalt found on the Earth. Two groups of achondrites are anomalous since they are younger than most achondrites and do not appear to come from the asteroid belt. One of these groups is from the Moon, and includes rocks similar to those brought back to Earth by the Apollo and Luna space missions. The other group is almost certainly from Mars and they will be discussed in more detail in Chapter 17. Both achondrites and chondrites are lumped together as 'stony' meteorites.

8.11 Conclusions

In this chapter we have undertaken a whirlwind tour of some of the characteristics of the Universe from the formation of the Universe itself and the origin of elements important to life through to the objects of the Solar System. All of this information is of importance to astrobiologists because, we could say quite simply, it puts the 'astro' into astrobiology. The nature of the Solar System and the bodies within it invite obvious questions, the most pertinent being: 'Is this typical in the Universe?' As we will see in Chapter 19, the distribution of planets, their sizes and characteristics are very varied around other stars, but there are aspects of our Solar System that do yield very general insights into the formation processes of potentially life-bearing planets. Above all, the journey we have taken provides a view on our cosmic home within which the elements required for life are synthesised. In the next chapter, we will take a closer link at the formation of carbon compounds in the Universe.

Further Reading

Books

Krauss, L.M., Dawkins, R.A. (2013) *Universe from Nothing: Why There is Something Rather than Nothing?* Atria Books, La Jolla.

Rees, M. (2003) *Our Cosmic Habitat.* Phoenix Books, Burlington.
Sagan, C. (2013) *Cosmos.* Ballentine Books, New York.

Papers

Basri, G., Brown, M.E. (2006) Planetesimals to Brown Dwarfs: What is a planet?. *Annual Reviews of Earth and Planetary Sciences* **34**, 193–216.
Burrows, A., Hubbard, W.B., Lunine, J.I., Liebert J. (2001) The theory of brown dwarfs and extrasolar giant planets. *Reviews of Modern Physics* **73**, 719–765.
Hoyle, F. (1946) The synthesis of the elements from hydrogen. *Monthly Notices of the Royal Astronomical Society* **106**, 343–383
Iben, I. (1967) Stellar evolution within and off the main sequence. *Annual Review of Astronomy and Astrophysics* **5**, 571–626.
Lopez, B., Schneider, J., Danchi, W.C. (2005) Can life develop in the expanded habitable zones around red giant stars? *The Astrophysical Journal* **627**, 974–985.
Raymond, S.N., Quinn, T.R., Lunine, J.I. (2004) Making other Earth's: dynamical simulations of terrestrial plant formation and water delivery. *Icarus* **168**, 1–17.
Sackmann, L.-J., Boothroyd, A.I., Kraemer, K.E. (1993) Our Sun. III. Present and Future. *Astrophysical Journal* **418**, 457–468.
Schuler, S.C., King, J.R., The, L.-S. (2009) Stellar nucleosynthesis in the Hyades Open Cluster. *Astrophysical Journal* **701**, 837–849.
Seeger, P.A., Fowler, W.A., Clayton, D.D. (1965) Nucleosynthesis of heavy elements by neutron capture. *Astrophyical Journal Supplement* **11**, 121–166.
Wallerstein, G., Iben, I., Parker, P., Boesgaard, A.M., Hale, G.M., Champagne, A.E., Barnes, C.A., Käppeler, F., Smith, V.V., Hoffman, R.D., Timmes, F.X., Sneden, C., Boyd, R.N., Meyer, B.S., Lambert, D.L. (1997) Synthesis of the elements in stars: forty years of progress. *Review of Modern Physics* **69**, 995–1084.

9

Astrochemistry – Carbon in Space

9.1 Astrochemistry: The Molecules of Life?

We now have knowledge of the structure of life and the basic molecules of which it is comprised. We also have a better understanding of how the elements from which life is constructed, including carbon, are synthesised in the fusion furnaces of stars.

With the benefit of these insights, in the next two chapters we will consider where the carbon-based molecules for life might have come from and how they might have been assembled to make the earliest organisms that ultimately led to the diversity of life that we know today. In these next two chapters we begin a deeper foray into linking several disciplines together – astronomy, chemistry and biology.

The investigation of how molecules are formed in space is encompassed by the field of astrochemistry. Astrochemistry is concerned with the production, distribution and fate of molecules in the Universe. The field of astrochemistry is very wide and as astrobiologists we do not need to know everything about chemistry in the Universe. Perhaps the most important question is whether diverse carbon compounds can be formed in space. Could some of the precursor carbon molecules for life have come from space and what is the complexity of carbon-based chemistry that we can achieve in the expanse of the **interstellar medium**? This question is important because it might tell us whether the chemistry from which terrestrial life is made could be universal. Is complex carbon chemistry a strange quirk of our little area of the Galaxy or are there good reasons to suspect that carbon compounds are everywhere being made? This chapter will explore some questions in astrochemistry with a special focus on carbon molecules.

9.2 Observing Organics

When space is viewed in the visible we see blackness – the familiar view we have of the night sky (when there are no clouds). It's not surprising therefore that we tend to think of space as a black, empty, rather inert void of nothingness. It also explains why for a very long time space didn't seem like a place of interest to chemists. There doesn't seem to be a lot going on there and the lack of chemicals in the diffuse stretches of interstellar space seem to preclude

Astrobiology: Understanding Life in the Universe, First Edition. Charles S. Cockell.
© 2015 John Wiley & Sons, Ltd. Published 2015 by John Wiley & Sons, Ltd.
Companion Website : www.wiley.com/go/cockell/astrobiology.

any reactions that would generate complexity. However, this view of space was definitively ended when we developed the ability to see space using infra-red and radio telescopes. In the infra-red, we see that space is full of clouds. In Figure 9.1 you can see the Orion Nebula in the visible on the left and in the infra-red on the right. The black void is revealed to have structure, huge swirling clouds of ionic, atomic and molecular material absorbing energy. Radio astronomy similarly shows us the presence of molecules in these regions. Although this material is diffuse, it is the domain of interstellar chemistry and within these regions surpisingly complex compounds are being formed.

9.3   In the Beginning

The early Universe after the Big Bang about 13.8 billion years ago was quite chemically simple and the elements available were minimal: hydrogen and deuterium, helium 3 and 4 and lithium. Chemistry during this time was limited to simple association and ion-molecule interactions. There was no possibility of prebiotic chemistry, let alone life. In this early stage the sorts of chemical reactions that could have occurred are shown below and you can see how limited they were compared to chemistry today:

$$H_2^+ + e^- \rightarrow H + H$$

$$HeH^+ + e^- \rightarrow H + He$$

$$H + e^- \rightarrow H^-$$

$$H^- + H \rightarrow H_2 + e^-$$

$$H^+ + H \rightarrow H_2^+ + \nu$$

$$H^+ + He \rightarrow HeH^+ + \nu$$

$$H^+ + Li \rightarrow LiH^+ + \nu$$

$$H^+ + H_2 \rightarrow H_3^+ + \nu$$

Once early stars formed, nuclear fusion within them led to **nucleosynthesis**. This process could not occur in the diffuse environments of space and so required the presence

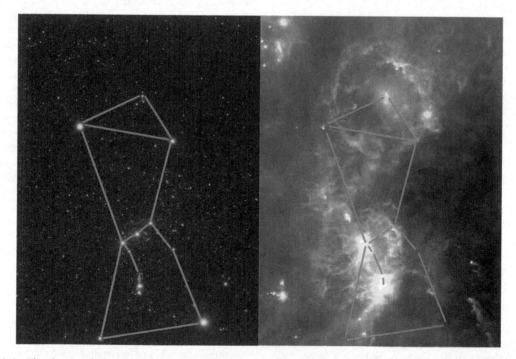

Figure 9.1   *The Orion nebula in the visible (left;* Source: Akira Fujii) *and infra-red (right;* Source: Infra-Red Astronomical Satellite).

Focus: Astrobiologists: Zita Martins

What do you study? I am an Astrobiologist. My research interests include the detection of biosignatures in space missions, and the possible contribution of meteorites and comets to the origin of life on Earth.

Where do you work? I am a Royal Society Research Fellow at Imperial College London.

What was your first degree subject? I have an MSc in Chemistry from Instituto Superior Técnico (Portugal) and a PhD in Astrobiology from the University of Leiden (The Netherlands).

What science questions do you address? 'How did life originate on Earth?' and 'Is there life anywhere else in the Solar System?'

What excites you about astrobiology? Astrobiology addresses the origin of life and the possibility of life existing elsewhere in the Universe, which are topics that have long fascinated human beings, including myself. Part of my research includes the analysis of meteorite samples which have not been extensively changed since the formation of the solar system 4.6 billion years ago. These samples are extraterrestrial and the oldest rocks you will ever hold in your hands. It can't get any more exciting than that!

of stars to generate the environments for fusion reactions, as we discussed in Chapter 8. These fusion reactions in low to high mass stars allowed for the formation of heavier elements with which carbon chemistry could occur.

9.4 Different Environments

As the Universe expanded and new generations of stars formed, so the diversity of environments in which molecules could be formed increased. Today there are an enormous variety of places where complex chemistry is occurring and in which carbon compounds are being formed. Let's take a brief tour and look at some of the characteristics of these environments.

9.4.1 Diffuse Interstellar Clouds

These are cold regions of space (90–100 K) in which there exist clouds of matter that are very diffuse (even more diffuse than a very strong laboratory vacuum) with about 10^8 m^{-3} of ions, atoms or molecules. The clouds are mostly atomic elements, but they contain simple diatomic molecules such as CO, OH, CH, CN and CH$^+$. The clouds are permeated by stellar UV radiation which is involved in inducing chemical reactions and ionising molecules. Figure 9.2 shows the diffuse clouds through which our

own Solar System is moving and some of the movements in the clouds themselves.

9.4.2 Molecular Clouds

Perhaps the most impressive places for chemistry are the Molecular Clouds. These are regions of space that are extremely cold (10–50 K). They are a little denser than diffuse interstellar clouds (about 10^{11}–10^{13} m^{-3} of ions, atoms or molecules) and the material is mostly molecular. About 100 different molecules have been detected so far. Some of these clouds are vast. A cloud with a mass of approximately 10^3 to 10^7 times the mass of the Sun is called a giant molecular cloud (GMC). These clouds are the most massive objects in the Galaxy, typically equivalent to about one million solar masses and have widths sometimes on the order of 150 light years across. Molecular clouds are the birth place of stars and within them many thousands of low and high mass stars can be formed. Some good examples of these clouds are the Orion Molecular Cloud (usually called the Orion Nebula), Sagittarius and the Eagle Nebula.

Giant molecular clouds are not penetrated very easily by optical and UV photons, which means that there is little ionisation there, accounting for the fact that most of the material is molecular and the dominant species is H$_2$.

A very nice example of a molecular cloud is B68 seen in Figure 9.3. B68 (Barnard 68) is towards the southern

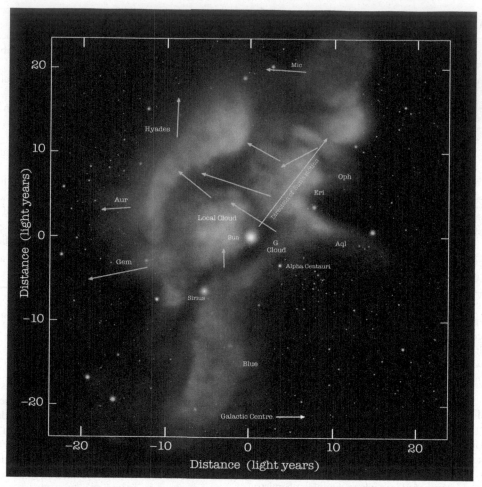

Figure 9.2 *The local interstellar clouds through which our Solar System is currently moving. The arrows indicate the movements within the diffuse clouds* (Source: NASA, Goddard).

constellation Ophiuchus at a distance of about 500 light years. The cloud on the left in the figure is imaged both in the visible and infra-red part of the spectrum. You can see how the cloud blocks out the light from behind it. On the right, in the infra-red, you can see the stars behind it showing that light gets through, illustrating the general principle that infra-red light can be used to probe the interior of molecular clouds.

9.4.3 Protoplanetary Disc

Some of the hottest places in the Universe are around newly forming stars. Within molecular clouds regions of gas begin to collapse under their own gravitational

attraction until they coalesce and spin by the conservation of angular momentum. In the centre of this newly forming accretion disc a star is formed as the material reaches sufficient pressures to ignite nuclear fusion. Around the star, material begins to collect and planets are formed (Chapter 8). Material from a great distance (to around 100 AU) is attracted towards the protostar. Figure 9.4 shows an illustration of the disc around TW Hydrae, an orange dwarf star approximately 176 light years away in the constellation Hydra.

Within the protoplanetary disc there is a large temperature gradient. The temperature of the outer disc is cold (10 K), but the temperature of the inner disc is ~100 K at 10 AU and ~1000 K at 1 AU. As well as hosting

(a)   (b)

Figure 9.3 The Dark Cloud B68. On the left it is seen in a composite visible/infra-red image showing how light is obscured by dust. On the right it is seen only in the infra-red in false colour showing how infra-red can be used to see through and into the cloud (Source: ESO).

Figure 9.4 An illustration depicting the disc that astronomers have detected forming around nearby star TW Hydrae approximately 176 light years away in the constellation Hydra (Source: NASA/JPL-Caltech/T. Pyle, SSC/Caltech).

high temperatures, the disc is also quite dense compared to diffuse or giant molecular clouds. Typical densities are $\sim 10^{16} – 10^{21}$ m^{-3}. Throughout the disc the physical conditions are also varied. Molecular gas is frozen on to dust grains in the outer disc, but ices evaporate in the inner disc. All of these variations create a nursery where all manner of compounds can be synthesised under different chemical conditions. The full diversity of chemistry that occurs throughout a protoplanetary disc is not fully known, but it seems clear that these are probably important places for complex astrochemistry, particularly as the material synthesised will be close to new planets being formed within the system. They provide a potential exogenous source of compounds for surface chemistry on new planetary bodies. This has special relevance for the formation and deposition of carbon compounds on habitable planets, as we'll see in the next chapter.

The production of organics within the disc has led to the notion of the soot line. Inwards of the soot line temperatures are too great for complex organic compounds to be formed and they will tend to be thermally degraded. Outside of the soot line temperatures are low enough for complex organic molecules to be formed.

9.4.4 Carbon-rich Stars

Some stars are carbon-rich and as a result they are very suitable places for the formation of complex carbon molecules. An example is IRC+10216 (CW Leo), seen in Figure 9.5, a star between 390 and 490 light years away

Figure 9.5   *IRC+10216 (CW Leo) – a carbon-rich star showing traces of its surrounding envelope* (Source: Izan Leao, the Very Large Telescope).

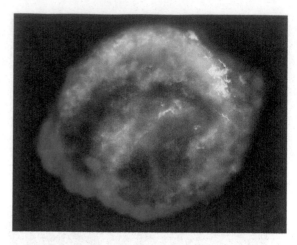

Figure 9.6   *Shock waves from supernova explosions. Here is shown a multiwavelength X-ray, infra-red and optical compilation image of 'Kepler's supernova' remnant, SN 1604 ('SN' means supernova), about 20 000 light years away* (Source: NASA).

(the name 'IRC' refers to Infra-Red Catalogue). A typical carbon star has an extended (ca. one light year) circumstellar envelope expanding at a velocity of $15 \, km \, s^{-1}$, in which chemical reactions can occur. About 60 molecules have been detected, mostly linear hydrocarbons. Near the photosphere of the star, which is the region within a star from which light is radiated, chemical reactions produce simple molecules such as CO, N_2, HCN and C_2H_2.

9.4.5   Shock Waves from Supernova Explosions and Other Astrophysical Violence

The universe is not a quiet place. Stars with sufficient mass explode as supernovae, generating intense and very violent shock waves. These rapid perturbations in physical conditions in the Universe are yet another way to create the sort of disequilibria which are good for driving chemical reactions.

Supernova explosions can generate supersonic shock waves with a speed of ~1 $km \, s^{-1}$. These shocks compress and heat the gas in the interstellar medium. There are a variety of different types of shocks (Figure 9.6). J-type shocks can cause the density of the gas to jump by four to

six times and generate gas temperatures of about 3000 K. As the shock wave passes, the gas cools quickly (in about a few tens to hundreds of years) and may increase in density further as it cools. The heat generated in these shocks can be used to drive endothermic chemical reactions (chemical reactions requiring energy) between neutral molecules, producing more complex molecules.

9.5   How are Compounds Formed?

So we see that there are a variety of environments where molecules can form and where they have been detected. Over 300 different molecules in total have been found to date.

But how can this chemistry happen given two essential properties of most places in interstellar space?

1. *Low temperatures*. When temperatures are low, there is a limited potential for thermal activation of chemical reactions. We are familiar with the **Arrhenius relationship**, which we will explore more in a later chapter. It tells us that the higher the temperature, the faster the rate of a chemical reaction. The extremely low temperatures of interstellar space limit chemical reactions and this problem would seem to apply to most places in the Universe.

We have already seen some exceptions to this idea. The interior of a protoplanetary disc or a supernova shock wave can generate very high temperatures, but certainly in giant molecular and diffuse interstellar clouds, the temperatures, generally less than 100 K, would suggest to us physical conditions not at all conducive to chemistry.

2. *Low pressures.* In a chemistry laboratory we are familiar with adding compounds together to start a reaction, but we are also familiar with the problems of dilution. If we add two reactive chemicals together that have been diluted in a large amount of water we observe that the reaction is either very slow or doesn't happen at all. If we want some spectacular results, we intuitively, as scientists, know that we need to add reagents together at some defined (and quite high) concentration that will encourage a chemical reaction. In other words, we need collisional activation.

Despite the apparent problems of low temperatures and pressures in space, chemistry can be driven by one thing we do have in these environments – high radiation.

Cosmic rays, which are comprised of protons, neutrons and ions of heavy elements such as iron can ionise hydrogen and helium. Photoionisation, for example with ultraviolet radiation from stars, can also generate carbon

ions that then lead to networks of complex chemistry. The ultraviolet radiation is primarily composed of the Lyman α photons (122 nm) from neighbouring stars or UV photons generated as the secondary product of galactic cosmic ray interactions with material in clouds.

By bombarding hydrogen, helium and carbon with cosmic rays or subjecting them to photoionisation we can produce ions in the gas phase as shown below:

$$H + cr \rightarrow H^+ + e^-$$

$$H_2 + cr \rightarrow H_2^+ + e^-$$

$$He + cr \rightarrow He^+ + e^-$$

$$C + \nu \rightarrow C^+ + e^-$$

where 'cr' means cosmic rays and ν is photoionisation by UV radiation. These ions can then go on to drive reactions that generate some surprisingly complex chemical reactions involving carbon, ultimately leading to cyanide, alcohol and nitrogenous carbon compounds. In Figure 9.7 some example reaction schemes are shown for the formation of carbon molecules starting with the irradiation of hydrogen and helium with cosmic rays.

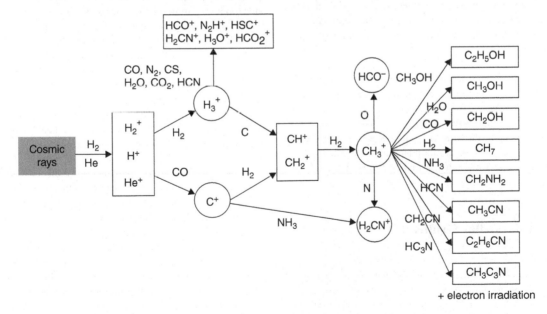

Figure 9.7   *Some examples of a wide variety of reactions in the interstellar medium driven by the irradiation of hydrogen and helium with cosmic rays. Arrows show reactants. Boxes and circles show products.*

Even the formation of water can occur in space, driven by cosmic ray interactions, but its pathways of formation are quite complex. One set of reactions is shown below.

$$H_2 + \text{cosmic irradiation} \rightarrow H_2^+ + e^-$$

$$H_2^+ + H_2 \rightarrow H_3^+ + H$$

$$H_3^+ + O \rightarrow OH^+ + H_2$$

$$OH_n^+ + H_2 \rightarrow OH_{n+1}^+ + H$$

$$OH_3^+ + e^- \rightarrow \mathbf{H_2O} + H; \; OH + 2H, \text{ etc.}$$

These reactions are intriguing to look at from an astrobiological point of view because you can see how by irradiating hydrogen with cosmic rays and then reacting it with simple oxygen ions we can end up with the solvent of life, formed in the interstellar medium. As we discussed in Chapter 3, water is abundant throughout the cosmos.

A remarkable example is **Quasar** APM 08279+5255, 12 billion light years away, which harbours a black hole 20 billion times more massive than the Sun ('APM' refers to Automatic Plate Measuring Facility, a facility that was used to examine astronomical image plates and which was used to find this object). The object has been shown to be producing vast amounts of water. The water, equivalent to 140 trillion times all the water in the world's oceans, surrounds the object.

These collected observations underline the fact that water as the solvent of life is perhaps not a remarkable and extraordinary thing, but a quite logical consequence of astrochemistry. We might predict, based on these observations, that water would be a likely solvent for life anywhere in the Universe just because it is so readily available, so hugely abundant and we know from empirical observation that it makes a very good solvent for life.

9.6 Interstellar Grains

Despite this impressive knowledge of the wide range of reactions that can occur in the gas phase in space, there is still a major problem. Gas phase chemistry just can't predict or explain the abundance of molecules that have been observed in space, particularly many complex molecules. Another mechanism for the production of

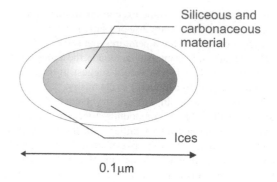

Figure 9.8 *A typical structure of an interstellar grain showing the core of siliceous and carbonaceous material surrounded by frozen volatiles (ices) on which chemical reactions can occur* (Source: Charles Cockell).

compounds must be occurring. It is now thought that in addition to gas phase reactions, interstellar silica and ice grains might be important locations for the formation of complex molecules.

Interstellar grains are small grains with silicate cores or carbon-rich cores with an icy coating. Figure 9.8 shows the structure of a typical interstellar grain. These grains provide three very important characteristics for astrochemical reactions: (i) They provide a surface on which reactants can collect to far higher concentrations than in the interstellar medium, (ii) They provide a surface on which reactants can be irradiated and/or come into contact with other reactants and (iii) products can collect on them and take part in further reactions.

A variety of reactions have been proposed that could occur in the icy rims of interstellar grains. There are three main types that we'll briefly consider here: .

1. *Eley-Rideal reactions.* Here only one molecule or atom binds to the surface of the grain and then reacts with a molecule or atom that remains in the gas phase to form a product that can then be desorbed from the surface (Figure 9.9).
2. *Langmuir–Hinshelwood reactions.* Here two molecules both bind to the surface of the grain and then come into contact, for example by diffusing across the surface, reacting and forming a product which can be desorbed (Figure 9.10).
3. *Hot atom reaction.* The hot atom reaction is not dissimilar to the Eley–Rideal reaction. In this case

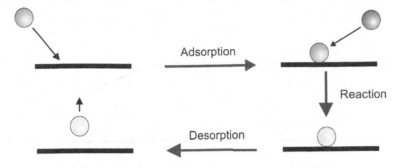

Figure 9.9 *Eley–Rideal reactions* (Source: Charles Cockell).

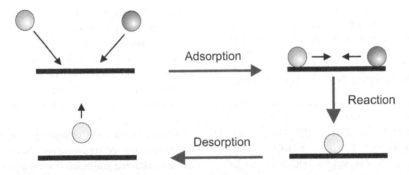

Figure 9.10 *Langmuir-Hinshelwood reactions* (Source: Charles Cockell and NASA).

an atom from the gas phase lands near an already absorbed atom and is able to move along the surface because it has high energy (hence 'hot atom'), thus reacting with it (Figure 9.11).

9.7 Forming Carbon Compounds

As we observed at the beginning of the chapter, of all the chemistries occurring in space, the formation of carbon compounds is of greatest interest to astrobiologists. By studying carbon compounds we can get insights into whether the building blocks of life, such as sugars and amino acids, could have been synthesised in space and subsequently delivered to early Earth and what the diversity of carbon compounds in space might be.

Once precursors are formed, then more complex carbon compounds, potentially the precursors to prebiotic molecules, can be produced. These reaction pathways (including some included in Figure 9.7) have various names associated with them. Here are just some of the variety of carbon reactions that can lead to more complex carbon molecules and particularly compounds with more than one carbon:

1. Carbon insertion

$$C^+ + CH_4 \rightarrow C_2H_3^+ + H \rightarrow C_2H_2^+ + H_2$$

2. Condensation

$$C_2H_2^+ + C_2H_2 \rightarrow C_4H_3^+ + H$$

3. Neutral–neutral reactions

$$C + C_2H_2 \rightarrow C_3H + H$$

4. Atomic insertion

$$N + C_3H_3^+ \rightarrow HC_3NH^+ + H$$

5. Radiative association

$$CH_3^+ + H_2O \rightarrow CH_3OH_2^+ + h\nu$$

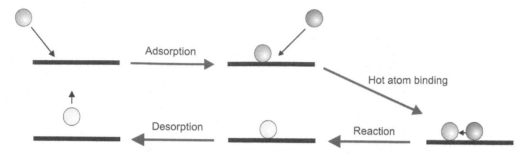

Figure 9.11   *Hot atom reactions* (Source: Charles Cockell).

Table 9.1   *A selection of the chemical formulae for just 25 gas phase carbon molecules observed in different regions in space providing a comparison to the more complex PAHs and other carbon compounds (Figure 9.12). These formulae illustrate the diversity of molecules and provide examples of carbon compounds containing the basic elements of life: C, H, N, O, P, S.*

CH_4	CO	CO_2	HC_8CN	C_3S
CH_3SH	C_2H_5CN	CSi	$(CH_2OH)_2$	$HNCO$
CH_3CN	CH_3COOH	CP	NH_2CN	C_2S
C_4H_2	HC_4CN	CH_3NH_2	$HCOOH$	SiC_3
CN	HC_6CN	CH_3CHO	C_3O	$HNCS$

The final pathways to compounds with multiple carbon atoms are still not fully understood and the diversity of carbon compounds observed in space has not been synthesised in the laboratory. However, the reactions above give us some idea of the potential chemical processes.

Formaldehyde was the first organic molecule to be detected in the interstellar medium. One of the most common molecules is CO, which is so abundant it is used to map interstellar clouds.

These reaction schemes generate some of the vast array of carbon molecules, from simple ones like CO to more complex compounds that have now been observed in space. In Table 9.1 you can see just some of the range of gas phase carbon-containing molecules that are thought to have been identified.

9.8   Polycyclic Aromatic Hydrocarbons

One important class of carbon compounds observed in space are the **polycyclic aromatic hydrocarbons** (PAHs). These are stable compounds, made mostly of carbon rings or aromatics (Figure 9.12). They make up about 10–15% of all the carbon in the galaxy. They are thought to exist in Diffuse Interstellar Clouds as they are generally resistant to UV radiation. They are very common in meteorites and assuming this is a result of synthesis in the protoplanetary disc, it might suggest widespread distribution of these compounds in planet and star-forming regions.

When transformed in chemical reactions, PAHs tend to lose their infra-red signature and the lack of their detection in interstellar ice grains has been suggested to be caused by the fact that they are engaged in complex chemical reactions on the surface of grains.

A role for PAHs in early biochemistry has not been demonstrated. Many of them are quite inert. However, under oxidative conditions, they can react to form **quinones** and other organic compounds that could be precursors to biological molecules. Quinones are involved in energy acquisition in life. PAHs are known to undergo photochemical conversion under UV irradiation. For example the PAH pyrene, when incorporated into phospholipid membranes, can release electrons after being irradiated. This has been suggested as a possible early light energy capture mechanism. PAHs could also act as molecular scaffolds for the formation of other molecules such as nucleic acids and so their inert nature might hide some potentially interesting roles for them in early biochemistry.

9.9   Even More Carbon Diversity

PAHs are complex molecules, but by no means the end of the story in terms of the sheer size of carbon macromolecules. Carbon 'buckyballs', including those with the structure C_{60}, were first observed in the nebula Tc-1 and

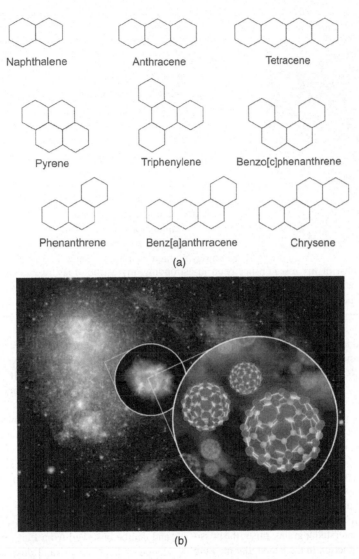

Naphthalene   Anthracene   Tetracene

Pyrene   Triphenylene   Benzo[c]phenanthrene

Phenanthrene   Benz[a]anthrracene   Chrysene

(a)

(b)

Figure 9.12 *(a) Some examples of polycyclic aromatic hydrocarbons showing how they are assembled from aromatic structures. (b) A figure depicting the discovery of buckyballs (C_{60}) in the Small Magellanic Cloud (Source: NASA).*

have been detected by NASA's Spitzer Space Telescope in the Magellanic Clouds (Figure 9.12). C_{60} buckyballs have a cage-like fused-ring structure which resembles a soccer ball, made of 20 hexagons and 12 pentagons. Carbon tubes, curious carbon onion structures made of **aromatics** and other intermixed and nonspecific carbon sheets and aromatic chains that could simply be described as 'soot' have been observed or are proposed to explain infra-red observations.

9.9.1 Prebiotic Compounds

As yet we do not have any definitive evidence of prebiotic compounds in the interstellar medium. There is a controversial claim for the detection of glycine in the interstellar medium, but this claim has not been sustained.

Astronomers at Copenhagen University have reported the detection of the sugar molecule glycolaldehyde, which can take part in the sugar-forming **formose reaction,**

Focus: Astrobiologists. Sun Kwok

What do you study? The synthesis of organic compounds in the late stages of stellar evolution

Where do you work? The University of Hong Kong, Hong Kong, China

What was your first degree subject? Physics

What science questions do you address? Using satellite-based telescopes, I observe the infra-red spectrum generated by the vibrational motion of complex organic molecules synthesised by old stars just before their deaths. By studying dying stars in the last phases of evolution, we can determine the sequence of chemical synthesis from simple molecules, such as acetylene, to benzene and then to complex organics with mixed aromatic-aliphatic structures. These complex organics are ejected into the interstellar medium and may have enriched the primordial solar system, with possible implications for the origin of life on Earth.

What excites you about astrobiology? I am most excited about astrobiology's interdisciplinary nature. Although my formal training was in physics and most of my research has been in astronomy, my astrobiology research prompted me to study organic chemistry, biochemistry, geology and planetary science. I believe that research that crosses traditional discipline boundaries has the greatest potential to lead to major breakthroughs in science.

---

which will be discussed in Chapter 11. The molecule was found in the binary system IRAS 16293-2422 located 400 light years from Earth (IRAS refers to the Infra-Red Astronomical Satellite).

The presence of amino acids, sugars and even nucleobases in meteorites suggests that these molecules might be formed in protoplanetary discs. This begs the obvious question of whether they could be found in other interstellar environments such as molecular clouds or carbon-rich stars and whether they are ubiquitous components of carbon chemistry in the Universe.

The detection of *iso*-propyl cyanide in the Sagittarius star-forming region near the centre of our Galaxy, presumed to be formed on interstellar grains, raises the possibility of amino acid production in the interstellar medium since the branched characteristic of this alkyl molecule is similar to the branched characteristics of amino acids found in meteorites (Chapter 11).

9.10   Comets

We can also gather data about organic material in the Universe from objects closer to home. Comets have long since been understood in the famous phrase of astronomer Fred Whipple (1906–2004) to be 'dirty snowballs'. They

exist in the **Oort cloud**, a spherically shaped reservoir of objects about $20\,000$–$100\,000$ AU away and in the toroidally-shaped **Kuiper Belt** (beyond the orbit of Neptune). The estimates of their number vary greatly There could be up to 10^{14} of them in the Oort cloud and potentially 1–10 billion objects more than 1 km in diameter in the Kuiper Belt. Their composition of water ice, organics and other volatiles, is confirmed by their low density which is typically on the order of 0.5–0.9 g cm^{-3}. Not all comets are the same and there may be diversity in the abundance of molecules to be found on different cometary bodies.

Comets find their way into the inner Solar System and can be observed from Earth with their characteristic double tails of plasma and dust as they evaporate under radiative heating from the Sun. Some of these cometary dust particles can be collected in the upper atmosphere as stratospheric Interplanetary Dust Particles (IDPs). Observations of Hale–Bopp and direct observations of Halley's comet have given scientists a fairly good idea of what they contain. CO and CO_2 are the most abundant ices, with sulfur-bearing species, formaldehyde, ammonia and other species on the order of 1% or less abundances. In the infra-red region, methane, ethane and acetylene have been observed. The presence of formaldehyde, H_2CO, a precursor of sugars, is also of significance to astrobiology.

Comets also contain a range of organic material. When the European Space Agency's Giotto spacecraft traveled past Halley's Comet in 1986 it detected grains of silicates and organic matter. These grains are sometimes referred to as CHON grains because of the apparently high abundance of molecules with these atoms. Confirmed molecules include formic acid, isocyanic acid, methyl cyanide and others. Dark material on comet nuclei may contain **amorphous** carbon material, representing complex organic material. The Halley's comet PUMA mass spectrometer instrument showed compounds with large molecular masses, up to at least a molecular weight of 160. Tentative interpretations have included benzene, toluene and, significantly, claims of adenine.

9.11   Chirality

The interstellar medium is not only the birthplace of new compounds, but may also provide the environment for other chemical features observed in life. One of the enduring mysteries of astrobiology, as we discussed in Chapter 3, is how **chirality** came about. Why does life predominantly use left-handed amino acids in its biochemistry?

Observation of the Orion nebula, which was mentioned earlier in the chapter, shows the presence of infra-red radiation which is circularly polarised – in order words rotated in one preferred direction. The presence of this radiation suggests that circularly polarised UV radiation may also exist. Different enantiomers of a molecule have a tendency to take up light of a particular polarised variety, hence the designations of L and D – which are a reference to the direction in which each one will tend to rotate light that is traveling with its waves aligned in one direction. If polarised UV radiation is generated from stars then it could be absorbed preferentially by one enantiomer selectively destroying it in the process, leaving an excess of the other enantiomer. This idea has not yet been proven.

Is it possible that the first slight enantiomeric excess was generated in the interstellar environment, eventually getting locked into life and resulting in the L-amino acids and D-sugars that we observe dominating terrestrial chemistry? Certainly a slight left-handed excess has been reported in amino acids in meteorites. We can imagine that if such an excess was incorporated into early life there would be a tendency for it to be magnified since life would be expected to operate better using only one consistently chiral form in all biochemical reactions. Biochemical reactions depend strongly on molecular recognition. Consistency in use of L- or D-amino acids in proteins and other molecules would tend to result in biochemistry where binding sites and inter-molecular recognition were also chirally consistent.

There are a variety of experiments that have successfully demonstrated chiral amplification. Certain protein helices tend to prefer one enantiomer over another when they are chemically polymerised. Catalytic reactions have been observed in which the product is more chiral than the reactant. Polymerisation rates of some compounds can be enhanced when there is a chiral preference. So in addition to biological evolution, there may be chemical means by which early chemistries with chiral excesses became purified in one enantiomer.

9.12   Laboratory Experiments

A developing area of astrochemistry is to perform experiments in the laboratory. These efforts make use of a vacuum chamber with a cold stage on which ices can be deposited. These ices, made of mixtures of defined composition, such as ammonia or water ices, are then irradiated with UV light or a variety of ions such as protons or heavy ions and the resulting products are analysed. The UV irradiation of ices composed of CH_3OH, NH_3 or CO results in the formation of the amino acid glycine, as well as acetamide, glyceramide and a variety of other carbon-based products. In the heavy fraction, PAHs and cyclic compounds have been detected.

Furthermore, ice irradiation experiments result in the production of hexamethylenetetramine (HMT), which under acid hydrolysis produces a range of amino acids such as alanine, proline and valine, although the extent to which these amino acids are produced in the interstellar medium or on comets is unknown.

Irradiation with charged particles including He^+, N^+ or Ar^+ ions similarly produces complex organic material referred to as ion-produced hydrogenated **amorphous** carbon. HMT is also synthesised under these conditions. The results of these experiments unsurprisingly depend on the ice compositions used, but broadly the conclusion can be made that the surfaces of interstellar grains and comets are

Debate Point: Replicating interstellar conditions in the laboratory

One of the challenges of astrochemistry is replicating interstellar conditions in the laboratory so that gas phase and interstellar grain chemical reactions can be examined in more detail (Figure 9.13). Scientists would like to know what compounds form, in what abundances and how they interact to lead to the complexity of compounds observed in space. Research has shown so far that when ices are irradiated under simulated interstellar conditions a variety of compounds are formed, including amino acids. Complications in this task include reproducing pressure and temperature conditions in the interstellar medium and the diverse radiation sources that might act on chemical species.

Figure 9.13 *Apparatus for irradiation and simulation of astrochemical environments in the laboratory. The image shows the main chamber at the W.M. Keck Research Laboratory in Astrochemistry, University of Hawaii* (Source: University of Hawaii).

What challenges can you list that would be relevant to attempting to understand astrochemical conditions using laboratory simulations? What technical solutions might there be and can you list the types of uncertainties that laboratory-based studies might cause when attempting to recreate conditions in the interstellar medium? What advantages are there to laboratory-based methods compared to attempts to directly observe these compounds?

Smith, I.W.M. (2011) Laboratory astrochemistry: gas phase reactions. *Annual Review of Astronomy and Astrophysics* **49**, 29–66.

environments where organic processing by radiation occurs, leading to complex organic material, some of which has prebiological significance. This confirms the general view that the interstellar environment is a place that gives rise to complex organic molecules that could have been deposited on the early Earth.

9.13   Observing these Molecules

How do scientists go about observing these molecules of interest in astrochemistry? It depends on what we want to observe. There are two approaches to detecting the molecules that we have been discussing – absorption

and emission spectroscopy. The principles behind these methods are discussed in Chapter 2.

Absorption spectroscopy. When atoms or molecules absorb radiation, they essentially 'rob' the light of energy at specific wavelengths, which is absorbed by a molecule, resulting in dips in the spectrum. The dips or troughs are quite specific to a given molecule and so can be used as a diagnostic feature of a molecule, a fingerprint if you will, of that molecule or collection of molecules. An example is shown in Figure 9.14 and shows the solids and ices in the molecular gas cloud around the massive protostar, NGC 7538 IRS9 (NGC refers to New General Catalogue).

You can see that we can tentatively identify a range of molecules in the cloud. The more powerful the instrument,

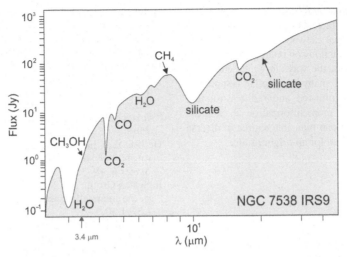

Figure 9.14 *An absorption spectrum for the molecular gas cloud NGC 7538 IRS9. The flux is given in the unit of jansky (Jy), a non-SI unit equivalent to 10^{-26} W m^{-2} Hz^{-1}. The absorption features are attributed to solids or ices, such as CO_2 and H_2O.*

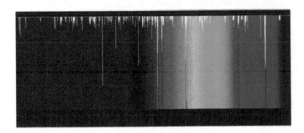

Figure 9.15 *An example of the absorption lines (shown as white lines) of the Diffuse Interstellar Bands across the spectrum (Source: NASA/P. Jenniskens and F.-X. Desert)*

the better the resolution and the finer the structure of the spectrum. The finer the structure, the more chance we have of resolving a variety of molecules and determining what they are.

A good example of a useful feature is the 3.4 μm absorption feature (Figure 9.14) which is characteristic of C–H bonds. It can be observed in the galactic centre, outside the galaxy and in meteorites as a fingerprint of the presence of organic carbon.

The most enigmatic features observed using absorption spectroscopy are the Diffuse Interstellar Bands (DIBs; Figure 9.15). Over 300 have been observed and they do not correlate with each other suggesting that they are made up of distinct molecules. The nature of the carriers of these absorption features has not been definitively characterised, but one likely candidate is PAHs. The

DIBs are complex and it has not proven easy to correlate them with laboratory experiments using interstellar ices and simulated irradiations. The strength of the DIBs is correlated with the extinction of light caused by dust and other matter which suggests that they are somehow linked to dust or ices.

Emission spectroscopy. Emission occurs when photons are produced by an energised gas that is exposed to radiation. The emission lines are again a type of fingerprint that is specific to particular molecules or atoms and can be used to identify the constituents. For example, emission bands at 3.3, 6.2, 7.7, 11.1, 12.7 μm wavelengths are attributed to C–H and C=C bonds in **aromatic** molecules.

These methods do have their limitations and optimal uses. Radio astronomy is blind to hydrogen, which is better detected in the UV radiation range. It is also not the method of choice for detecting molecules in the solid phase (including compounds on interstellar grains). Infra-red spectroscopy picks up many molecular structures and was used to detect PAHs. It can also pass through interstellar grains and so is useful for detecting silica, ices and other components of interstellar material. However, it is a poor method for detecting molecular nitrogen.

9.14 Conclusions

Once thought to be an inert emptiness, outer space is now known to be a place with diverse environments capable of generating complex compounds, including many carbon compounds. Unlike conventional chemistry in a laboratory on the Earth, which is primarily driven by high concentrations and thermal activation, in space cosmic rays, UV radiation and reactions on interstellar grains are thought to be just some of the ways in which chemistry can occur. The nature of many interstellar compounds remains enigmatic, but infra-red and radio astronomy are being used to gather more information about these molecules. They may be one plausible source of diverse carbon compounds required for an origin of life.

Further Reading

Books

De Duve, C. (1995) *Vital Dust: Life as a Cosmic Imperative*. Basic Books, New York.

Kwok, S. (2013) *Stardust: The Cosmic Seeds of Life*. Springer, New York.

Papers

Belloche, A., Garrod, R.T., Müller, H.S.P., Menten, K.M. (2014) Detection of a branched alkyl molecule in the interstellar medium: *iso*-propyl cyanide. *Science* **345**, 1584–1586.

Bettens, R.P.A., Herbst, E. (1997) The formation of large hydrocarbons and carbon clusters in dense interstellar clouds. *Astrophysical Journal* **478**, 585–593.

Bohme, D.K. (1992) PAH and fullerene ions and ion/molecule reactions in interstellar circumstellar chemistry. *Chemical Reviews* **92**, 1487–1508.

Burgh, E.B., France, K., McCandless, S.R. (2007) Direct measurement of the ratio of carbon monoxide to molecular hydrogen in the diffuse interstellar medium. *Astrophysical Journal* **658**, 446–454.

Ehrenfreund, P., Irvine, W., Becker, L., Brucato, J.R., Colangeli, L., Derenne, S., Despois, D., Dutrey, A., Fraaije, H., Lazcano, A., Owen, T., Robert, F., ISSI study team. (2002) Astrophysical and astrochemical insights into the origin of life. *Reports on Progress in Physics* **65**, 1427–1487.

Herbst, E., Chang, Q., Cuppen, H.M. (2005) Chemistry on interstellar grains. *Journal of Physics: Conference Series* **6**, 18–35.

Inglesias-Groth, S. (2004) Fullerenes and buckyonions in the interstellar medium. *Astrophysical Journal* **608**, L37–L40.

Kaiser, R.I. (2002) Experimental investigation on the formation of carbon-bearing molecules in the interstellar medium via neutral-neutral reactions. *Chemical Reviews* **102**, 1309–1358.

Marty, B., Alexander, C., Raymond, S.N. (2013) Primordial origins of Earth's carbon. *Reviews in Mineralogy and Geochemistry* **75**, 149–181.

McBride, E.J., Millar, T.J., Kohanoff, J.J. (2013) Organic synthesis in the interstellar medium by low-energy carbon irradiation. *Journal of Physical Chemistry* **117**, 9666–9672.

Snow, T.P., McCall, B.J. (2006) Diffuse atomic and molecular clouds. *Annual Review of Astronomy and Astrophysics* **44**, 367–414.

Tielens, A.G.G.M. (2008) Interstellar polycyclic aromatic hydrocarbon molecules. *Annual Reviews in Astronomy and Astrophysics* **46**, 289–337.

10

The Early Earth (The First Billion Years)

Learning Outcomes

➤ Understand hypotheses about the formation of the Earth, its early oceans and atmosphere.
➤ Understand some of the chronological sequence of events that occurred in the early history of the Earth.
➤ Understand some of the evidence that tells us about environmental conditions on the early Earth.
➤ Describe some of the environmental conditions in which life might have emerged.
➤ Describe some of the implications of these conditions for early life.
➤ Understand why these conditions would have challenged early life.

10.1 The First Billion Years of the Earth

In previous chapters we grasped the universal context of where the elements for life came from and how stars and planets form. We investigated how one source of carbon compounds for life could be the interstellar and newly formed protoplanetary environments. We are now in a position to focus on the Earth, consider its formation and the conditions on its early surface which provided the setting for emergence of life. What were conditions like on the early Earth and how did the earliest environments develop to a point where they were suitable for the persistence of life? What might have been some of the consequences of the early Earth environment for life?

In this chapter we will try to put these questions into perspective.

Broadly we'll consider the first billion years from when the Earth was formed 4.56 billion years (or giga annum, **Ga**) ago to 3.5 Ga (billion years) ago when the first evidence for life appears in the rock record. This period of Earth history covers the Hadean eon (4.56–4.0 Ga ago) and part of the Archean (which runs from 4.0 to 2.5 Ga). We will explore the geological history of the Earth in greater detail in Chapter 13. This is an immensely important time, the time when the Earth formed, when the first continents and the oceans came into existence, when the origin of life occurred and when life proliferated to a sufficient biomass to leave detectable traces of itself in the rock record.

10.2 The Earth Forms and Differentiates

When the Earth formed in the protoplanetary disc (Chapter 8), it was a homogenous mass of molten rock. There were three major heat sources that were important at this time:

1. Collisions involved the transfer of kinetic energy into heat. They occurred as the Earth was being formed from early planetesimals. As material pummelled the surface during the formation of the Earth in the early protoplanetary disc, so the planet was heated.
2. The gravitational forces that compressed material as more of it gathered into the newly forming Earth generated heat. Impacts of giant planetesimals would also have contributed to compression.

Astrobiology: Understanding Life in the Universe, First Edition. Charles S. Cockell.
© 2015 John Wiley & Sons, Ltd. Published 2015 by John Wiley & Sons, Ltd.
Companion Website : www.wiley.com/go/cockell/astrobiology.

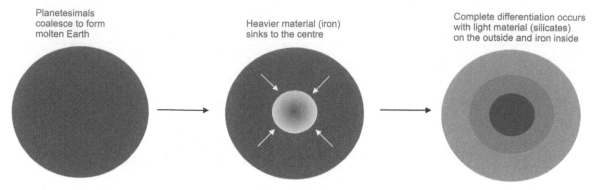

Planetesimals coalesce to form molten Earth

Heavier material (iron) sinks to the centre

Complete differentiation occurs with light material (silicates) on the outside and iron inside

Figure 10.1   *The differentiation of the Earth (Source: Charles Cockell).*

3. The radioactivity of elements would have contributed to internal heat energy. The three major elements that release energy from radioactive decay in planetary materials are uranium (^{235}U and ^{238}U), potassium (^{40}K) and thorium (^{232}Th).

These three sources of energy would have kept the planet molten. However, eventually, as energy dissipated, particularly collision and compression energy, the planet cooled down. Within the first 50 million years after accretion, temperatures within the Earth cooled to the melting point of iron. In a process, called global chemical **differentiation** (Figure 10.1), the heavier elements, including iron, began to sink down into the core of the Earth, while the lighter elements such as silica floated up towards the surface. Iron began to solidify and coalesce in the centre of the planet as an iron core. This process was helped by the property that liquid iron and liquid silicates are immiscible. Like oil and water they cannot mix and so will tend to separate.

Global differentiation was completed by about 4.3 Ga ago resulting in a composition and structure we are familiar with today (Figure 10.1). The interior of the planet contains a solid iron–nickel core surrounded by a liquid iron outer core. The core accounts for about a third of the mass of the Earth. Above this is the **mantle** and between this and the surface is the **crust**. It is within the crust that the biosphere resides. The crust is primarily formed of **silicates** (Chapter 3; Figure 3.15).

10.3   The Formation of the Moon

At about the same time as the Earth was forming, the Moon formed. In the early years of planetary sciences, there were several competing theories for how this could have happened. Each idea had certain predictions associated with it that can be tested:

1. **The common origin hypothesis.** As the planetesimals accreted from the protoplanetary disc, some of the material coalesced and formed the Earth, eventually differentiating into the structure of the planet we see today, and some of it formed the Moon.
2. **The fission hypothesis.** The early Earth rotated so fast that some of the material flew off and formed the Moon.
3. **The captured object hypothesis.** The Moon is another body entirely that was captured by the Earth's gravitational attraction early in the history of the Solar System.

The Moon is an interesting and unusual object and some of these characteristics are worth pondering. No other planetary body has such a large moon relative to its size (except Pluto). It has a diameter 27% and a mass 1.2% of the Earth's. The Moon has only a very small iron core, but it has a bulk density about the same as the Earth's mantle, suggesting a compositional similarity. The density of the whole Moon is less than the total bulk density of the Earth. The Moon is highly depleted in volatile elements and has a similar oxygen isotope composition to the Earth. All of these factors, taken together, imply that the Earth and Moon probably have a shared history.

These data seem to rule out some of the original theories. If the Moon and the Earth came from exactly the same protoplanetary material then we would expect their bulk density to be the same. The different sizes of body would have caused differences in differentiation and therefore the density in different regions of both bodies, but the overall

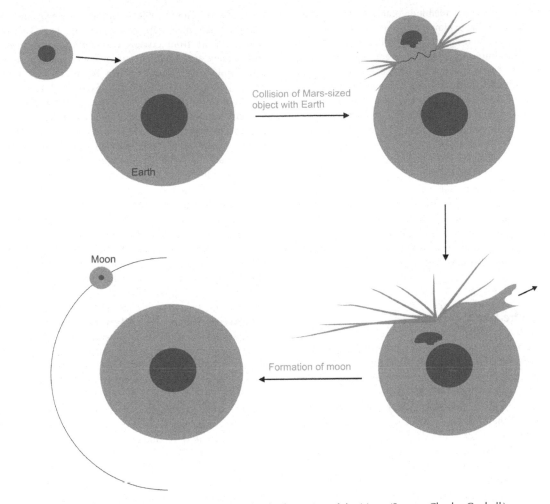

Figure 10.2 *The giant impact hypothesis for the formation of the Moon* (Source: Charles Cockell).

density should be nearly identical. Instead, the Moon has a lower density. The alternative theory, the fission hypothesis, is not supported by computer models which suggest that from a gravitational point of view, it is unlikely. The Earth would have had to be spinning implausibly fast. The geological structure of the Moon does not support the idea that it is an alien body captured by the Earth. This is in contrast to Mars's moons, Phobos and Deimos, and Neptune's moon, Triton, which are thought to be captured objects. Some of the moons of Saturn and Jupiter are also likely to be captured objects, so this is certainly not an irrelevant theory with respect to explaining how moons are formed around some planets.

To account for the data, the **Giant Impact Model** predicts that around 10–20 million years after the iron

in the Earth sank to the centre of the Earth to form the core, a planet about the size of Mars (sometimes rather romantically called Theia, after the Greek goddess who was the mother of the moon goddess, Selene), collided with Earth (Figure 10.2). Part of the material from the collision remained in orbit around the Earth.

By the process of collision and accretion, this orbiting material coalesced into the Moon. This theory explains why the bulk density of the Moon is slightly lower than the Earth, because the giant impactor would have stripped off lighter material from near the surface (the early mantle) and flung it into orbit around the Earth.

This idea was first proposed about 30 years ago, but it took calculations by modern high-speed computers to prove the feasibility of the idea. Today it is widely accepted as the most plausible hypothesis for the formation of the Moon. It explains many of the characteristics of the Moon, including its lower bulk density than the Earth and its similar oxygen isotope composition.

Today the Moon is receding on account of tidal interactions with the Earth. Its orbital distance from Earth is increasing at 3.8 cm year^{-1}. One manifestation of this interaction is a slowing of the Earth's rotation. Studies of coral growth rings show that, in the Devonian, 370 million years ago, there were 400 days in a year, implying a day lasting about 22 h. In the Hadean, the day might have been ~10 h long. Clearly this changing day length has implications for microbial cycles on early Earth linked to day length (such as the period of photosynthetic activity) and much later in Earth history, the cycles of animals linked to the lunar cycle. This is just one example of the astrobiological implications of a large moon. Later in the book, other ramifications will be explored.

10.4   The Early Oceans

Once the Earth began to cool and differentiate, the early oceans could begin to condense and form.

There have been a number of hypotheses to explain where the water to form the Earth's oceans came from. One theory is that water is constantly being delivered by **comets** and therefore, over time, the volume of Earth's oceans has been increasing. If it is assumed that water in comets survives collisions with the Earth, then one estimate (that $2.2–8.5 \times 10^{21}$ kg of water has been delivered to the Earth over its lifetime based on theoretical impact rates through time) would suggest that at least three times the present ocean volume has arrived over time. The theory is intriguing, but generally not supported by evidence, as we will see.

Other people hypothesised that comets impacting the Earth were a major source of water that contributed to the formation of the oceans, but that this process occurred early on and very rapidly when there were many of these objects.

A way to determine the origin of the Earth's water is to look at the ratio of two forms of hydrogen in the water: (i) 'normal' hydrogen (H), containing one proton and one electron and (ii) deuterium (D), containing one neutron as well as one proton and one electron. Hydrogen and deuterium are **isotopes**. We will examine how we can measure isotopic values in Chapter 13 in more detail. For now, it is sufficient to know that the ratio of these two isotopes can be measured and we could compare

that ratio to different objects in the Solar System to see if any of them match up. Fortunately, we have D/H ratio measurements from comets collected by spacecraft in situ (Halley and 67P/Churyumov–Gerasimenko) and by astronomical observations (Hale–Bopp and Hyakutake). They show that the comet ratio is much higher, about two to three times higher, than the ratio in Earth's oceans (which is 1.55×10^{-4}, or the **Standard Mean Ocean Seawater** or SMOW value). These data suggest that the Earth's oceans did not predominantly come from comets. These data also refute the idea that the Earth's oceans are the result of continuous cometary delivery over time. Although comets may not have provided the dominant source of water for the oceans, they probably did bring in a range of volatiles to contribute to the early inventory, including CH_4, CO, NH_3, N_2 and others.

So what objects did provide the water? Studies of the D/H ratio in meteorites, particularly **carbonaceous chondrites**, a type of meteorite with a water composition up to 22% by mass, show values of about 0.2 to 7.3×10^{-4}, similar to the terrestrial oceans. These objects come from the **asteroid** belt between Mars and Jupiter and many examples of them have been collected on the Earth. Mass calculations based on the isotopic data suggest that comets would only have provided about 10% at most of the early Earth's oceans, most of it coming from asteroidal material.

One leading hypothesis today is therefore that the water probably came from the accretion of asteroidal planetesimals containing water and the water was incorporated into the Earth when it formed. It is thought that this water then degassed from the planetary interior and condensed on the surface. We know that even today primordial materials are being degassed. The study of seawater from deep-ocean vents shows that it is enriched in ^{3}He, a primordial isotope of helium, compared to the more common ^{4}He. Light noble gases such as neon and helium have an isotopic abundance much more similar to the composition of the Sun suggesting that deep within the Earth primordial volatiles are today being outgassed to the surface. The water is thought to have been released from the interior in a catastrophic degassing event early in the history of the Earth as the Earth was differentiating and compressing.

Not all of this water was retained. The total inventory delivered during accretion was almost certainly higher than is present in the Earth's oceans today. Some of this water would have been lost during reactions with ultraviolet (UV) radiation on the early Earth which would have split water into its constituents, hydrogen and oxygen. The oxygen would have oxidised surface rocks and gases in the atmosphere; the hydrogen, because it is light, is lost to space. The on-going collision of asteroids at this time and the period of bombardment would also have caused **impact erosion** of the atmosphere and loss of some of this water.

At what time was the Earth's surface cool enough for oceans to form? Certainly one event that frustrated the permanent formation of the oceans would have been the Moon-forming impact. Estimates of the energy released during this event suggest that it would have completely vapourised the oceans (and a good amount of silicate rock as well). The result of this impact was an atmosphere between 300 and 500 times the present atmospheric pressure, and composed of water and carbon dioxide. Within a few million years the atmosphere cooled down and the surface of the Earth became solid once more. Once the temperature reduced to $\sim$600 K, in a very short space of time (on the order of $\sim$1000 years) the water would have precipitated out, such that by 4.4 Ga ago, oceans had formed on the surface of the Earth. By $\sim$4.3 Ga ago it is likely that the temperatures of the oceans were low enough (i.e. below the upper temperature limit for life; see Chapter 7) to support life. Nevertheless, the oceans were probably warmer than today and could have been somewhere between 50 and 80 °C.

There is direct evidence for oceans on the early Hadean Earth. Scientists have investigated ancient **zircons**. Zircons are zirconium silicates ($ZrSiO_4$). They are very tough and remarkably resilient to erosion. They are generally small crystals of about 0.1–0.3 mm in size, but they are ubiquitous in most rock types. Furthermore, the zirconium in zircons can be replaced by other elements such as uranium, which can be used to date rocks in **radiometric dating** (Chapter 13). Zircons were successfully extracted from rocks obtained from Jack Hills in Australia. These zircons are some of the oldest minerals on Earth. The Acasta Gneisses in northern Canada, which preserve a record of **granites** exposed to **metamorphism**, are another rare rock outcrop that dates to the Hadean $\sim$4.2 Ga ago.

Dating the zircons of the Jack Hills shows them to have an age of 4.4 Ga. The oxygen isotopic values of these zircons have also been measured (they are found to have $\delta^{18}O$ values of 5–8‰. You might like to review this statement when this notation has been introduced in the next chapter). The isotope values are thought to suggest that the lava from which the zircon crystallised contained material that had reacted with liquid water. These oxygen isotope values suggest low-temperature alterations (i.e. rock reactions in ocean water). Earth's surface was apparently cool enough for oceans to form at 4.4 Ga.

As well as being warmer, the chemistry of these early oceans was probably very different than today. High

partial pressures of CO_2 (pCO_2) dissolved in water would have caused the pH to be something on the order of pH 4.8–6.5, compared to today's pH, which is $\sim$8.0. Although we do not have chemical compositions for the very earliest oceans, 3.5 Ga old **fluid inclusions** from rocks in Australia and modelling studies suggest that the ocean at $\sim$3.5 Ga ago was much more concentrated in sodium and chloride (about twice modern day seawater) and enriched in bromine. Calcium concentrations could have been 20 times present-day values. We do not have direct chemical measurements for ocean water earlier than this.

Some of the best and oldest fully formed rocks that attest to the presence of oceans during the early Archean come from the Akilia and Isua regions in Western Greenland. Radiometric dating (which we will explore in Chapter 13) shows the Akilia and Isua rocks to be about 3.87 and 3.81 Ga old, respectively. Many features of these rocks suggest that they formed in contact with liquid water. They contain limestones (carbonates) that generally form in aqueous environments as a result of the abiotic precipitation of calcium carbonate (or in the recent history of the Earth from the carbonate skeletons of sea creatures, such as **foraminifera**). The rocks contain sandstones, which are essentially a large quantity of broken-down **quartz** (SiO_2) crystals and other impurities. Quartz crystals are produced when rocks are **weathered** or broken up by wind and water. Their presence in the Isua rocks suggests that continental rocks (essentially land masses) were being weathered away at this time and the quartz material was then deposited in shallow water or in the oceans. Finally, the rocks also contain lava that has a 'pillow'-like shape (**pillow lava**), the shape that lava tends to take up when it is erupted into water and instantly solidifies, quenching into rounded giant blobs. These rocks also suggest that the oceans had formed by this time. Rocks in Pilbara, Western Australia, dated to 3.5 Ga ago have been investigated and contain evidence for ancient carbonates and sulfates, minerals again associated with liquid water environments and possibly shallow coastal-like sediments.

10.5 The Early Crust

Before the first land masses formed, the Earth would have been covered by a magma ocean. It remains highly controversial how much continental land area was then formed on the Hadean Earth when the oceans came into existence and when life arose. It is debated at what rate continental land area was formed through to the

Figure 10.3 *An artist's impression of the early continents of the Hadean Earth* (Source: Science Photo Library).

present day. Some estimates place the Hadean land area as 10–15% of today's continents. However, very little Hadean rock record remains to provide constraints on this estimate. Most of the land masses on the very early Hadean Earth were probably basalt volcanic islands that formed over plates and volcanic hot spots. It is thought that an early form of **plate tectonics** had been established in the Hadean and would have been responsible for recycling crust at a fast rate, possibly continuously recycling all of the surface material in ~100 000 years into the mantle. The different temperature gradient from the surface into the mantle of the Hadean Earth may have resulted in a very different plate tectonic regime than the one we observe today (Figure 10.3). We will discuss plate tectonics in more detail in Chapter 13. Certain types of rocks such as **komatiites**, which are formed at high heat fluxes and are rare today, would have been present. By about 4 Ga ago, the Earth had cooled further and a stable plate tectonic regime had probably been established. Nevertheless, later into the Archean, the rocks that survive to the present day have evidence for **hydrothermal** veins, suggesting hot water flowing through the rocks, consistent with a picture of higher heat flow on the early Earth.

10.6   The Early Atmosphere

Immediately after its formation, the Earth is thought to have had a thin atmosphere composed primarily of helium and hydrogen gases. The Earth's gravity could not hold these light gases and they easily escaped into outer space. Today, H_2 and He are of low abundance in our atmosphere (about 5.5×10^{-5} and 5.2×10^{-4}%, respectively). For the next several hundred million years, volcanic out-gassing began to create a thicker atmosphere composed of a wide variety of gases that were part of the original volatile inventory when the Earth formed. Some of the gases that were released were probably similar to those created by modern volcanic eruptions (Figure 10.4). It is thought that greater than 85% of this atmosphere was produced by the early catastrophic degassing that also contributed to the formation of the oceans, the rest through "continual" degassing.

The early atmosphere would have been **reducing**, containing constituents such as sulfur dioxide (SO_2) and hydrogen sulfide (H_2S) produced by volcanic outgassing. Carbon dioxide (CO_2) and carbon monoxide (CO) would have been outgassed and produced from cometary impacts or reactions between impactors and the atmosphere. Methane (CH_4) would have been produced by degassing in the early stages and maybe augmented by microbial **methanogenesis**. The atmosphere would have contained nitrogen gas (N_2), up to a bar, similar to present-day levels, and derivatives of this atmospheric gas such as nitrogen oxides, produced by lightning and impacts. Ammonia (NH_3) would have been present from primordial outgassing, production at hydrothermal vents, and the reaction of nitrogen oxides with reduced iron (Fe^{2+}) in the oceans, although its lifetime in the atmosphere is short. It could have persisted for a long time if the atmosphere had hydrocarbon haze layers that would have protected it from destruction by UV radiation. Hydrogen (H_2) would have been outgassed, eventually being lost to space in addition to the hydrogen produced by the photolysis of water. The oxygen so produced would have reacted in the atmosphere with reduced gases or with surface minerals to form oxidised compounds. As the environment became more oxidised, CO_2 would have dominated over reduced gases and the atmosphere would have become only weakly reducing. Throughout the Hadean this CO_2 might have been consumed in **carbonatisation** reactions with volcanic rocks (whereby the gas is converted into carbonate minerals).

By human standards, this early atmosphere was very poisonous. It contained almost no oxygen. In Chapter 14, we will look in more detail at the evidence that tells us that the early atmosphere had very low O_2 concentrations.

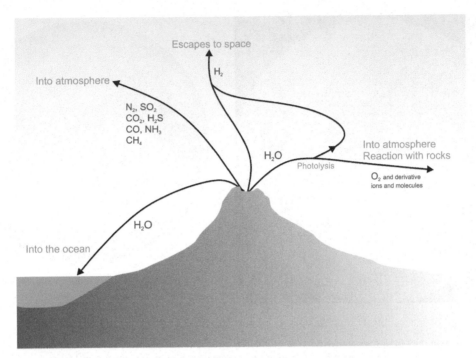

Figure 10.4 *Some gases produced by volcanic eruptions would have contributed significantly to the atmosphere of the very early Earth (Source: Charles Cockell).*

10.7 The Temperature of the Early Earth

During the Hadean, the Sun is thought to have been at least 30% less luminous than it is today. The lower solar flux would have significantly reduced solar **insolation**. It was recognised a long time ago that this causes a problem in explaining the presence of liquid water on the planet at that time. This is known as the **Faint young Sun paradox**. Although the higher heat flux from the interior of the planet would have gone some way to warming the surface (particularly during the early years after accretion), there remains a problem in explaining how liquid water was maintained throughout the Hadean and early Archean. One possibility is that greenhouse gases such as CO_2 and CH_4 could have compensated, raising planetary surface temperature. CO_2 concentrations could have been about 50–500 parts per thousand (compared to ~400 ppm today) and CH_4 concentrations could have been 100–1000 ppm (compared to the concentration today of 1.7 ppm). Another possibility is that NH_3, a very strong greenhouse gas, could have warmed the surface. Ammonia is rapidly destroyed by UV radiation, but organic hazes in the early atmosphere would have protected it. It could have been that a number of these atmospheric constituents acted together to warm the early Earth.

10.8 The Late Heavy Bombardment

During the early history of the Solar System between about 4.1 and 3.9 Ga ago, there is evidence for heavy bombardment from asteroidal material in the form of giant basins on the inner planets. As a result of plate tectonics, most of this evidence has been erased from the surface of the Earth. However, large impact basins on the Moon, Mercury and Mars provide strong evidence for this period of bombardment. Examples of them on the Moon and Mars are shown in Figure 10.5. In Figure 10.6 a graph provides information on the extent of these early craters and the similar profiles of size against surface coverage for the inner planets.

The dating of rocks from the Moon (Chapter 13) shows that the majority of these basins were formed just

-8000   -4000   0   4000   8000   12000
elevation meters

(a)   (b)

Figure 10.5   *Giant impact basins are evident on the Moon as dark 'mare' regions (left) and on Mars (right), such as Hellas Planitia, a 2300-km diameter crater shown in the southern hemisphere on a topographic map of Mars acquired by the Mars Orbiter Laser Altimeter (MOLA) instrument on the Mars Global Surveyor spacecraft (Source: NASA).*

after 4 Ga ago. We also know that these basins would have required quite substantial objects to form them. For example, the Mare Orientale on the Moon (about 330 km in diameter) would have required a 100-km impactor to have formed it.

In summary, the late heavy bombardment is recognised as a time of much higher impact flux than today (Figure 10.7).

It is worth pointing out here that direct age dating of lunar rocks enables us to date the period of bombardment because we are fortunate to have access to pieces of the Moon. However, for planetary bodies where we do not have rocks from a location of interest, we can use the evidence of bombardment itself to age date the surfaces. Given a certain assumed impact flux over time, taking into account spikes like the heavy bombardment, we can count the number of craters on a planetary surface and thus work out how old that surface is. **Crater age dating** has been remarkably successful in enabling planetary scientists to date the surfaces of other planetary bodies.

A question of interest to astrobiologists is whether this bombardment was the tail-end of a higher impact flux associated with the remnants of the early Solar System being swept up or something more fundamental – a real 'spike' in the frequency of impacts?

One theory is that it was a genuine increase in the impact flux. The idea behind this hypothesis is that initially the Solar System had a large population of icy objects located beyond Saturn. They were in stable orbits around the Sun for several hundred million years. However, over time, gravitational interactions between Saturn and Jupiter caused perturbations in the orbits of the gas giants, pushing Uranus and Neptune outwards. As these planets migrated, their gravitational interactions began to scatter the remaining planetesimals into the inner Solar System. A small fraction impacted the Moon and rocky planets, making immense craters. Calculations suggest that the bombardment would have lasted less than 100 million years.

Impacts on the early Earth would have had profound consequences for the environment. At the site of impact, rock, water and, if present, organisms would be vaporised. Deeper rock would be heavily shocked, fractured or melted at high pressures and temperatures. Giant tsunamis, if the impact was into the oceans, would generate enormous waves around coastal areas (although without

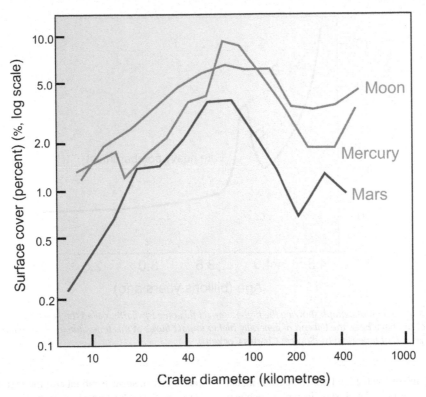

Figure 10.6 *Planetary bodies in the inner Solar System preserve a record of large impacts on their surface. The graph shows that the surface coverage of craters of a given diameter is similar for the Moon, Mercury and Mars, suggesting that they were subjected to a similar bombardment early in Solar System history. The significantly lower number of craters on Mars is partially accounted for by the presence of an atmosphere as well as active surface erosion in its early history. The Earth is not shown since most (but not all) craters have been eroded or destroyed by plate tectonics making such a plot unreliable. The surface of Venus has been reworked with new lava since heavy bombardment and so an ancient surface no longer exists with which to do this analysis. The axes are logarithmic scales.*

Debate Point: Age dating using craters

A common method to date the surface of planetary bodies is to count the craters on them or in a region of interest. If we have some idea of the rate of cratering in the Solar System we can then work out how many craters we would expect and compare it to the number observed. Thus, the age of the surface can be determined. Newer surfaces have fewer craters, older surfaces more. How reliable do you think this method is? What sort of uncertainties might there be? A lack of definitive and very accurate data on impact fluxes through all of geological time might be one uncertainty. In young terrains where there are very few craters, the statistics might be questionable, particularly if we are trying to date a small area. Some craters might overlay others (saturation cratering) obscuring the true number of craters. Secondary craters could be produced from material thrown out from primary cratering events. List and discuss uncertainties that might exist with this method. How would you overcome some of these problems using remote (telescopic or orbital) information?

Hartmann, W.K., Neukum, G. (2001) Cratering chronology and the evolution of Mars. *Space Science Reviews* **96**, 165–194.

McEwan, A.S., Bierhaus, E.B. (2006) The importance of secondary cratering to age constraints on planetary surfaces. *Annual Review of Earth and Planetary Sciences* **34**, 535–567.

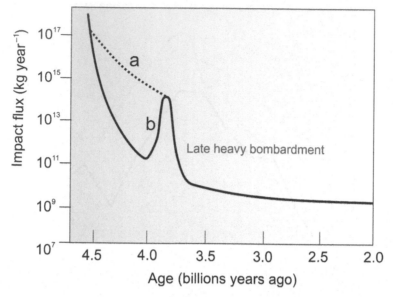

Figure 10.7 *A simple schematic graph showing the higher impact flux on early Earth, called the late heavy bombardment. The high impact flux could have been the tail end of generally higher impact fluxes at this time (line a), or a real peak in fluxes over a relatively short period of time (line b) (Source: Charles Cockell).*

multicellular organisms on the land, the biological impact may have been minimal). Thick dust in the atmosphere would have cut out sunlight for months or years, depending on the scale of the event, potentially cooling the planet. Aerosols caused by eruptions and vaporisation would have remained in the atmosphere for decades, changing atmospheric chemistry.

Even after the late heavy bombardment there is evidence that impacts were still, intermittently, a challenge for early life. Small sphere-shaped rocks, presumed to have been formed by molten rock flying through the atmosphere and solidifying into spheres like raindrops, are thought to represent evidence for early impacts. They are known as **tektites**. Layers of them have been found in the Barberton rocks of South Africa and the Pilbara rocks of Australia (both 3.5–3.2 Ga old).

10.9 Implications of the Early Environment for Life

We have looked at some of the features of the early Earth and the conditions on the planet. What would be the implications of these conditions for life? In some ways, the early Earth was like the present-day Earth. Oceans would have provided habitats for life and early continents offered the possibility of surface habitats for the first microorganisms. Perhaps, through a hydrological cycle and the evaporation of the oceans and subsequent precipitation, there would have been freshwater bodies on land that would have provided homes for certain microorganisms. However, there are many characteristics about early Earth in its first billion years that were very different. The atmosphere lacked oxygen and the impact flux was greater. These two differences we will explore in more depth.

The lack of oxygen (O_2) in the atmosphere would have meant that there was no ozone (O_3) shield. Today, oxygen reacts with solar UV radiation in the stratosphere and forms ozone. Although this layer of gas, if compressed into a single layer, would only be 3 mm thick, it is a strong absorber of short wavelength ultraviolet radiation (Figure 10.8). It cuts the UV radiation off at about 295 nm. Without an ozone shield wavelengths as short as 200 nm could have reached the surface of the early Earth, with shorter wavelengths than this screened by CO_2. The significance of this difference is that these shorter wavelengths cause more biological damage. In Figure 10.8 you can see the **action spectrum** for DNA damage. The action spectrum is a measure of the biological damage caused by particular wavelengths of radiation. You can see that at short wavelengths of about 200 nm, the damage to DNA is about six orders of magnitude greater than at say 340 nm

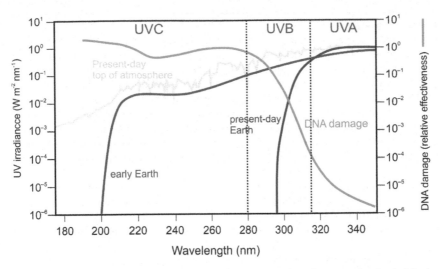

Figure 10.8   *A diagram showing the putative UV flux on the early Earth (red) and present-day Earth (blue) compared to the extraterrestrial flux at the top of the Earth's atmosphere today (yellow). Also shown is the sensitivity of the molecule, DNA, to different wavelengths of UV radiation (green) illustrating the disproportionately greater damage at short wavelengths.*

in the longer wavelength UV region of the spectrum. We can multiply the action spectrum by given light spectra and then integrate across the whole wavelength range to get the **biologically effective irradiance**. This is given by:

$$E_{\text{eff}} = \sum_{\lambda=200\,\text{nm}}^{\lambda=400\,\text{nm}} S_\lambda E_\lambda \Delta\lambda \qquad (10.1)$$

where E_{eff} is the biologically effective irradiance, S_λ is the spectral irradiance (the intensity of light being received at a given wavelength) and E_λ is the value of the action spectrum at a given wavelength. These two quantities are multiplied together at different wavelengths across the spectrum and summed (denoted by the sum symbol, Σ). Here the calculation is shown from 200 to 400 nm, which is thought to be the UV radiation wavelength range on the surface of the early Earth. The wavelength range can be modified for the present-day Earth or other environments.

Calculations such as these suggest that in the worst case (with no other absorbers such as organic hazes in the atmosphere) damage to biomolecules such as DNA on the early Earth would have been about a thousand times higher than today.

You will notice from Figure 10.8 that UV radiation is categorised into different wavebands. UVC radiation is from 100 to 280 nm (here only the part relevant for the early Earth is shown), UVB radiation is from 280 to 315 nm and UVA radiation is from 315 to 400 nm (again, only a part of this range is shown in Figure 10.8).

The good news for life is that UV radiation is quite easily screened. Particles of soil, iron, rock and other substrates rapidly attenuate UV radiation so that protected beneath these layers, the radiation would have been reduced quite significantly. Nevertheless, it would have been a significant factor in preventing, or at least challenging, anything attempting to colonise the exposed surface of the early Earth. The true extent of UV radiation also remains largely unknown. If there was a higher dust loading in the early atmosphere or, as some people have suggested, some sort of hydrocarbon smog generated from the UV irradiation of methane and volatile organic compounds, then the biologically effective irradiance may not have been as much as a thousand times greater.

The high impact flux on the early Earth has similarly been suggested as a factor that would have challenged early life. Very large impacts have been suggested to have sterilised the planet or vapourised the top few hundred metres of the oceans by generating atmospheric temperatures in excess of 2000 K. Impacts would have produced large hydrothermal systems on the early Earth. It has been pointed out that the late heavy bombardment could have caused a **bottleneck** for evolution, accounting for the preponderance of hyperthermophiles at the base of the phylogenetic tree of life. Frequent boiling of the oceans could generate an intense selection pressure on early organisms, favouring those with heat-loving characteristics. The genetic evidence for a deep hyperthermophilic ancestry for life is a matter of intense

---

Debate Point: Radiation – good for life?

Both UV and ionising radiation are generally considered to be bad for life. One of their effects is to cause mutations within DNA. UV radiation, for example, can generate *oxygen free radicals* and these can indirectly cause damage to DNA. There are a variety of forms of UV radiation-induced damage including thymine (pyrimidine) dimers, which are covalent linkages between thymine bases in DNA. There are a variety of biochemical pathways and molecules that have evolved in life to remove DNA lesions, including the enzyme, photolyase, a light-induced enzyme that splits pyrimidine dimers and excision repair, a pathway which removes a section of damaged DNA and resynthesises the strand. It is clear from these pathways that there is a strong selection pressure to remove radiation-induced damage. However, could higher UV radiation also act to accelerate evolution by increasing the mutation rate? Could radiation, at some certain level, be good for life? What do you think? Was radiation bad or good for early life, or both?

Pfeifer, G.P., You, Y.H., Besaratinia, A. (2005) Mutations induced by ultraviolet light. *Mutation Research* **57**, 19–31.

Todd, P. (1994) Cosmic radiation and evolution of life on Earth: Roles of environment, adaptation and selection. *Advances in Space Research* **14**, 305–313.

---

discussion (Chapter 7). It also now seems probable that these early impacts were not globally sterilising. Thus, the role of early impacts in modifying evolution is an open question.

Despite these differences and the additional challenges that early life would have faced, we know that by about 3.5 Ga ago the Earth was differentiated, oceans had formed and early land masses were in place. Many of the habitats available for life in the oceans and on the land masses existed. It is within this environment that early life originated, prospered and left its evidence in the rock record. In the next chapter we will examine ideas about the origin of life. We will explore these early life forms and the evidence for their existence in Chapter 12. In Figure 10.9

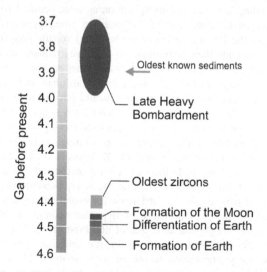

Figure 10.9 *Some of the major events in the first billion years of Earth history.*

you can see a summary of the first billion years of the Earth's history.

10.10   Conclusion

The first billion years is the time during which some of the most biologically significant events in our planet's history occurred. The evidence suggests that the surface became suitable for life soon after the accretion of the planet, frustrated by the Moon-forming impact and early bombardment. Oceans were formed from early degassing of volatiles and within a few hundred million years of its formation it seems that the Earth had many of the habitats that we are familiar with. Despite these parallels, the early Earth had some characteristics very different from today. High impact flux, an **anoxic** atmospheric composition very different from today, high UV radiation and a high heat flow through the crust are just a few of the differences that would have substantially influenced the availability of energy and the distribution of habitats for life. With this context in place, we are now in a position to examine how life might have originated in these environments and then to investigate how scientists go about looking for evidence of life in the early rock record.

Further Reading

Books

Dalrymple, G.B. (1991) *The Age of the Earth*. California: Stanford University Press, Stanford.

Lunine, J.I. (1999) *Earth: Evolution of a Habitable World*. Cambridge University Press, Cambridge.

Papers

Albarède, F., Ballhaus, C., Blichert-Toft, J., Lee, C.T., Marty, B., Moynier, F., Yin, Q.Z. (2013) Asteroidal impacts and the origin of terrestrial and lunar volatiles. *Icarus* **222**, 44–52.

Altwegg, K., Balsiger, H., Bar-Nun, A., Berthelier, J.J., Bieler, A., Bochsler, P., Briois, C., Calmonte, U., Combi, M., De Keyser, J., Eberhardt, P., Fiethe, B., Fuselier, S., Gasc, S., Gombosi, T.I., Hansen, K.C., Hässig, M., Jäckel, A., Kopp, E., Korth, A., LeRoy, L., Mall, U., Marty, B., Mousis, O., Neefs, E., Owen, T., Rème, H., Rubin, M., Sémon, T., Tzou, C.-Y., Waite, H., Wurz, P. (2014) 67P/Churyumov–Gerasimenko, a Jupiter family comet with a high D/H ratio. *Science* doi: 10.1126/science.1261952

Halliday, A.N. (2008) A young Moon-forming giant impact at 70–110 million years accompanied by late-stage mixing, core formation and degassing of the Earth. *Philosophical Transactions of the Royal Society* **366**, 4163–4181.

Harrison, T.M. (2009) The Hadean crust: evidence from >4 Ga zircons. *Annual Reviews of Earth and Planetary Science* **37**, 479–505.

Hazen, R.M. (2013) Paleomineralogy of the Hadean Eon: a preliminary species list. *American Journal of Science* **313**, 807–843.

Hopkins, M., Harrison, T.M., Manning, C.E. (2008) Low heat flow inferred from >4 Gyr zircons suggests Hadean plate boundary interactions. *Nature* **456**, 493–496.

Kasting, J.F. (2005) Methane and climate during the Precambrian era. *Precambrian Research* **137**, 119–129.

Kasting, J.F., Zahnle, K.J., Pinto, J.P., Young, A.T. (1989) Sulfur, ultraviolet radiation and the early evolution of life. *Origins of Life and Evolution of the Biosphere* **19**, 95–108.

Schaefer, L., Fegley, B. (2010) Chemistry of atmospheres formed during accretion of the Earth and other terrestrial planets. *Icarus* **208**, 438–448.

Valley, J.W., Peck, W.H., King, E.M., Wilde, S.A. (2002) A cool early Earth. *Geology* **30**, 351–354.

Wood, B.J., Walter, M.J., Wade, J. (2006) Accretion of the Earth and segregation of its core. *Nature* **441**, 825–833.

Yang, X.Z., Gaillard, F., Scaillet, B. (2014) A relatively reduced Hadean continental crust and implications for the early atmosphere and crustal rheology. *Earth and Planetary Science Letters* **393**, 210–219.

Zahnle, K., Schaefer, L., Fegley, B. (2010) Earth's earliest atmosphere. *Cold Spring Harbour Perspectives in Biology* **2**, a004895.

11

The Origin of Life

Learning Outcomes

➤ Understand the history of ideas about the origin of life.
➤ Understand the types of carbon in the interplanetary environment and its potential delivery to early Earth.
➤ Understand some plausible pathways to the production of organic matter on early Earth.
➤ Describe the RNA world and some of its purported features.
➤ Describe the putative nature of early cells.
➤ Discuss the controversies in ideas about the pathways for the origin of life.

11.1 Early Thoughts on the Origin of Life: Spontaneous Generation

We have investigated how carbon molecules can be synthesised in the interstellar medium, suggesting that carbon chemistry is widespread in the Universe. With a knowledge of astrochemistry, an understanding of the basic atomic and molecular structures of life and an insight into the environmental conditions in the first billion years of Earth history, we can proceed to ask one of astrobiology's most fundamental questions – how and where did life originate? This question is the subject of this chapter.

To begin with, it is worth turning to some history to understand past ideas about life and how people thought it was begat from the abiotic world. One line of thinking, 'materialism', was accepted by the ancient Greeks: that physical processes in the Universe are accounted for by the structure of matter. This line of thinking was most forcefully developed by the atomists, such as Democritus (ca. 460–370 BC), who believed that all matter was composed of particles. For them, the soul, which was a kind of fire made from atoms, was part of living things. Similar ideas were developed by ancient Indian and Islamic thinkers. This school of thought was a departure from previous ideas which invoked more superstitious agents such as spirits, gods and forces that imbued matter with certain properties.

Once chemical elements had been discovered, however, it became increasingly difficult to describe, from a physical basis, the part of life that somehow made it different. In the early years of the European Enlightenment in the seventeenth century the concept of 'vitalism' regained ground: that life was infused with a vital force that transforms dead into living things.

Alongside these changing ideas ran the influential notion of **spontaneous generation**, a two-millennia-old concept whereby **abiotic** material could be transformed into living matter. This process occurred by the infusion of abiotic material with 'vital heat' through the combination of environmental conditions to which matter was subjected. Spontaneous generation explained how mice emerged from wheat and maggots from meat. Many sixteenth century texts provide examples of strange experiments whereby wheat husks mixed with old underwear in a jar could be used to generate mice.

Astrobiology: Understanding Life in the Universe, First Edition. Charles S. Cockell.
© 2015 John Wiley & Sons, Ltd. Published 2015 by John Wiley & Sons, Ltd.
Companion Website : www.wiley.com/go/cockell/astrobiology.

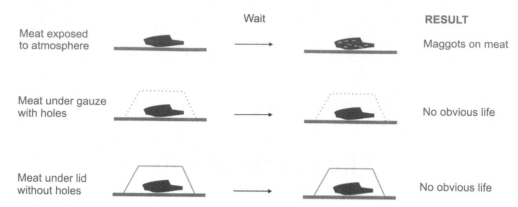

Figure 11.1 A schematic showing Francesco Redi's elegant experiment to demonstrate that spontaneous generation was not responsible for the formation of maggots in meat (Source: Charles Cockell).

Francesco Redi (1626–1697), an Italian physician and naturalist, was one of the first people to challenge spontaneous generation with an elegant experiment using meat. He showed that if meat was covered by a lid, no maggots would form. But he also showed that if it was covered by gauze that had holes to allow in the 'vital force', but were too small to allow in flies, maggots would still not be generated. His control experiment was meat exposed fully to the atmosphere, in which maggots appeared. His experiment (Figure 11.1) seems absurdly simple today, yet at the time it elegantly showed that a vital force did not account for the appearance of maggots in meat.

These data provided some of the first insights into the fact that maggots must be produced by a vector of some sort, determined to be flies.

However, although spontaneous generation was disproven for flies, it wasn't going to disappear that easily. Turbevill Needham (1713–1781), a well-respected scientist of his time, reported the famous Mutton-Gravy experiment. He had taken some of his left-over gravy and heated it in a fire. Transferring this into vials with stoppers he reported how the gravy, after it had been sealed from the outside world, teemed with microscopic life, regardless of whether it had been heated or not. Thus, he surmised, spontaneous generation was proven. The organic matter of the gravy had been infused with a 'life-force'.

We now know that it was likely that his gravy either became contaminated with microbes in the air after it was taken from the fire to be poured into the vials or that it contained spore-forming bacteria (Chapter 7).

Italian scientist Lazaro Spallanzani (1729–1799), who is more famously known for his pioneering studies on the regeneration of organs in animals like frogs, was the first to take a good shot at the theory. He did the same experiments as Needham using broths, but he was more careful. He put wetted seeds used to make his broths in the vials first, sealed them and then heated them to kill anything in them. In this way particles from the air could be stopped from contaminating the infusions. In vials heated for short periods, large organisms died very quickly (we now think these must have been **protozoa**, which are large amoebae. They are not tolerant of heat). He noticed that organisms of the 'lower class', as he called them, could tolerate heat for many minutes until even they no longer moved. These were almost certainly bacteria. He had in this important experiment shown that microbes are differentially affected by heat, some being more susceptible than others. He finally showed that if his vials were heated for long enough then they could be turned, in his words, into 'an absolute desert'. He had demonstrated the concept of sterilisation.

Now you might think that this remarkable set of experiments would finally end the idea of spontaneous generation, but not so. By sealing his vials, so his critics said, Spallanzani had denied the organic matter contact with the atmosphere and hence any 'life-force'. Remarkably it would be 38 years before an Austrian scientist made an experimental apparatus to take this new criticism on.

One way to provide air to a broth without introducing microbes is to heat the top of the tube where the air gets in. The simple addition of an open tube heated with a flame underneath to sterilise it allowed Theodore Schwann (1810–1882) to provide clean air to his boiled vials. In 1837 he published his results in the *Annals for Physics and Chemistry* and they were conclusive.

One could become somewhat exasperated, because still the theory of spontaneous generation would not go away. Others had difficulty reproducing Schwann's experiments and after boiling and allowing through heated air, they still found life emerging in their vials. This inconsistency in the results fuelled the continuity of the debate.

Schwann's observations were confirmed by Frenchman Charles Cagniard de la Tour (1777–1859), who also performed experiments on how yeast grew. In 1838, in the *Annals of Chemistry and Physics*, he reported on some fundamental discoveries of importance to astrobiology and the limits of life.

In his work he made the first observations that yeast could survive freezing – a notable difference from many higher animals and a fact that would later become important for understanding how life can survive in permafrost and in polar environments. This observation still looms large in considerations of how life might survive in freezing extraterrestrial environments, if it is there. He deserves the recognition of almost certainly being the first person to show that microbes can grow without oxygen. He showed that yeasts could cause fermentation in an atmosphere of carbon dioxide. His observations did not stop there. He noticed that yeast does not die in the absence of water. Thus, he had documented freezing tolerance, desiccation tolerance and oxygen-free growth of microbes.

Despite the wonderful observations that were being made with yeast, the **germ theory** was not going to be accepted without a little bit of further struggle. German scientist Justus von Liebig (1803–1873) was not impressed. As a chemist he was convinced that all reactions attributed to these new-fangled organisms were really chemical and you could explain them without having to invoke biology. Writing in 1839 he asserted bravely 'The ability to cause fermentation ... does not depend on an effect brought about through contact with the yeast'. This view would permeate his writings and lectures. His insistence that chemistry was all that there was to fermentation would muddy the waters. Like a lot of science though, he was not entirely wrong.

In the late 1890s, Eduard Buchner (1860–1917) did in fact show that if you took the extract from yeast you could get sugar to ferment. We now know that he had liberated the enzymes needed for fermentation, even if the yeast cells themselves were dead. In some ways he confirmed Liebig's ideas; he had shown that chemical reactions were sufficient for fermentation, it's just that they were chemical reactions from living things. Buchner might be described as having contributed to founding the science of biochemistry – the science of studying chemical reactions that occur within life. However, Liebig's focus purely on chemistry as the cause of fermentation would weaken those seeking to get acceptance of germ theory for a while.

It is at this stage in the story that the now famous Frenchman Louis Pasteur (1822–1895) comes onto the stage. He had a mind of wonderful clarity when it came to planning experiments. He began to publish observations on the growth of yeast. He would later pioneer the process of Pasteurisation – where the rapid and short-lived heating of milk could kill off microbes without changing its taste, and so help preserve it. As his scientific momentum got going he began to challenge some of the fundamental beliefs of his day. There was no greater irritant to him and his fellow scientists than spontaneous generation.

His response to this centuries old question was an experiment of ingenious simplicity. He invented his famous swan-neck flasks (Figure 11.2). These flasks contained a variety of liquids including yeast extracts, pepper water and urine. He heated them until they boiled for several minutes. Then he cooled them. The swan neck flasks were a very simple alteration of the flasks used by Schwann. The swan necks prevented the aerially transported entities from entering the broth. In contrast, flasks in which the swan neck was removed were rapidly contaminated and grew these microbes. This experiment showed that broths were contaminated by organisms in the air and that life could not spontaneously be generated in a sterile medium. He concluded his remarkable paper in which he reported the results with a sentence of extraordinary clarity and importance – 'There exist continually in the air organised bodies which cannot be distinguished from true germs'. With this sentence microbiology had truly been propelled to centre stage in medicine and our understanding of the environment. Pasteur's experiment was a response to a French Academy of Sciences prize in 1859 that challenged scientists to disprove spontaneous generation.

Despite the end of spontaneous generation as an idea for the formation of life in a continuous process on the Earth, we know that it must have happened once. When life first arose there was a transition from abiotic chemistry to life. So the problem of spontaneous generation still hasn't really gone away. However, to avoid confusing it

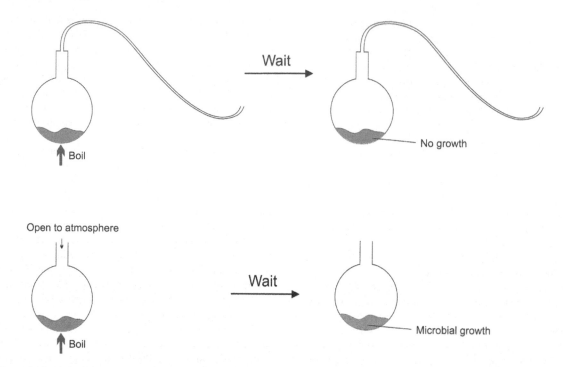

Figure 11.2   *Louis Pasteur's swan-neck flasks and his experiment to disprove spontaneous generation. The swan necks prevent microbes from dropping directly into the flasks after the broth has been heat-sterilised. Flasks with open necks become colonised and microbial growth occurs* (Source: Charles Cockell).

with the old obsolete ideas of spontaneous generation, this transition is often referred to today as abiogenesis. How did simple molecules come together to form more complex ones and ultimately a self-replicating organism?

11.2   Some Possible Ideas for the Origin of Life

There are a number of over-arching theories to account for the origin of life. One is that it came from somewhere else (**panspermia**). Of course this doesn't solve our problem since it pushes the problem off to somewhere else. Even if we were to find evidence that this was the case, we would then be faced with the problem that we would have to find out where and in what environment the origin of life occurred. If it happened on some distant planet around another type of star then we are going to find it very difficult to replicate or study the possible environment for the origin of life. Nevertheless, the possibility that life can be transferred between planets will be discussed later.

Another possibility is that it was created by some higher order being or entity, that could be a God, if you subscribe

to such ideas, or an alien race. However, although I won't spend time in this book attempting to refute these ideas, they are not very valuable, because they are experimentally intractable. They still don't escape the question of how the first molecules came together to form self-replicating organisms that were the ancestors to life on the Earth, whoever or whatever decided to do this.

So regardless of what our view is on the process and environment for the origin of life, the only experimentally tractable approach to the problem at the current time (and taking into account that we think such an environment requires an available liquid) is to assume that it occurred in an Earth-like environment and then use our knowledge of the early Earth and organic chemistry in the Universe to attempt to suggest pathways to this origin. This is the approach this chapter will discuss.

11.3   The Synthesis of Organic Compounds on the Earth

The question of how life arose has intrigued biologists for a long time. Famously in 1871, Charles Darwin

(1809–1882) wrote a private letter to his friend Joseph Hooker (1817–1911) in which he said: 'But if (and Oh! what a big if!) we could conceive in some warm little pond, with all sorts of ammonia and phosphoric salts, light, heat, electricity etc., present, that a protein compound was chemically formed ready to undergo still more complex changes, at the present day such matter would be instantly devoured or absorbed, which would not have been the case before living creatures were formed'.

Darwin's speculation was significant because he had identified that the origin of life required a specific environment that would provide an energy source and the appropriate environment for chemical syntheses to occur. He also, intriguingly, was the first person to point out that an origin of life might be difficult to find today because of the presence of abundant life (heterotrophs) that would be likely to eat the organic material being produced.

In the early twentieth century, Aleksandr Oparin (1894–1980) and John Haldane (1892–1964) suggested independently that the ocean might have been a giant prebiological reactor, with complex organics being synthesised as the aqueous environment was bombarded with UV radiation. Oparin proposed that organic aggregates, which he called **coacervates**, were formed and they developed to do successively more complex chemistry. Eventually they resembled early cells. This Oparin–Haldane hypothesis set the stage for early ideas on the origin of life.

The first successful experiment to try to answer the question of whether organics could be synthesised in an early Earth-like environment was carried out by Stanley Miller (1930–2007) and Harold Urey (1893–1981) in 1953. This experiment used an artificial early Earth atmosphere to investigate what would happen when the mixture was reacted in the presence of an electrical discharge.

The experimental set up (Figure 11.3) was quite simple. A container with water simulates the early ocean. It is heated so that water vapour enters the artificial

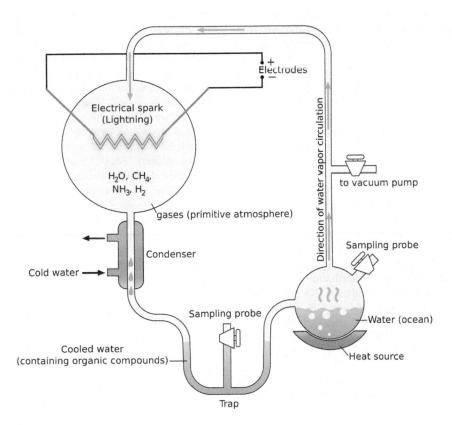

Figure 11.3 *A schematic showing the apparatus involved in the Urey-Miller experiment* (Source: wikicommons, YassineMrabet).

Table 11.1 *Just a few of the compounds generated in the Urey–Miller experiment by sparking CH₄, NH₃, H₂O and H₂.*

Compound	Yield (%)
Glycine	2.1
Glycolic acid	1.9
Alanine	1.7
Lactic acid	1.6
N-Methyl alanine	0.07
α-Amino-*n*-butyric acid	0.34
α-Aminoisobutyric acid	0.037
β–Alanine	0.76
Succinic acid	0.27
Aspartic acid	0.024
Glutamic acid	0.051
Iminodiacetic acid	0.37
Formic acid	10
Acetic acid	0.51

'atmosphere' where it is subjected to an electrical discharge in a mixture of methane, ammonia and hydrogen. The vapour from this process is then cooled using a condenser and collected in a trap at the bottom of the apparatus. The result was the formation of a diversity of molecules, including amides, carboxylic acids and, crucially, amino acids such as alanine, aspartic acid and others (Table 11.1). Absent were the sulfur-containing amino acids, but this was not surprising as the Urey–Miller experiment did not contain sulfur-containing gases (e.g. H_2S). When H_2S is included, the sulfur-containing amino acids are produced. The aromatic amino acids (tryptophan, tyrosine, phenylalanine and histidine) were not produced, which might be explained by the difficulty of synthesising ring structures in the gas phase reactions in the Urey–Miller experiments. It might suggest that these amino acids came from other reactions or environments.

The mixture that Urey and Miller used was very reducing and today we think the early Earth atmosphere was not so reduced. The atmosphere contained CO_2 but probably had lower concentrations of H_2 and CH_4 compared to the experiment, and higher concentrations of N_2. A good yield of organic compounds from gas phase chemistry generally requires a CH_4 to CO_2 ratio greater than 1. More modern experiments, however, have confirmed the idea that amino acids can be produced in artificial early Earth atmospheres that are more oxidised. Experiments using CO_2, N_2 and H_2O in the atmosphere (with iron present and

calcium carbonate that buffers the pH to prevent amino acid destruction from acidity) produce amino acids. Although these experiments successfully demonstrate the idea that simple chemical compounds, when supplied with energy, could be reacted into compounds of prebiological significance, it is now also thought that other locations (such as hydrothermal vents, as we will discuss later) may have been favourable sites for prebiotic chemistry in addition to gas phase reactions in the atmosphere.

The results of the Urey–Miller experiment actually bear striking resemblance to the data from the rest of the Solar System. The lack of organic molecules in the atmospheres of Venus and Mars, which are dominated by CO_2 and N_2, is in stark contrast to the hydrocarbon haze of Saturn's moon, Titan. High concentrations of methane play a major role in atmospheric chemistry on this moon and, under irradiation, produce a variety of compounds including HCN, ethane, propane, benzene and more complex hydrocarbons whose characteristics have not yet been identified.

The Urey–Miller experiment focused on electrical discharges, but energy could also have been provided by UV radiation, geothermal heating, impact shock and cosmic rays. The greatest *continuous* source of energy on the early Earth, as it is today, would have been sunlight. Many experiments have been conducted since Miller's day using short wavelength UV light and different initial gas mixtures to examine the yields of organic compounds under the theoretical conditions of the early Earth. They have shown the production of organic molecules.

Another source of energy on the early Earth was that associated with shock pressures induced by asteroid and comet impacts. Shock experiments in the laboratory have been shown to produce HCN, aldehydes and amino acids about a million times more efficiently than UV radiation. Infrequent large impacts may have generated large quantities of prebiotic materials. Similarly to the Urey–Miller experiment, impact shocks are far more efficient at producing prebiotic compounds such as amino acids in reduced atmospheres compared to more oxidised atmospheres.

11.3.1   Possible Reaction Pathways

Nowadays extensive organic chemistry experiments have yielded some quite satisfactory ideas about how the earliest molecules or 'building blocks' of life might have arisen. Hydrogen cyanide (HCN), thought to have been an important component of the inventory of early Earth molecules (and it has been detected in space), can react to form a number of important compounds. For example, in the **Strecker synthesis**, named for Adolph Strecker (1822–1871) a German chemist who studied amino acids, the reaction of HCN with ammonia and an aldehyde, followed by hydrolysis, leads to the formation of an amino acid as shown below:

$$R\text{-CHO (aldehyde)} + HCN + NH_3$$

$$\rightarrow R\text{-CH}(NH_2)\text{-CN} + H_2O \qquad (11.1)$$

Followed by hydrolysis:

$$R\text{-CH}(NH_2)\text{-CN} + H_2O \rightarrow R\text{-CH}(NH_2)\text{-CONH}_2 \quad (11.2)$$

$$R\text{-CH}(NH_2)\text{-CONH}_2 + H_2O$$

$$\rightarrow R\text{-CH}(NH_2)\text{-COOH (amino acid)} + NH_3 \quad (11.3)$$

The problem with this reaction is that it requires the full range of aldehydes to form the corresponding different amino acids required for life. It has yet to be shown that these reactions can occur under natural environmental conditions at significant yields. Nevertheless, Strecker synthesis provides a plausible pathway and could be one route for the formation of amino acids in the protoplanetary disc and in material that eventually becomes incorporated into carbonaceous meteorites such as **carbonaceous chondrites**.

Chemical reactions have been identified that could have provided early sugars. The polymerisation of formaldehyde (HCHO), analogously to HCN, can lead to the formation of sugars through the **formose reaction**. The reaction begins with two formaldehyde molecules condensing to make glycolaldehyde, which further reacts with another formaldehyde molecule to make glyceraldehyde. An **isomerisation** of glyceraldehyde forms dihydroxyacetone, which can react with glycolaldehyde to form the sugar, ribulose and through another isomerisation, the crucial sugar for nucleic acids, ribose. Dihydroxyacetone can also react with formaldehyde to produce the sugars, tetrulose and aldoltetrose. These reactions require high concentrations of formaldehyde and alkaline conditions. The abundance of these on early Earth is not known. Although formaldehyde can be formed in the gas phase (such as, e.g., through the production of radicals generated by the action of short-wavelength ultraviolet radiation on CO_2 and H_2O), it is thought that land-based environments, where high concentrations of formaldehyde could accumulate to drive the formose reaction, are a more likely location.

Hydrogen cyanide (HCN) is a rather ubiquitous participant in prebiotic reaction sequences and not only does it play a role in proposed pathways of amino acid and sugar formation, but it may have been involved in the production of nucleobases. HCN can condense with other HCN molecules to form diaminomaleonitrile. This molecule can react with yet more HCN to form adenine. The concentrations of HCN may need to have been high for yields to be large, but conceivably this could have occurred in evaporated water bodies or even seasonally frozen ponds in which freezing would drive the concentration of compounds in inter-grain spaces. Reactions of HCN in ammonia can produce guanine and cyanoacetylene, another CN-containing compound that is a precursor to cytosine. Mixed with phosphates and sugars and exposed to dehydration reactions, these components can yield numerous nucleosides and nucleotides, the building blocks of the nucleic acids.

Yet another means to achieve polymerisation and subsequent hydrolysis into smaller prebiological molecules is to return to the Urey–Miller type environment. When mixtures of gases, particularly reduced gases, are exposed to UV radiation or ionising radiation, one notable product is a brown-orange-reddish substance called 'tholins' by Carl Sagan (1934–1996) and Bishun Khare (1933–2013), after the ancient Greek word for muddy (*tholos*). Tholins are a variety of highly polymerised organic molecules of undefined composition (they vary depending on irradiation conditions and initial gas composition), which can hydrolyse (react in water) to yield a variety of amino acids and other compounds.

One question is how nitrogenous compounds would have been formed on the early Earth. The nitrogen-nitrogen bond in N_2 gas is very strong and not easily pulled apart. In the prebiotic era, before biological nitrogen fixation, how was such a transformation achieved? There are a number of possible solutions. Lightning can split N_2. In an early CO_2-dominated atmosphere, it is suggested that 3×10^9 to 10^{10} moles of NO could have been formed each year. The production of nitrogen oxides can also occur in impact events as the energy imparted to the atmosphere during atmospheric entry and during collision can raise local temperatures to ∼1500 K, sufficient for these reactions. Ammonia may also have been available from the primordial inventory of volatiles. It seems therefore, that even in the absence of biological nitrogen fixation, there are routes to the production of fixed nitrogen compounds that would then be able to play a role in prebiotic chemistry.

Debate Point: Is the origin of life inevitable?

Wherever there are suitable hydrothermal sites or other environments for the formation of organic compounds, is an origin of life inevitable? This question is a critical one for astrobiology because if it is inevitable then the statistical chances of life elsewhere are very high. If it is an unusual event, then life may be rare. If the origin of life is unusual it would mean that even if the number of habitable planets was enormous, the statistical chances of life may not be dramatically improved. Discovering more habitable planets may mean little for the probability of finding life if a crucial reaction required for the assembly of a component of life is a chemically extraordinarily rare event and required a very specific environment. What do you think? Is the origin of life inevitable and, when it does occur, how long do you think it needs? It is not expected that you can find a definitive answer to this question! However, a discussion about it can stimulate a variety of in-depth discussions on the reactions required for the origin of life.

11.4   Delivery from the Extraterrestrial Environment

We saw in Chapter 9 that complex organic compounds can be formed in the interstellar environment in a variety of locations from molecular gas clouds to protoplanetary discs. Evidence that prebiological compounds could have come from space and been delivered to the surface of the early Earth is supported by the study of meteorites.

One of the most intensively studied of these meteorites is the Murchison meteorite (Figure 11.4), a type of **carbonaceous chrondrite**, a fragment of an asteroid from the asteroid belt between Mars and Jupiter containing material from the earliest Solar System.

Carbonaceous chondrites have been shown to contain amino acids (Table 11.2), including those found in biological systems on Earth such as proline and valine and over 50 non-protein amino acids, including norvaline and α-aminoisobutyric acid. Approximately 70 different

Table 11.2   *Some examples of carbon compounds identified in carbonaceous chrondrites (from Sephton, 2002).*

Compound	Concentration (ppm; unless specified)
Macromolecular material	1.45%
Amino acids	60
Purines	1.2
Pyrimidines (uracil and thymine)	0.06
Ketones	16
Aldehydes	11
Alcohols	11
Aliphatic hydrocarbons	12–35
Aromatic hydrocarbons	15–28
Sulfonic acids	67
Phosphonic acids	1.5
Carboxylic acids	372.3
Sugar-related compounds	~60

amino acids have been characterised with a total abundance of about $10-60\,\mu g\,g^{-1}$ meteorite. They are part of the soluble carbon fraction of meteorites, which makes up about 30% of the organic carbon in these bodies, the rest of it accounted for by highly complex refractory compounds including PAHs that were discussed in Chapter 9. The contribution of amino acids to the total organic carbon in meteorites is nevertheless low.

The structure of these amino acids tells us something about their mode of formation. The majority of biological amino acids are α-amino acids, meaning that the amine group (NH_2) in the amino acid is on the α carbon (Figure 11.5). However, only 65% of extraterrestrial amino acids in meteorites are in the α form. The remainder are in the β, γ and δ positions. These observations suggest that amino acids were synthesised in the early protoplanetary disc by broadly non-selective methods,

Figure 11.4   *The Murchison meteorite, which fell to Earth in 1969 in the town of Murchison, Australia* (Source: U.S. Department of Energy).

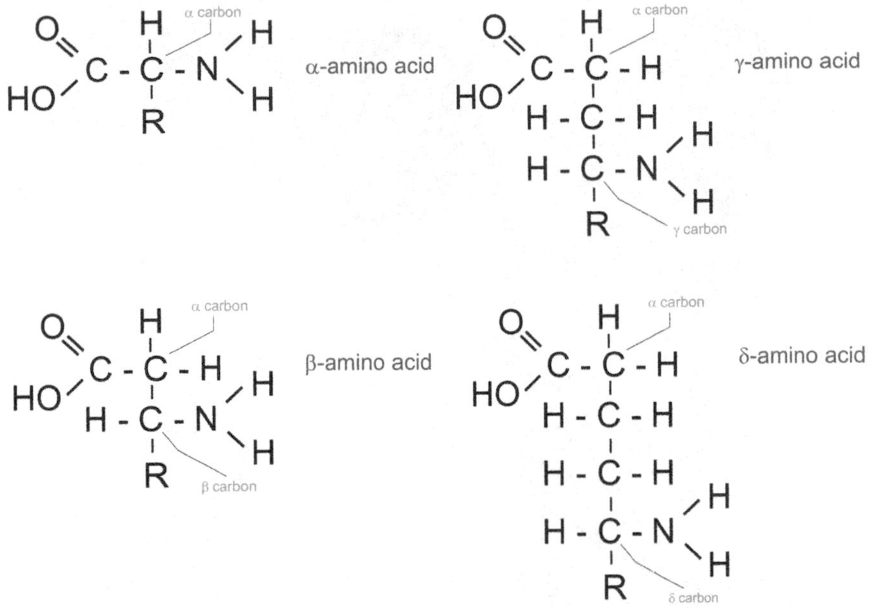

Figure 11.5 The structure of α-, β-, γ- and δ-amino acids (Source: Charles Cockell).

such as the Strecker synthesis discussed earlier, leading to the vast range of compounds that are observed within the meteorites. An enduring question is why life has used only a portion of these amino acids and what factors led to their selection, if indeed an origin of life had available the full diversity of these extraterrestrial amino acids.

Attempts have been made to find peptides in meteorites. Glycine–glycine peptides have been reported, but the lack of other peptide bonds between other amino acids that have been detected (for example between glycine and alanine) suggests that Gly-Gly peptides were formed by dehydration reactions rather than co-polymerisation during their formation. No proteins have yet been detected in meteorites.

The amino acids within meteorites have also been suggested to have an enantiomeric excess of L-amino acids, similarly to terrestrial life. To avoid the ubiquitous problem of the contamination of the meteorites with organics in the terrestrial environment, Pizzarello and Cronin (2000) published a paper in which they studied the α-methyl amino acids in meteorites that are rare in the biosphere. They found that in meteorites there was an enantiomeric excess of L-amino acids of between 1 and 9%, suggesting that amino acids in meteorites could have provided the chiral enrichment in early life that was amplified by chemistry and biology.

In addition to amino acids, nucleobases have been detected in meteorites, for example guanine and adenine, suggesting that the molecules of the genetic material could also have come from space.

A remarkable number of other compounds have been found in meteorites, including carboxylic acids, sulfonic and phosphonic acids, amines, amides, alcohols and sugars.

The data that have been gathered to date suggests that exogenous sources of building blocks could have helped the origin of life and that there are a number of pathways by which the earliest molecules might have been available for early cells to be assembled.

The extent of this delivery could have been significant. At about 4 Ga ago it is estimated that the delivery of organic material could have been about 10^5–10^6 kg year^{-1} from comets, about two orders of magnitude less for meteorites, but 10^8 kg year^{-1} or more from interstellar dust, with this quantity declining as the impact flux declined. The total estimated extraterrestrial delivery of all material to the Earth by 3.9 Ga ago could have been on the order of 10^{16} to 10^{18} kg of material (compare this to the total organic carbon in life on Earth today, which is on the order of 6×10^{14} kg). The vast majority of this delivery peaks at the size range of about 100 μm as Interplanetary Dust Particles (IDPs). Once the size range reaches centimetre

Figure 11.6 *RNA can fold into complex structures, such as here seen for a tRNA molecule from yeast. The inset shows the 2-D schematic of its shape* (Source: wikicommons, Yikrazuul).

to metre scales the objects are in the realms of meteoritic material.

11.5 The RNA World

A question in the origin of life remains. How did early molecules come together into a self-replicating organism? One clue that was suggested in the 1980s and led to the award of the 1989 Nobel Prize to Sidney Altman and Thomas Cech was the result of their discovery that ribonucleic acid, RNA (Chapter 3), can fold in such a way as to become a biological catalyst, like an **enzyme**. Their discovery led to the idea of the **RNA World**, a prebiological world dominated by RNA.

RNA is generally found as a single-stranded molecule, but it can form double-stranded RNA when the base pairs are complementary, leading to the formation of bulges, loops and hairpins and resulting in the three-dimensional folding of the one-dimensional RNA

molecule (Figure 11.6). tRNA is an example of such a molecule (Chapter 4).

Altman and Cech showed that small fragments of RNA can catalyse chemical reactions, later shown to include their own replication. These fragments can be as small as five base pairs. These **ribozymes** (Figure 11.7) may be a relic from a time when RNA dominated biochemistry and could reflect the first biochemical reactions without proteins.

Ribozymes are chemically reactive because they have a reactive hydroxyl group at the 2′ position in the ribose sugar, hence the name, ribonucleic acid, as opposed to *deoxy*ribonucleic acid for DNA which lacks this oxygen. This relative reactivity, and therefore in some sense instability, of the RNA molecule is one reason why it is proposed as a precursor to DNA.

As with all origin of life theories, a perennial question is how these RNA molecules became sufficiently concentrated to carry out chemistry and eventually replication into large numbers of ribozymes in small volumes that could then go on to do more complex chemistry. One

number of experiments in different RNA sequences and conformations (shapes) whereby the fastest replicating ones would have outcompeted the ones nearby. A type of chemical Darwinian evolution might have got going whereby every RNA molecule that replicated faster than the others would have dominated the local RNA pool and thus have been the template from which further mutations would have occurred.

The existence of RNA and DNA in the modern world should not prejudice us to discount the possibility of other types of nucleic acid that could have preceded RNA and DNA and have long since disappeared. A dizzying array of nucleic acids has been synthesised. Some of them have some plausible advantages to RNA and may therefore have been precursors to RNA. One such is PNA (peptide nucleic acid) in which the bases are attached to a polyamidic backbone. It has been shown to form double helices and can serve as a template for RNA synthesis. It is not chiral and so avoids the need for chiral selectivity that would be required by the architecture of life as we know it. TNA (threose nucleic acid) contains threose, a sugar with a simple four-membered ring. TNA can bind to DNA and RNA showing its compatibility with modern genetic material. However, the sugar is prone to instability. Other suggestions have included HNA (hexitol nucleic acid) and ANA (altritol nucleic acid), which are capable of forming helices and/or binding with RNA, DNA or short strands of nucleotides. The biochemistry of these alternative structures is still not fully understood, but even if they did not exist on the early Earth, they offer model systems with which to investigate how early biochemical systems might have evolved.

It has not been lost on biochemists that many of the nucleobases are found in multiple locations in biochemistry. For example, ATP is used both as a molecule for transporting energy and as a building block for DNA assembly (Chapter 4). Adenine turns up in a variety of **coenzymes** and precursors in the biosynthesis of molecules. Even guanine is to be found in the energy-rich molecule, guanosine triphosphate (GTP), which plays a role, like ATP, in some reactions of energy transfer. It seems very likely that the pervasive presence of nucleobases and nucleotides throughout biochemistry reflects an early stage in chemical evolution when energy production and information storage emerged from the same suite of molecules.

A question that remains unresolved in astrobiology is: which came first, metabolism or genes? As protein synthesis, at least today, requires RNA, then one can imagine that RNA came first, with proteins being added to the RNA scaffold and resulting in more complex

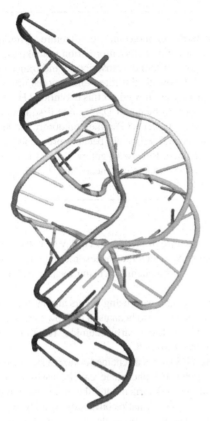

Figure 11.7 *A ribozyme. The hammerhead ribozyme which, in a manner similar to proteins, can catalyse reversible cleavage and joining reactions at specific sites within RNA molecules* (Source: wikicommons, William G. Scott).

solution is the binding of the RNA to mineral or clay surfaces such as the clay, montmorillonite. Surfaces can act as a concentration mechanism for the molecules and provided the binding strength to the material is not too high, they can peel away from the surface after replication. Essentially the surface becomes a scaffold to hold RNA in place while it codes for replica RNA molecules. Clays can themselves act as catalysts for chemical reactions and have been shown to assist the assembly of monomers into long chains of nucleotides. The second way to concentrate the RNA is within vesicles, which will be discussed in the next section.

Naturally, unaided by modern nucleic acid repair mechanisms found within cells, the **mutation** rate within the RNA would have been quite high, but on the flip side, these errors would have acted to generate a vast

---

Debate Point: What came first?

A perplexing question in the origin of life is what came first in terms of molecular architecture? Proteins are three-dimensional structures that carry out important functions. Yet at some point the information to make them was encoded in a one-dimensional linear strand of nucleic acid to achieve the system of transcription and translation we observe today. How did this happen? Did proteins evolve first? If they did, how did this three-dimensional information get encoded into a one-dimensional molecule and what intermediate processes allowed for this evolution? How would proteins have been replicated? To evolve in the other direction also poses problems. If DNA came first, how did the base pairs evolve to encode the three-dimensional information required for proteins? Did they randomly produce lots of protein structures until some of them happened to replicate faithfully and ultimately became prevalent? Discuss what we know about proteins and nucleic acids and their interrelationships. Review Chapters 3 and 4 if necessary. Consider the question: which came first, nucleic acid or protein, or did they emerge in concert, and if so, how?

Joyce, G.F. (2002) The antiquity of RNA-based evolution. *Nature* **418**, 214–221.

---

replicating entities. The ability for RNA to both carry out enzyme-catalysed reactions and store information makes it a suitable molecule as a candidate early system for self-replicating biochemistry. In contrast, although proteins have catalytic activity, it is difficult to see how they could replicate themselves and how they would pass on information to future generations of proteins that could be subject to evolutionary pressures and improvements.

11.6   Early Cells

At some point, biochemistry must have become enclosed in cell membranes. Membranes, as we saw in Chapter 4, perform the function of concentrating molecules within a small space and thereby enhancing chemical reactions. Without membranes, molecules would have a tendency to become diluted in water bodies. As we also discussed in Chapter 4, membranes are formed from **amphiphilic** molecules that are hydrophobic at one end and hydrophilic at the other (fatty acids or lipids). When added to water, fatty acids, that exhibit these characteristics, have a tendency to spontaneously arrange in such a way that they form lipid bilayers or more simple micelles, which are spheres with the hydrophilic heads pointed outwards and hydrophobic tails pointing inwards. The property that amphiphiles have of forming membranes accounts for the structure of organism membranes today.

It is a relatively simple matter to demonstrate the formation of vesicles, even using quite primitive materials. Carboxylic acids such as nonanoic and decanoic acid, extracted from the Murchison meteorite, have been shown to spontaneously form membrane structures when added to water. The fact that long chained-carboxylic acids from meteorites can form membranes might suggest that the structure of membranes today broadly reflects the mechanism by which the earliest membranes might have enclosed early biochemistry using locally available compounds.

Simple metabolisms using sugars and proteins in artificial vesicles have also been used to show how membranes could have enclosed early chemical reactions, whether RNA or protein-based.

Figure 11.8 shows a scheme for how a simple molecule such as glucose-1-phosphate can be assimilated across a membrane and turned into the long-chain molecule, starch, through enzymatic conversion by a glycogen phosphorylase. The large molecules are trapped in the vesicle and they become concentrated. The breakdown of these molecules by an amylase to the two-unit sugar, maltose, results in a product that is free to leave the cell. This simple type of experiment illustrates the general principle of resources entering a cell, waste products leaving and complex molecules accumulating inside a cell.

In systems analogous to that shown in Figure 11.8, experiments have been conducted in which an RNA polymerase was encapsulated within vesicles. The provision of ADP externally to the vesicles was sufficient for it to enter the vesicles and drive the production of RNA polymers inside the vesicles. **Polymerase Chain Reaction** has also successfully been accomplished in artificial vesicles. All of these experiments demonstrate the principle of encapsulation – quite simple vesicles made from lipids can act as a method of concentrating reactants sufficiently to allow for localised chemistry to occur in addition to providing a barrier to the external environment.

These experiments also demonstrate the principle that simple cellular structures can allow for growth and, in particular, nucleic acid experiments show the possibility for primitive replication of information in early cells.

Focus: Astrobiologists: Dave Deamer

What do you study? Self-assembly processes related to the origin of cellular life.

Where do you work? University of California, Santa Cruz.

What was your first degree subject? Chemistry.

What science questions do you address? How are self-assembly processes at the molecular level related to the origin of cellular life?

What excites you about astrobiology? Years ago I discovered that certain soap-like molecules called amphiphiles were present in the Murchison meteorite and formed cell-sized vesicles in water. If such amphiphilic compounds were available on the prebiotic Earth they could have assembled into the membranes required for primitive cells. We have now found that amphiphiles not only assemble into membranes, but the self-assembled structures can promote non-enzymatic synthesis of polymers like RNA. After being synthesised, the polymers are encapsulated in the vesicles to form protocells. These are not alive, but instead represent an early stage of evolution toward the first forms of life.

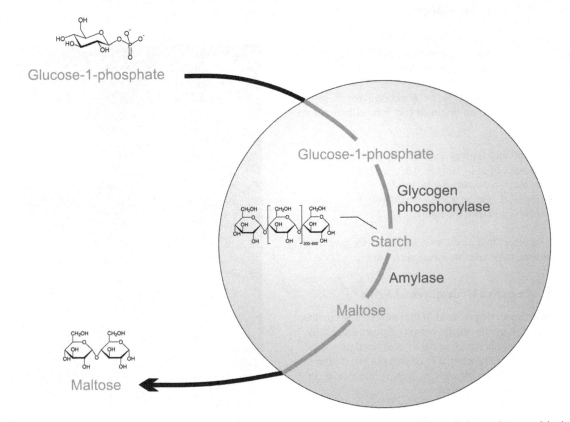

Figure 11.8   *Simple metabolisms could have occurred in enclosed membranes, illustrated here with the production of the long chain carbohydrate, starch, occurring within an enclosed early cell (Source: Charles Cockell).*

Lipids possess a variety of intriguing properties that are reflected in modern cells. Once vesicles are formed they tend to concentrate hydrophobic substances within them, offering a concentration mechanism and a means by which other molecules with useful functional properties can be incorporated into the structure. The exterior of lipid vesicles has been shown to be able to bind sugar molecules, leading to a carbohydrate 'coat', similar to a cell wall. These early cells could have been much smaller than modern prokaryotic cells because of their simple biochemical reactions and lack of bulky molecules like ribosomes, perhaps almost an order of magnitude smaller ($\sim$0.1 μm in diameter) than typical modern prokaryotes.

Once structures and early molecules became encapsulated, Darwinian evolution would have operated on the whole cell rather than on isolated molecules, resulting in selection pressures for the improvement of the whole cell. At this point, life had truly transitioned into a self-replicating entity defined by cellular processes.

11.7   Where did it Happen?

If we assume that life originated on the Earth, then one fundamental question is: where could this have occurred? There are a variety of hypotheses on possible locations for the origin of life.

For any environment to be conducive to chemical reactions needed for an origin of life, it must have three characteristics:

1. An available energy source to drive chemical syntheses.
2. A means of concentrating molecules.
3. A physical environment conducive to complex molecules and their assembly.

Several places that have these characteristics have been hypothesised to be suitable locations for life.

11.7.1   Deep Sea Hydrothermal Vents

Near mid-ocean ridges, cold water travels through the crust into the geothermally heated regions around the ridges and is expelled back into the oceans at high temperatures. During this process of hydrothermal circulation in the crust many chemical changes and interactions occur, which today include, in sequence:

1. Cold seawater sinks down through the crust.
2. O_2, K, Ca, SO_4 and Mg are removed from the fluid during interactions with rocks.
3. Na, Ca and K from the crust enter the fluid.
4. At high temperatures (350–400 °C), Cu, Zn, Fe, and H_2S from the crust dissolve in the fluids.
5. Hot and acidic fluids with dissolved metals rise up through the crust.
6. The hydrothermal fluids mix with cold, O_2-rich seawater. Metals and sulfur then combine to form metal-sulfide minerals, giving them their characteristic name of **black smokers** (Figure 11.9).

As the vents host geochemical disequilibria from these circulating fluids, in the present day they are home to a huge abundance of microorganisms, many of them carrying out chemoautotrophic metabolisms. On the early Earth, one hypothesis is that these geochemically active regions would have facilitated an origin of life.

The hydrothermal vents have been the focus of the hypothesis known as the 'iron–sulfur' world. This hypothesis assumes metal catalysis of the formation of organic molecules from simple inorganic gases and compounds such as CO, CO_2, HCN and H_2S using iron–sulfur minerals within the vent. At high temperatures (between

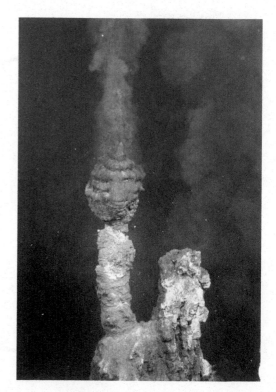

Figure 11.9   *A black smoker hydrothermal vent at the Brothers volcano, Kermadec Arc, 400 km off the coast of New Zealand* (Source: NOAA).

about 150–300 °C) carbon monoxide and hydrogen, for instance, can be reacted in **Fischer–Tropsch** syntheses using iron and nickel catalysts to produce organic molecules such as alkanes and other petroleum-related compounds (indeed, these pathways are used in industrial production of synthetic lubricants and oils). However, the notion of prebiotic syntheses at vents goes beyond the production of alkanes.

The original proposal, which was put forward by Günter Wächterhäuser, described the formation of primitive membranes from FeS (mackinawite), allowing for a semi-permeable barrier across which a proton motive force could have been established with the movement of electrons across the membrane, through primitive iron-sulfur clusters, such as **ferrodoxin**-like entities.

The reaction schemes proposed for an origin of life are quite speculative at the current time and although some of them are based on reaction pathways shown in the laboratory, they have not been directly demonstrated in hydrothermal vents.

For example, the following reaction is theorised at the vents:

$$CO + H_2S + 2H_2 \rightarrow CH_3SH + H_2O \qquad (11.4)$$

The methyl thiol (CH_3SH) produced is in turn converted catalytically with carbon monoxide in the presence of nickel and iron sulfides at ~100 °C (shown both theoretically and experimentally) to the methyl thioester of acetic acid:

$$CH_3SH + CO \rightarrow CH_3COSCH_3 + H_2S \qquad (11.5)$$

Reaction of the CH_3COSCH_3 so produced with H_2O yields acetic acid in a simple reaction proposed to represent the first stages of carbon fixation in early chemoautotrophic organisms:

$$CH_3COSCH_3 + H_2O$$
$$\rightarrow CH_3COOH(\text{acetic acid}) + CH_3SH \qquad (11.6)$$

The reduction of CO at high (>200 °C) temperatures and elevated pressures (~2000 bar) has also been shown to generate pyruvate, which it is proposed would provide the substrate for the precursors of the types of biochemical pathways discussed in Chapter 5, such as the reductive citric acid cycle.

The surfaces of iron–sulfur particles have also been proposed to provide an environment for the cyclisation reactions of HCN and related molecules (e.g. C_2H_2 and CH_3C_2H) to yield relatives and precursors to the nucleobases required for genetic material.

There is some circumstantial evidence that makes the idea of the hydrothermal origin of life attractive. Iron–sulfur proteins, including ferrodoxin, are ancient. They are ubiquitous, being found in various enzymes, including those involved in electron transfer (Chapter 5). They include NADH dehydrogenase, hydrogenases, various cytochromes, reductases, nitrogenase and the photosynthetic apparatus. These iron–sulfur clusters are suggested to be a relic of these ancient interactions between mineral surfaces and early biochemistry.

The processes proposed for hydrothermal vents have encouraged the debate about whether metabolism or nucleic acids came first. The potential for early metabolic reactions in vents has given support to the idea that basic metabolisms were first established and then encoded later in genetic material.

11.7.2 Land-based Volcanic Pools

Land-based environments share many of the characteristics of deep-sea hydrothermal systems. Heat from geothermal gradients, complex inorganic chemistry and mineral surface reactions (Figure 11.10) could have occurred in analogy to ocean hydrothermal systems. One obvious difference is that small volcanic pools may have localised chemicals in a smaller space than hydrothermal vents where molecules not retained within the vent structure will be diluted into ocean water. Land-based volcanic pools offer yet another form of geothermal gradient.

11.7.3 Impact Craters

During the collision of an asteroid or comet, an enormous thermal pulse is delivered into the local environment which can generate circulating hydrothermal systems. These hydrothermal systems are somewhat akin to volcanic hydrothermal systems. However, some features of impact craters stand out as different to volcanic hydrothermal systems and might make them suitable for the origin of life chemistries.

Impacts occur into all rock types (Figure 11.11). They are indiscriminate and so, unlike volcanic hydrothermal systems that are found in the specific chemistries of ocean crust environments such as volcanic rifts, impacts can establish hydrothermal systems in every rock type available. Thus, different hydrothermal systems produce different clays, secondary minerals and porosities, generating a whole diversity of experiments in the origin of life. They cool down at different rates, depending on the size of the impactor, therefore generating a range of hydrothermal systems of different sizes and with different cooling rates, again generating variety in rates

(a) (b)

Figure 11.10 *(a) Land-based geothermal (volcanic) pools, such as these pools in Kamchatka, Russia which are just over 1 m in length have been discussed as locations for organic syntheses. (b) Here, Dave Deamer from the University of California, Santa Cruz, samples them to study their potential for prebiotic chemistry* (Source: Dave Deamer, University of California Santa Cruz).

Figure 11.11 *The modern day Pingualuit crater in northern Quebec, Canada, containing a lake. The 3.4-km diameter crater is 1.4 million years old. No present-day crater contains an active, hot hydrothermal system.* (Source: NASA).

This idea is supported by data on the optimal temperatures for chemical reactions suggested for deep-sea hydrothermal vents. For example, at hydrothermal vents different compounds are predicted to be formed at different temperatures with ketones and alkenes preferentially formed at ~220, alcohols at ~150 and organic acids at ~60 °C. As an impact hydrothermal system cooled and went through these successive temperature regimes, from higher to lower temperatures, so different compounds could have been formed.

There are no craters on the Earth today with active hydrothermal systems, but there are craters that contain water bodies that give us a tantalising view of what such environments might have looked like (Figure 11.11).

11.7.4 Beaches

Beaches provide a mechanism for concentrating organic compounds by successive tidal motion over rocks. Energy in this system could come from seawater compounds or UV radiation impinging on the surface.

The notion of chemical reactions occurring in beach environments has been taken up in specific schemes for peptide and protein formation. One theory postulates that free amino acids on the early Earth (perhaps generated through the Strecker synthesis described earlier or delivered by meteoritic material) would have reacted with cyanate (OCN^-) ions to form N-carbamoyl amino acids. These would have dried out on beaches within or on the surface of rocks. In this dry phase they react with

of chemical reactions and the complexity of compounds. Impact hydrothermal systems have a defined lifetime as the crater cools. A crater with a diameter of about 5 km can sustain a hydrothermal system for several thousand years; a crater of 200 km diameter can host a hydrothermal system that persists for hundreds of thousands of years. We can imagine going from an early very hot stage in which simple prebiotic compounds could be formed to later stages where lower temperatures would be conducive to more complex molecules in different regions of the impact structure.

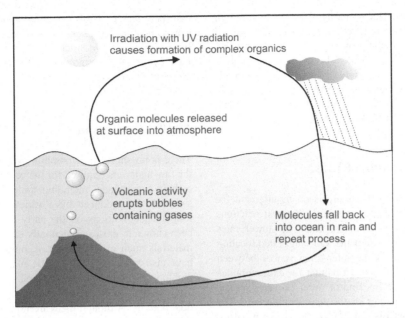

Figure 11.12   *Bubbles as a vehicle for prebiotic synthesis* (Source: Charles Cockell).

nitrogen oxides produced by lightning. This results in the formation of N-carboxy anhydrides (NCA) which either break down to amino acids, or, when the tide rises and submerges the NCA in neutral seawater, causes it to form polypeptides. This process might be repeated to generate proteins.

11.7.5   Bubbles

An intriguing hypothesis for the origin of complex molecules involves bubble formation in the oceans. In this scenario, simple organic compounds are trapped in bubbles from volcanoes, which are liberated from the ocean surface into the atmosphere (Figure 11.12). In the atmosphere they react with UV radiation in sunlight to form complex organic compounds which then rain back into the oceans and the cycle is repeated.

Over the course of several cycles, this irradiation and entrapment in bubbles provides both energy for the formation of complex compounds and a concentration mechanism (the bubbles) for more complex materials, and eventually protocells, to form. The proposed cycle bears some similarity to the way in which methane is processed into complex hydrocarbons high in Titan's atmosphere, as we shall see later (Chapter 18). Titan may be a large-scale organic chemistry laboratory for studying some of these processes (even if not life itself).

11.7.6   The Deep Sub-Surface

Subsurface life today constitutes a considerable portion of the biosphere. Some researchers have suggested that this environment would have been suitable for an origin of life. Energy would have been provided by **geothermal** heat. The concentration of molecules would have occurred within the pore spaces and fractures of rock deep underground, eventually allowing a variety of chemistries to be 'tested out' in diverse sub-surface environments, heat flows and chemical solution compositions. Eventually life would have moved to surface environments once it had evolved in the deep sub-surface.

11.7.7   Mineral Surfaces

Throughout these discussions there has been mention of mineral surfaces. Minerals have one major attraction – they can catalyse and concentrate. Some clays (such as montmorillonite) have been shown to catalyse the polymerisation of nucleotides into RNA polymers, and are therefore suggested as potential catalysts in an RNA world. Minerals such as pyrite have a strong propensity to bind cations, such as phosphate, and can therefore act to concentrate reactants. Low concentrations of phosphate are assumed on early Earth because of the low solubility of the primary phosphate containing mineral, **apatite**. The possibility that minerals would have concentrated phosphate on their surfaces, acting to create locally high concentrations for prebiotic reactions, has been suggested to give them a potential role in the origin and early evolution of life. Minerals have been shown to catalyse reactions that give higher yields of prebiotic compounds than non-mineral-containing experiments. For example borate minerals enhance the production of the sugar, ribose, from precursor formose-type reactants.

11.8 A Cold Origin of Life?

It is worth considering that there is no requirement for an origin of life to happen quickly. Chemical reactions do occur in cold environments, albeit at slower rates than hydrothermal environments. One suggested location for an origin of life is the inter-grain spaces between ice crystals, which today are a habitat for extremophiles (Chapter 7). Here, the freezing of water excludes solutes and produces salty water in the inter-grain space. The eutectic point of NaCl is $-21.2\,°C$ at 23.3% salt by mass, thus liquid water at below $0\,°C$ can exist in ice sheets. As the water freezes and excludes the salts, the concentration of organic molecules can occur. The energy for complexification to occur might come from geothermal energy (although at low flux) into the environment. The hypothesis has the problem that high salt concentrations are generally deleterious to complex molecules. However, other scenarios where prebiotic chemistry occurs in relatively fresh water, but at temperatures above freezing (but still in cold environments), have also been proposed.

The obvious problem is that chemical reactions follow an **Arrhenius relationship** whereby, as temperatures drop, chemical reaction rates drop off (about twofold for each drop of $10\,°C$). However, if the Earth was clement for an origin of life at about 4.4 Ga ago and life did not emerge until a few hundred million years after this, perhaps there was plenty of time for a cold origin of life.

11.9 The Whole Earth as a Reactor?

There is a tendency for scientists to favour particular reactions and environments as plausible locations for the origin of life. However, it is conceivable that the entire Earth was a giant pre-biotic reactor with organic chemistry occurring in many different places. Many of the hypotheses listed above need not be mutually exclusive. Organic chemistry may have been occurring in all of these environments. Nevertheless it still remains the case

that there was probably one environment where the first self-replicating organisms that gave rise to the diversity of life with its common biochemical architecture arose. The Last Universal Common Ancestor (LUCA) must have appeared somewhere.

11.10 Conclusions

There is no shortage of possible pathways for providing the raw materials required for the origin of life. Prebiotic carbon compounds, including amino acids, have been shown to be produced by a variety of plausible environmental conditions on the early Earth and they have been shown to exist in potentially exogenously delivered materials such as meteorites. A range of environments have also been suggested to be plausible locations for prebiotic synthesis and the assembly of early organics. They include hydrothermal systems, impact craters and tidal regions. Although there are now a diversity of both reaction sequences and environments that have been identified, it still remains one of the most significant questions in astrobiology to identify how and where the first cells emerged.

Further Reading

Books

Cairns-Smith, A.G. (1990) *Seven Clues to the Origin of Life: A Scientific Detective Story.* Cambridge University Press, Cambridge.

Deamer, D. (2012) *First Life: Discovering the Connections Between Stars, Cells and How Life Began.* University of California Press, Oakland.

Hazen, R. (2005) *Genesis: The Scientific Quest for Life's Origins.* National Academies Press, Washington, D.C.

Oparin, A.I. (1938) *The Origin of Life.* Macmillan, New York.

Papers

Bada, J.L. (2013) New insights into prebiotic chemistry from Stanley Miller's spark discharge experiments. *Chemical Society Reviews* **42**, 2186–2196.

Chakrabarti, A.C., Breaker, R.R., Joyce, G.F., Deamer, D.W. (1994) Production of RNA by a polymerase protein encapsulated within phospholipid vesicles. *Journal of Molecular Evolution* **39**, 555–559.

Chyba, C., Sagan, C. (1992) Endogenous production, exogenous delivery and impact-shock synthesis of organic molecules: an inventory for the origin of life. *Nature* **355**, 125–132.

Deamer, D.W. (1997) The first living things: a bioenergetics perspective. *Microbiology and Molecular Biology Reviews* **61**, 239–261.

Gilbert, W. (1986) The RNA World. *Nature* **319**, 618.

Huber, C., Wächterhäuser, G. (1997) Activated acetic acid by carbon fixation on (Fe,Ni)S under primordial conditions. *Science* **276**, 245–247.

Joyce, G.F. (2002) The antiquity of RNA-based evolution. *Nature* **418**, 214.

Sczepanski, J.T., Joyce, G.F. (2014) A cross-chiral RNA polymerase ribozyme. *Nature* doi:10.1038/nature13900.

Lane, N., Allen, J.F., Martin, W. (2010) How did LUCA make a living? *BioEssays* **32**, 271–280.

Martin, W., Baross, J., Kellet, D., Russell, M.J. (2008) Hydrothermal vents and the origin of life. *Nature Reviews Microbiology* **6**, 801–814.

Miller, S.L. (1953) A production of amino acids under possible primitive Earth conditions. *Science* **117**, 528–529.

Mulkidjanian, A.Y., Bychkov, A.Y., Dibrova, D.V., Galperin, M.Y., Koonin, E.V. (2012) Origin of first cells at terrestrial, anoxic geothermal fields. *Proceedings of the National Academy of Sciences* **109**, E821–E830

Pizzarello, S. (2007) The chemistry that preceded life's origins: a study guide from meteorites. *Chemistry and Biodiversity* **4**, 680–693.

Pizzarello, S., Cronin, J.R. (2000) Non-racemic amino acids in the Murray and Murchison meteorites. *Geochimica et Cosmochimica Acta* **64**, 329–338.

Ricardo, A., Szostak, J.W. (2009) Life on Earth. *Scientific American*, **2009**, September, pp. 54–61.

Russell, M.J., Hall, A.J. (1997) The emergence of life from iron monosulphide bubbles at a submarine hydrothermal redox and pH front. *Journal of the Geological Society* **154**, 377–402.

Schwartz, A. (2007) Intractable mixtures and the origin of life. *Chemistry and Biodiversity* **4**, 656–664.

Sephton, M.A. (2002) Organic compounds in carbonaceous meteorites. *Natural Product Reports* **19**, 292–311.

Wächtershäuser, G. (1990) Evolution of the first metabolic cycles. *Proceedings of the National Academy of Sciences* **87**, 200–204.

12

Early Life on Earth

Learning Outcomes

➤ Describe the environment in which life emerged.
➤ Be able to calculate isotope fractionation using the delta notation and understand its use in investigating geological and biological processes.
➤ Describe the methods for searching for life on early Earth, including isotopic and morphological approaches.
➤ Describe the evidence for life on the early Earth.
➤ Understand and describe some of the problems in interpreting evidence of life on the early Earth.
➤ Describe some of the inferred metabolisms of life on early Earth.
➤ Understand the applications of these principles to the search for life elsewhere.

12.1 Early Life on the Earth

One of the most compelling challenges in astrobiology is finding evidence for the earliest life on the Earth. We are motivated to address this because it might help us to understand how quickly the origin of life occurred once habitable conditions became available. Furthermore, by finding such evidence we are in a better position to attempt to define the environmental conditions in which the earliest organisms lived and ultimately make assessments about the conditions required for early life on other planetary bodies.

The evidence for the earliest life is highly controversial and still debated. Much of the debate surrounds the oldest fossils and whether they contain evidence of life or non-biological processes. In the previous two chapters we discussed the formation of the Earth and the environment in which life originated and began to proliferate. It covered the Hadean **eon** (4.56–4.0 Ga ago). In this chapter we will look at the putative types of life that grew on the Earth at this time and the evidence for this life. The timespan that considers the biological and geological context for the emergence of life on the Earth includes the Hadean and the early **Archean**. The Archean **eon** ran from 4.0 to 2.5 Ga ago.

In this chapter the focus is on understanding the principles that emerge in the controversy about early life on Earth and the problems that lie in the use of any given method for detecting early life. In fact, it would not be surprising if between the time this book was written and the time you are reading this, new evidence has been presented and old evidence refuted. This is the process of science, but it is happening particularly quickly in studies of early life on Earth. Nevertheless, we can identify timeless principles and general complications in interpreting evidence in the rock record for life long ago.

12.2 Early Life – Metabolisms and Possibilities

In Chapter 10 we discussed some of the earliest rock record and how it shows us the presence of liquid water on the early Earth. These same rocks are the material from which astrobiologists seek evidence for the earliest life on the Earth. Despite the presence of these rocks in some diverse locations on the Earth, which provide a useful way to probe for early life, we lack a good rock record from the first ~10% of Earth's history. Nevertheless, a remarkable quantity of information has been gained about the prospects for early life from a small amount of material.

As we saw in Chapter 10, although the early Earth was a very different place from today in some regards (lack of oxygen, higher heat flow, higher UV radiation), in many respects it resembled the present-day Earth, particularly with respect to potentially available habitats for life. When I say **habitat**, I am talking about the physical space for life. Of course, the very different chemistry of the early Earth would have created very different opportunities and constraints on energy sources available, nutrients, trace elements and so on. In other words, the niches would have been different, where **niche** generally means the particular energy and resources that an organism uses, rather than the physical space that it occupies.

Nevertheless, some of the habitats available to early life could have included the open oceans, hydrothermal vent-like environments in the oceans, the surface of early volcanic land masses, the margins or shallow waters around the continents (inter-tidal regions), freshwater pools of water formed on continental crust from precipitation, the deep sub-surface (both in ocean sediments and in the crust) and so on.

As we explored in Chapter 6, the characteristics of the Last Universal Common Ancestor (LUCA) are not precisely known. The inferred hyperthermophilic nature of this organism based on phylogenetic evidence remains a point of discussion. Deep-branching (ancient in the phylogenetic tree of life) hyperthermophiles may reflect a real evolutionary provenance, an evolutionary artefact of the phylogenetic tree that does not represent the true nature of the earliest organisms, or some type of sequencing bias.

Some researchers favour the suggestion that the earliest organisms on the Earth were autotrophic. The reductive citric acid cycle or acetyl-CoA pathways, which are widespread in the domains bacteria and archaea, and ubiquitous in the deepest branches, may have been an early means by which CO_2 from the atmosphere was assimilated. Review Chapter 5 if necessary at this point to refresh your memory of the different organisms and how they gain energy. Some uses of elements in biology, particularly metals found in iron–sulfur proteins and transition metal-containing enzymes found in energy pathways, might be vestiges of a more intimate link between early biochemistry and mineral surfaces of metal-rich environments.

It is also supposed that H_2 could have played an important role in early metabolisms. Hydrogen would have been exhaled at hydrothermal vents and produced by **serpentinisation** reactions whereby water reacts with minerals such as **olivine**, resulting in the formation of **serpentine** and hydrogen.

The three reactions that constitute the process of serpentinisation can be summarised as below. In Reaction 1, hydrogen is generated when fayalite (the Fe-rich end member of **olivine**) reacts with water. The silica produced can react with forsterite (the Mg-rich end-member of olivine) to form the mineral serpentine (Reaction 2). In Reaction 3, serpentine is formed by reaction of water with forsterite without hydrogen production.

Reaction 1

Fayalite + water → magnetite + aqueous silica + hydrogen

$$3Fe_2SiO_4 + 2H_2O \rightarrow 2Fe_3O_4 + 3SiO_2 + \mathbf{2H_2}$$

Reaction 2

Forsterite + silica + water → serpentine

$$3Mg_2SiO_4 + SiO_2 + 4H_2O \rightarrow 2Mg_3Si_2O_5(OH)_4$$

Reaction 3

Forsterite + water → serpentine + brucite

$$2Mg_2SiO_4 + 3H_2O \rightarrow Mg_3Si_2O_5(OH)_4 + Mg(OH)_2$$

Hydrogen concentrations in the atmosphere in a mildly reducing state could have been ~0.1% or greater. The evolution of the capacity to use hydrogen as an electron donor would have opened up the possibility of chemoautotrophic iron and sulfate reduction (using oxidised iron and sulfate ions as the electron acceptors, respectively). It would also have made possible **methanogenesis**, whereby hydrogen is coupled to CO_2 as the electron acceptor, to generate energy. Some deep branching bacteria such as *Aquifex* and *Sulfurihydrogenibium* are capable of using hydrogen as an electron donor and nitrate and iron as electron acceptors.

Sulfur-based metabolisms seem to have an ancient heritage and may reflect a time in the early history of the Earth with more hydrothermal activity. However, low concentrations of sulfate inferred for the Archean oceans (perhaps less than 2.5 µM) may have limited the possibilities for sulfate reduction in the ancient oceans.

Many of the organisms that oxidise sulfur compounds prefer to use sulfite (HSO_3^-) as the electron acceptor, a compound readily formed from the reaction of SO_2 (from

Figure 12.1 *Hot springs in Yellowstone National Park are host to microbial communities that use hydrogen and sulfur metabolisms that could have been present on the early Earth. The foreground is about 2 m across.*

volcanic gases) with water. As we saw in Chapter 5, sulfate requires initial ATP investment to activate into a form in which it can be used. Thus, sulfite reduction, which does not require this step, is more energetically favourable and could have been the precursor metabolism. The genes for this process (dissimilatory sulfite reductase, *dsr*) are found in at least five separate groups in the bacteria and archaea. Of the archaeal groups, the Korarchaeota are found in sulfur-containing hot springs in Yellowstone National Park (Figure 12.1). The Crenarchaeota are dominated by organisms that have sulfur-based metabolisms and many of them, such as *Thermoproteus*, are hyperthermophilic. Some of them use sulfur as the electron acceptor and hydrogen as the electron donor. Organisms, such as *Pyrococcus*, which occupy the other branch of archaea, the Euryarchaeota, are anaerobic hyperthermophiles that oxidise elemental sulfur. Some of the similarities in the metabolic pathways of the sulfur metabolisers and methanogens suggest there could be an evolutionary link.

Organisms capable of using organic compounds on the early Earth may have existed. Many deep branching bacteria and archaea are capable of fermentation, suggesting that this could have been an early trait. These early fermenters would have used dead chemoautotrophs as a source of food or even organic material produced in non-biological syntheses or delivered from space by some of the processes that were discussed in Chapter 9 and 11.

An enduring question in astrobiology is: which came first, heterotrophy or autotrophy, or did they emerge at roughly the same time? As the evidence described above shows, this is an unresolved question. However, the availability of inorganic redox couples and the presence of organic carbon, both from chemoautotrophs and exogenously delivered organics, would suggest that early Earth was habitable with respect to both modes of metabolism.

In addition to fermentation, the ability to link the use of organics as electron donors to compounds such as sulfate and iron in anaerobic respiration would have opened up new possibilities in energy extraction from the environment. Organisms such as *Archaeoglobales* are capable of chemoautotrophic growth on hydrogen or anaerobic respiration using organic compounds, both coupled to sulfate reduction. There are a variety of deep-branching microorganisms capable of anaerobic respiration using sulfate suggesting that anaerobic respirations were an early evolutionary innovation. By analogy, similar points have been made about anaerobic respiration using ferric iron as the electron acceptor. Although in an oxygen-poor world most of the iron would have been present as Fe^{2+}, UV radiation could have **photooxidised** some of this to Fe^{3+}, or Fe^{3+} could have been produced by the activity of ancient microbial anaerobic iron oxidisers.

One major metabolic conundrum is the evolution of photosynthesis. It seems plausible to imagine that the earliest photosystems may have resembled the simple light-driven pumps in organisms like the *Halobacterium*. These halophiles use **bacteriorhodopsin**, a simple light-driven pump, to create a proton motive force that can do work. The later evolution of bacteriochlorophylls would have allowed for simple cyclic electron transport chains in early anoxygenic photosynthesisers. These organisms would have used cyclic phosphorylation to make ATP.

It has been suggested that long-wavelength absorbing bacteriochlorophylls may even have begun as infra-red detectors, being used by organisms living near hydrothermal vents to detect heat (**thermotaxis**), later being applied to the acquisition of visible light for energy production. The phylogenetic affiliation of early phototrophs also remains enigmatic. There have been no identified photosynthetic archaea, suggesting that photosynthesis evolved first in the bacterial domain, with green non-sulfur bacteria, purple bacteria and heliobacteria all suggested as candidate ancestors of the first photosynthetic organisms. Once the bacteriochlorophylls had evolved, then photosynthetic organisms could have begun to evolve to exploit the full range of thermodynamically plausible electron donors, including reduced iron (photoferrotrophy).

At some momentous time, organisms developed the ability to couple photosystems to the use of water as an electron donor. The two photosystems in cyanobacteria are thought to have evolved from an ancestral photosystem I, which bears similarities to the photosystem in green sulfur bacteria and other anoxygenic photosynthesisers, followed by the evolution of photosystem II, perhaps by gene duplication. The fact that certain cyanobacteria retain the capacity to carry out photosynthesis with H_2S suggests support for this scenario, with these organisms retaining their early capacity for anoxygenic photosynthesis. The implications of this new-found capacity to tap into the immense availability of liquid water as a source of electrons to drive electron transfer reactions cannot be overestimated. In Chapter 14 we will explore the consequences of this single event in greater detail. For now it is sufficient to point out that it would have given early cyanobacteria the ability to colonise diverse habitats and to spread widely, leaving anoxygenic photosynthesisers confined to places where there was local availability of reduced iron, sulfide, hydrogen and organic compounds. Cyanobacteria would have increased the production of carbon in the biosphere by about three orders of magnitude with enormous implications for the quantity and diversity of carbon-containing compounds that would have acted as electron donors available to anaerobic respirers, and later on a planetary scale, aerobic respiration.

The evolution of oxygenic photosynthesis would have had important implications for microorganisms. It would have enabled oxygen, even if only locally, to have driven aerobic respiration and it would have opened up chemoautotrophic pathways using oxygen as the terminal electron acceptor, such as iron oxidation, nitrification using ammonia and nitrite as electron donors and sulfide oxidation using oxygen. During this time atmospheric oxygen concentrations would have remained low as the oxygen in the atmosphere was mopped up by reduced gases, as we will see in Chapter 14. We can conceive of 'oxygen oases' on the early Earth (although one should be careful not to be too **anthropocentric** about this, because the term 'oases' belies the fact that the oxygen would have been lethal to **obligate** anaerobic organisms at that time, generating oxidative stress from **oxygen free radicals**).

One of the most profound consequences of oxygenic photosynthesis would have been the potential for the spread of aerobic respirers. We cannot say for certain that oxygenic photosynthesis was required for their evolution since abiotic, localised trace sources of oxygen from water **radiolysis** in the deep sub-surface cannot be ruled out as a source of oxygen for this metabolism. Certainly oxygenic photosynthesis would have greatly expanded the extent and distribution of oxygen in the biosphere.

One characteristic of these early cells may have been that they were a lot less specialised than present-day organisms. It has been proposed that early organisms may have engaged in horizontal gene transfer much more promiscuously than present-day organisms. Instead of thinking about the earliest organisms as giving rise to a definitive last common ancestor, we should instead consider early cells to have been some sort of collective, whereby information was being shared widely, leading to a whole variety of experimental metabolisms. Some of these experiments were more successful than others and would have become more specialised as they evolved into discrete organisms.

12.3   Isotopic Fractionation

One way in which we can search for evidence of this past life that might have lived in early habitats is to look for isotopic evidence. **Isotopes** are atoms in two or more forms of the same element that contain equal numbers of protons, but different numbers of neutrons in their nuclei. As was briefly alluded to in Chapter 10, isotopes have many other uses in astrobiology quite apart from looking for life, for example determining the ratios of deuterium to hydrogen in ocean water and comparing this to meteorites and comets to attempt to determine where Earth's oceans originally came from.

Many elements of low atomic weight (including those used extensively in life) have two or more **stable isotopes**. Some examples are shown below with their atomic mass numbers shown as superscripts.

Hydrogen – ^1H; deuterium, ^2H, is sometimes written as D
Carbon – ^{12}C, ^{13}C
Nitrogen – ^{14}N, ^{15}N
Oxygen – ^{16}O, ^{17}O, ^{18}O
Sulfur – ^{32}S, ^{33}S, ^{34}S, ^{36}S

The different numbers of neutrons confer on elements different chemical properties. The lighter isotope has a higher molecular vibrational frequency, thus it tends to form a weaker bond and so it is slightly more reactive. As molecules with the lighter isotope react faster the light isotope tends to become concentrated in the products of reactions. This chemical process whereby isotopes are separated or enriched in products is called **isotopic fractionation**. For most processes (we will encounter an exception later in the book), this fractionation is proportional to the mass of the isotopes and so it is sometimes referred to as **mass-dependent isotope fractionation**. The principle of this is shown in Figure 12.2 for ^{12}C and ^{13}C.

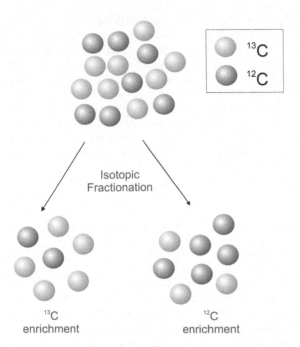

Figure 12.2 *Isotopic fractionation illustrated with the two stable carbon isotopes. A variety of physical and chemical processes can drive fractionation.*

There are a number of ways in which isotopic fractionation can occur in a chemical environment. For example systems that are open, in other words where products can escape, such as reactions that produce gases, can be effective at fractionating isotopes.

Isotope fractionation is also generally greater at lower temperatures. Small differences in mass are less important when all molecules have very high kinetic and vibrational energies (at high temperatures), but when a reaction is cooled down then the difference between the reaction rates of the light and heavy isotope become greater.

12.3.1   Carbon Isotopes

One element used to investigate the isotopic composition of early life is carbon. As we have already seen on a number of occasions, carbon is the major 'backbone' element of biological molecules and is therefore a ubiquitous potential marker for life. Carbon has two stable isotopes: ^{12}C, which makes up about 98.9% of carbon, and ^{13}C, which makes up about 1.1% of carbon on the Earth. There is also the radioactive isotope ^{14}C, which is unstable and decays to ^{14}N. It has a very low abundance

(about one part in a trillion). ^{14}C is used in the carbon dating of archaeological artefacts.

Following on from the previous section, it shouldn't surprise you to learn that biology tends to 'prefer' the lighter isotope, ^{12}C. In other words, chemical reactions mediated by life tend to become enriched in ^{12}C over ^{13}C because of the slightly greater reactivity of ^{12}C.

The question we might ask ourselves is: how can we actually measure this fractionation? There is an internationally agreed system with which to do this.

12.4   Measuring the Isotope Fraction: The Delta Notation

We need some way to express isotope fractionation which creates an internationally recognised norm and an easy way of comparing different samples. The 'delta' notation is the most important way of specifying the degree of fractionation. For carbon isotopes, for instance, the degree of fractionation is defined as:

$$\delta^{13}C = \left\{ \left[\left(\frac{^{13}C}{^{12}C} \right)_{sample} - \left(\frac{^{13}C}{^{12}C} \right)_{std} \right] \div \left(\frac{^{13}C}{^{12}C} \right)_{std} \right\} \times 1000‰$$

There are several things to notice about this equation. We first measure a ratio between the two carbon isotopes ($^{13}C/^{12}C$) for our sample of interest. It is the ratio of the heavier isotope over the lighter one. The ratio for a standard (std) is subtracted from this value and this value is expressed as the fraction of the same standard ratio. For any given element the standard is a rock or material that has been internationally agreed by the scientific community to be the standard of choice. In Table 12.1 you can

Table 12.1 *Typical isotope standards.*

Name of standard	Isotope ratio provided by standard	Value of standard
Standard mean ocean water (SMOW)	$^{2}H/^{1}H$	0.00015575
	$^{18}O/^{16}O$	0.0020052
Pee dee belemnite (PDB)	$^{13}C/^{12}C$	0.0112372
	$^{18}O/^{16}O$	0.0020672
Canyon diablo troilite (CDT)	$^{34}S/^{32}S$	0.045005
Air	$^{15}N/^{14}N$	0.003676

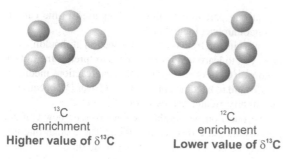

Figure 12.3 *An enrichment in the heavier isotope of an element leads to higher delta values, an enrichment in the lighter isotope leads to lower delta values expressed with respect to the heavier isotope.*

see a variety of these for different elements. For some elements there is more than one standard that can be chosen so when isotope values are published it is important to state the standard used.

The final isotope fractionation is also expressed in parts per thousand and given the somewhat curious symbol ‰, which means **per mil** or 'per thousand'. A value of +10‰

means that the sample is enriched in the heavier isotope by 10 parts per thousand relative to the standard.

The delta notation means that a sample containing less of the heavy isotope compared to a lighter one will yield a lower isotope ratio and a lower delta value. If the ratio is lower than the standard, then the isotopic value will be negative. This principle is shown for carbon in Figure 12.3.

For carbon, typical values of $\delta^{13}C$ range from ~0 ± 2‰ for seawater and limestones to −31‰ in petroleum (which is essentially the remains of dead organisms) to −70‰ for methane produced by microorganisms, illustrating the general principle that, as biological fractionation increases the content of ^{12}C, so the $\delta^{13}C$ becomes more negative (lower).

Photosynthetic carbon (organic carbon) produced by organisms such as cyanobacteria, algae and plants is depleted in ^{13}C and has typical values of $\delta^{13}C$ of approximately −10 to −30‰. All autotrophic organisms (organisms that get their carbon from CO_2; Chapter 5) will tend to produce fractionation of carbon. In Figure 12.4 you can see a range of carbon isotope fractionation values. Unfortunately, many of the carbon fixation pathways used in autotrophs produce quite similar fractionations so it

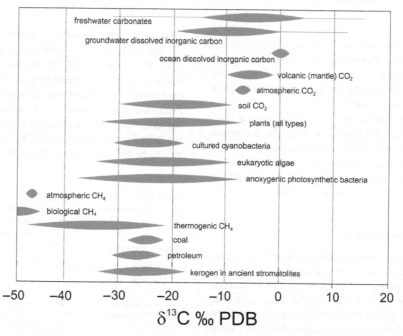

Figure 12.4 *A range of carbon isotope fractionation values for non-biological and biological carbon compounds expressed with respect to the Pee Dee Belemnite (PDB) standard (Table 12.1). 'Thermogenic' methane refers to methane produced by non-biological heat-driven break-down of organic compounds to methane.*

is not readily possible to use ancient carbon isotopic fractionation values to infer the types of biochemical pathways or metabolisms that might have been present.

12.5   Sulfur Isotope Fractionation

Another commonly studied set of isotopes in the rock record are sulfur isotopes. The stable sulfur isotopes are ^{32}S, ^{33}S, ^{34}S, ^{36}S. The fractionation between ^{32}S and ^{34}S is the most studied. In analogy to carbon, biology prefers the lighter isotope, ^{32}S, as it reacts slightly faster than the heavier isotope ^{34}S.

Sulfur isotope fractionation in the natural environment is largely due to sulfate-reducing bacteria (Chapter 5), which you'll remember are anaerobic bacteria that oxidise organic matter (or hydrogen) using sulfate, SO_4^{2-}, as the electron acceptor. When organisms are limited in sulfate, fractionation from the source is quite small, but when they have plenty of it (>1 mmol) then fractionation values as great as −45‰ have been measured. Fractionation values also vary between species.

SO_4^{2-} is reduced to H_2S during the microbial activity, which becomes enriched in ^{32}S because of isotopic fractionation. When the H_2S reacts with minerals to form

sulfides (such as iron sulfide), then the signature of sulfur fractionation is preserved in the minerals of the rock and it can be detected later. Thus, low values of $\delta^{34}S$ in sulfidic rocks imply the presence of life.

12.6   Using Ancient Isotopes to Look for Life

We can look at ancient carbon in rocks laid down through Earth history and see what the isotopic fractionation patterns and values are. This could be used to tell us whether ancient carbon found in the earliest rocks on the Earth is of biological origin.

Ancient organisms are heated and pressurised when rocks undergo **metamorphism** as they are buried. During this process the carbon is heated and pressurised and turns into **kerogen**, a generic term for complex undefined carbon compounds. Under intense heating and pressure this kerogen itself may be turned into **graphite**.

In Figure 12.5 you can see a whole range of carbon isotope measurements that have been made for rocks of different ages. The Figure shows two values, the values at the top ($C_{carbonate}$) are the isotopic values for

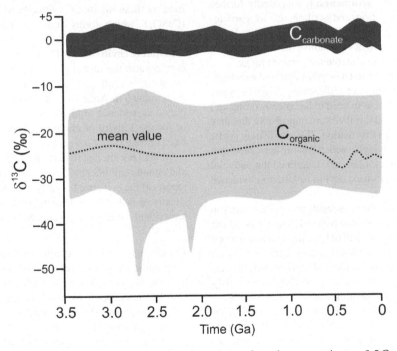

Figure 12.5   *Ranges of carbon isotope fractionation values through time from the present day to ~3.5 Ga ago in the inorganic fraction ($C_{carbonate}$) and organic fraction ($C_{organic}$) of rocks showing evidence for biological fractionation through time.*

non-biological carbonate rocks, precipitated in watery environments. Unsurprisingly, without biology, we see no significant carbon isotope fractionation. The values in the bottom part of the graph ($C_{organic}$) are the values obtained from carbon that is presumed to have once been organisms (e.g. kerogen).These materials do show isotopic fractionation (a depletion of ^{13}C) throughout time, which is taken as evidence that organisms (autotrophs) have been at work through Earth history, taking up CO_2, preferentially using ^{12}C and thereby causing a ^{13}C depletion and a negative $\delta^{13}C$ value.

Evidence for early life on Earth has been presented from the ancient Akilia rocks from Greenland (which are greater than 3.8 Ga old). The carbon (graphite) within them was found to be depleted in ^{13}C, which was reported as evidence that organisms had preferentially taken up ^{12}C and left a signature of their presence. This was presented as evidence for life just over 3.85 Gyr ago. Other data, which involved analysing graphite within phosphate minerals in the rocks (**apatite**) showed remarkable fractionation values with a mean value $-37‰$, which, given losses of light carbon through processes such as metamorphism, would give isotopic values that could be interpreted to be caused by organisms such as methanotrophs that use methane as an electron donor (Chapter 5). The rocks were interpreted to be **sedimentary**, supposedly formed by chemical precipitation and settling out as particles from seawater. They were critical indicators of early life because they were thought to establish the existence of a liquid hydrosphere in a habitable temperature range.

However, re-mapping and new petrologic and geochemical analyses do not support a sedimentary origin for some of the rocks. They appear instead to be igneous (volcanic) rocks such as lavas and it is therefore improbable that they hosted life at the time of formation. The apatite may not be as old as originally thought, which has further called into question the evidence for life and underlined the potential problems with using isotope data in isolation as evidence for life.

Evidence has also been sought in another outcrop of Greenland rocks, the Isua rocks. These rocks, about 3.7–3.8 Ga have been claimed to have light carbon isotope values in carbonates and ancient sedimentary rocks (with a mean of ca. $-15‰$), indicative of life. However, the carbonates are thought to have been produced more recently during the metamorphism of the ancient volcanic rock.

More recent contamination is always a potential problem with finding evidence of life in rocks. As we saw in Chapter 7, many organisms live within rocks (such as chasmoendoliths) where they gain energy and nutrients and protection from extremes experienced outside the rocks. A period of 3.8 billion years is a very long time and in that time we have very little idea of how the rocks have been colonised at different stages at which they have been exposed to potential microbial colonists.

The Greenland rocks remain controversial and they illustrate just one example of the exchange and counter-exchange that occurs between scientists trying to resolve these issues.

Isotopic studies of carbon from rocks in the Pilbara of Australia (3.5–3.2 Ga old) and the Fig Tree groups in South Africa (~3.3 Ga old) similarly show evidence of isotopic excursions ($\delta^{13}C$) of up to $-32‰$ that some have claimed to be biological and by others to have been caused by alteration processes since the rocks were laid down.

Evidence for life has been sought in other isotopic fractionations. For example, fractionation in sulfur isotopes caused by biological sulfate reduction is a favoured means to attempt to infer a biological process. Such signatures are reported in 3.47 Ga old rocks from the North Pole region of Western Australia (it has nothing to do with the geographical north pole!). Sulfur isotope fractionation values in pyrite grains of up to $-21.1‰$ and a mean of $-11.6‰$ suggest that biological sulfate reduction occurred in these sediments when they were being formed and could be evidence of life at that time. The sulfate used by these microbes is thought to have been **gypsum** ($CaSO_4$), which forms at low temperatures (less than ~60 °C), suggesting that the microbes grew in a clement, **mesophilic** environment. The evidence, if correct, further corroborates the idea that sulfate reduction is an ancient metabolism. Nitrogen isotope fractionations (negative $\delta^{15}N$ values) in Archean rocks have been variously interpreted as evidence of nitrogen fixation on the early Earth or for the activity of chemoautotrophic organisms using fixed nitrogen.

We will not attempt to delve into detail on the arguments and counter-arguments for particular rocks and particular reports of the evidence for early life. What is important for a textbook is that we understand some basic principles that are evident from the foregoing. Some main points are:

1. One has to be very careful to distinguish recent contamination from a truly ancient signature of life.
2. The signature has to be within a geological context that makes sense for life. Finding isotopic signatures of life in something that was once molten lava makes it a questionable signature for life. We expect to find these signatures in rocks such as sedimentary rocks associated with clement aqueous environments where life could have grown.

3. These attempts to find life illustrate the difficulties and controversies of finding a signature even with the well-equipped laboratories that we have on the Earth. It places into perspective the difficulties of finding isotopic or other evidence of life in extraterrestrial environments with even smaller sample sizes that are collected and returned to the Earth or using equipment sent to a planetary body that must fit on a spacecraft.

12.7 Morphological Evidence for Life

Another way to look for evidence of ancient life is to search for shapes that resemble fossilised cells, the basis of life (Chapter 4). We can look for tell-tale signs of spirals, filaments and other shapes that suggest cellular structures consistent with the sorts of morphologies we find associated with cells.

12.7.1 How are Microorganisms Fossilised?

Work in the laboratory shows that microorganisms do fossilise and their shapes can therefore be preserved. Much work has been done to simulate this process.

The **extracellular polysaccharide** (EPS) produced by organisms is generally negatively charged (by functional groups such as sulfates or phosphates). These negatively charged groups tend to bind cations such as calcium. The effect of this binding is to increase the concentrations of these ions locally to a point where the precipitation of minerals can occur on the cell membrane or wall (**biomineralisation**). In the case of calcium, raised concentrations of this element at the surface of the cell can encourage calcium carbonate precipitation. This can occur in shallow water environments, for example. The organisms are then preserved by the minerals formed around them.

Another route to fossilisation is by silicification (silica-containing rocks, or **cherts**, are one favoured material in which to look for evidence of early life). Silica is a molecule that consists of oligomers of SiO_2. Raised concentrations of silica will cause it to precipitate around the cell with the effect of essentially entombing the organism in the mineral. The silica molecules are negatively charged. Silica can bind to the surface of the cell by different means including:

1. Electrostatic interactions with positively charged sites on the cell surface (for instance ammonium ions).
2. Cation bridging, for example mediated by an iron hydroxide molecule which sits between a negatively charged phosphate group in the cell EPS and the negative charge on the silica molecule, and
3. Hydrogen bonding between a hydroxyl group on the cell surface (in the EPS for instance) and the oxygen in the silica molecules.

These processes could occur where there are high concentrations of silica, for example around the edges of a volcanic hot spring.

Microorganisms can induce mineral precipitation through their metabolic activity. For example, during photosynthesis, CO_2 is consumed around a cell, which would otherwise mix with water forming carbonic acid (H_2CO_3). This causes an increase in the pH around a cell, increasing the stability of carbonates and causing carbonate mineral precipitation around a cell. However, in the case of silica precipitation, it is thought to be entirely passive, with silica polymerisation caused by the binding of silica to the cell, an abiotic chemically-driven precipitation.

The minerals so formed will eventually get heated and pressurised when they are formed into rocks as the material in which they reside gets buried. The silica will turn into **cherts** and the calcium carbonates into limestones. During this process the carbon in the cells is also pressurised and heated and turned into **kerogen** or **graphite**. Alternatively, the cells might be **permineralised**, whereby they are replaced by minerals, but the morphology can be preserved intact.

These processes can be simulated in the laboratory and they show that there are plausible means by which microorganisms can be preserved in the rock record. We can also see at once that these processes, as they occur over many millions to billions of years, can also preserve inorganic materials and structures that either look like organisms, or through geological processes, experience carbon isotope fractionations and other chemical alterations that can confound researchers looking for life.

12.7.2 Evidence for Fossil Microbial Life

A lot of morphological evidence has been presented for early fossilised microbial life preserved by the processes we have just discussed. William Schopf famously presented photomontages of inferred **microfossils** from rocks ranging in age from 0.7 to 3.5 Ga. Some of them are from the Warrawoona Group of the Pilbara Craton in West Australia and are ~3.5 Ga old. They were interpreted to be early cells with striking resemblance to the filamentous, branched structures of modern-day cyanobacteria with segmentation (**septation**; Figure 12.6)

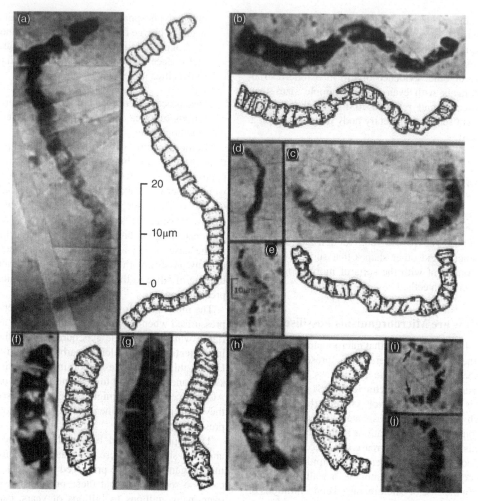

Figure 12.6 *Some of the first suggested early Earth microfossils described by William Schopf.* (Source: Schopf 1993). Reproduced with permission of American Association for the Advancement of Science.

However, the morphological evidence has since been questioned by Martin Brasier (1947–2014) and colleagues. Rather than being in an ancient sedimentary rock, the microfossils were claimed to be associated with a hydrothermal vein created by the interaction of hot water in the ancient rock that was emplaced more recently. Temperatures within the vein could have been over 200 °C, above the known upper temperature limit for life. Furthermore, the fossils have been claimed to be quite non-uniform in shape, more irregular than biological fossils. Instead, they have been suggested to be the result of organic synthesis in non-biological **Fischer–Tropsch reactions**, also seen in deep-ocean hydrothermal vents.

Other morphological evidence has been presented. Microbial mat-like structures and rod and filament-shaped cells have been reported from the ancient volcanic rocks of the Barberton and Pilbara and are thought to document life in shallow water environments. Many of these features were preserved when the cells were replaced by silica solutions or carbonate minerals.

Some of the most convincing evidence from the Archean (although not the early Archean when life first arose) are microfossils in the 2.55 Ga **cherts** and carbonate formations in the Transvaal Supergroup in South Africa. These are made of tubes and sphere-shaped structures of about 2–3 μm size or less.

Focus: Astrobiologists: Nicola McLoughlin

What do you study? Textural and chemical evidence of microbial life, or so-called biosignatures from some of the earliest rocks on Earth. I am interested in when and where the earliest life forms emerged, the first microbial metabolisms, and how these microorganisms co-evolved with the Earth's environment.

Where do you work? I conduct fieldwork on ancient Archean rocks mainly in the Barberton Greenstone Belt of South Africa and the Pilbara Craton of West Australia, because these locations contain some of the world's best preserved and earliest rocks. Then back at the Centre for Geobiology in Bergen, Norway, we employ microscopic imaging techniques to seek putative biosignatures in the samples. We then use various geochemical tools at a range of specialist laboratories across the world to measure the elemental and isotopic signatures in these rocks.

What was your first degree subject? Natural Sciences (Geology)

What science questions do you address? When did life on Earth first emerge, and what are the oldest reliable traces of life in the rock record? How did these earliest life forms make a living, what metabolisms did they employ? Was there an early sub-seafloor biosphere, and were these microorganisms true endoliths, that is rock boring? What is the oldest verifiable microfossil, and what chemical techniques can we use to understand its primitive metabolism? Can we distinguish biogenic from abiogenic stromatolites using tools that could be employed by a Martian rover to confirm the biogenicity of this type of laminated sedimentary structure?

What interests you about astrobiology? The opportunity to tackle fundamental questions about what is life, how do we detect life, and what environments does life inhabit? The challenge of exploring unknown worlds and habitats, combined with the refreshing interdisciplinary nature of this field. There are always lively debates about what is robust evidence of life, plus a real pioneer spirit when for example, new data is reported from Martian rovers. Astrobiology is limitless and really makes you think in new ways – there is never a dull day!

Moving back further in time, microfossils that include long filaments from a hydrothermal set of rocks ~3.23 Ga old have been reported from the Sulfur Springs Group of rocks in the Pilbara Craton, Australia. The filaments, which are made of pyrite are narrow (~2 μm or less) and very long (~200–300 μm). They are not hollow, but have carbon coatings.

There are other suggested lines of evidence for microfossils in the early Archean and you are encouraged to read the most recent papers to see where debate about any particular set of fossils is.

Morphological evidence poses some similar challenges to investigating isotopic evidence, including:

- The problem of dealing with ancient samples that have long since been pressurised and heated during rock deformation, making the signature of morphology difficult to discern.
- The problem of ensuring that the fossils are associated with a plausible geological context for life.
- Ensuring that the fossils are not abiotic.

One problem with morphology is that you can create microfossil-looking structures (so-called **biomorphs**) by

chemical means. Morphology on its own is at best an ambiguous indicator of biogenicity.

Filament-looking structures have been produced by some researchers in the laboratory using only chemical solutions such as $BaCl_2$, Na_2SiO_3, NaOH. When the solutions are added together, inorganic aggregates precipitate. In silica-carbonate solutions, filamentous materials are precipitated like the ones shown in Figure 12.7 which appear very much like the ancient fossils reported from the rock record. To add to the complexity of this situation, these inorganic precipitates can even be shown to bind organic carbon, so that when they are **lithified** or preserved in hydrothermal fluid by **silicification**, they could easily appear to be the organic remains of organisms.

These complications have led some in the community to propose criteria by which all suggested microfossils should be judged – a sort of minimum set of criteria to accept that they could be of biological origin. One set of criteria is that putative ancient fossils should:

1. Be in rocks that are shown to be sedimentary and have undergone very low levels of geological alteration (metamorphism).
2. Be made of kerogen.

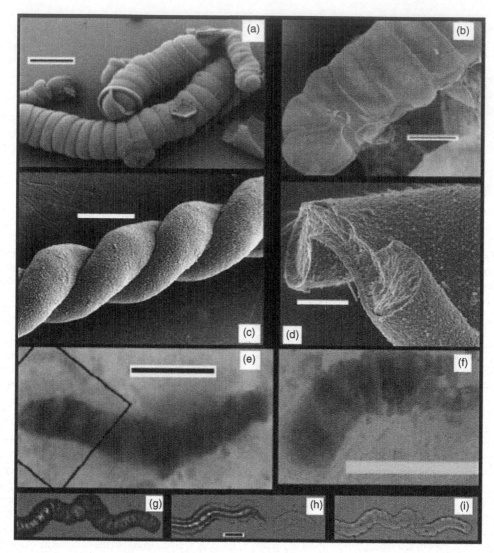

Figure 12.7 A variety of abiotic precipitates produced by chemical reactions that look like fossil life. (a–d) Scanning electron microscopy images of filaments. (a, b, d) Filaments containing silica and barium carbonate. (c) Barium carbonate crystal aggregate after dissolution of silica in mild alkaline solution. (d) Silica skin, coating the exterior of the aggregates. (e, f) Filamentous structures found in the Warrawoona chert, Australia for comparison. (g, h, i) Optical micrographs of synthetic filaments, showing the progressive dissolution of the solid interior of the filaments in dilute ethanoic acid, leaving a hollow silica membrane whose morphology is that of the original filament. Scale bars in (a, b) 40 μm, in (c) 1 μm, in (d) 4 μm, in (f–i) 40 μm. (g–i) are at the same magnification. (Source: Garcia-Ruiz et al. 2003. Reproduced with permission of American Association for the Advancement of Science).

3. Exist with other fossils (and not just be an isolated occurrence).
4. Be of a size at least as great as the minimum known size for viable cells.
5. Be hollow (suggesting a cellular origin).

Some of these criteria may be difficult to demonstrate and we should not rule out the possibility of finding credible fossil evidence of microbes that have been completely replaced by minerals and therefore have no kerogen and are not hollow. However, attempting to set up well-defined criteria that can be systematically followed provides a way to compare data and assess weak points in the interpretation.

Although there is much controversy, new methods, approaches and samples are allowing astrobiologists to home in on the most reliable samples. The existing work shows that chemical signals without morphology cannot reliably demonstrate life and that finding the morphological evidence of cellular life is an important quest in investigating early life on Earth.

12.7.3 Stromatolites

Life could potentially manifest itself at microscopic scale as microfossils as we have just seen. Another type of morphological feature of life is **stromatolites**. These are features that can be observed at the macroscopic scale and they are found in some locations on the present-day Earth. Shark Bay stromatolites in Australia, for instance, result from the interaction between microbes and the physical and chemical environment (Figure 12.8). The cyanobacteria trap sediment with a film of **extracellular polysaccharide** (EPS; Chapter 4) that each cell secretes. The polysaccharide binds the sediment grains that are made of calcium carbonate. However, there is a problem, because the cyanobacteria need sunlight to grow and the sediment will block it out. Many of the cyanobacteria that make up the stromatolites have the ability to move (glide) and they will tend to move towards the light and thus their growth keeps pace with the accumulating sediment. Shark Bay stromatolites grow in this way.

Another way in which stromatolites can grow is by the cyanobacteria forming their own carbonate framework. Photosynthesis tends to make the water around the cell alkaline and an alkaline pH is more favourable for the precipitation of carbonate minerals compared to neutral pH.

The growth of stromatolites is very slow. At Shark Bay, they grow less than 1 mm year^{-1}. Domed structures can grow over 30 cm above the water and take hundreds of years to grow to this size. Others are intermittently covered by water, growing as the tide rises and falls daily. Yet others are submerged permanently, growing at depths of 3.5 m. While these stromatolites look like rocks, examined under a microscope they can be seen to teem with life.

Although one might think that these large structures would be easier to identify than the microfossils discussed earlier, their remains in the rock record nevertheless are disputed. Some early putative stromatolites are from the Strelley Pool Formation, near Nullagine, Pilbara, Australia. The putative stromatolites are about 3.43 Ga old (Figure 12.9). Stromatolites are also observed in the Fig Tree Group rocks in South Africa which are about 3.3 Ga old.

Figure 12.8 *Present-day stromatolites growing in Shark Bay, Australia. Each one is about 0.5 m across* (Source: Jon Clarke).

Figure 12.9 *Putative early stromatolites (seen here as wavy rock textures) from the Strelley Pool Formation, near Nullagine, Pilbara, Australia* (Source: Jon Clarke).

Debate Point: Stromatolite structures without biology?

Are layered structures in the rock biological? As you read this chapter and the next one, you might like to consider some of the factors that might make this question a reasonable one to ask. Even without a detailed knowledge of geology, list some of the natural processes that you can think of that would lead to layers or laminations that could ultimately be preserved in the rock record. Given what you have learned elsewhere in this chapter, describe some of the methods that you might use to find out whether a stromatolite is produced by biological processes. If you were handed a sample of a layered rock and given an open-ended budget to buy any type of analytical equipment you wanted, describe a sequential set of experiments that you would perform that could be used to find out whether the sample ever contained life.

Grotzinger, J., Rothman, D.H. (1996) An abiotic model for stromatolite morphogenesis. *Nature*, **382**, 423–425.

The Australian stromatolites are made up of wavy textures with laminations up to 200 μm thick. They have carbonate minerals within them, similar to modern stromatolites. They form conical, columnar and dome-shaped structures. Domed structures are also observed in the Fig Tree stromatolites. However, the laminations are isopachous (of the same thickness), which might suggest mineral growths rather than biofilms.

One potential problem to be resolved with these stromatolites is that they existed at a time when oxygenic photosynthesis may not have existed and so they might not have been formed by cyanobacteria. However, there are anoxygenic photosynthesisers that could have played a role in stromatolite formation. In many present-day **microbial mats** we find anoxygenic photosynthesisers taking up residence in well-defined bands beneath the surface. Such organisms might well have played a role in forming microbial mats and stromatolites on the early Earth.

12.8 Biomarkers

Some microorganisms produce compounds that are quite specific to biology and even particular metabolisms. For example **hopanoids** are lipid compounds that are thought to be involved in the rigidification of cell membranes. When they degrade they leave hard to degrade carbon structures. One class of compounds, the 2α-methyl-hopanes, are associated with cyanobacteria and have been found in 2.6–2.5 Ga old shales and iron formations in the Hamersley Group of rocks in Western Australia.

Biomarkers offer excellent potential since they can allow for specific metabolisms to be identified. One of their weaknesses is their tendency to be destroyed by the high temperatures (greater than ~300 °C) encountered during metamorphism. In addition, the potential for biomarkers to be mobilised and introduced into rocks by later geological fluids means their age and origin needs to be carefully evaluated. New discoveries are showing that a variety of hopanoids are produced by a much wider diversity of organisms than previously believed, so some of them may not be as specific as originally envisaged. They have also been difficult to convincingly identify in the oldest Archean rocks.

All of these methods discussed in this chapter for identifying ancient life should be regarded as mutually reinforcing. Biomarkers, along with isotopic and fossil evidence, offer yet another way to attempt to corroborate the presence of life in ancient rocks.

12.9 The Search for Extraterrestrial Life

The search for ancient life on the Earth not only reveals information about the emergence of life on the Earth, but it provides us with the basis to improve methods to look for life elsewhere. Some of the difficulties in the search for extraterrestrial life are evident from this chapter. There is much controversy about morphological and isotopic evidence for life, even when we have access to complex laboratories and analytical methods. Therefore, you can get some idea of the difficulty of applying these methods to small samples returned from locations such as Mars or analysed *in situ* using robots. However, a particularly valuable set of lessons are to be learned for developing robust criteria for assessing evidence of preserved microscopic life. By testing criteria for fossil life on terrestrial samples, particularly those with strong evidence for life, we develop a process for testing the biogenicity of putative fossil life elsewhere.

12.10 Conclusions

We have explored some of the plausible metabolisms for early life. The deep branching nature of many chemoautotrophs suggests that this metabolism was established early. However, fermentations and anaerobic respiration could also have developed early using indigenous organics, particularly the organics derived from chemoautotrophs and phototrophs.

The methods used to search for life broadly fall into some key categories: morphological, isotopic and biomarker evidence. All of them have their challenges, not least the confounding effects of signatures produced by abiotic processes. Nevertheless, it is possible to develop assessment criteria that allow scientists to examine putative evidence of early life on Earth and therefore develop robust frameworks for examining early rocks for signs of life. These same methods will be used to search for life in locations such as Mars.

Further Reading

Books

Brasier, M. (2013) *Secret Chambers: The Inside Story of Cells and Complex Life*. Oxford University Press, Oxford.

Knoll, A.A. (2004) *Life on a Young Planet: The First Three Billion Years of Evolution on Earth*. Princeton University Press, Princeton.

Schopf, J.W. (1999) *The Cradle of Life*. Princeton University Press, Princeton.

Wacey, D. (2009) *Early Life on Earth: A Practical Guide*. Springer, New York.

Papers

Allwood, A.C., Walter, M.R., Kamber, B.S., Marshall, C.P., Burch, I.W. (2006) Stromatolite reef from the early Archean era of Australia. *Nature* **441**, 714–718.

Brasier, M.D., Green, R., McLoughlin, N. (2004) Characterization and critical testing of potential microfossils from the early Earth: The Apex 'microfossil debate' and its lessons for Mars sample return. *International Journal of Astrobiology* **3**, 139–150.

Brasier, M.D., Green, O.R., Jephcoat, A.P., Kleppe, A.K., Van Kranendonk, M.J., Lindsay, J.F., Steele, A., Grassineau, N.V. (2002) Questioning the evidence for Earth's oldest fossils. *Nature* **416**, 76–81.

Buick, R. (2008) When did oxygenic photosynthesis evolve?. *Philosophical Transactions of the Royal Society B* **363**, 2731–2743.

Crewe, S.A., Paris, G., Katsev, S., Jones, C., Kim, S.-T., Zerkle, A., Nomosatryo, S., Fowle, D.A., Adkins, J.F., Sessions, A.L., Farquhar, J., Canfield, D.E. (2014) Sulfate was a trace constituent of Archean seawater. *Science* **346**, 735–739.

Garcia-Ruiz, J.M., Hyde, S.T., Carnerup, A.M., Christy, A.G., Van Kranendonk, M.J., Welham, N.J. (2003) Self-assembled silica-carbonate structures and detection of ancient microfossils. *Science* **302**, 1194–1197.

House, C.H., Schopf, J.W., McKeegan, K.D., Coath, C.D., Harrison, T.M., Stetter, K.O. (2000) Carbon isotope composition of individual Precambrian microfossils. *Geology* **28**, 707–710.

Knoll, A.H., Javaux, E.J., Hewitt, D., Cohen, P. (2006) Eukaryotic organisms in Proterozoic oceans. *Philosophical Transactions of the Royal Society B* **361**, 1023–1038.

McLoughlin, N., Brasier, M.D., Perry, R.S., Wacey, D., Green, O.R. (2007) On biogenicity criteria for endolithic microborings on Early Earth and beyond. *Astrobiology* **7**, 10–26..

Noffke, N., Eriksson, K.A., Hazen, R.M., Simpson, E.L. (2006) A new window into Early Archean life: Microbial mats in Earth's oldest siliciclastic tidal deposits (3.2 Ga Moodies Group, South Africa). *Geology* **34**, 253–256.

Orange, F., Disnar, J.R., Gautret, P., Westall, F., Bienvenu, N., Lottier, N., Prieur, D. (2012) Preservation and evolution of organic matter during experimental fossilisation of the hyperthermophilic archaea *Methanocaldococcus jannaschii*. *Origins of Life and Evolution of Biospheres* **42**, 587–609.

Rasmussen, B. (2000) Filamentous microfossils in a 3,235-million-year-old volcanogenic massive sulphide deposit. *Nature* **405**, 676–679.

Schopf, J.W. (1993) Microfossils of the Early Archean Apex Chert: new evidence of the antiquity of life. *Science* **260**, 640–646.

Schopf, J.W., Kudryavtsev, A.B., Agresti, D.G., Wdowiak, T.J., Czaja, A.D. (2002) Laser-Raman imagery of Earth's earliest fossils. *Nature* **416**, 73–76.

Shen, Y., Buick, R., Canfield, D.E. (2001) Isotopic evidence for microbial sulphate reduction in the early Archaean era. *Nature* **410**, 77–81.

Westall, F., Cavalazzi, B., Lemelle, L., Marrocchi, Y., Rouzaud, J.-N., Simionovici, A., Salomé, M., Mostefaoui, S., Andreazza, C., Foucher, F., Toporski, J., Jauss, A., Thiel, V., Southam, G., MacLean, L., Wirick, S., Hofmann, A. l., Meibom, A., Robert, F., Défarge, C. (2011) Implications of in situ calcification for photosynthesis in a 3.3 Ga-old microbial biofilm from the Barberton greenstone belt, South Africa. *Earth and Planetary Science Letters* **310**, 468–479.

13

The History of the Earth

Learning Outcomes

➤ Know the basic internal structure of the Earth.
➤ Understand plate tectonics and some of its consequences for life.
➤ Describe the rock cycle.
➤ Describe and understand how relative and absolute age dating of rocks is accomplished.
➤ Be able to use basic radiometric dating principles to calculate the age of materials and understand the principle of isochron dating.
➤ Understand the concept of geological time and how it is divided.
➤ Describe some basic biological features of certain geological time periods.

13.1   The Geological History of the Earth

Having learned about the process of the formation of the Earth, conditions on the earliest Earth and investigated some of the evidence for early life, we can now put this into a wider geological context by taking a step back and looking at the Earth over its whole history. How do we measure and examine geological time? This chapter will explore the geological history of the Earth and some of the methods that astrobiologists use to understand when major biological events occurred. The principles and observations elaborated in this chapter provide us with an understanding of one planet that we know is inhabited, the Earth. They are directly applicable to the study of other planetary bodies.

13.2   Minerals and Glasses

Before we begin our journey into the history and composition of the Earth, we should learn something about minerals. Whether you are concerned with the composition of the Earth, Mars, the core of Jupiter's moon, Europa, or you are thinking about the composition of a distant exoplanet, you will come across minerals. Minerals are generally inorganic substances that have an ordered atomic structure and can be defined with a **stoichiometric** chemical formula.

As we saw in Chapter 3, silicon makes a poor element with which to construct biological molecules because it has a habit of binding to oxygen to form very stable silicates. However, these are the materials from which planetary bodies are constructed and their diversity is sufficiently great to allow for a remarkable array of solid materials.

A favourable configuration of binding between silicon and oxygen is a tetrahedron (SiO_4). Once we have this structure, we can put it together like toy blocks, into a wide range of configurations. At this stage you should refer to Figure 3.15 in Chapter 3, where you can see the structure of some silicate minerals in the context of a wider discussion about the chemistry of silicon compounds. You can see how small changes in the ordering of silicon in minerals have dramatic consequences for their structure and type.

We could keep the silica tetrahedra isolated. In this case, the oxygen atoms, which are negatively charged, will tend to bind to metal cations, which cross-link the isolated tetrahedra. This class of minerals is called the island silicates or nesosilicates. In the **olivines**, which are one example of this class, the tetrahedra are cross-linked by Mg^{2+} or Fe^{2+} ions. The Mg^{2+} and Fe^{2+} have roughly the same radii and the same charge, so they tend to be interchangeable in binding to the silica tetrahedra. The result is that we get an olivine series from forsterite (Mg_2SiO_4) to fayalite (Fe_2SiO_4) and a variety of compositions in between, depending on how much Mg^{2+} or Fe^{2+} there is in the melt from which the mineral is being formed.

If we string the tetrahedra together in a single chain, we produce the class of minerals called the **pyroxenes**. We can go one step further and attach these chains together to form double chained silicates. These are the **amphiboles**. And if we go even further and attach the chains together to form large two dimensional sheets of silica tetrahedra, we form the **sheet silicates** or **phyllosilicates**, such as clays and micas. Yet another configuration is to form three-dimensional structures from these tetrahedra. These give rise to the framework silicates or tectosilicates, including the **feldspar** and **quartz** groups of minerals. You can think of the silicate **minerals** as just a whole series of two- and three-dimensional configurations of silica.

The diversity of minerals within these basic structures is accounted for by the addition of a wide variety of cations such as Mg^{2+}, Fe^{2+}, Ca^{2+}, K^+, which fit within the cavities formed in the structures generating different types of minerals, as we discussed for the olivines. In the framework silicates, the silicon atoms can be replaced by aluminium atoms in different ratios, leading to a diversity of minerals, such as the variety of chemical compositions found in the **feldspars**.

Minerals are not limited to the silicates. There is a vast diversity of non-silicate minerals such as carbonates (e.g. limestone; $CaCO_3$) and sulfides (e.g. fool's gold or pyrite, FeS_2), which play an important role in Earth system processes and, in the case of carbonates for instance, in the preservation of the past evidence for life.

All of these minerals come together to form rocks. A **rock** merely describes a solid lump of planetary material. It can be comprised of individual minerals. An example of a rock is basalt (Figure 13.1) which contains olivine, pyroxenes and feldspars.

A rock need not contain minerals. If the material has been rapidly quenched, for example lava erupting into water, so that crystals of particular minerals have not had time to form, it could be a homogenous material or a **glass**. In glass the silica and cations are evenly distributed through the material, but in a non-ordered or non-crystalline structure. A good example of a glass is obsidian (Figure 13.1), a silica-rich volcanic glass that

(a)   (b)

Figure 13.1 (a) Basalt, an example of a rock containing minerals. (b) Obsidian, a silica-rich volcanic glass.

was used by ancient people to make cutting implements, because the glass can be fractured to produce very sharp edges.

13.3 Types of Rocks

Minerals or glasses come together to form broadly different types of rocks. We recognise three main types of rocks, which are described below.

13.3.1 Igneous Rocks

Igneous rocks are produced from the cooling and solidification of **magma**. The rocks can be formed beneath the surface of the Earth in which case we refer to them as **intrusive** rocks (or plutonic rocks). If they are formed on the surface, they are called **extrusive** rocks (or volcanic rocks).

Within magma, as different minerals have different melting points they will tend to crystallise out at different temperatures, a process called **fractional crystallisation**. The exact sequence and type of minerals formed (and thus the type of rock) is complex and depends upon changing magma chemistry. This sequence of mineral crystallisation is called the Bowen's reaction series after Norman Bowen (1887–1956), a petrologist who carried out experiments on mineral formation. Conversely, as we heat an igneous rock, different minerals will melt at different temperatures, leading to **partial melting**.

There are over 700 types of igneous rocks known, all produced at different temperatures and conditions, hence the type of igneous rock yields information on the magma from which it was formed and the conditions deep in the Earth. Some igneous rocks, such as basalt and basaltic glass, provide nutrient-rich habitats for life in the deep oceans and on land.

Rocks that cool slowly, particularly intrusive varieties, tend to have plenty of time to grow large crystals and so are coarse-grained, such as **granite**. Extrusive rocks that erupted onto the surface, such as basalt, tend to become finely grained because they cool much quicker. Basalt glass is a common product of the eruption of volcanic rocks at divergent plate boundaries where lava erupts into the oceans or in places like Iceland, where lava erupts into and below glaciers.

Igneous rocks also turn out to be economically important, being sources of important ores for elements such as tungsten and uranium. These are examples of **incompatible elements** that have either an inappropriate charge or ionic size to be incorporated into the mineral chemical structure. They are excluded from the mineral crystals and become concentrated as ores in veins at a late stage of rock solidification.

We can systematically characterise igneous rocks according to their silica content. Rocks with very high silica are called felsic and include a volcanic glass such as obsidian (Figure 13.1) and its crystalline counterpart, rhyolite. As we go through intermediate rocks, we arrive at **mafic** rocks, which are volcanic rocks such as basalt with a silica content of 45–52%. **Ultramafic** rocks include **peridotite** from which the mantle is made and have silica contents of less than 45% (Figure 13.3).

13.3.2 Sedimentary Rocks

Sedimentary rocks are rocks formed from the deposition of material at the surface of the Earth, such as in water bodies, through the action of wind, rain and glaciers. Although sedimentary rocks are enormously important for human activity since they affect decisions on civil engineering and the integrity of buildings, they only account for about 8% of rocks in the crust, the rest being igneous and metamorphic rocks. Good examples of sedimentary rocks are chalk (calcium carbonate) and mudstones (made of at least 50% silt or clay-sized particles) and clays, which are a type of **phyllosilicates** and are the altered weathering products of rocks.

Sedimentary rocks can be broadly subdivided into several types. Clastic sedimentary rocks are those formed in moving water bodies such as rivers or streams and usually contain clays and rock fragments. Biochemical sedimentary rocks are those formed from biological material (such as coal formed from dead plant matter). Chemical sedimentary rocks are rocks formed from chemical reactions, for example **cherts**, formed from the precipitation of silica out of solution in a supersaturated solution of silica.

Sedimentary rocks are of huge significance to astrobiologists because they form strata, which provide information on the age of rocks and especially the environments of their deposition. As sediments tend to accumulate organic material and dead organisms, they often preserve the biological evidence of past times, from macroscopic features such as dinosaur bones to microbial microfossils.

Astrobiologists are particularly interested in the process of **diagenesis**, which describes how sedimentary rocks change over time from processes such as pressurisation and heating. Diagenesis will determine whether signatures

Facilities Focus: Studying rocks and planetary materials

We can examine the composition of rocks with a wide diversity of methods. One very powerful approach is to use synchrotron radiation (Figure 13.2). A synchrotron is a type of particle accelerator whereby a particle beam is accelerated and guided by magnetic fields into a circular route, allowing for the development of high energy beams in relatively small spatial areas. When particles such as electrons are rapidly circulated through magnetic fields they give off radiation across the whole electromagnetic spectrum. By diverting some of this radiation off from the main ring (or storage ring as it is usually called), it can be channelled into areas outside the ring and used to carry out experiments. Synchrotron radiation can be used in geosciences to probe the structure and orientation of metal ions within rocks and minerals, to investigate materials under intense temperature and pressures inside planets, to study the effect of microbes on rock structures, to map elements including metals at very small scales within samples, for example in lunar rocks or meteorites, and to study the distribution of carbon in fossils, as well as other applications. It finds use in a whole range of applications where a focused beam of radiation with a known wavelength can be used to probe the structure of matter. In particular, such facilities are useful to astrobiologists because they can be used to study materials returned from other planetary bodies in sample return missions when sample quantities are generally very small.

Figure 13.2   *The Diamond Synchrotron Source in Oxford (UK), which generates beams of radiation with a distribution from X-rays to the infra-red. The facility produces electrons of 3.0 GeV that are kept in a 561.6 m circumference storage ring and then fed into experimental 'hutches' in which given wavelengths can be used to study materials. For example X-rays are used to study the oxidation state and structure of metals in minerals produced by bacteria or in ancient rocks from the Moon or Earth.*

of life can be preserved and may provide insights into why signatures of life have been lost if they are thought to have been originally present.

13.3.3   Metamorphic Rocks

Metamorphic rocks are formed when rocks are transformed, typically when they are heated to temperatures greater than about 200 °C and pressurised to values that exceed about 1000 bar (100 MPa). This process can occur at any time rocks are buried, but it is particularly prevalent at plate boundaries, for example at convergent boundaries where **subduction** exposes rocks to intense temperatures and pressures beneath plate boundaries. These large-scale changes to rocks are referred to as regional metamorphism, in contrast to contact (or thermal) metamorphism, when rocks are altered in some localised process such as during contact with hot magma.

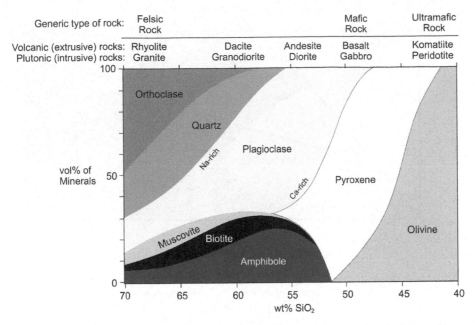

Figure 13.3 *An example classification of different types of igneous rocks according to their content of different minerals. The content of minerals is correlated with the overall percentage of silica (x axis).*

Not all rock minerals change state at the same temperature. For example, olivines and quartz are quite stable to high temperatures, such that the mineral assemblages found in metamorphic rocks can be used to ascertain information about the temperature and pressure conditions they were exposed to. The different types of metamorphic rocks and the temperature and pressure conditions with which they are associated are shown in Figure 13.4. The rocks are often a complex assemblage of different minerals, for example greenschist is so named because it contains green-coloured minerals such as serpentine. The different types of metamorphic conditions, grouping given temperatures and pressures, are referred to as **facies**.

As the rocks recrystallise and deform, they often form very striking banded patterns, known as foliation. In other rocks, the high temperatures cause chemical reactions to occur in which minerals may be completely replaced by other elements and compounds. The process of **metasomatism** is chemical alteration by hydrothermal or carbonaceous fluids. Metasomatism can also occur by rock–melt interaction.

Metamorphism has a great deal of relevance to astrobiologists because recrystallisation, deformation and metasomatism have a usually detrimental effect on the preservation of organics and biosignatures in rocks. In general it destroys them or makes them harder to detect and potentially introduces contamination. These problems were discussed in Chapter 12.

13.4 The Rock Cycle

All three of the different rock types can interchange. The best way to summarise these changes is to recognise a **rock cycle** that describes the different types of rocks produced on the Earth and their fate in the Earth system. The rock cycle is a pictorial representation of events leading to the origin, destruction, change and reformation of rocks (Figure 13.5).

The rock cycle shows how these rock families are interrelated and can be derived from one another. We can follow just a few of these pathways. For example extrusive igneous rocks, once solidified, may be subjected to wind and water erosion depending on their location. The material produced will eventually be transported into rivers and other water bodies to contribute to sedimentary rock. Sedimentary rock will eventually be buried, ultimately becoming metamorphic rock, and, if buried deep enough, it will melt into magma and can eventually become igneous rock. Metamorphic rocks, too, when exposed on

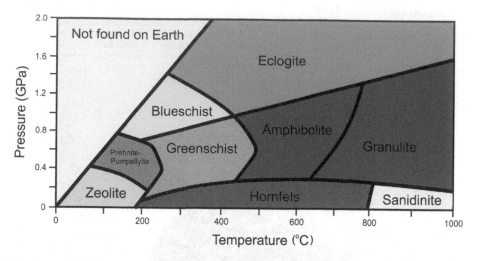

Figure 13.4   *The names of different types of metamorphic rocks showing the pressure (given as giga-Pascal, GPa) and temperature regimes at which they are formed. The separation between the rock types is gradual, although shown here as black lines for clarity.*

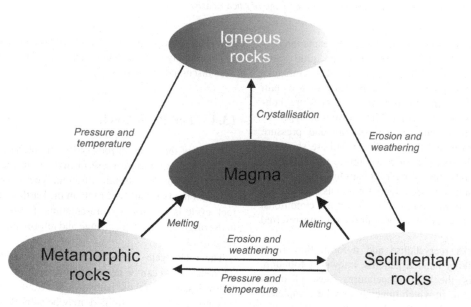

Figure 13.5   *The rock cycle showing the different pathways of rock types and their interchanging fates (Source: Charles Cockell).*

Debate Point: Anthropic rocks.

In an intriguing paper, geologist James Underwood proposed that geologists need to introduce a new class of rock called Anthropic Rock. Recognising the huge quantity of synthetic rock being made by people, including concrete and even synthetic basalt, Underwood proposed that Anthropic Rocks could be used to classify this diversity of anthropogenic rocks. These rocks, like any others, eventually weather and contribute to new sedimentary rocks, and will, given enough time, eventually be subducted. They are therefore part of the rock cycle. What do you think? Do such rocks deserve a new category and what similarities and differences to existing rock categories do you think they have?

Stone, E.C. (1998) From shifting silt to solid stone: the manufacture of synthetic basalt in ancient Mesopotamia. *Science* **280**, 2091–2093.

Underwood, J.R. (2001) Anthropic rocks as a fourth basic class. *Environmental and Engineering Geoscience* **7**: 104–110.

---

the surface of the Earth are eroded by water and wind to contribute to the production of sedimentary rocks and of course, sedimentary rocks can be broken down by wind and rain to contribute to yet new sedimentary rocks. This highly interconnected cycle has been going on since the Earth was first formed. As we saw in Chapter 10, the evidence of sedimentary rocks in the rock record from ~3.8 Ga ago suggests that the rock cycle, with all three major types of rocks, was well established on the Hadean and Archean Earth. You might like to reconsider the rock cycle in Chapter 17, when the aqueous history of Mars is discussed.

13.5 The Composition of the Earth

Once the Earth had formed and differentiated, an internal structure developed similar to what we observe today (Figure 13.6). Our picture of the internal structure of the Earth comes from a variety of sources. Thermodynamic modelling can give us ideas about how we would expect certain geological materials to behave at large scales. Seismic tomography allows us to directly measure seismic waves passing through the Earth's interior, whose speed and deflections depend upon the density, composition and flow characteristics of the material they are passing through. Studies of the moment of inertia of the Earth, which focus on how the Earth spins and what mass distribution is required to explain its spin, tell us about the distribution of density within the Earth. Geological studies of rocks exposed on the surface of the Earth and erupted from the interior give us a great deal of information about the crust.

The centre of the Earth is made of a solid (inner) core surrounded by a liquid (outer) core which is responsible for generating the magnetic field. The core is comprised of ~85–90% iron and nickel and the remainder an unknown mix of volatiles (H, C, N, O, S or Si). These other elements are required to account for the dual solid/liquid characteristics of the core. The temperature of the solid core is estimated to be ~5400 °C. The core comprises about one third of the mass of the Earth, but the volume is only 16% of the planet.

Above the core is the mantle. The core–mantle boundary is at a depth of 2900 km. The mantle is itself split into the upper and lower mantle with a ~250 km transition zone between them. The upper mantle is primarily made of **peridotite**, which is a dense, coarse-grained igneous rock, consisting mostly of the minerals olivine and pyroxene. Between about 400 and 650 km depth, olivine is not stable and transitions into high-pressure phases (such as the minerals wadsleyite and ringwoodite). Below 650 km depth the minerals are dominated by the silicate perovskites, a set of high-pressure calcium and magnesium iron silicates. The mantle temperature ranges from about 600 to 900 °C at the interface with the crust to 4000 °C at the interface with the core.

Above the mantle is the crust. The transition between the crust and the mantle occurs in a region known as the Moho (or Mohorovičič) discontinuity after Croatian seismologist, Andrija Mohorovičić (1857–1936) who first used seismology to see this transition.

13.6 The Earth's Crust and Upper Mantle

The crust of the Earth is where biological activity occurs. It occupies ~1% of the volume of the Earth. At the current time we do not know the maximum depth to which life penetrates in the crust. If we assume that the upper temperature limit for life is 122 °C (Chapter 7) and the **geothermal gradient** (the increase in temperature with depth in the Earth) is something on the order of 15 °C

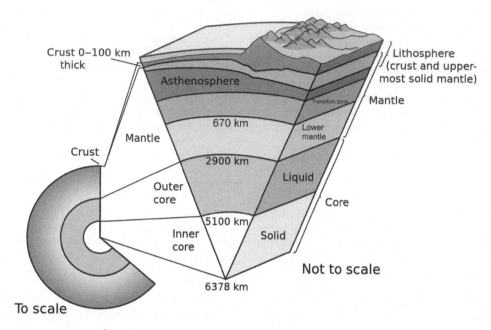

Figure 13.6 *The internal structure of the Earth* (Source: USGS).

km^{-1} in oceanic crust depending on the location (it is about 25 °C km^{-1} in continental crust), then the maximum depth of life is probably about ∼10 km.

We can recognise two different types of crust. The continental crust has a density of about 2.7 g cm^{-3} and contains about 10% **granite**. The crust also contains sedimentary rocks (about 8%), and a variety of other metamorphic and igneous rocks. The continental crust is about 20–90 km thick. The ancient cores of the continental crust are referred to as **cratons**.

The oceanic crust is thinner than the continental crust (about 5–10 km thick) with a density of 3.0 g cm^{-3} and it is composed mostly of **basalt**, a common extrusive igneous rock, and **gabbro**, a group of dark, mafic intrusive igneous rocks. The oldest ocean crust is about 180 million years old in the northwest Pacific.

The Earth's crust has a very different composition to the bulk properties of the whole Earth (Figure 13.7) primarily because of the differences between the silicates in the crust and the iron in the core.

The crust is part of the **lithosphere**, which includes the crust and part of the solid upper mantle. Below the lithosphere is the **asthenosphere**, which is part of the upper mantle and behaves plastically, which means that it flows.

13.7 Plate Tectonics

The lithosphere is not a continuous piece of rock. Instead, it is broken into individual pieces or plates. The plates move over the asthenosphere. The asthenosphere is characterised by convection cells. The formation of convection cells within the asthenosphere is accounted for by the behaviour of rocks when they are heated. As they get hotter the rocks have a lower density than cooler rocks. The result is that they will tend to rise, while cooler rocks sink, generating a circulating pattern. A misconception is to think that these convective cells entirely drive the movement of the plates. That would be a logical line of thinking. However, this does not seem to be the case and we will come back to why this is so towards the end of the section.

It is important to understand a major conceptual point here. The rocks within the mantle are solid. This may seem counterintuitive. However, solid rocks, when heated sufficiently, become ductile and flow, albeit very slowly, like a fluid. This allows for the convective movement of the rocks within the Earth's interior.

The molten lava that erupts onto the surface of the Earth is rock that has been melted locally in magma chambers. The lower pressure nearer the surface of the Earth enables hot solid rock to melt. This is decompression melting.

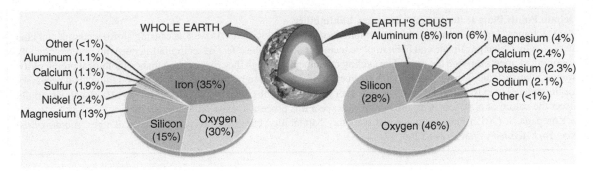

Figure 13.7 *The composition of the crust of the Earth compared to the whole Earth, showing evidence of* differentiation.

To understand this, review the phase diagram for a typical substance shown in Figure 2.18, Chapter 2. You can see that the reduction of pressure at a given temperature can cause material to cross the melting curve from solid to liquid. The process of rock melting is also called **partial melting** because only some of the minerals in a rock at any given temperature and pressure conditions may melt depending on its composition. The observation of liquid rock erupting onto the surface of the Earth should therefore not confuse you into thinking that it is an eruption from an essentially liquid layer under the crust.

Plate tectonics is not a universal phenomenon. On Mars, which has a smaller volume than the Earth, heat loss was more rapid and plate tectonics ceased early on, if it ever really became properly established. The lack of plate tectonics accounts for the vast shield volcanoes where volcanic eruptions remained in a fixed location on the surface. An example is Olympus Mons, which is 22 km high and about 600 km wide (Figure 13.8). On Venus, the relatively uniform distribution of craters over the surface argues against active plate tectonics. The lack of plate tectonics on Venus, despite the fact that it is a similar size as the Earth, is attributed to the lack of water in the crust, driven off by the runaway greenhouse effect which heats the surface of the planet to just over 460 °C. The paucity of water in the crust makes it stronger and less able to fracture and move. Nevertheless, it is still possible that convective circulation occurs deep within Mars and Venus, but if it does occur it does not connect with the surface in the same way as it does on the Earth.

There are 12 major plates on the Earth, depending on how we define them, and many smaller plates (Figure 13.9).

The idea that the crust might be moving and might be made of plates was a revolutionary theory. It was first

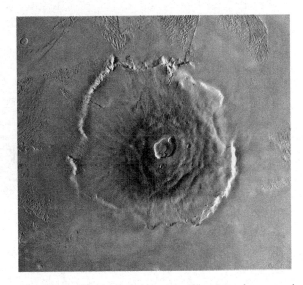

Figure 13.8 *The lack of plate tectonics on Mars has caused the formation of vast shield volcanoes as lava erupts in one place. This is an orbital image of Olympus Mons, a 22-km high, 600-km wide volcano.*

indirectly suggested by Alfred Wegener (1880–1930) in the early years of the twentieth century. He proposed the theory of continental drift in 1912. The observation that continental outlines seemed to fit together and the discovery of similar fossil plants in South America and Africa suggested the idea that the continents were once joined. However, there was no obvious way in which such enormous forces to move entire continents could be derived and Wegener's suggestions took at least 50 years to become mainstream science. The forces that were originally suggested to drive plate tectonics, such as tidal forces, were shown to be too weak to

---

Debate Point: Plate tectonics and planetary habitability

Plate tectonics clearly has many important implications for the biosphere. As well as causing shifting continental land masses, which change climate and habitat, plate tectonics also has the effect of recirculating nutrients and biologically important elements through the crust. The cycling of elements through the crust might enhance biological productivity. Some scientists believe that plate tectonics is necessary for a planet to be habitable. What do you think? Is it necessary for the process to exist for a planet to host life over geological time periods? How does the size of a planet potentially affect the nature of plate tectonics and would we expect this process to exist on extrasolar planets?

Korenaga, J. (2012) Plate tectonics and planetary habitability: current status and future challenges. *Annals of the New York Academy of Sciences* **1260**, 87–94.

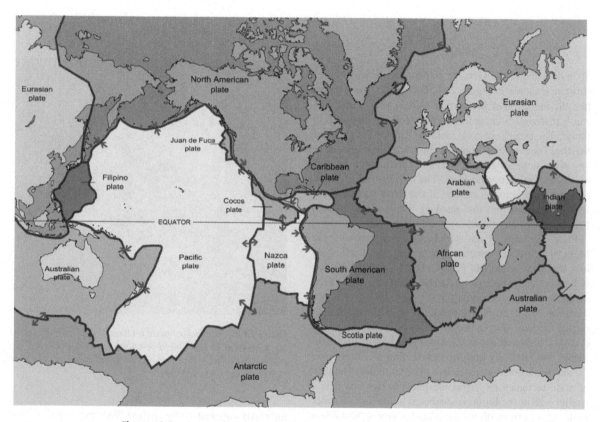

Figure 13.9   *A map showing plates and major plate boundaries on the Earth.*

account for movement, explaining the slow uptake of the idea.

Wegener's ideas were confirmed by several lines of evidence in addition to the observations of similar fossils on both sides of oceanic divides. The ceaseless and careful mapping of the ocean floor in the mid-twentieth century, partly motivated by a desire to improve maps for submarines, revealed enormous ridges down the middle of the oceans. These were the evidence for ocean spreading regions from which new crust was emerging, creating the conveyor belt that shifts the continents apart. Equally as compelling was the discovery of alternating magnetic fields preserved in the iron-containing minerals (**magnetite**) in rocks on the ocean floor. These rocks preserve the record of the Earth's magnetic field that experiences periodic reversals. As the rocks are produced from ocean spreading centres, solidify and begin their journey from the ridges, so the flipping back and forwards of the magnetic field is preserved in characteristic magnetic bands in the spreading rock – evidence that the Earth is in movement, that new crust is being synthesised and driving apart plates.

Plate tectonic theory describes the movements of these plates (Figure 13.10) and the processes that occur within them. Plate boundaries are marked by seismic and volcanic activity. At plate boundaries plates diverge, converge or slide sideways past each other. Let's look at each of these plate boundary types.

Convergent boundaries exist where plates slide towards one another and create a **subduction** zone, where one plate is pushed under another (e.g. the Cascade Mountains in the USA). Where more dense oceanic crust collides with continental crust it is driven under the continent. As the plate is subducted it carries with it volatiles such as water. The intense pressures at depth squeeze the water out and the water mixes with the mantle peridotite above it. Water has the effect of lowering rock melting temperature. Thus the rock is melted, causing the formation of lines of volcanoes or **volcanic arcs**. The region of continental crust along such plate boundaries in the oceans is called **continental shelf**. Where continent to continent collision occurs, ocean basins can be closed off and mountain building can occur. For example, the Himalayas formed from the collision of the Indian and Eurasian plates. In some instances, these collisions can cause ocean crust and mantle to be forced up onto the continents. This material so deposited is called an **ophiolite**.

Divergent boundaries exist where the plates spread apart. In the oceans these spreading mid-ocean ridges (or sea-floor spreading regions) generate volcanic activity and seismic activity and are the locations of the formation of new ocean basins. The melting that occurs at these boundaries is decompression melting, caused by the lowering of pressure as material moves towards the surface of the Earth. As well as in the ocean (e.g. Mid-Atlantic Ridge), plate spreading can also occur within continental crust (one debated example is the Eastern African Rift).

Transform boundaries occur where two plates slide past one another. The San Andreas Fault in California is such an example. Seismic activity can be intense along such boundaries.

How are these plate movements linked to the underlying mantle convection? One hypothesis is related to the structure of the plates themselves. Where plates subduct, the rock is pressurised and becomes denser. This dense material has a tendency to pull the plate downwards. At the

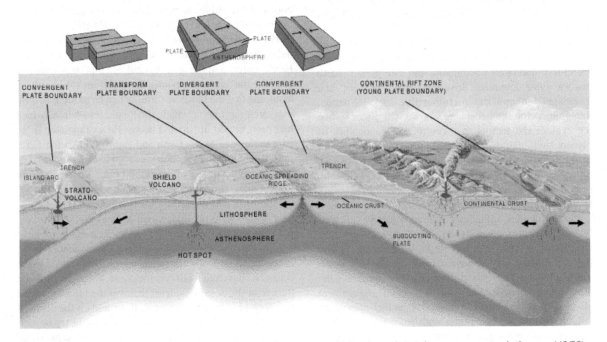

Figure 13.10 *Interactions between plates and the different types of plate boundaries that are recognised.* (Source: USGS).

other end of the plate, for example, in a mid-ocean ridge, the plates are pushed apart by the injection and expansion of the magma formed by decompression melting during rifting. These factors mean that the plates tend to move into the mantle through the subduction zone. They essentially pull themselves into the mantle, apparently largely independently of convection beneath them.

However, this cannot be the only explanation for plate movement since there are plates that are not linked to subduction zones, such as the North American Plate. There is still much debate about the involvement of tidal forces and there are adherents to the idea that if large convection cells do not drive plate movements, plate movement might be driven by smaller scale movements in channels beneath the lithosphere.

Plate tectonics is not the only mechanism for generating surface geological activity such as volcanism. Rising plumes of rocks within the mantle melt as they approach the surface and are subjected to lower pressures, generating a hotspot, which, like a welding torch, can burn its way through the crust and generate local volcanic activity within a plate. Perhaps the best illustrated example of this is the Hawaiian island chain (Figure 13.11). Here a hotspot has erupted lava into the oceans over tens of millions of years. As the Pacific plate has moved over the essentially static hotspot, so a series of islands has been generated along the plate running, for example, from Kauai which is about six million years old to the modern day Big Island of Hawaii.

Plate tectonics provides a framework for interpreting many aspects of Earth on a global scale, relating many seemingly unrelated phenomena and interpreting Earth history. Plate tectonics drives mountain building and associated **igneous** and **metamorphic** rock formation, thus influencing the geosphere. As we will discuss when we explore the factors that control the habitability of planets later, plate tectonics also influences heat transfer in the interior of the Earth and may therefore have an influence on the dynamo that generates the magnetic field, responsible for protecting the atmosphere from being sputtered away by the **solar wind**.

Plate tectonics influences the arrangement of continents, affecting solar heating and cooling, and thus winds and weather systems. Rapid plate spreading and hotspot activity may release carbon dioxide and affect global climate. Thus the system of plates influences the atmospheric composition. The continental arrangement affects ocean currents, the rate of spreading affects the volume of mid-oceanic ridges and hence sea level. The placement of continents may contribute to the onset of ice ages. Thus the hydrosphere is influenced.

Finally of course, the movement of continents creates corridors or barriers to migration, results in the formation of new ecological niches and the transport of habitats into more or less favourable climates, thus influencing the biosphere.

13.8   Dating Rocks

We have seen how the process of plate tectonics plays a fundamental role in the geological history of the Earth and how the rock cycle describes the way in which rock types can change structure and form, transforming from one type to another. We have also seen how sedimentary rocks can preserve evidence for life; yet **diagenesis** (e.g. through metamorphism) can destroy those signatures. However, a major question in geology is how we can date the Earth and the rocks of which it is composed. Only when we can date rocks can we put the major biological and geological changes that have occurred on the Earth over time into perspective. The methods developed to do this are directly applicable to the study and dating of other planetary bodies.

Prior to modern geology, Archbishop James Ussher (1581–1656) calculated the age of the Earth at 6000 years. He noted that his calculations were made based on the books of the Bible, namely Genesis, and pinpointed the origin of the Earth to be 26 October 4004 BC.

This crude attempt might be viewed with some humour today, but it was the first attempt to place an age on the Earth. Later, a more geophysical attempt was made to estimate the age of the planet. Lord Kelvin (1824–1907), assuming that the Earth was similar to a large lump of iron, measured the rate at which small spheres of iron, which had been heated up, cooled down. He extrapolated these data to the size of the Earth and estimated that the Earth must be about 20–40 million years old to be as cool as it is. This elegant estimate of course did not take into account the complex formation history of the Earth, its internal composition and the presence of radioactively-generated heat, which were not understood at that time.

Other empirical attempts were made using geological approaches. They included looking at the amount of salt in the ocean. If we know the rate at which salt is added to the ocean and how much salt is in the ocean, we might estimate the age of the oceans. At least that was the idea. We could also look at sediment thickness. If we add up the depth of the thickest sediments we can find we could try to estimate the rate of deposition and arrive at some approximation of the age of the Earth. Using these methods, people

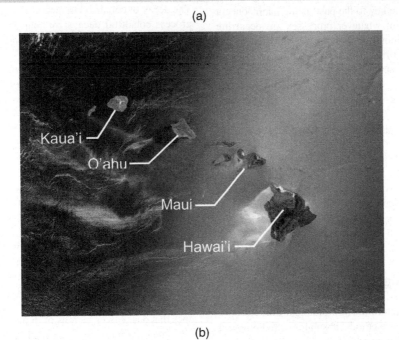

Figure 13.11 *The Hawaiian island chain. An example of a hotspot creating volcanic activity along a plate that is moving above it, shown (a) diagrammatically and (b) as a satellite photograph of the main islands (Source: USGS).*

arrived at ages of a few hundred million years. It should be fairly evident why these two approaches were unreliable. The first method makes huge assumptions about the constancy of ocean composition over time, the second method depends upon which sediments we look at. As we already know, there are very few rocks that date back to the early Earth and, where they do occur, they are not embedded in a sedimentary sequence that continues uninterrupted to the present day.

Despite the early inaccurate attempts to date the Earth, we certainly can use the present to understand the past. The principle of **uniformitarianism** was elaborated by geologist James Hutton (1726–1797) in the late 1700s, although the term was coined by polymath William Whewell (1794–1866). It was further advanced by geologist Charles Lyell (1797–1875) in his work, *The Principles of Geology*.

Hutton realised that most sedimentary layers were deposited from gradual, day to day processes. He realised that it took a long time to form these rocks. This was far different from what others believed prior to this time.

'The present is the key to the past' is the main concept of uniformitarianism: whatever processes are occurring today (volcanism, mountain building, earthquakes, sedimentation) also occurred in the past and probably at the same (or very comparable) rates. We can use an understanding of these processes observed today to investigate past geological processes and their rates.

Thus, by examining layers of sedimentary rock, geologists developed a time scale for dividing up Earth history. The early attempts to date rocks were the right idea; there was just a need for more accurate methods to get an absolute date on samples.

In the next sections, we will learn how geologists use both radiometric techniques and relative age dating to infer the ages of rock units and determine the divisions of the geologic time scale.

13.9   Age-dating Rocks

How do geologists determine how old rocks are? Here we explore the two basic methods: absolute (or radiometric) and relative age-dating of rocks.

13.9.1   Absolute Dating of Rocks

You'll remember that in Chapter 12 we examined some of the evidence for life on the early Earth and in that chapter we discussed how some elements form isotopes (the form of an element that has a different number of neutrons). Some of these elements are radioactive and decay to form daughter products. These elements are **radioisotopes** (isotopes that spontaneously decay, giving off radiation). Radioisotopes decay at a constant exponential rate and this rate of decay is measured by the **half life** (Figure 13.12). The half life is the time it takes for one-half of the radioactive material to decay. The process by which the decay occurs varies between elements.

In the case of **alpha decay**, the element releases a helium nucleus (two protons and two neutrons) and changes into a new element. Uranium-238 decay to lead-206 involves reactions in which alpha particle release occurs.

In **beta decay**, a neutron turns into a proton with the release of an electron, again changing the element (e.g. the decay of carbon-14 to nitrogen-14). The **atomic number** is increased by one, but the **atomic mass number** is not changed.

During **gamma radiation decay**, a gamma ray is emitted, but the number of nuclear particles does not change. The decay of cobalt-60 to nickel-60 is an example. In this process cobalt-60 decays by beta decay to an activated nickel-60. The nickel-60 activated nucleus then emits two gamma rays.

Radioisotopes may decay to form a different radioisotope. Thus, there may be a series of radioactive decays before a stable isotope is finally formed. For example, the decay of uranium-238 to lead-206, one important decay process used to date rocks, involves no less than eight alpha and six beta decays. The stable isotope is called the 'daughter' formed from decay of the radioactive 'parent' isotope (and any radioactive intermediates).

Radioisotopes are trapped in minerals when they crystallise and they will decay through time forming stable isotopes that themselves are trapped within the mineral.

Determining the ratio of parent isotope to daughter product reveals the number of half lives that has elapsed since the mineral was formed. Fortunately, for people interested in dating materials, there is a whole variety of radioisotopes that have different decay times, from ones that have half lives on the order of billions of years to ones that decay over periods of thousands of years.

We can formally write an equation to describe this process. The relative amount of an isotope at a time, t, given as n, can be expressed in relation to the amount at the start (n_0). As the decline is exponential then:

$$n = n_0 e^{(-\lambda t)}, \tag{13.1}$$

where λ is a constant.

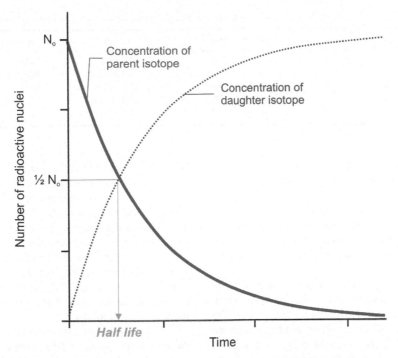

Figure 13.12 *A graphical illustration of the decay of a radioisotope over time. The half life is the time it takes for one-half of the radioactive isotope to decay according to the exponential decline in activity (Source: Charles Cockell).*

However, we know that after one half life $n = n_0/2$ and $t = T$, where T is the half life (the time it takes for n to drop to one-half the original value).

Rearranging Equation (13.1) with these conditions gives the important equation for dating materials using the decay of radioisotopes:

$$n = n_0\exp(-\,0.693t/\text{T}) \qquad (13.2)$$

As an example of the use of this equation, let's consider the age of the Solar System using the decay of ^{40}K to ^{40}Ar, which has a half life of 1.25 **Ga** (Table 13.1). We know that today the ratio of ^{40}Ar to ^{40}K in certain extraterrestrial materials is 11.5 to 1. This can be measured.

We can calculate the amount of ^{40}K, n_0, in the original material. It is simply given by: (the concentration of ^{40}K today) + (the concentration of ^{40}K that has decayed).

This is equal to: (the concentration of ^{40}K today) + (11.5 times the concentration of ^{40}K today).

The latter number is derived from the measured ratio. In other words, the ^{40}K in the original material equals 12.5 times the ^{40}K concentration today.

Using Equation (13.2), $n = n_0$ exp $(-0.693t/\text{T})$ gives us: $1 = 12.5$ exp $(-0.693t/\text{T})$; and using a half life, T,

of 1.25×10^9 years, this gives us an age of the Solar System of 4.56 billion years. These data are consistent with studies of many materials, including lunar rocks and a variety of meteoritic materials.

Some examples of common isotopes used in age dating are shown in Table 13.1.

We can choose an isotope and its daughter and measure the concentration of both to determine how old a rock is (assuming that when the rock was laid down it began with a certain concentration of radioactive element).

All of these elements have different uses. The decay of ^{87}Rb to ^{87}Sr, ^{238}U to ^{206}Pb and ^{40}K to ^{40}Ar are suitable for investigating the ages of ancient rocks that were laid down on the early Earth or other planetary bodies.

By contrast, the short half life of carbon-14 makes it appropriate for carbonaceous materials a few thousands of years old or less (Table 13.1). Carbon-14 is constantly being produced in the atmosphere (at 9–15 km height) from the action of neutrons produced by the collision of **cosmic rays** with the atmosphere. The neutrons react with nitrogen-14 to produce carbon-14. Once the carbon gets incorporated into organic material (and is therefore no longer being produced) it starts to decay. Carbon-14 dating

Table 13.1   *Some examples of common isotopes used to date rocks with their half lives and decay type.*

Parent isotope in rock	Daughter isotope	Half life	Type of decay
Rubidium-87 (^{87}Rb)	Strontium-87 (^{87}Sr)	47 Ga	Beta
Uranium-238 (^{238}U)	Lead-206 (^{206}Pb)	4.47 Ga	Alpha (8), beta (6)
Potassium-40 (^{40}K)	Argon-40 (^{40}Ar)	1.25 Ga	Beta
Uranium-235 (^{235}U)	Lead-207 (^{207}Pb)	704 Ma	Alpha (7), beta (4)
Carbon-14 (^{14}C)	Nitrogen-14 (^{14}N)	5730 yr	Beta

(**carbon dating**) is extremely useful for archaeology, but not so useful for long-term geological studies.

In general, we can say that radiometric age dating is useful to about 10 half lives for any system. Beyond this, the decay is so extensive that the results become ambiguous.

Many of these radioisotopes complement each other and can be used on the same rock materials to corroborate measured ages.

Radiometric age dating was first used to determine the age of the Earth by geologist Arthur Holmes (1890–1965). He realised that the decay of uranium-238 to lead-206 with a half life of 4.47 Ga could be used to determine the age of the Earth. His original estimate was about 3.0 Ga. His estimate was off because the early calculations assumed that all the lead within the sample had come from radioactive decay of uranium-238 and there was none there to begin with. However, this assumption is problematic because rocks on Earth, particularly old ones, have been metamorphosed, exposed to other rocks and so on. They often contain a variety of elements, including elements with radioactive and non-radioactive isotopes, when they are formed. This complication was ultimately overcome by looking at lead isotope concentrations in meteorites, which represent the unaltered materials of the early Solar System. By examining how much lead there is in undifferentiated meteorites, scientists could determine how much would have been incorporated into the earliest rocks. These provide a control on the dating of rocks on the Earth. The revised estimates gave a value for the age of the Earth that we are familiar with today.

13.9.1.1   *The Isochron Method of Age Dating*

Holmes's work was pioneering since it eventually led to the **isochron** method of age dating. In isochron dating, as well as measuring the parent and daughter isotopes for the radioisotope we are using (e.g. for very old lunar rocks, we might measure the parent, ^{87}Rb, and its daughter decay product, ^{87}Sr; see Table 13.1), we also measure

an additional isotope of the daughter element (^{86}Sr) that is not the product of radioactive decay and whose concentration is therefore stable over geological time scales. Since the chemistry of the non-radioactive isotope (^{86}Sr) is similar to the daughter product of radioactive decay (^{87}Sr), we would expect the same relative amounts of each to have been incorporated when the rock first formed and before radioactive decay started in the rock. Thus, the non-radioactive isotope (^{86}Sr), provides us with a way to correct for the presence of some of this element (Sr) in the rock when it formed.

We plot on one axis the ratio of the concentration of the radioactive parent (^{87}Rb) to the concentration of the non-**radiogenic** isotope of its daughter product (^{86}Sr). On the other axis we plot the ratio of the concentration of the daughter product to the non-radiogenic isotope of the daughter (^{87}Sr/^{86}Sr). The value for the non-radiogenic isotope of the daughter product of ^{87}Rb decay (^{86}Sr) essentially normalises the values to take into account how much of that type of element (Sr) was in the rock from the beginning in our different samples before decay began. This plot provides a straight line from which the age can be derived (Figure 13.13) from the gradient. The younger the rock, the more the gradient will tend to a horizontal line. How is this explained? Let us consider this in more detail.

Consider the rock when it first forms (Figure 13.13b). The ratio of the daughter product (^{87}Sr) to its non-decay isotope (^{86}Sr) will be the same in all fresh rocks since there has been no decay to produce any more ^{87}Sr beyond that originally incorporated. The similar chemistry of both isotopes (in this case ^{87}Sr and ^{86}Sr) means that similar amounts are incorporated into any given sample at the beginning, hence the constant ratio on the y axis. By contrast, the parent isotope (^{87}Rb) is chemically different from the non-radiogenic isotope of the daughter product (^{86}Sr) and so the quantity of both is not correlated and may be different in different samples, leading to different ratios along the x axis and hence the horizontal line (Figure 13.13b).

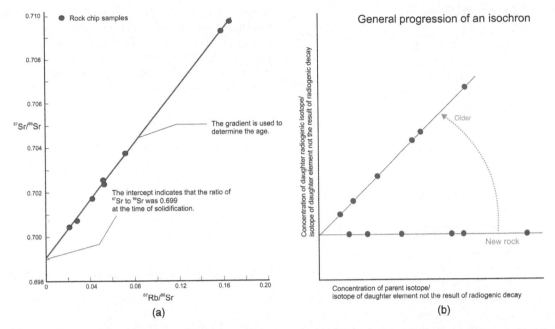

Figure 13.13 (a) An isochron for a lunar rock. This sample is a dunite, an igneous rock of ultramafic composition (sample number 72417) collected by the Apollo 17 mission. Different chips of the rock are analysed. Its age is calculated using the rubidium–strontium system. The gradient of the graph gives an age of 4.47 ± 0.1 Ga. (b) A graph showing the general principle of how an isochron curve develops as a rock ages after initial solidification (Source: Charles Cockell).

What happens as time progresses? The ratio of the concentration of the parent radioisotope (^{87}Rb) to the non-radiogenic isotope of the daughter product (^{86}Sr) gets smaller (x axis) since the concentration of the non-radiogenic isotope of the daughter product (^{86}Sr) stays the same, whilst the ^{87}Rb decays. By contrast, the concentration of the daughter of the radioisotope (^{87}Sr) increases (because of radioactive decay) whilst the concentration of the non-radiogenic isotope of the daughter product (^{86}Sr) stays the same, hence the ratio on the y axis increases. Overall the gradient of the line increases (Figure 13.13b).

A more formal method of understanding this approach is to write the equation for the isochron method:

$$\frac{D + \Delta P_t}{D_i} = \frac{\Delta P_t}{P - \Delta P_t}\left(\frac{P - \Delta P_t}{D_i}\right) + \frac{D}{D_i}$$

where D is the initial concentration of the daughter isotope (^{87}Sr in the example here), D_i is the concentration of the isotope of the daughter product which is not the result of radioactive decay (which is assumed to be at a constant rate; ^{86}Sr in the example here), P is the initial concentration of the parent isotope (^{87}Rb in the example here), and ΔP_t is the total amount of the parent isotope which has decayed by time t.

The equation presented, although it looks complex, is useful because it contains values that can be directly measured. $P - \Delta P_t$ is the concentration of parent isotope measured at the time of the study. $D + \Delta P_t$ is the concentration of the daughter isotope at the time of measurement and D_i is the concentration of the isotope of the daughter product which is not the result of radioactive decay at the time of measurement.

It is the two values $\frac{D + \Delta P_t}{D_i}$ (the relative concentration of daughter isotope and its isotope that is not the result of radioactive decay; ^{87}Sr/^{86}Sr in Figure 13.13) and $\frac{P - \Delta P_t}{D_i}$ (the relative concentration of parent and the isotope of the daughter that is not the result of radioactive decay; ^{87}Rb/^{86}Sr in Figure 13.13) that we plot to obtain an isochron graph.

The slope of the graph, $\frac{\Delta P_t}{P - \Delta P_t}$, is equal to $(e^{\lambda t} - 1)$, where t is the time since the rock was formed and λ, the proportionality constant, is given by $\ln 2/T$, where T is the half life. Thus, we can use the slope to determine the age of the samples.

13.9.1.2 Extinct Radioisotopes

A particularly clever use of radioisotopes is dating the Earth's differentiation. When the Earth accreted from the protoplanetary disc, it had incorporated within it the radioactive nuclide ^{182}Hf (hafnium). This isotope has a short half life of nine million years, which means that within 90 million years (10 half lives) it was effectively gone. These elements are known as **extinct radionuclides** as they are no longer found on the Earth. They turn out to be very useful. ^{182}Hf is an element that prefers to be in rocks such as those found in the mantle. It is called a lithophile element. Its decay product, ^{182}W (tungsten), however, which is a siderophile element, prefers to be mixed in the iron of the core. If the Earth had differentiated quite quickly (in less than 100 million years) then Hf would have been trapped in the mantle before it had all decayed. The tungsten produced from its decay would also have been trapped in the mantle. We would expect, today, to find enrichment in ^{182}W in the silicate rocks of the mantle and crust compared to the non-radiogenic tungsten isotope ^{183}W. By contrast, if core formation happened very slowly we would expect the hafnium to have mostly decayed before the core formed, giving a chance for its daughter isotope, ^{182}W, to move into the core. Enrichment in ^{182}W is indeed found in the Earth's silicates, showing that core formation occurred quite quickly. The magnitude of the enrichment suggests that differentiation was occurring about 30 million years after the Earth was formed.

One question you might wonder is where these early extinct radionuclides such as ^{182}Hf came from in the first place? They were produced in neutron capture reactions in stars and supernovae (Chapter 8) just prior to the formation of the Solar System, presumably in the star-forming region in which our Solar System was formed. Meteorites also provide a source of products of extinct nuclide decay, telling us about the environment and the materials from which the Solar System formed.

13.9.2 Relative Dating

Another way to date rocks is by comparing them to other sequences of rocks that have a known age (perhaps determined by radiometric age dating as described in the last section). The principles of relative age dating were elaborated by the Danish geologist Nicolas Steno (1638–1686). By comparing rock units to decipher their age relative to one another we can date rocks. To do this we need to recognise some basic geological relationships between rocks. There are many ways to interpret rocks and examine their interrelationships, often made complicated by multiple episodes of re-working, plate tectonics and so on. Nevertheless, some basic ideas can be used to try to make sense of what we see in the rock record. Let's examine just some of the principles which apply to rocks, particularly strata of sedimentary rocks (Figure 13.14).

13.9.2.1 The Principle of Superposition

The principle of *superposition* recognises that in general any given rock layer is younger than the ones below it (the oldest on the bottom and the youngest on the top). Assuming that there has not been large scale deformation or re-working of the rock strata, then this principle applies to rocks where sediments, lava or other rocks have been sequentially layered on top of one another over time.

13.9.2.2 The Principle of Original Horizontality

Settling under gravity generally tends to make sediment layers (and igneous rock eruptions such as lavas) settle in a horizontal form. If layers are folded by rock deformation then the principle of *original horizontality* tells us that the episode of deformation probably occurred after the rocks formed.

13.9.2.3 The Principle of Crosscutting Relationships

If we have a series of rock strata and then find that another feature has cut across those strata, then it is reasonable to assume that this event occurred after the rock strata were laid down. Such a feature could be a fault or a dyke (an intrusion of magma or sediment into a rock). Any feature that cuts across rocks is younger than the youngest rock that is cut (Figure 13.14).

13.9.2.4 The Principle of Faunal Succession

Organisms have evolved and gone extinct through time and therefore the fossil content of rocks changes in a systematic way, reflecting evolutionary changes. Sometimes paraphrased as 'organisms within rock units change with time', if we know the evolutionary sequence of organisms [perhaps through studies of phylogenetic trees (Chapter 6) or fossil evidence from another locality] we can place the rocks into the correct chronological sequence using the fossil record.

None of the above principles are mutually exclusive and by applying a number of them to a given rock

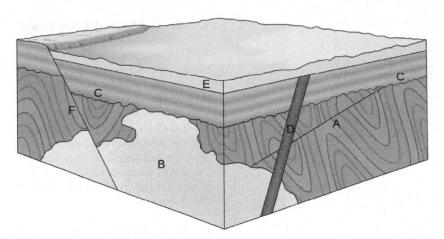

Figure 13.14 *Diagram illustrating some principles of relative age dating of rocks showing in particular the principle of cross-cutting relationships. These relations can be used to give structures a relative age. Zone A is a folded rock strata cut by a fault. The fault must have occurred after the rock was folded. Zone B is a large intrusion of rock cutting through Zone A, which must have occurred after Zone A was folded. Zone C is erosion cutting off Zones A and B on which rock strata were deposited, which must have happened after Zones A and B were formed. Zone D is a volcanic intrusion cutting through Zones A, B and C, which means is it younger than these units. Zone E is even younger rock strata (overlying Zones C and D). Zone F is a very recent fault cutting through Zones B, C and E.* (Source: wikicommons, Woudloper).

stratigraphy, it is possible to ascertain the age of any given strata of interest.

13.9.3 Unconformities

Another tool which geologists use to understand the relationship of rocks to one another is by looking at 'unconformities'. An unconformity is an erosional surface that separates two quite distinct rock masses showing that they were not laid down continuously, but that there was a hiatus or delay in between them. The study of unconformities and their ages can be guided by the principles that we have just examined. Some of the most famous of these features are 'Hutton's Unconformities', a series of unconformities across Scotland first identified by James Hutton as examples of changes in rock formations across time. These features led him to his ideas on **Uniformitarianism**. Here it is worth outlining some of the types of unconformities that geologists recognise on Earth, and on other planetary bodies.

13.9.3.1 Nonconformity

If rocks in a horizontal sequence were eroded down to igneous or metamorphic bedrock and then followed by subsequent deposition of sedimentary layers, then a nonconformity may result (Figure 13.15).

13.9.3.2 Angular Unconformity

If the rocks above and below an unconformity have different orientations then it shows that there was a period of deformation which rotated the rock, followed by erosion, and then renewed deposition. It is the easiest of the three types to recognize because the units are at an angle and truncated with the units above them (Figure 13.15).

13.9.3.3 Disconformity

In the disconformity, sedimentary rocks in a nearly horizontal fashion are eroded. The erosion profile is then covered by subsequent sedimentary deposition. This process shows that there was a period of erosion and then renewed deposition in nearly horizontal layers. It is the most difficult of the unconformities to recognise because the units are nearly horizontal and only a small discontinuous layer can be observed (which could be a soil horizon; Figure 13.15). It is different from a nonconformity, whereby sedimentary rock are overlain over an eroded igneous bedrock.

Mars provides some spectacular examples of unconformities such as the layers of lava sitting on top of clays and sulfates observed by the Curiosity rover at Mount Sharp in Gale Crater (Figure 13.16).

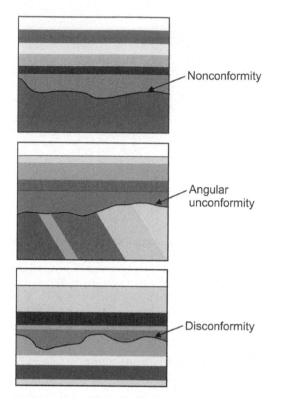

Nonconformity

Angular
unconformity

Disconformity

Figure 13.15 *Different types of unconformity used in the relative dating of rocks* (Source: Charles Cockell).

Figure 13.16 *An unconformity on Mars (Mount Sharp, Gale Crater) is shown by the white dots in the diagram. Lava is thought to be above the white dots, and ancient sediments below* (Source: NASA).

13.10 Geological Time Scales

We have seen how rocks can be laid down. By considering the ideas behind the relative age dating of rocks and the formation of unconformities, we bring to our attention the long time-spans over which these processes occur. But what are these time periods? Although the Earth's geological history is punctuated by sudden catastrophic changes (such as earthquakes and volcanoes), geological processes are in general very slow. Here are just some examples of the rates of change of various types of geological features given as the typical distances over which movements occur during one year:

- Cutting of the Grand Canyon – 0.06 cm
- Opening of Atlantic Ocean along the mid-ocean ridge – 2.8 cm
- Movement of the San Andreas Fault – 5 cm
- Uplift of the Alps – 0.05 cm

Nevertheless, over millions and billions of years, these changes accumulate and along with biological changes, they can be used to define geological time periods as we will see when we discuss mass extinctions in a later chapter.

13.11 The Major Classifications of Geological Time

Geological time scales were developed in the 1800s from the relative dating of rocks. More recently, radiometric techniques have allowed scientists to determine the absolute ages of units with greater accuracy. Many of the names relate back to localities in England (e.g. the Devonian from Devonshire) as Victorian English scientists had a particular fascination with fossil hunting and geological classification.

Geological times, in some sense like the Linnean classification of species, are organised in a hierarchy (Figure 13.17).

The largest subdivision of geological time is an **Eon**. The Hadean, Archean, Proterozoic and Phanerozoic are

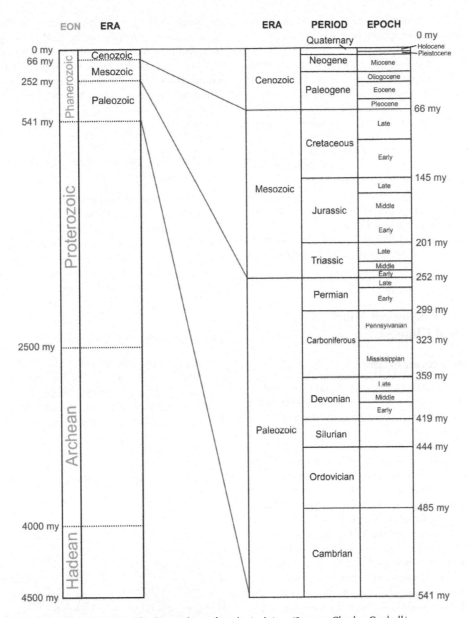

Figure 13.17 *The hierarchies of geological time (*Source: Charles Cockell*).*

the four major eons. Sometimes the Hadean, Archean and Proterozoic are lumped together in a supereon called the **Precambrian**.

At the next level down, the eons are segmented into **Eras**. A good example would be the division of the Phanerozoic Eon (in which you and I live) into the Paleozoic (the oldest era), the Mesozoic and the Cenozoic, which is the youngest era and again, the one in which you and I live.

Eras themselves are further subdivided into **Periods**. In the Phanerozoic the oldest period is the **Cambrian**. The most recent period of the Phanerozoic is the Quaternary.

Finally, we can split these periods yet further into **Epochs**. The Quaternary period is itself split into two epochs, the Pleistocene and the Holocene.

In summary, you and I live in the Holocene Epoch of the Quaternary Period of the Cenozoic Era of the Phanerozoic Eon.

13.12   Some Geological Times and Biological Changes

13.12.1   The Precambrian

As we discussed earlier, geological time periods are often defined by changes in the rock record. For example, the end of the Cretaceous is defined by a marked change in the rocks that is associated with a mass extinction, as we shall discuss in Chapter 15. Unsurprisingly, therefore, geological time periods are to some extent correlated with the major changes in biology occurring on the Earth, since environmental changes which are reflected in changes in the rock record are often accompanied by major changes in the dominant life forms on the planet. It is worth making some observations about some of the major geological time periods with respect to the classification scheme we have just looked at and how they correlate to biology.

Geological time begins with the Precambrian, the supereon that covers the Hadean, the Archean and the Proterozoic. Precambrian time covers approximately 88% of Earth's history. During this time the Earth's biota was exclusively microorganisms.

One enormously influential change that occurred during the Proterozoic was the Great Oxidation Event, a rise in oxygen ~2.4 Ga ago that radically altered the planet for good. We'll spend the next chapter looking at this in more detail. Despite this significant perturbation, we can characterise the Precambrian as a period of microbial domination of the Earth. As was discussed in earlier chapters, we don't know how and when the various microbial metabolisms

that were discussed in Chapter 5 emerged. They could all have arisen very rapidly on the early Archean Earth and then spent the next two billion years refining and evolving, but generally not changing much. Alternatively, the Precambrian might have been a long period of continual innovation, not just in metabolisms, but also in quorum sensing, microbial communication and other facets of the microbial world.

13.12.2   The Phanerozoic: The Rise of Animals and Complexity

As we transition from the Proterozoic (the last eon of the Precambrian) into the Phanerozoic at 541 million years ago, major changes in the Earth's biology begin to occur. The Phanerozoic begins after the Precambrian and is generally the time associated with the rise and widespread existence of multicellular organisms. The Phanerozoic is split into three Eras, each of which has its own characteristic geology and biology.

13.12.2.1   The Paleozoic Era

First there is the Paleozoic Era (541–252 million years ago). Paleozoic means 'ancient life'. The Cambrian, the first period of this Era, began with the breakup of the world-continent **Rodinia** and closed with the formation of a new world continent, **Pangaea**, as the Earth's plates came together once again.

During this time the first multicellular organisms become visible in the fossil record. These fossils were first observed in the Ediacaran Hills in the Flinders Ranges of Southern Australia and they are enigmatic (Figure 13.18).

Figure 13.18   *The enigmatic creatures and body plans of the Ediacaran* (Source: wikicommons, Ryan Somma).

They display a diversity of radial, frond-like and linear worm-like forms, suggesting the possibility that this was a time when many symmetrical body plans were being experimented with. The Ediacaran fauna, appearing at about 575 Ma ago, are almost certainly not the first multicellular eukaryotes on the Earth, but they are the first fossil evidence for an array of animal forms. They range in size from a few centimeters to over a meter and are followed about 20 million years later by the first evidence for organisms with bilateral body plans.

The Ediacaran fauna give way to the organisms of the Cambrian explosion. About 542 million years ago a remarkable increase in the fossil record of animal life occurs, partly attributable to the evolution of skeletons which improved the preservation of these organisms in the rock record. This expansion in fauna begins with the Small Shelly Fauna (SSF) that comprises a variety of small shelled worms and other organisms. They are the precursor to a large expansion in the fossil fauna. The Chengjiang biota, Yunnan Province, China and the slightly younger Burgess Shale in Canada (Figure 13.19), although not the only locations of these fossils, preserve some of the best examples of Cambrian fauna.

The Cambrian Period is sometimes known, very colloquially, as the Age of the Trilobites, because of the ubiquity of these arthropods, which inhabited a vast number of marine environments (Figure 13.20). There are at least 15 000 known species from the fossil record. They dominated marine habitats until the end of the

Figure 13.20 *Trilobites. Ubiquitous marine denizens of the Cambrian. Here is shown a species called* Paradoxides *that lived ~500 million years ago. This specimen is about 10 cm long.*

Permian when they were destroyed by the end-Permian mass extinction. There were many other groups that are represented in the Cambrian fauna, including the echinoderms (which today include the sea urchins and starfish) and sponges. The Cambrian also saw the emergence of our own phylum, the Chordata, of which vertebrates (chordates with backbones) are a sub-phylum. Small tadpole-like creatures set the stage for the emergence of the wide diversity of vertebrates that would one day colonise the entire planetary surface.

The extinction of some of these forms of life that defines the Cambrian–Ordovician boundary was followed by a continued increase in the diversity of marine life, sometimes called the Great Ordovician Biodiversification Event (GOBE) that follows the Cambrian explosion.

During the Paleozoic, early land plants developed. The earliest of plants are thought to have evolved from algae. Originally living in ponds and around water bodies on the continents, periodic desiccation would have provided a selection pressure for early plants to have developed thick cells walls, desiccation resistant early leaves and roots to acquire water and nutrients from beneath the surface. These structural innovations were requirements for the permanent transfer of multicellular organisms from water to land. The earliest of these plants include *Cooksonia* (Figure 13.21), a cylindrical plant that grew from a few millimetres to a few centimetres tall and first appeared in the rock record 425 Ma ago. The first evidence for roots in the fossil record is at ~400 Ma ago. The Rhynie Chert of Scotland, which preserves exquisite fossils from ~410 Ma ago, contains at least seven types of plants and suggests a growing diversity of plants since the appearance of

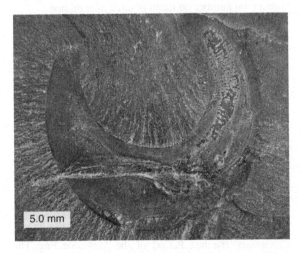

Figure 13.19 *An example Cambrian fossil (*Ottoia prolifica*), a type of marine invertebrate, from the Burgess Shale* (Source: wikicommons, Wilson 44691).

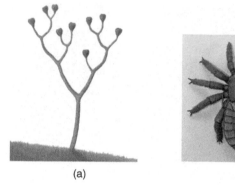 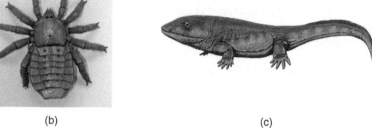

(a)   (b)   (c)

Figure 13.21   *The invasion of land. By the Carboniferous, plants, insects and early tetrapods had invaded the land. (a) An artist's impression of one of the earliest plants (Cooksonia). (b) A model of the eight-legged invertebrate trigonotarbids (Source: Jason Dunlop). (c) An artist's impression of an early tetrapod, Pederpes.*

Cooksonia. Plants were almost certainly not the first denizens of the land masses. Exceptionally preserved fossils of the Doushantuo Formation (635–551 Ma old) at Weng'an, South China, suggest the presence of **lichens** on the land masses in the Precambrian and Cambrian. Microbial mats and free-living biofilms of microorganisms on rocks were likely already growing on the surface of the continents before this time during the Precambrian. However, plants were probably the first multicellular organisms to substantially invade the land.

Rapidly following the plants, the first insects were invading the land masses. Fossil evidence found in rocks from the Lake District, England suggests a millipede invasion of land ~470 Ma ago and by 410 Ma ago, fossils in Shropshire, England suggest that spiders, giant scorpions and centipedes had colonised land. The extinct trigonotarbids, eight-legged animals that had a more beetle than spider-like appearance (Figure 13.21) were a dominant invertebrate of the time. By ~335 Ma there is conclusive evidence that the insects had taken to the skies, with fossils of dragonflies and other orders of aerial insects.

Around 350 million years ago the extent of plant and insect invasion was so great that the Earth was covered in enormous and biologically productive forests. These Carboniferous (359–299 Ma ago) forests were eventually preserved in shallow water, giving rise to coal which drove the Industrial Revolution. The forests were dominated by mosses, ferns and horsetails (flowering plants did not emerge until much later, becoming globally widespread about 120 million years ago). Some of the carboniferous plants were vast, including the 35 m high *Lepidodendron.*

The Carboniferous forests were home to vertebrates as well as insects and plants. Around ~370 Ma ago, fossil evidence from Elgin, Scotland shows that the first fish-like amphibians were on land. By ~345 Ma, genuine **tetrapods** with short tails, such as the 1 m-long *Pederpes* (Figure 13.21) were walking on land. The first tetrapods were an important evolutionary innovation since they were the precursors of reptiles and mammals.

This rise of land biota is significant because these organisms would eventually dominate life on the Earth. Although today the oceans cover 71% of the Earth, there are estimated to be about four times as many eukaryotic species on land compared to the marine environment, with a total number on Earth of circa nine million.

The end of the Paleozoic is marked by the largest mass extinction in history, the end-Permian extinction, which wiped out approximately 90% of all marine animal species including the trilobites and 70% of land animals.

13.12.2.2   *The Mesozoic Era*

The Paleozoic is followed by the Mesozoic (252–66 Myr ago; *Mesozoic* means 'middle life'). At the beginning of this era the continents were joined as Pangaea, but it began to break up around the middle of this era (~225 million years ago). The climate during this Era can generally be characterised as warm, with hot and dry conditions during the Triassic (the first Period of the Mesozoic), and warm tropical conditions running through into the Cretaceous, the last Period of the Mesozoic Era. The discovery of dinosaur fossils in Antarctica is testament to the fact that warm conditions at times stretched to the poles.

Reptiles became the most abundant animals because of their ability to adapt to the drier climate of the Mesozoic Era. Skin helped to maintain body fluids and embryos laid in eggs with shells protected against desiccation. The earliest reptiles probably emerged in the mid-Carboniferous from organisms such as *Hyalonomus*, a 30 cm-long lizard-like creature that laid eggs. By providing a liquid environment enclosed within a shell, the evolution of eggs enclosing offspring marked the true independence of animals on land. These organisms would diversify into the Mesozoic.

Eventually, the diversification of reptiles became so great that the Mesozoic is also known as the 'Age of the Dinosaurs' (Figure 13.22). The dinosaurs were part of the great lineage of the archosaurs. Today the archosaurs are represented by crocodiles and birds. The first small dinosaurs appeared in the Triassic. *Eoraptor*, for example, was an agile, small 1 m-high dinosaur (Figure 13.22). Larger and more abundant dinosaurs appeared in the Jurassic Period and went on to dominate into the Cretaceous. Some of the largest of these were the sauropods such as *Brachiosaurus* (Figure 13.22), which reached lengths of over 20 m.

Alongside the sauropods evolved the therapods, including *Deinonychus* (Figure 13.22), a fast human-sized creature with an enormous claw on its hind leg, probably used for slashing prey. The therapod lineage would give rise to the *Archaeopteryx*, the oldest bird.

Other reptiles were also important during this time including the winged pterosaurs (which were not of the dinosaur lineage, but separate winged reptiles). Some of the pterosaurs, such as *Quetzalcoatlus* (Figure 13.22) achieved wing spans of up to 15 m.

Throughout the Mesozoic the most successful marine reptiles were the Plesiosaurs and Ichthyosaurs with streamlined bodies and paddles for efficient movement through the water (Figure 13.22). However, the entire marine realm experienced a revolution in diversity, with the emergence of new plankton groups, new invertebrates such as the ancestors of crabs and lobsters, and the appearance of modern fish, including sharks.

Small mammals appeared during this Era (Figure 13.22) such as the dog-like *Thrinaxodon*. They were warm-blooded animals and hair covered their bodies. They belonged to the cynodonts (meaning 'dog teeth') in the clade Cynodontia and were successful at rummaging in the undergrowth, living on the edge of ecosystems dominated by reptiles. They had differentiated teeth types and ribs only at the front of their thorax characteristic of mammals (unlike reptiles). They would be the progenitors of all modern mammals.

The main plant life of this time was **gymnosperms** or plants that produce seeds but no flowers, for example Pine Trees. Flowering plants (**angiosperms**) appeared midway through the Era and achieved rapid increases in diversity. Today they dominate plant life.

The Mesozoic Era ended with a mass extinction event 65.5 million years ago. Many groups of animals, including the dinosaurs, disappeared suddenly at this time. This change in the rock record defines the end of the Cretaceous and the end of the Mesozoic Era.

Many scientists believe that this event was caused by a comet or asteroid colliding with the Earth. The effect would be a vast cloud of dust that filled the air, blocking out sunlight. The plants died from cold and a lack of light for photosynthesis and the herbivores subsequently died. The carnivores that ate the plant-eaters died. Many animals today, including the mammals, are descendants from the survivors of this extinction event. We will explore the evidence for this scenario in more detail in Chapter 15.

13.12.2.3 *The Cenozoic Era*

The Mesozoic Era is followed by the Cenozoic Era (66 Myr to the present). The climate in this Era has been in general warm and mild, interspersed in the last few million years by ice ages. Marine animals such as whales and dolphins evolved in this Era. Mammals began to increase in abundance and evolve adaptations that allowed them to live in many different environments, on the land, air and the sea, niches vacated by the reptiles. Grasses increased in diversity and abundance and provided a food source for grazing animals.

Many mountain ranges formed during the Cenozoic Era including the Alps in Europe, the Himalayas in India and the Rocky Mountains in the United States. The growth of these mountains may have helped to cool down the climate as Ice Ages occurred late in the Cenozoic Era (in the Quaternary Period). It is suggested that the mountain ranges would have provided a large area of fresh rock surfaces on which rock weathering reactions occurred, which consume the greenhouse gas CO_2 (Chapter 16), thus causing cooling. As the climate changed, the animals had to adapt to the rise and fall of the oceans caused by melting glaciers. This era is sometimes called the 'Age of Mammals' (Figure 13.23). The Cenozoic Era sees biology evolve into shapes and forms that we are familiar with today, including the evolution of modern humans, which we will return to in Chapter 21.

Figure 13.22   *Artists' impressions of some of the reptiles and a mammal of the Mesozoic.* (a) Eoraptor resto (Source: wikicommons, Conty). (b) Brachiosaurus altithorax (Source: wikicommons, Nobu Tamura). (c) Quetzalcoatlus (Source: wikicommons, public domain). (d) Deinonychus antirrhopus (Source: wikicommons, Michal Maňas). (e) Plesiosaurus dolichodeirus (Source: wikicommons; Adam Stuart Smith). (f) It would remiss not to remember that this was the Era when the first mammals, albeit small, emerged. This image shows the dog-like Thrinaxodon (Source: wikicommons, Nobu Tamura).

Figure 13.23 *An image depicting some of the forms of mammals of the Miocene epoch (23.030–5.333 million years ago) of the Cenozoic (66 million years ago to the present) and illustrating more generally a world in which mammals were now dominant after the fall of the reptiles. The large animals in the centre left are the extinct chalicotheres, a group of herbivorous mammals related to the horses and camels (ungulates).*

Figure 13.24 *Cenozoic biology. A tool-building ape emerges capable of calculating orbital dynamics and leaving the planet on which it evolved (Source: NASA).*

The tool building ability of humans (Figure 13.24) and the changes wrought on planet Earth through industrial activity has been suggested to be so dramatic that it has even been proposed that a new geological epoch should be defined: the **Anthropocene**.

13.13   Conclusion

The internal structure of the Earth is such that the surface of the planet is made up of plates that move over geological time periods. The movement of plates has consequences for the geosphere, hydrosphere, atmosphere and biosphere. The way in which different rock types are altered by plate tectonics and other processes such as weathering can be described by the rock cycle. Although plate tectonics does not necessarily occur on other planets, the rock cycle describes the fate of rocks in many ways, for example, erosion and sedimentation, that we would

---

Debate Point: The anthropocene

Geological time periods are usually defined by changes in the rock record. Often these changes are accompanied by large changes in the biota on the planet, perhaps one of the best examples being the end-Cretaceous extinction that wiped out dinosaurs and many other land and marine organisms. With about seven billion of us on the planet and our industrial activities now changing the composition of the planetary atmosphere, it has been proposed that we should define a new epoch in geological time: the Anthropocene. This Epoch would follow on from the Holocene. Do you think this is justified? Is it consistent with the factors that define changing geological epochs, eras and eons in the past history of the Earth? Can you list some of the major factors that have changed on the Earth since the rise of human civilisation and that have been caused by it? Do they merit a new epoch?

Crutzen, P.J. (2002) The geology of mankind. *Nature* **415**, 23.

Steffen, W., Persson, A., Deutsch, L. *et al.* (2011) The Anthropocene: from global change to planetary stewardship. *Ambio*, doi: 10.1007/s13280-011-0185-x

---

expect to occur on other planetary bodies, particularly locations with liquid water. A variety of methods can be used to date rocks, both absolutely and by using relative methods. The division of geological time into hierarchies of increasing detail is somewhat akin to phylogenetics in biology and allows for a systematic ordering of geological knowledge. These approaches are just as applicable to other planets as they are on the Earth. In recent years, some scientists have suggested that the effect of human industrial activity on planet Earth is so great that a new Epoch should be defined as the Anthropocene.

Further Reading

Books

Canup, R.M., Righter, K. (2000) *The Origin of the Earth and Moon.* University of Arizona Press, Tucson.

Hazen, R.M. (2013) *The Story of Earth: The First 4.5 Billion Years, from Stardust to Living Planet.* Penguin Books, London.

Papers

Burke, K. (2011) Plate tectonics, the Wilson cycle, and mantle plumes: geodynamics from the top. *Annual Reviews of Earth and Planetary Science* **39**, 1–29.

Favilli, F., Egli, M., Brandova, D., Ivy-Ochs, S., Kubik, P., Cherubini, P., Mirabella, A., Sartori, G., Giaccai, D., Haeberli, W. (2009) Combined use of relative and absolute dating techniques for detecting signals of Alpine landscape evolution during the late Pleistocene and early Holocene. *Geomorphology* **112**, 48–66.

Hazen, R.M., Papineau, D., Bleeker, W., Downs, R.T., Ferry, J.M., McCoy, T.J., Sverjensky, D.A., Yang, H. (2008). Mineral evolution. *American Mineralogist* **93**, 1693–1720.

Hawkesworth, C.J., Kemp, A.I.S. (2006) Evolution of the continental crust. *Nature* **443**, 811–817.

Karson, J.A. (2002) Geologic structure of the upper-most oceanic crust created at fast- to intermediate-rate spreading centers. *Annual Reviews in Earth and Planetary Sciences* **30**, 347–384.

Pisani, D., Poling, L.L., Lyons-Weiler, M., Hedges, S.B. (2004) The colonization of land by animals: molecular phylogeny and divergence times among arthropods. *BMC Biology* **2**, doi:10.1186/1741-7007-2-1.

Rosing, M.T., Bird, D.K., Sleep, N.H., Glassley, W., Albarede, F. (2006) The rise of continents – an essay on the geologic consequences of photosynthesis. *Palaeogeography, Palaeoclimatology, Palaeoecology* **232**, 99–113.

Valentine, J.W., Jablonski, D., Erwin, D.H. (1999) Fossils, molecules and embryos: new perspectives on the Cambrian explosion. *Development* **126**, 851–859.

Wagner, G.A., Krbetschek, M., Degering, D., Bahain, J.-J., Shao, Q., Falguères, C., Voinchet, P., Dolo, J.-M., Garcia, T., Rightmire, G.P. (2010) Radiometric dating of the type-site for *Homo heidelbergensis* at Mauer, Germany. *Proceedings of the National Academy of Sciences* **107**, 19726–19730.

Walker, J.D., Geissman, J.W., Bowring, S.A., Babcock, L.E. (2012) Geologic Time Scale v. 4.0. *Geological Society of America*, doi: 10.1130/2012.

Watanabe, Y., Martini, J.E.J., Ohmoto, H. (2000) Geochemical evidence for terrestrial ecosystems 2.6 billion years ago. *Nature* **408**, 574–578.

14

The Rise of Oxygen

Learning Outcomes

➤ Explain how we know that the concentration of oxygen on the early Earth was low.

➤ Understand the major sources and sinks of oxygen on the Earth.

➤ Describe some of the theories for why sudden rises of oxygen occurred through Earth history.

➤ Explain the concept of the Snowball Earth and how these events occur and are reversed.

➤ Describe the consequences of the rise of oxygen for life.

➤ Explain some of the ideas linking the rise of oxygen to the emergence of multicellular life and ultimately intelligence.

14.1 Dramatic Changes on the Earth

There is a tendency, given the short span of human life, to regard the Earth's history as almost unchanging. However, throughout the planet's history there have been some dramatic events. Mass extinctions, the colonisation of land, the rise of animals – all of these events either count as major evolutionary transitions or perturbations to the Earth system. Crucially, some of these changes have been caused by life itself.

Each one of these changes could occupy a textbook in itself, but here we are interested not just in describing a variety of the major events that can shape a biota

co-evolving with its planet. We would like to identify particular changes that could have implications for the emergence of life on any planet in the Universe and represent profound ways in which a biota can influence the environmental conditions on a planet.

Perhaps one of the most significant environmental changes that has occurred throughout the history of the Earth is the rise in atmospheric oxygen. About 2.4 billion years ago (the **Great Oxidation Event**) and again about 700 million years ago, there was a dramatic rise in oxygen in the Earth's atmosphere leading to levels we are familiar with today.

As we discussed in Chapter 10, the early Earth had an atmospheric composition that was dominated by a range of gases such as CO_2, N_2, H_2, CH_4, SO_2, H_2S outgassed from the primordial inventory of volatiles. The concentrations of oxygen at this time were negligible. But at some point in Earth history oxygenic photosynthesis evolved which, as we explored in Chapter 5, has the feature of being able to use water as an electron donor to drive energy (ATP) production with the concomitant release of oxygen as a waste product.

Today, the concentration of oxygen in the Earth's atmosphere is 20.95% and this gas is the electron acceptor in aerobic respiration, which powers almost all multicellular organisms, including the one writing this textbook. That last comment is not a flippant and frivolous aside. If we make the assumption that there is a cause–effect relationship in that statement, then we are saying that the rise of oxygen is necessary for the rise of intelligence on this planet and potentially others.

Astrobiology: Understanding Life in the Universe, First Edition. Charles S. Cockell.
© 2015 John Wiley & Sons, Ltd. Published 2015 by John Wiley & Sons, Ltd.
Companion Website : www.wiley.com/go/cockell/astrobiology.

Quite apart from intelligence, aerobic respiration powers most of the productive chemoheterotrophic biosphere and, without it, the biomass on the planet would be smaller. You will remember from Chapter 5 that anaerobic modes of metabolism typically generate about ten times less energy than aerobic respiration.

These points bring us at once to our major question: how did we get from the low oxygen atmosphere of the early Earth to the present-day planet with its hugely productive aerobic respiring flora and fauna? What factors caused this transition and what were the consequences for the planet and life? The rise of oxygen has sometimes been colloquially referred to as the greatest pollution event of all time. This is somewhat unfair because the organisms (cyanobacteria) behind the rise of oxygen hardly knew that they were polluting the atmosphere, but it nonetheless sufficiently captures the sheer importance of this event.

In this chapter, we'll explore this change in more detail as a key example of the changes in the conditions for life on a planet and co-evolution of life and its planetary environment.

14.2 Measuring Oxygen Through Time

How do we know that oxygen concentrations have changed through time? There are a number of ways that scientists can measure past atmospheric gases. One way is to extract gases from ice cores and measure the concentration. This has been successfully accomplished for the study of CO_2 in Antarctic ice cores. However, these can only take us back hundreds of thousands of years. We don't have ices from the early history of the Earth.

Instead, scientists rely on **geological proxies**. Geological proxies are changes in the characteristics of rocks or minerals or compounds that can tell us something about the environment in which they were formed. In short, a proxy is a measurement of some sort that provides information about something we cannot measure directly. We look at the rock record to get proxies for oxygen abundance further back in Earth history.

These geological proxies suggest that the Earth has experienced at least two major rises of oxygen throughout its history (and other changes in concentration since these two major rises, which we will discuss). The first change occurred about 2.4 Ga ago. Before that time oxygen concentrations in the atmosphere were negligible, probably much less than 10^{-5} times the present atmospheric level

(and potentially as low at 10^{-15} times present atmospheric level). The first rise in the Paleoproterozoic (the first era of the Proterozoic, from 2.5 to 1.6 Ga ago) took oxygen concentrations to a small percentage of present-day levels. A second rise in oxygen at about 700 Ma ago occurred in the Neoproterozoic (the last era of the Proterozoic, from 1 Ga to 541 Ma ago). Oxygen concentrations rose to at least 10% of the present-day value, with a rise or rises after that leading to its present-day value. Nearer the present, in the Paleozoic, there is evidence for a rise in oxygen concentrations to above present-day values ($\sim$30%). In Figure 14.1 you can see a schematic showing the major patterns of atmospheric oxygen concentrations over the history of the Earth.

14.2.1 Minerals that Form at Low Oxygen Concentrations

One very effective way to infer oxygen concentrations is to look at the prevalence of minerals whose formation is associated with quite specific oxygen concentrations. Although oxygen concentrations are very variable in different environments (even on present-day Earth there are plenty of underground environments with no or little oxygen) the general trends of the abundance of different minerals can provide important insights on global conditions.

An example of a mineral that tends to be formed at low oxygen concentrations (Figure 14.2) is pyrite (iron sulfide, FeS_2), which is generally formed at an oxygen concentration less than 0.1% of the Present Atmospheric Level (or PAL). Note that the term PAL is not the absolute percentage of oxygen, but a percentage of the present atmospheric level, which is 20.95%. Therefore 10% PAL would be an oxygen concentration of 2.1% (care must be taken because oxygen levels can sometimes be stated at absolute concentration or as PAL). Uraninite (uranium oxide, which is mainly UO_2) is another example. It forms in oxygen concentrations less than 0.01% PAL. Siderite (iron carbonate; $FeCO_3$) forms at 0.001% PAL or less. In particular, we can look at the concentrations of these oxides in **detrital** minerals, which are minerals deposited in sediments. They reflect the conditions in the atmosphere to which the sediments are exposed. In modern rivers, pyrite, uraninite and siderite are completely oxidised during transport and so are absent in detrital sediments.

All of these minerals are more common in sediments before about 2.4 Ga ago and suggest general conditions of low oxygen concentrations in the atmosphere at that time.

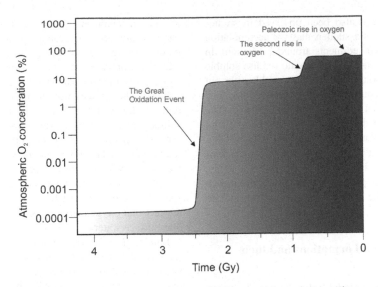

Figure 14.1   *Schematic showing the history of terrestrial atmospheric oxygen through time. The graph shows the two major rises in oxygen and an oxygen pulse in the Paleozoic. The y axis is a logarithmic scale. The value before ~2.4 Ga ago is an upper limit* (Source: Charles Cockell).

Figure 14.2   *Minerals whose formation is favoured in low oxygen concentrations below Present Atmospheric Levels (PAL;* Source: wikicommons, Rob Lavinsky).

14.2.2   Changes in the Oxidation State of Elements

We can also look at the **oxidation state** or redox state of minerals through time as a proxy for oxygen concentrations.

Some elements are soluble in only one oxidation state. For example, iron and manganese are much more soluble when they are in a reduced form (Fe^{2+} and Mn^{4+}, respectively) compared to their more oxidised form (Fe^{3+} and Mn^{6+}, respectively). Uranium and molybdenum, by contrast, are much more soluble when they are in their oxidised forms (U^{6+} and Mo^{6+}), compared to their reduced form (U^{4+} and Mo^{4+}).

It is clear that at the Great Oxidation Event there were dramatic changes to iron. Ancient soils or **paleosols** older than about 2.4 Ga old are significantly depleted in iron relative to those younger than 1.8 Ga. This observation is explained by a change in the oxidation state of iron between 2.4 and 1.8 Ga ago. In water before 2.4 Ga ago iron was mainly in its soluble reduced form (Fe^{2+}) and was more readily leached from soils. After the Great Oxidation Event, in its more insoluble (Fe^{3+}) form, it was less readily leached from soils.

'Red beds', which are iron-rich sediments derived from oxidised iron or hematite (Fe_2O_3), dramatically increase in abundance after ~2.3 Ga, suggesting widely oxygenated conditions.

Similar features have also been observed in studies of manganese. There is a sudden massive accumulation of manganese in ocean sediments around 2.2 Ga ago. In seawater of circumneutral pH, O_2 would oxidise soluble Mn^{2+} to produce insoluble Mn^{4+} that would settle out in the sediments, accounting for the observations. The chemistry suggests that water became oxygenated. The result of this massive deposition was highly significant for humans. The Kalahari manganese field is 11×50 km in extent and ~50 m thick. It is the location for much of the manganese which is mined in South Africa today, showing how ancient deposits laid down as a consequence of the oxidation of the atmosphere now have economic significance.

14.2.3 Banded Iron Formations and their Isotopes

A very important line of evidence for the rise of oxygen about 2.4 Ga ago is Banded Iron Formations (BIFs). BIFs are distinctive rocks that contain bright red bands of iron oxides in between layers of **chert** (silica-rich rock; Figure 14.3).

The process of formation of the BIFs remains poorly understood. However, they are thought to be caused by the precipitation of iron oxides from seawater. Early waters in an oxygen-poor atmosphere would have had iron in its very soluble reduced form (Fe^{2+}). Every now and again, this iron was oxidised to iron oxides (containing iron in its Fe^{3+} state), which are not very soluble and would have precipitated out of the water column to form the BIFs.

What processes caused this? Local oxygenation of water, perhaps caused by seasonal blooms of cyanobacteria producing oxygen from photosynthesis, is suggested

Figure 14.3 *A banded iron formation showing the layers of cherts (grey/black) interspersed with layers of iron oxides (red;* Source: wikicommons, Woodloper).

as one mechanism. Chemoautotrophic iron-oxidising bacteria are also suggested as another candidate for oxidising iron and producing insoluble iron oxides. An additional process may have been caused by a rise in oxygen forming sulfates from the reaction of oxygen with sulfides. The sulfates washed into the oceans and were reduced by sulfate-reducing bacteria. The biological sulfide that was formed then bound to reduced iron and formed iron sulfides in the deep oceans. BIFs are rarely found after 1.8 Ga ago, showing that the oxygenation of the oceans was a slow delayed process, as the rise of oxygen is thought to have occurred about 600 million years before the end of BIF formation. This could reflect the fact that oxygenation of the deep oceans took longer than surface water.

Banded iron formations also record isotopic evidence for changes in oxygen concentrations. Some elements, when they are oxidised, not only change oxidation state, they also become isotopically fractionated. When the element, chromium, is oxidised from Cr^{3+} to Cr^{6+} it can become fractionated, causing an increase in the $^{53}Cr/^{52}Cr$ ratio. Fractionated Cr isotope data indicate the accumulation of Cr^{6+} in ocean surface waters approximately 2.8–2.6 Gyr ago, evidence of a possible transient elevation in surface ocean oxygenation before the first great rise in atmospheric oxygen.

14.2.4 Sulfur Isotope Fractionation

In the previous sections we focused on the formation and preservation of minerals at low oxygen concentrations and changes in oxidation states of elements as proxies for oxygen concentrations in aqueous environments. Here we explore in more detail the use of isotopes as proxies, which can be used to constrain both aqueous and atmospheric oxygen levels.

Sulfur has become an important element for investigating past oxygen levels. Sulfur exists in four stable isotopes (^{32}S, ^{33}S, ^{34}S and ^{36}S). In Chapter 12 we saw how ^{32}S and ^{34}S can be used to search for evidence of life in the ancient rock record. In particular, the fractionation, or preferential use of ^{32}S compared to ^{34}S, by organisms results in a higher concentration of ^{32}S in sulfides produced from biological processes such as microbial sulfate reduction.

However, sulfur fractionation can only be achieved if sulfate concentrations reach a concentration of at least 1 mM (one millimole). The lack of significant $^{34}S/^{32}S$ sulfur fractionation in Archean rocks suggest that sulfate concentrations were quite low and this in itself suggests that the surface of the Earth was not very oxidising, since oxidised conditions are required to produce sulfate ions

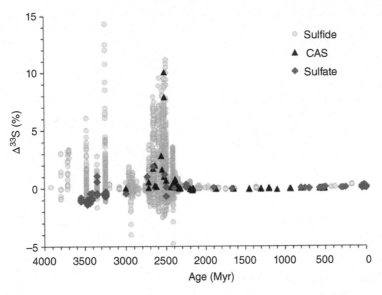

Figure 14.4 *Mass independent fractionation (MIF) of sulfur over time, showing the disappearance of the $\Delta^{33}S$ signal after ~2.4 Ga ago. The graph shows the sulfur fractionation measured in different types of sulfur compounds including sedimentary sulfides (grey circles), sulfates (red diamonds), and carbonate associated sulfate (CAS), which is sulfates found in carbonate rocks [black triangles; from Claire et al. (2014)]. Note that 'Big delta' ($\Delta^{33}S$) is the deviation of $\delta^{33}S$ from fractionation expected in proportion to mass.*

(SO_4^{2-}). By 2.3 Ga ago, sulfide minerals with strong fractionation indicative of biological activity can be found, suggesting that the concentration of oxygen had risen, causing sulfate concentrations in the oceans to rise.

The fractionation observed in ancient rocks is usually proportional to the relative masses of the isotopes. The reason for this is that the mass of the isotopes affects their chemical reactivity and the strength of their bonds, creating small differences, such as slightly more reactive bonds, in light isotopes (Chapter 12). This **mass-dependent fractionation** is the typical way to look for life. For example, in the case of sulfur, mass-dependent fractionation would lead to a prediction that given a certain degree of fractionation, we would expect the fractionation between ^{33}S and ^{32}S to be half that between ^{34}S and ^{32}S.

However, there is another way in which isotopic fractionation can occur and that is by a process that fractionates the sulfur isotopes independently of their mass (**mass-independent fractionation**). One example of a process that can yield mass independent fractionation is the light-driven splitting (**photolysis**) of SO and SO_2 gases produced by volcanoes, specifically by short-wavelength ultraviolet radiation.

For SO_2 to be photolysed, short-wavelength UV radiation must penetrate deep into the atmosphere. As we

discussed in Chapter 10, in the absence of oxygen in the atmosphere, ozone does not form in the stratosphere. On the present-day Earth, ozone absorbs UV radiation below about 295 nm wavelength but, on the early Earth, in the absence of ozone, UV radiation down to 200 nm could have penetrated into the atmosphere, allowing for mass independent fractionation of sulfur gases.

Once oxygen accumulated in the atmosphere it would have had two effects: (i) it would react with sulfurous gases, forming sulfuric acid, which would rain out of the atmosphere, thus clearing the atmosphere of sulfur gases and (ii) it would generate an ozone shield that would block short-wavelength UV radiation. The presence of mass-independent fractionation in sulfur isotopes before ~2.4 Ga ago, but the disappearance of this signature about 2.4 Ga ago is interpreted to be evidence of the appearance of oxygen in the atmosphere (Figure 14.4).

14.3 Summarising the Evidence for the Great Oxidation Event

Accepting some uncertainty in the dates when the various changes occurred, there is actually a remarkable agreement amongst a variety of evidence that we have just examined (and other evidence besides). The evidence suggests an abrupt rise in oxygen occurs at about 2.4 Ga ago (Figure 14.1). This marks the Great Oxidation Event, an important transition in the Earth's geochemical environment.

Although the corroborated evidence for rises in oxygen is impressive and self-consistent, it raises the question of why this rise in oxygen occurred. What factors led to an apparently dramatic, and rapid, rise in geological terms?

To understand some of the hypotheses for how this change occurred, we must first understand the sources and sinks of oxygen.

14.4   The Source of Oxygen

The primary source of free oxygen was cyanobacterial oxygenic photosynthesis (Figure 14.5).

Oxygen gas is produced as these organisms used water as the electron donor (Chapter 5):

$$6CO_2 + 6H_2O \rightarrow C_6H_{12}O_6 + 6O_2 \qquad (14.1)$$

Photosynthesis is thought to have arisen between 3.5 and 2.7 Ga, but the exact timing is difficult to pin down. Crucial to this discussion is the likelihood that oxygenic photosynthesis evolved well before the Great Oxidation Event, implying that prior to the event the oxygen produced by cyanobacteria must have been somehow mopped up before it was capable of causing a large scale increase in atmospheric oxygen concentrations. This view

is supported by a number of lines of evidence for 'oxygen oases' existing well before the large-scale rise in atmospheric oxygen. Examples include 2.8-Ga shallow water limestones in Canada that attest to localised oxygenated water at that time and biomarkers for cyanobacteria in rocks older than 2.4 Ga.

14.5   Sinks for Oxygen

There are pathways by which oxygen can be removed from the atmosphere. These 'sinks' for oxygen mop it up from the atmosphere before it has a chance to build up. Two major processes are:

1. Oxidation of dead organic material. The reaction pathway shown above for photosynthesis is reversible in two ways. The reverse reaction is the process of aerobic respiration whereby organics are oxidised using oxygen as the electron acceptor by organisms to produce energy (Chapter 5), thus mopping up oxygen. The reaction can also be reversed when organic material from dead organisms oxidises abiotically in an oxygen-containing atmosphere. In both cases, oxygen is consumed.

2. Reaction with reducing elements and compounds. Oxygen reacts with reduced gases such as hydrogen and hydrogen sulfide exhaled from volcanoes (Figure 14.6) and reduced elements such as ferrous iron (Fe^{2+}) in water bodies. In the process it is removed from the atmosphere.

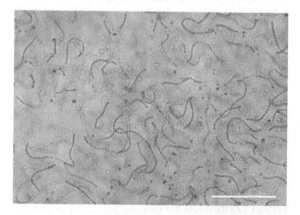

Figure 14.5   *Cyanobacteria. The primary suspects responsible for the large-scale increase in atmospheric oxygen. Here is shown a species of the filamentous, segmented genus* Nostoc. *The scale bar is 100 µm (Source: wikicommons, Gibon).*

Figure 14.6   *Volcanoes are one source of reduced compounds (gases) that would have depleted atmospheric oxygen in early Earth history. This impressive eruption is from Mount Redoubt, Alaska, on 21 April 1990 (Source: USGS).*

14.6   Why did Oxygen Rise?

In the previous sections we saw that there are both sources and sinks of oxygen. We can surmise that the rise of oxygen during the past history of the Earth must be caused by a change in the balance between sources and sinks.

Much of the oxygen produced by the very first photosynthesisers might have been mopped up by volcanic gases in the atmosphere and reduced elements such as ferrous iron in the oceans. In other words, oxygenic photosynthesis might have been around some time before the Great Oxidation Event, but the oxygen just didn't have a chance to accumulate in the atmosphere.

There have been a variety of proposed mechanisms that could have led to discrete rises in oxygen. Let's examine a few of these ideas:

1. ***There was a change in the redox state of gases being exhaled from the mantle.*** A continent stabilisation period is thought to have occurred around 2.7 Ga. At this time it has been suggested that volcanism went from being mainly submarine to a mix of land-based and submarine. Submarine volcanism is more effective at producing reducing gases and so in turn more effective at mopping up oxygen. If its contribution declined then less oxygen would have been consumed and it could have begun to build up. However, one problem with this theory is that the difference in the production of reduced gases does not seem to be able to account for the scale of the rise in oxygen.

2. ***There was a change in marine nutrient supply.*** Phosphorus is an important nutrient for cyanobacteria, the organisms responsible for all oxygenic photosynthesis prior to plants. Phosphorus is often lost from water by adsorption onto iron oxides. However, as cyanobacteria locally increased the availability of oxygen in the oceans and caused the water to become more oxygen-rich, the iron would have precipitated out (as Banded Iron Formations). Eventually, when the iron became more depleted it would have been less capable of pulling phosphorus out of the water column. The availability of phosphorus would have increased, spurring a rise in photosynthetic productivity and thus enhancing the source of oxygen in a feedback process.

3. ***The rise in oxygen may have been caused by a decrease in atmospheric methane levels.*** Methane has a tendency to react with oxygen and remove it from the atmosphere. The reduction in methane production has been hypothesised to be linked to a cooling Earth. As the mantle cooled it erupted less nickel-rich rocks. Ni is a key metal co-factor in a number of **enzymes** involved in biological **methanogenesis**. As its availability decreased, the methanogens could not be as productive, reducing the atmospheric concentrations of methane available to react with oxygen, allowing oxygen concentrations to increase. Problems exist with the idea since the Ni decline seen in the rock record may not be enough to trigger a methane collapse.

4. ***There was a period of enhanced organic burial.*** If more carbon was buried then there would be a net build-up of oxygen since it would not have a chance to react with the organic matter. However, the carbon would eventually be returned into the atmosphere as CO_2 when the organic material was subducted and heated, so although this mechanism can provide a short term possibility of enhancing oxygen concentrations, it is difficult to see how it could cause a long-term irreversible increase in oxygen.

5. ***There was a switch between two feedback stabilised steady states driven by the interactions between oxygen and UV radiation.*** UV radiation has a tendency to destroy oxygen and thereby decrease its lifetime in the atmosphere. Once atmospheric oxygen concentrations reached a threshold level, UV shielding by ozone became effective. This increased the atmospheric lifetime of oxygen, causing oxygen levels to increase yet further in a positive feedback effect. The Great Oxidation Event marks the switch between the low oxygen and higher oxygen steady states.

6. ***The Earth became irreversibly oxidised as the lighter gas, hydrogen, was lost to space.*** During the early history of the Earth, the planet lost its initial inventory of light hydrogen gas into space. Most hydrogen atoms in gases that existed after that time, including those in hydrogen itself and methane, ultimately were produced from the breakdown of water (H_2O). As these gases were lost to space or broken down into their constituents, resulting in the loss of hydrogen to space, there was therefore a net build-up of oxygen on the planet, which was heavier and did not so readily escape to space. Thus, the Earth became irreversibly oxidised.

We do not know the answer. All of these ideas (and there are others besides) are the subject of active scientific debate. None of them have to be mutually exclusive and it is possible that a number of these factors, and others, combined to create a situation where the Earth flipped from one steady state of low oxygen to another with higher levels of oxygen.

Furthermore, it is almost certainly the case that the same exact factors were not responsible for the rise in oxygen ~2.4 Ga ago and the rise ~700 Ma ago. Here we have focused on the Great Oxidation Event, but similar mechanisms in the changes in sources and sinks of oxygen would have driven the second major rise in oxygen.

The regulation of oxygen at nearly constant levels over long periods of time between these rises is thought to be driven by different mechanisms. Increasing levels of oxygen are likely to increase the oxidative weathering of pyrite (iron sulfides) and organic matter, leading to more consumption of oxygen and hence stabilising concentrations in a negative feedback effect. Higher oxygen concentrations may also enhance aerobic respiration which will consume more oxygen, thus providing another negative feedback on oxygen concentrations. At high levels of oxygen, since the emergence of plants, forest fires would be more common, consuming oxygen in reactions with organic matter.

14.7   Snowball Earth Episodes

What is in little doubt is that these changes in oxygen concentration would have had profound consequences for the environment of the Earth. One of the most dramatic climatic changes that has been linked with both major increases in atmospheric oxygen concentration is the **Snowball Earth** episodes (Figure 14.7).

Around the first rise in oxygen there is evidence in the rock record for substantial glaciations. At 2.45–2.22 Ga ago in the Paleoproterozoic Era, the **Huronian** interval

of glaciations are recorded by a number of features. The rocks of that period contain dropstones, which are rocks that drop out from floating ice sheets as they melt, introducing large stones into otherwise fine-grained deep-water sediments. Powerful glaciers leave scratch marks on rocks, caused by ice and rocks within glaciers tearing away at rock over which they are moving. Mixtures of rocks, called tillites, are also a feature of glacial environments and are found in the rock record.

Perhaps one of the most telling pieces of evidence for global glaciation is the presence of features of glaciation in rocks that were at the equator when they were formed. When sediments form, they trap mineral grains such as magnetite (Fe_2O_3) that are magnetic. These tiny bar magnets orient themselves according to the direction of the Earth's magnetic field. The magnetite found in the Paleoproterozoic rocks lies almost parallel to the rock layers, which implies that the Earth's magnetic field was horizontal to the Earth's surface. This occurs near the equatorial regions (nearer the poles, the magnetic field feeds through the surface into the core of the Earth and is nearly perpendicular to the ground). Thus, the evidence suggests glaciated conditions near equatorial regions.

The evidence for global glaciations in the Paleoproterozoic is repeated in the Neoproterozoic with little (if any) evidence for glaciations in the two billion years in between. The Neoproterozoic glaciations (two to four of them between 715 and 610 Ma ago), ending with the Marinoan glaciation ~610 Ma ago, coincide with the time of the second major rise in oxygen.

There are a number of possible reasons for these glaciations. A decrease in methane, which was discussed above linked with the rise in oxygen, would have reduced the greenhouse effect. Methane is about 20 times more

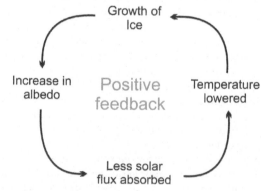

Figure 14.7   *Left: An artist's impression of a Snowball Earth* (Source: Chris Butler/SPL). *Right: The general scheme of the positive feedback effect in the climate system that results in the onset of a Snowball Earth.*

effective than CO_2 as a greenhouse gas and its reduction might have caused the Earth to cool, triggering glaciations. Yet another cause could have been the configuration of the continents. If they were localised to the equator, as is suspected, then they could have caused intense tropical rains, which might have reduced atmospheric carbon dioxide concentrations in the **carbonate–silicate cycle**, in which CO_2 is consumed in reactions with rocks, thereby reducing the greenhouse effect.

The onset of glaciations could have caused runaway cooling. Russian scientist Mikhail Budyko (1920–2001) calculated that, if the Earth was covered in glaciers at 30° latitude and higher, the effect would be that the Earth would reflect more sunlight than it absorbs, causing further cooling. The equatorward advance of glaciers would reflect yet more sunlight in a positive feedback effect (Figure 14.7) until the Earth was completely frozen over.

This dramatic and sobering scenario raises the obvious question – how did the Earth escape from this catastrophe, as it clearly did for us to be around to think about it? The answer lies in the build-up of carbon dioxide. Volcanic eruptions would have caused carbon dioxide to build up in the atmosphere. However, with very little ocean and rocks exposed under the global ice cover to react with it or absorb it, the concentrations would have risen to a point where an intense greenhouse warming would have set in, melting the glaciers and returning the Earth to its pre-glaciated state. If this hypothesis is correct then we would expect to find evidence for large layers of carbonate rocks, formed when the high concentrations of carbon dioxide in the atmosphere were precipitated in minerals at the end of the glaciation. These **cap carbonates** are indeed found in the rock record (Figure 14.8). Their carbon isotopic values, similar to carbon dioxide exhaled by volcanoes, supports the idea that a greenhouse episode ended the Snowball glaciations.

A conundrum raised by scientists is how the Earth's biota survived such precipitous changes in the global environment. Despite global glaciations, many habitats would remain. Deep-sea habitats would have existed as the oceans would not have completely frozen. The ice layers would have dramatically reduced light for photosynthesis and, if they were on the order of kilometres thick, completely eliminated it. However, we know of many productive photosynthetic communities that live on the surface of ice sheets. In the High Arctic, microbial mats live on the surface of ice sheets, gaining energy from sunlight and in the process generating organic carbon which feeds many heterotrophic organisms that live within the mats. On the surface of many polar glaciers are found **cryoconite holes**, which are small circular

Figure 14.8 *Cap carbonates are evidence for high concentrations of CO_2, which would have been responsible for a greenhouse episode ending the Snowball Earth. This is a Neoproterozoic outcrop of cap carbonate at the Doushantuo Formation at Jiulongwan, China, between other rock formations (Fm; Source: International geobiology conference).*

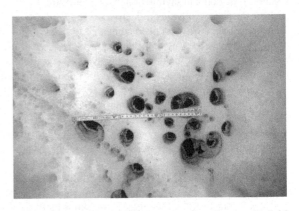

Figure 14.9 *Cryoconite holes on the Juneau Icefield, Alaska. Just one type of microbial community that can be successful on the surface of ice. Each hole is several centimetres in diameter and is filled with sediment and microorganisms (Source: Nozomu Takeuchi).*

depressions containing wind-blown sediments, cyanobacteria and a whole assemblage of microbes. The dark sediments absorb sunlight because of their dark brown or black colour, melting the ice and creating an aqueous habitat in which the microbes can grow. Cryoconite holes (Figure 14.9) can be just a few centimetres in diameter.

Even more remarkable are organisms that live within ice sheets or permanently ice-covered lakes of the Antarctic.

Debate Point: Life on a present-day Snowball Earth

A Snowball Earth seems intuitively to be a rather catastrophic episode. Ice cover would have cut down light penetration into the oceans and reduced temperatures in the upper zone of the ice-covered oceans. It would also have reduced gaseous interchange between the oceans and the atmosphere. Nevertheless, we can imagine that many habitats, such as deep-ocean hydrothermal vents, would have remained relatively unchanged. Or would they? As a thought experiment, consider the effect of a Snowball Earth on present-day Earth and its atmosphere, oceans and biosphere. What cycles and feedbacks would you expect to change? Consider the relative abundance of different habitats for life and how and why these habitats would alter. You might like to reconsider this question when you have completed Chapter 16.

Corsetti, F.A., Olcott, A.N., Bakermans, C. (2006) The biotic response to Neoproterozoic snowball Earth. *Palaeogeography, Paleoclimatology, Palaeoecology* **232**, 114–130.

In a similar way to the cryoconite holes, the microbes form dark matted communities within the ice. Provided they are not buried too deep, enough sunlight penetrates into the ice to be absorbed by the microbes, which warm up and melt the ice around them, resulting in a watery inclusion within the ice in which they can grow. Even if we did assume a completely ice-covered Earth, we can see many examples of microbial communities on the Earth today that would have inhabited the surface of this icy world. There is little doubt in saying though, that under these conditions, the photosynthetic productivity of the Earth would have been dramatically reduced compared to today. Without wide open expanses of ocean or productive coastal regions to provide a habitat for organisms living within the **photic zone**, the regions available for photosynthetic organisms could have been limited.

Although this is a rather diminished view of the biosphere, it is not certain that the Earth was completely ice-covered. More recent modelling results suggest that the tropics and equatorial regions may not have been completely ice-covered, leaving a region of equatorial water ice-free. Budyko's cataclysmic predictions, although prescient at the time he conceived them, were perhaps not played out in full. Nevertheless, the Snowball Earth episodes remain some of the most dramatic examples of the link between the changing biota on a planet and planetary scale environmental conditions.

14.8   Other Biological Consequences of the Rise of Oxygen

The Snowball Earth events would have been influential in determining the distribution of life, but the rise in oxygen itself would have had profound implications. Oxygen is reactive and in combination with UV radiation or other ionic species it produces reactive oxygen chemical species such as **oxygen free radicals** (Figure 14.10). Free radicals can oxidise membrane lipids, tear cofactors from proteins and oxidise key biological molecules such as proteins and nucleic acids. There are a number of enzymatic pathways that have evolved to cope with this stress, including *Superoxide Dismutase*, a ubiquitous enzyme which catalyses the quenching of superoxide anions (the precursors to damaging hydroxyl radicals) into oxygen and hydrogen peroxide. Although oxygen had some beneficial consequences, not least the fact that aerobic respiration can yield at least ten times as much ATP than anaerobic metabolisms, anaerobic organisms on the surface of the Earth that had evolved to live on an Earth with low concentrations of atmospheric oxygen had three options: evolve, go into hiding or go extinct.

The evolutionary machinery to deal with high oxygen and its free radicals was almost certainly in place before the first rise in oxygen. Localised production of oxygen by cyanobacteria before the first major rise in atmospheric oxygen would have produced oxygen oases in which aerobic organisms could have evolved. It would have provided a selection pressure for the evolution of mechanisms to deal with reactive oxygen species a long time before the rise in atmospheric oxygen. The large-scale rise in atmospheric oxygen would probably have presaged a spread of these oxygen adapted organisms into a greater number of surface habitats.

The rise in oxygen and general oxygenation of the Earth's environment would have changed the abundance of electron donors and acceptors for redox reactions. Sulfate and fixed nitrogen compounds (e.g. nitrate and nitrite) would have been more widely sustained and available, expanding the spatial range and abundance of chemoheterotrophic reactions available. Concomitantly, however, methane, hydrogen and other reduced compounds would have become less available, reducing the

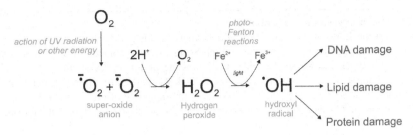

Figure 14.10 *A schematic showing some of the reactions involving oxygen that produce biologically damaging radicals, including hydroxyl radicals, whose production can be catalysed by iron and light in 'photo-Fenton reactions'. The rise in atmospheric oxygen would have increased the prevalence of these reactions* (Source: Charles Cockell).

spatial abundance of chemoautotrophic reactions, at least in habitats directly exposed to the atmosphere.

14.9   Oxygen and the Rise of Animals

One of the most dramatic consequences of the rise of oxygen is thought to have been the emergence of more complex organisms taking advantage of the possibility of aerobic respiration. As we saw in Chapter 5, aerobic respiration can generate much more energy (**ATP**) than anaerobic modes of metabolism. With this greater energy availability it is supposed that larger organisms, and more complex organisms, became possible.

Near the Great Oxidation Event in the fossil record, preserved remains of small complex organisms begin to appear. Fossils of spiral-shaped organisms (*Grypania*) are found in shales aged 1.9 Ga. They are probably algae. Fossils of acritarchs, organic fossils that are possibly the remains of eukaryotic organisms with a diameter about 50 μm, are found in Siberian shales from ~2.1 Ga ago. Other 1.85 Ga-old fossils of acritarchs have been found in China. Small, symmetrical fossils 1.7 Ga old are found in localities in China. Spiny acritarchs are found in rocks from ~1.5 Ga in Australia. These organisms were not the large multicellular differentiated organisms that we are familiar with today. However, they suggest the possibility of forms of organisms that have greater energy available for growth and complexity and this could be a consequence of the first rise in oxygen and the greater biomass sustained.

The second, Neoproterozoic, rise of oxygen is coincidental with the rise of large animals. Enigmatic moulds of tubular and frond-shaped organisms which lived during the Ediacaran period have been found in rocks between 585 and 542 Ma. As we saw in the previous chapter, they were first discovered in Australia's Ediacara Hills. In Chapter 13 we also saw that at the beginning of the Cambrian period (542 Ma) hard-bodied remains burst into the fossil record. The Ediacaran fauna disappear and the numbers of fossils preserved increases enormously. The **biodiversity** of animal life increases. This is the 'Cambrian Explosion'. At this time, organisms evolved the ability to precipitate minerals used for skeletons and hard shells, which are more easily preserved, accounting for these observations.

The link between the second rise of oxygen and the emergence of animals could be related to the availability of sufficient oxygen to drive energy intensive multicellular structures. A concentration of about 10% is thought to have been sufficient to allow for animal respiration.

One further factor must be considered and that is the movement of energy from one level in a food chain to the next. As energy is acquired by organisms and passed to a predator in the next **trophic level**, so an ecological chain is established. Aerobic respiration can support more trophic levels than anaerobic respiration because of its capacity not only to generate more energy, but generally to more efficiently produce energy and biomass compared to anaerobic respiration. It seems likely that this contributed to greater complexity in ecosystems, since aerobic respiration allowed for more energy accumulation at each trophic level and therefore longer food chains and successively larger organisms at each level.

An obvious question is why organisms would have tended to get larger. This has interested evolutionary biologists for decades. One way to view this tendency is to consider the fact that larger organisms, in addition to other possible benefits, would have been more capable of overwhelming prey. If the energy to get larger exists

---

Debate Point: The rise of oxygen on other planets

The rise of oxygen on the Earth at the Great Oxidation Event was clearly contingent on a variety of factors, not least the sources and sinks of oxygen. Despite these complexities, we can consider how planetary size might affect a hypothetical rise of oxygen elsewhere. Consider the following argument: A smaller planet might have lower abundances of reduced gases to act as a sink for oxygen. The sink would therefore become more rapidly depleted and O_2 concentrations might rise sooner, potentially leading to a more rapid rise of multicellular life and intelligence. Do you agree with this argument? What other factors about a planet might modify the hypothetical history of oxygen we have seen on the Earth?

McKay, C.P. (1996) Time for intelligence on other planets. In: *Circumstellar Habitable Zones* (Doyle, L.R., ed.), Travis House, Menlo Park, pp. 405–419.

Catling, D.C., Glein, C.R., Zahnle, K.J., McKay, C.P. (2005) Why O_2 is required by complex life on habitable planets and the concept of planetary 'oxygenation time'. *Astrobiology* **5**, 415–438.

---

through aerobic respiration, then selection pressures resulting from competition between organisms will have a tendency to allow for larger forms, an evolutionary 'arms race'. Predation would have been an important factor. Trilobite fossils preserve the injuries of attacks by one of the Cambrian's top predators, *Anomalocaris* (a model of this formidable early creature is shown in Figure 14.11), a precursor to arthropods, which provides evidence for the interactions between predator and prey at that time.

Once large multicellular organisms emerged, then there was an inevitability that the process of mutation would 'explore' the available parameter space of size resulting in a range of organism sizes (that, of course, is not to say that organisms cannot also become more diminutive if there are evolutionary selection pressures that favour smaller size).

Yet another factor which could have played a role in the exploration of new body plans and forms is the poor

regulation of genes. Today, Hox genes (one group of the homeobox genes) are the genes that control body plans and symmetry in animals. Presumably, during the early rise of animals these would have been less well regulated and complex. Some scientists have suggested that weak organisation and regulation of these genes might have enabled life to explore the parameter space of body organisations in geologically short periods of time.

Some researchers have suggested that the rise of animals could be related to other environmental factors. For example, forced into proximity by the global glaciations, cooperative cell structures, eventually differentiated cell structures, could have emerged. Whilst these sorts of ideas are intriguing, they are difficult to experimentally test. It seems reasonable to assume, given what we know about the lower energy yield of anaerobic respiration, that oxygen was probably one major environmental change that allowed for the rise of complex organisms.

14.10   Periods of High Oxygen

The chapter has focused on the majority of Earth history when oxygen was lower than today. However, there have been times when oxygen concentrations were higher than today. The changing balance between oxygen sinks and sources, which has caused the rises in oxygen witnessed through time, has also caused periods of elevated O_2 compared to the present-day.

Around 300 Ma ago, oxygen levels rose ~30% and this high oxygen concentration is thought to explain the episode in insect giganticism observed in the rock record at that time. Invertebrates, such as insects, which primarily rely on diffusion or primitive forms of abdomen contraction to obtain their oxygen, would have been able to grow to larger sizes with the greater availability of the gas.

Figure 14.11 *A model of* Anomalocaris. *An early Cambrian predator that provides evidence for a predator–prey 'arms race', which could be one explanation for the increase in animal size in the Cambrian. The organism was about 1 m long* (Source: wikicommons).

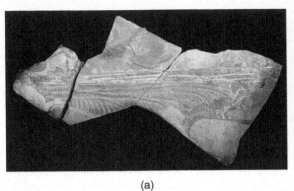

(a)　　　　　　　　　　　　　　　　　(b)

Figure 14.12 *Giant insects are thought to bear witness to a period when oxygen concentrations were higher than today. (a) A* Meganeura *fossil wing from a griffenfly (ancestor to the dragonflies), about 50 cm long. (b) These giant insects are illustrated with a model showing one of these impressive creatures on a log about 25 cm in diameter. (Source: wikicommons, Ghedoghedo and GermanOle).*

Fossil imprints of *Meganeura*, a griffenfly (ancestor to the dragonflies) with a 1-m wingspan, as well as other evidence, bear witness to this period (Figure 14.12).

The reasons for the rise of oxygen are uncertain. One possibility is that as the giant forests of the Carboniferous were buried, so the carbon locked up in them was buried, causing a net build-up of oxygen from photosynthesis. This in itself could have been caused by the lack of organisms able to degrade the structural materials of plants, **lignin** and **cellulose**, at that time.

14.11　Conclusions

Throughout the history of the Earth, life has been influenced by major geological changes, such as the shifting position of the continents. However, life itself has wrought changes on the Earth environment, with implications for the geosphere and the subsequent opportunities for life. In this chapter, we looked at the history of oxygen in the Earth's atmosphere as one important example of geosphere–biosphere coupling. We saw how, by using geological proxies, geochemists have been able to reconstruct a history of oxygen. By correlating this information to geochemical and biological observations in the rock record, scientists are beginning to understand how this coupling worked. We find that, in addition to being linked to radical changes in the Earth's chemistry, rises in oxygen may also be behind some of the most important evolutionary transitions in our planet's history, most notably the appearance of animals, ultimately leading to intelligence. The chapter invites obvious questions, which include: Was this rise inevitable? Would it always occur on other planets and with the same general patterns? What does the history of oxygen tell us about the prospects and probability of large animal-like organisms on other planets and therefore the likelihood of the rise of intelligence?

Further Reading

Books

Canfield, D.E. (2014) *Oxygen A Four Billion Year History*. Princeton University Press, Princeton.

Lane, N. (2002) *Oxygen: The Molecule that Made the World*. Oxford University Press, Oxford.

Papers

Anbar, A., Duan, Y., Lyons, T.W., Arnold, G.L., Kendall, B., Creaser, R.A., Kaufman, A.J., Gordon, G.W., Scott, C., Garvin, J., Buick, R. (2007) A whiff of oxygen before the great oxidation event. *Science* **317**, 1903–1906.

Berner, R.A., Beerling, D.J., Dudley, R., Robinson, J.M., Wildman, R.A. (2003) Phanerozoic atmospheric oxygen. *Annual Reviews of Earth and Planetary Sciences* **31**, 105–134.

Canfield, D.E., Poulton, S.W., Narbonne, G.M. (2007) Late Neoproterozoic deep-ocean oxygenation and the rise of animal life. *Science* **315**, 92–95.

Catling, D.C., Zahnle, K.J., McKay, C.P. (2001) Biogenic methane, hydrogen escape, and the irreversible oxidation of the early Earth. *Science* **293**, 839–843.

Catling, D.C., Claire, M.W. (2005) How Earth's atmosphere evolved to an oxic state. *Earth and Planetary Science Letters* **237**, 1–20.

Claire, M.W., Kasting, J.F., Domagal-Goldman, S., Buick, R., Stueeken, E., Meadows, V. (2014) Modeling the signature of sulfur mass-independent fractionation produced in the Archean atmosphere. *Geochimica et Cosmochimica Acta* **141**, 365–380.

Dudley, R. (1998). Atmospheric oxygen, giant Paleozoic insects, and the evolution of aerial locomotor performance. *Journal of Experimental Biology* **201**, 1043–1059.

Farquhar, J., Bao, H.M., Thiemens, M. (2000) Atmospheric influence of Earth's earliest sulfur cycle. *Science* **289**, 756–758.

Frei, R., Gaucher, C., Poulton, S.W., Canfield, D.E. (2009) Fluctuations in Precambrian atmospheric oxygenation recorded by chromium isotopes. *Nature* **461**, 250–253.

Guo, Q., Strauss, H., Kaufman, A.J., Schröder, S., Gutzmer, J., Wing, B., Baker, M.A., Bekker, A., Jin, Q., Kim, S.-T., Farquhar, J. (2009) Reconstructing Earth's surface oxidation across the Archean-Proterozoic transition. *Geology* **37**, 399–402.

Hoffmann, P.F., Schrag, D.P. (2002) The Snowball Earth Hypothesis: testing the limits of global change. *Terra Nova* **14**, 129–155.

Holland, H.D. (2006) Oxygenation of the atmosphere and oceans. *Philosophical Transactions of the Royal Society* **361**, 903–915.

Knoll, A.H. (2011) The multiple origins of complex multicellularity. *Annual Review of Earth and Planetary Sciences* **39**, 217–239.

Lenton, T.M., Watson, A.J. (2000) Redfield revisited: 2. What regulates the oxygen content of the atmosphere? *Biogeochemical Cycles* **14**, 249–268.

Schirrmeister, B.E., de Vos, J.M., Antonelli, A., Bagheri, H.C. (2013) Evolution of multicellularity coincided with increased diversification of cyanobacteria and the Great Oxidation Event. *Proceedings of the National Academy of Sciences* **110**, 1791–1796.

Stolper, D.A., Revsbech, N.P., Canfield, D.E. (2010) Aerobic growth at nanomolar oxygen concentration. *Proceedings of the National Academy of Sciences* **107**, 18755–18760.

15

Mass Extinctions

Learning Outcomes

➤ Understand what mass extinctions are.

➤ Know that there are five mass extinctions in the Phanerozoic and be able to point out some of their characteristics.

➤ Describe some of the purported causes of mass extinctions.

➤ Be able to explain some of the major lines of evidence for the role of an asteroid impact in the Cretaceous–Paleogene extinction.

➤ Understand the nature of controversy in science using evidence for mass extinctions as an example.

➤ Explain why some scientists describe the current changes in the Earth system as a sixth mass extinction.

15.1 Extinction

The evolution of life on the Earth has not been a continuously gradual and steady process. As we saw in the last chapter, life itself can precipitate momentous global-scale changes to the environment. The Great Oxidation Event, essentially resulting from the evolution of oxygenic photosynthesis, had enormous implications for life. In an example such as this we see how life is a driver of environmental changes. Life, however, has sometimes been very much a passenger in the co-evolution of the Earth–Life system. A large-scale asteroid impact can cause huge perturbation to the biosphere and is imposed from outside the Earth, with life a witless victim of these changes (although it is not beyond the means of an intelligent organism to avert these threats, as we shall see later).

Whether life is a driver or a passenger, these changes have made the large-scale evolutionary trends of the planetary biosphere a haphazard affair. Sudden changes not only create new opportunities for life, but they have also destroyed a great diversity of **taxa**. Throughout the history of Earth, the biosphere has been subjected to episodes of **mass extinction**. We'll explore some of these extinctions, the proposed causes of them and their significance for understanding the tenure of life on a planet.

15.2 What is Extinction?

An extinction occurs when the last individual of a species dies out. However, unfortunately, this isn't necessarily the definitive end. It can occur long before that. An extinction is described as a **Functional Extinction** when individuals remain, but the odds of sustainable reproduction are low. In other words, the species is effectively extinct even though individuals remain. A functional extinction can occur because a small population of an organism may become so inbred that it becomes genetically unhealthy and unable to produce viable offspring.

One iconic, and perhaps sobering, example of human-caused extinction is the Passenger Pigeon or Wild Pigeon (*Ectopistes migratorius*; Figure 15.1). In the nineteenth century this bird, which occupied the forests of North

Figure 15.1 *A taxidermy specimen of the Passenger or Wild Pigeon* (Source: wikicommons, Ltshears).

America, was so numerous that its numbers were estimated to be between three and five billion. Its flocks were so large and extensive that they blocked out sunlight and covered hundreds of square miles as they migrated. Sold as meat, their intensive and industrial-scale hunting led to massive population declines in the late 1880s. The destruction of forests for wood reduced their habitat until eventually, by the late nineteenth century, their population had reduced to below its minimum viable size. The pigeon had essentially become functionally extinct. The last passenger pigeon in the wild was shot at Babcock, Wisconsin, United States, in September 1899. However, the last Passenger Pigeon on Earth, named Martha, died alone at the Cincinnati Zoo at about 1:00 p.m. on 1 September 1914.

Natural extinctions can occur when the environment of a species changes faster than the species can adapt, when a species' adaptations are no longer sufficient in allowing it to acquire and compete for resources. Extinctions can be local, widespread, or global. A good example of a local extinction would be the Grey Wolf (*Canis lupus*) present today in some regions of North America, but made extinct in England probably during or just after the reign of King Henry VII (1485–1509).

Extinction is the fate of most species on the Earth. Probably about 99.9% of all species that have ever existed have gone extinct. Most of these have disappeared because of local extinctions or conditions that have pushed a single species beyond the brink, a so-called species extinction. An **extinction event** refers to episodes that are not global, but may affect multiple species. An example of

such an event is the end-Pleistocene extinction in the northern hemisphere ~11 000 years ago which caused the extinction of many large mammals including the woolly mammoth (*Mammuthus primigenius*). It may have been caused by climate change and/or human hunting.

However, within the rock record is evidence for episodes when much more widespread extinction occurred. These episodes are called **mass extinctions**.

A mass extinction is an extinction of many species on the planetary scale. They usually imply extraordinary and rapid planetary changes and affect all kinds of life (e.g. plants and animals, equatorial and polar organisms, large and small animals, carnivores and herbivores). One appropriate definition of a mass extinction was given by evolutionary biologist, David Jablonski as 'any substantial increase in the amount of extinction (i.e. lineage termination) suffered by more than one geographically widespread higher **taxon** during a relatively short interval of geologic time, resulting in an at least temporary decline in standing diversity'.

The possibility of mass extinctions was recognised by French geologist, Baron Georges Cuvier (1769–1832) who pointed out that giant bones found in ice age deposits did not correspond to any living organism. Until recent times, gradual geological change was considered the dominant force. Catastrophism was considered interesting, but rather sensationalist. Today, we recognise that some mass extinctions are associated with catastrophic events.

Although mass extinctions have punctuated Earth history, there is no strong evidence to suggest a regular pattern. In the 1980s, claims were made of periodicity in extinction events, with a favoured theory being that extinction occurs every 26 million years. One suggestion was that it was caused by some distant companion to the Sun, called Nemesis, which perturbed the inner Solar System, causing asteroid and comet impacts, at regular intervals. Other scientists wondered whether periodic extinctions might be linked to the Solar System's 250 million year journey once round the Milky Way Galaxy, with changes induced by the Earth's periodic passages through the spiral arms of the galaxy. These theories have fallen away in favour of the view that mass extinctions are triggered by diverse causes that do not have any fixed periodicity.

Mass extinctions can occur slowly (a gradual mass extinction) or very quickly (a sudden mass extinction). The mechanisms underlying any given extinction will determine how quickly the process occurs. Following any mass extinction is a **recovery interval** during which environmental stability and increasing organism diversity (**adaptive radiation**) occurs.

The organisms affected by these events are themselves classified into groups. **Hold-over taxa** are those taxa that survive the extinction, the particular representatives of groups that make it through the extinction. **Disaster taxa** are those organisms that briefly proliferate and take advantage of the disturbed conditions. They are the opportunistic groups that dominate shortly after a mass extinction. **Progenitor taxa** are those organisms that go on to form the phylogenetic ancestors of the newly dominant groups that rise to prominence in the longer-term after an extinction.

15.3   Five Major Mass Extinctions

There are five mass extinctions that are recognised in the Phanerozoic. Table 15.1 shows each of these extinctions and some estimates of the magnitude of extinction at family, genus and species level. Extinctions are generally named after the end of the geological period in which they occurred or after the two periods between which they occurred.

The values of the magnitude of extinction at different levels of biological hierarchy shown in Table 15.1 are estimates and it is important to understand why they are uncertain. They are based on the fossil record, which is incomplete. This incompleteness is caused by poor preservation of many organisms and the fact that environments in which preservation occurs represent only a small proportion of environments on the Earth at the time. However, relative comparisons before and after an extinction event are probably quite accurate since the preservation of skeletons would not necessarily be expected to change.

Defining absolute abundance of particular **taxa** is more difficult. Uncertainties arise because the numbers depend on the classification of fossils and how they are assigned to families, **genera** and species which is difficult because many of the fossils are representatives of long-extinct organisms. The numbers vary depending on the way in which these uncertainties are dealt with mathematically. Nonetheless, the presence of major losses in species diversity in the fossil record is not in doubt and the relative severity of these events can be quite well quantified. The end-Permian extinction is the most severe of the five major mass extinctions of the Phanerozoic.

These numbers also illustrate that extinction rates are lower at higher hierarchies of biological classification (Chapter 6). If you contemplate this you can understand why. To eliminate an entire family would require the extinction of many species across different genera and in different areas of the world, so generally it is more difficult to make extinct successively higher levels of the biological hierarchy. This is reflected in the statistics.

The five major mass extinctions occur against a backdrop of increasing diversity of life on planet Earth (Figure 15.2). Despite large reductions of diversity during these episodes, following the extinction, the diversity of life has, within a few tens of millions of years or sooner, reached pre-extinction levels. Extinctions have the effect of vacating **niches**, which are then occupied by new species following extinction. This may explain why diversity can, at least in the context of geological time periods, rapidly recover.

Perhaps most famous of these empty niche opportunities is the demise of the dinosaurs at the Cretaceous–Paleogene (K–Pg) **boundary**, which vacated niches for the mammals. Up until then, although they evolved roughly at the same time as the reptiles, most mammals were small or nocturnal. They made a living at the edges of a biosphere dominated by reptiles. Following the Cretaceous, with

Table 15.1   *The five major extinctions of the Phanerozoic. Some example estimates of the magnitude of extinction at different levels of biological hierarchy are shown.*

Extinction	Age (Mya)	Family extinction (%)	Genus extinction (%)	Species Extinction (%)
Cretaceous–Paleogene extinction	~65.5	17	40	75
End Triassic extinction	~201	23	50	80
Permian–Triassic extinction	~252	57	56	80–90
Late Devonian extinction	~372	19	57	75
Ordovician–Silurian extinction	~444	27	57	86

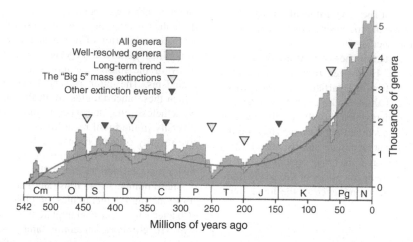

Figure 15.2 *The changing diversity of genera of marine organisms during the Phanerozoic showing all genera and those that are well resolved (well established as separate genera). As well as the major five extinctions (The 'Big 5') of the Phanerozoic, smaller extinctions ('Other Extinction Events') that nevertheless exhibit reductions in diversity in the rock record are also shown (Source: wikicommons, Albert Mestre).*

most of the available niches for large organisms vacated, small rat-like mammals were successful. Mammals eventually were able to take over the newly available ecological space on land, sea and in the air, for example, vying with giant carnivorous birds in North and South America. The rise of the mammals would eventually lead to us.

15.4   Other Extinctions in Earth History

It is easy to become transfixed by the five major mass extinctions. They are, after all, sobering. However, life on Earth has experienced other more minor extinctions that are not fully understood, nor their causes identified. Changes in environmental conditions, including some of the causes we will discuss in the next section, mean that life on Earth is being constantly exposed to perturbations which have the potential to push species beyond their limits.

You will also notice that our discussion is exclusively focused on the Phanerozoic. The reason for this bias is that we have little evidence that microbes are subject to mass extinctions since their remains are generally poorly preserved in the rock record. Large-bodied organisms fossilised in rocks provide a better record with which to assess mass extinctions. Exceptions are microorganisms such as the **foraminifera**, which make calcium carbonate shells

and are found in the rock record. They were subject to extinctions at the K-Pg boundary, suggesting that microorganisms do suffer during planetary-scale environmental changes.

We do know of periods in Earth history before the Phanerozoic when there have been dramatic environmental changes. Some of these we have discussed already – the **Late Heavy Bombardment, Snowball Earth** episodes and the **Great Oxidation Event**. It might be reasonable to expect that even though the fossil record evidence is poor, these could have been periods of extinction for groups of prokaryotes and early eukaryotes that were very specialised or had small geographical distributions affected by changes in environmental conditions.

15.5   Causes of Mass Extinction

What are the causes of mass extinction? There are several leading reasons, which we will discuss for some of the major mass extinctions in the following sections. Here some general comments are made to introduce them.

Volcanic causes. Large-scale volcanic eruptions have the potential to release CO_2 into the atmosphere, increasing the greenhouse effect and thereby causing a dramatic warming period. Warming can potentially cause a variety of knock-on effects including causing thermal stress to organisms in water bodies and changes in habitat ranges. The melting of **methane gas hydrates** by volcanic activity would cause the release of methane into the atmosphere, itself causing an increase in greenhouse warming. We will discuss volcanic causes in more detail in this chapter.

Sea-level change. Sea-level changes alter the area of continental shelf or land available for colonisation. It is known that the diversity of species is correlated to the physical area of habitat available. The more habitat that is available, the more species can be accommodated. This is known as the **species–area relationship**. A reduction of the available area of habitat in sea and on land caused by sea-level change can therefore cause extinction. Changes in sea-level can be linked to warming episodes whereby the melting of ice caps would increase sea-level, illustrating how some extinction mechanisms may be linked.

Asteroid and comet impacts. The Earth has been pummelled by asteroids and comets throughout its history. Large impacts have the potential to loft dust into the atmosphere, cutting out sunlight, shutting down photosynthesis and thereby disrupting food chains, amongst other effects. We will discuss this mechanism in more detail in this chapter.

Supernova explosions and gamma ray bursts. This is one of the more unusual suggested causes of mass extinctions. We live in a rather quiet neighbourhood in the Galaxy. However, it is estimated that about once every 200–300 million years, as the Solar System undergoes its 250 million year journey around the Galaxy, a supernova explosion could occur close enough to the Earth (~30 light years) to cause the depletion of the ozone layer and affect climate through ionisation of the atmosphere and the perturbation of atmospheric chemistry. Certainly these events do occur in our galaxy. In 1604, a supernova 20 000 light years away was observed (Kepler's supernova; Chapter 9, Figure 9.6). It was the most recent event that could be observed with the naked eye. Gamma ray bursts, which are intense and very distant bursts of radiation (possibly caused by supernova explosions or collisions between neutron stars), could cause changes in atmospheric conditions if they occurred close enough. The major problem with assigning these events as causes of mass extinctions is the difficulty of linking the events with changes in the rock record. Attempts have been made to search for iron-60 and beryllium-10 in ice cores. These short-lived isotopes are formed in supernova explosions and so elevations in their concentrations would suggest the presence of a close event. Beryllium-10 peaks have been

claimed for the polar ice record, such as one suggested peak at 40 000 ago. An iron-60 peak at 2.8 Ma ago in the deep ocean crust was reported. However, the breadth of these peaks, their frequency and the inability to find a candidate explosion by astronomical methods make these claims doubtful.

Biological changes. Life itself could perpetuate extinction. As the biosphere changes, so organisms better equipped to persist can out-compete other organisms to extinction. Alternatively, new organisms can change the environmental conditions to such an extent that other organisms are driven to extinction. The **Great Oxidation Event** (Chapter 14) is an example of how an evolutionary innovation in metabolism (the use of water as an electron donor in oxygenic photosynthesis) caused global scale environmental changes by increasing the concentrations of oxygen in the atmosphere. Some surface-dwelling **anaerobic** microorganisms might have been driven to extinction. As we'll see for the end-Devonian extinction, the evolution of forests is one event that could have caused extinction.

Multiple effects. None of the events described above are mutually exclusive and it is possible that some extinction events are a combination of these mechanisms, all of which would cause stress to the biosphere.

Let's examine some mass extinctions and their proposed causes.

15.6   The End-Cretaceous Extinction

There can be little doubt in saying that of all the Phanerozoic extinctions the Cretaceous–Paleogene (K-Pg), which was once named the Cretaceous–Tertiary (K-T) extinction (the Paleogene is informally known as the lower Tertiary), attracts the greatest interest. It is also called the end-Cretaceous extinction. It receives the most attention partly because it is the most recent and so the fossil record is the best preserved. Perhaps this is unwarranted since its scale of devastation did not match the end-Permian extinction. Its association with the demise of the dinosaurs might account for some of its infamy, but it also caused the extinction of oceanic **taxa** (including plankton groups) and many land animals.

However, the extinction is instructive because the suggestions for the causes of this extinction have varied across almost all of the possible mechanisms briefly described in the last section. It is therefore worth examining this extinction first. One of these suggested causes – meteor impact – has strong evidence associated with it.

Debate Point: What If... The K-Pg Extinction Didn't Happen?

'What If?' type thought experiments are not always useful scientifically, but they can have a role in stimulating us to think about things differently and to consider how much about our Universe is contingent on specific circumstances and how much about the Universe is inevitable. One such *gedanken experiment* (thought experiment) is what would have happened if the K-Pg extinction had not occurred. Would the dinosaurs have risen to intelligence? Perhaps the most well-known of these speculations is Dale Russell's dinosauroid (Figure 15.3). Russell was curator of vertebrate fossils at the National Museum of Canada and speculated on an evolutionary path taken by the Troodontid dinosaurs. Russell noticed that there had been an increase in the encephalisation quotient or EQ (the relative brain weight when compared to other species with the same body weight) among the dinosaurs. Russell, who discovered the first Troodontid skull, pointed out that its EQ was six times higher than that of other dinosaurs and that it might have had greater cognitive capacities than most dinosaurs. He speculated that the language of an intelligent dinosaur, if it had had chance to evolve, would have sounded like bird song. The intelligent dinosaur would have three fingers on each hand and large eyes. Some people think this whole thought experiment is too *anthropocentric*. What do you think? Is there scientific value in such speculations? Russell's thought experiment raises yet another question: Is the rise of intelligence inevitable?

Figure 15.3 *Dale Russell's intelligent dinosaur shown alongside a model of a troodontid,* Stenonychosaurus inequalis. *Inset: a close-up of the head and shoulders* (Source: Canadian Museum of Nature, Ottawa).

Russell, D.A., Séguin, R. (1982) Reconstruction of the small Cretaceous theropod *Stenonychosaurus inequalis* and a hypothetical dinosauroid. *Syllogeus* **37**, 1–43.

Was it an asteroid impact? There is a great deal of evidence for an asteroid impact at the K-Pg boundary and this evidence takes various interconnected forms. The evidence for asteroid impact at the end-Cretaceous is one of the great scientific detective stories of all time and so it is worth describing this information.

Geologists Luis Alvarez (1911–1998) and his son Walter published work in the 1980s showing evidence for a spike in the heavy element, iridium, at the K-Pg boundary in the rock record (Figure 15.4). It was many times greater than the background concentration. Iridium is associated with asteroidal material and although the Earth had a primordial inventory of the element, it would have sunk into the Earth during **differentiation** and accumulated in the deep interior. A spike in its concentration on the surface of the Earth is interpreted to be the introduction of the element from space. Although one might also envisage an increase in iridium from volcanic activity, which would bring it from the subsurface, this isn't thought to be the best explanation given other evidence for an impact.

Other signs include the presence of **shocked quartz** at the boundary (Figure 15.5a). When quartz is subjected to intense pressures associated with the shock wave from an asteroid or comet, the material becomes filled with 'planar deformation features', which are linear features. These features are found in quartz at the K-Pg boundary.

During an impact, rock is melted and ejected into the atmosphere. As it transits the atmosphere, it has a tendency to solidify to form spheres (just like rain drops,

molten rock will form a sphere as this shape minimises the surface area of the material) or sometimes dumbbell shapes. These objects land on the ground, forming layers of glassy **tektites** (Figure 15.5b). The presence of large numbers of these at the boundary is yet further evidence for an impact. The tektites have the chemistry of limestones and sulfates (not volcanic rocks), which is evidence they were formed from an object impacting with sedimentary rock, not from a volcanic eruption.

There is further evidence for impact. The presence of soot and charcoal at the boundary has been suggested to be caused by global wildfires triggered by impact heating around the world.

The environmental effects of such an impact would have been devastating. The evidence suggests that a rock of approximately 10-km diameter collided with the Earth. The impact energy would have been equivalent to about 100 million megatons of the explosive, TNT (a curious unit commonly used by impact scientists). During the impact, dust would have been lofted into the atmosphere shutting out sunlight or seriously reducing it (to ~20% of normal levels) for several months to several years, thus interrupting photosynthesis and causing a short-term period of intense cold. In the specific case of this boundary, the vapourisation of sulfate deposits would have generated sulfuric acid droplets high in the atmosphere, themselves contributing to cooling by forming highly reflective aerosols. Models suggest that the dust cloud could also have perturbed the hydrological cycle, potentially preventing rain for many months. Herbivores would have gone without fresh food and died, eventually disrupting other layers of the food chain, such as carnivores, dependent upon them (Figure 15.6). In summary, the effects of an impact are thought to be an 'impact winter', a period of cold, dark and dry conditions. Further stress could have been imposed onto the biosphere by destruction of the ozone layer from reactions of chemicals injected into the upper atmosphere, potentially causing transient increases in UV radiation. Wildfires caused by heated material launched across the surface of the Earth would have added yet another layer of calamity.

Finally, the fossil evidence following the event is intriguing and consistent with an impact winter scenario. A 'fungal spike' is observed at the boundary, potentially suggesting what might simply be described as a rotting biosphere, resulting from dead and decaying biomass. Following this is a marked increase in fern spores. The fern spike suggests that large areas were covered in ferns, which often take advantage of newly opened areas and would be the first vegetation to grow in places where forests and other vegetation had died.

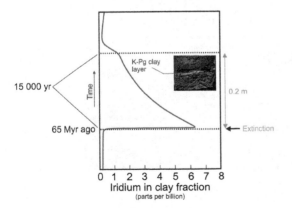

Figure 15.4 *An iridium spike at the Cretaceous–Paleogene boundary in the rock record suggests the arrival of an asteroid. The data shown here are from clays found in limestones. Inset is the K-Pg clay layer from Drumheller, Alberta, Canada, one layer corresponding to the K-Pg boundary in which an iridium anomaly is observed.*

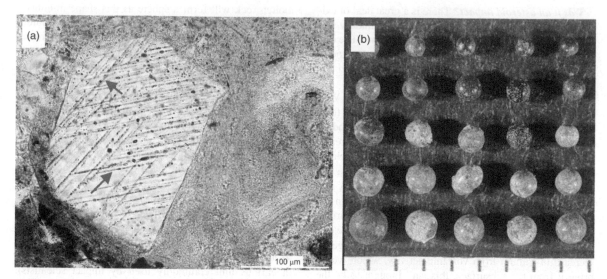

Figure 15.5   Evidence for geological changes at the K-Pg boundary associated with an impact event. (a) A piece of shocked quartz (centre left) containing planar deformation features. They are linear features within the crystal; examples are highlighted with red arrows (Source: Martin Schmieder). (b) Small tektites from the K-Pg boundary. The bottom scale bar is in millimetres (Source: Geological Survey of Canada).

Figure 15.6   The dinosaurs face an impact winter (Source: BBC).

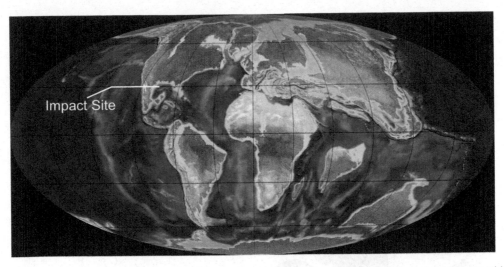

Figure 15.7 *The site of the Chicxulub impact crater in Mexico, proposed collision site of the asteroid responsible for the end-Cretaceous extinction, shown on a map with the putative pattern of continents at that time.*

Further support for the impact hypothesis was provided by the claim that the site of the impact had been identified. The Chicxulub crater (180–300 km in diameter; Figure 15.7) was discovered in the Yucatan Peninsula, Mexico in 1978 by a geophysicist working for a Mexican Petroleum company, Petróleos Mexicanos. They did not release the data for commercial confidentiality reasons, but it was "rediscovered" by geologist Alan Hildebrand in 1991. Its location in the rock record coincides with the K-Pg boundary and the diameter of the crater is consistent with a 10-km rock that would have been required to cause global-scale extinction.

It is difficult to find direct evidence of the impactor itself. Most of the rock would have been vapourised during the intense heat and pressures generated during the impact itself. However, a possible fragment, 100-trillionth of a gram, was reported to have been found in a drill core from the boundary in the northwest Pacific.

Was it sea level change? In the Cretaceous the sea level was high and the land masses were separated (despite the proximity of continents to one another). These separated continents meant that there was lots of continental shelf area for shallow marine organisms. However, in the early Paleogene the sea level was lower, the continents were further apart and there may have been less shelf area for shallow marine organisms. The lowering of sea level would therefore have caused a loss of continental shelf area, for example the loss of interior seas such as Western Interior Seaway of North America. If at the same time the land masses became arid, as is suspected, then there may

have been globally less habitat available for life, causing extinction.

Was it volcanism? During the end of the Cretaceous there is evidence for large-scale volcanic activity associated with rising heat plumes from the mantle (**flood basalts**). The Deccan Traps in India (so called because the eroded layers of rocks look like stairs, which in Swedish is the word 'trappa') record eruptions of more than 500 000 million km^3 of basaltic lava over about five million years (Figure 15.8). These flood basalts entail the eruption of large amounts of lava over prolonged time periods and could have had profound consequences for life. Sulfur from the volcanic eruptions could have formed sulfuric acid aerosols in the atmosphere, resulting in short-term cooling. Emission of CO_2 from the eruptions would have caused longer-term global warming. Disruption to food availability caused by these radical changes to the environment might have caused mass extinctions. It's quite possible that volcanism created stress in the biosphere, which was pushed over the edge by the asteroid impact. The consensus today is that the asteroid impact hypothesis has strong support, but that other contributions to extinction cannot be ruled out.

Volcanism has been proposed for other mass extinctions. Massive flood basalts in Siberia are associated with the end-Permian extinction and these Siberian Traps have been proposed as a contributor to major climate changes. There appears to be correspondence between some (but not all) mass extinctions and major volcanic episodes (Figure 15.9).

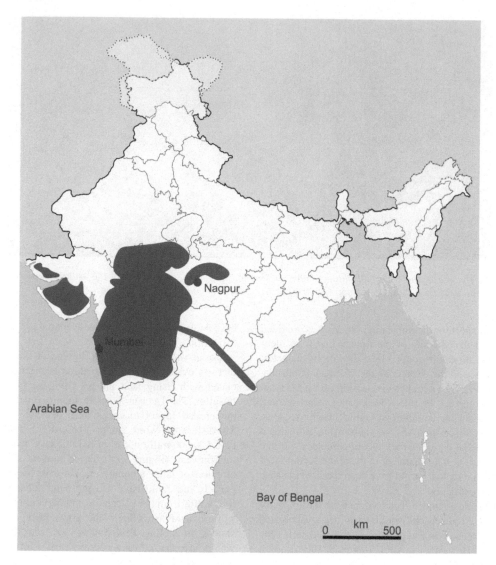

Figure 15.8 *The Deccan traps (in red) erupted towards the end of the Cretaceous and can be found across a wide area of present-day India. They are evidence for large-scale volcanic eruptions.*

15.7 The Other Four Big Extinctions of the Phanerozoic

15.7.1 End-Ordovician Mass Extinction

At the end of the Ordovician (~444 Ma ago) a range of organisms met their fate (Ordovician–Silurian extinction). Groups of corals, trilobites, brachiopods (marine animals with shells that have an upper and lower shell) and cephalopods (which include the squid and octopus) died.

One possible cause could be cooling. At the time the continents were gathered at the southern polar region, which might have triggered widespread glaciations both on the continents and in continental shelf areas.

Perhaps one of the most intriguing suggestions for a cause of the extinction is a gamma ray burst or a supernova explosion (Figure 15.10). One predicted effect of these

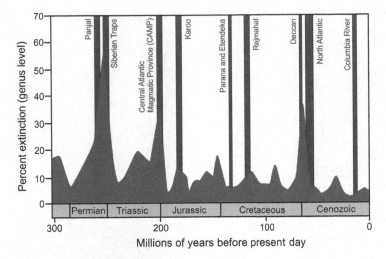

Figure 15.9 *A correspondence between some (but not all) extinctions (red) and major volcanic eruptions (including flood basalts; blue vertical bars) has been suggested as an indication of volcanically-induced extinctions. The exact role and extent of volcanic activity in Phanerozoic extinctions remains controversial.*

Figure 15.10 *The remnants of a supernova explosion. This one is SN 1987A, at a safe distance of 168 000 light years from the Earth. Have any others caused extinctions? (Source: ESO).*

events is the formation of aerosols in the upper atmosphere from ionising radiation interactions with the atmosphere. These aerosols would cool the Earth and the evidence of cooling at the end of the Ordovician has been suggested to be evidence for such an event.

15.7.2 Late Devonian Mass Extinction

The Late Devonian extinctions (~372 Ma ago) claimed a range of groups including representatives of the trilobites,

brachiopods, corals, cephalopods and fish. The changes are thought to be attributed to at least two major extinction episodes or possibly even a series of smaller-scale extinctions.

Like the Ordovician extinction, the late Devonian one was associated with the onset of glaciation and a number of causes have been suggested. One leading contender is the evolution of the first forests, which might have increased the drawdown of CO_2 from the atmosphere in photosynthesis, reducing the greenhouse gas concentrations and cooling the Earth. A side effect of the spread of forests would have been the break-up of soils and rocks by their extensive root systems. Up until this time plants were limited to surface-dwelling small plants that penetrated no more than a few centimetres into the soil. The massive new release of nutrients from soils might have caused rivers and ocean environments to have become anaerobic as oxygen was used up by organisms, itself causing extinction.

Cooling might also be caused by an impact event and although this has been suggested for the end-Devonian extinctions, the evidence is not nearly as persuasive as the end of the Cretaceous.

15.7.3 The Largest of all Mass Extinctions: The End-Permian Extinction

The end-Permian extinction (~252 Ma ago) (or Permo-Triassic Extinction) brought an end to over 90% of

Figure 15.11   *The trilobites. Just one group destroyed by the end-Permian extinction. This specimen is ~3 cm long.*

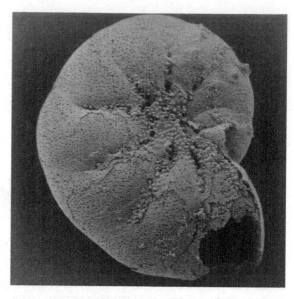

Figure 15.12   *A foram showing its calcium carbonate shell. This specimen is about 0.5 mm in diameter* (Source: wikicommons, Hannes Grobe).

species. It is the largest extinction that has occurred on the Earth. It destroyed trilobites that had survived all previous environmental challenges since the early Cambrian (Figure 15.11) and it eliminated major ecosystems including coral reefs and forests.

Like the end-Cretaceous extinction, it coincided with the eruption of flood basalts. The Siberian Traps cover two million km^2 of present-day Siberia and the volume of lava produced could have been up to four million km^3, erupted possibly in the space of one million years or less. It remains controversial whether the flood basalts were the only extinction mechanism or a contributor.

Methane produced by methanogens and/or the melting of methane gas hydrates, which would have released methane into the atmosphere, would further have contributed to global warming. Sulfur dioxide erupted from the flood basalts is thought to have mixed with rainwater, producing sulfuric acid, which rained out as acid rain, helping to destroy the forests. There is widespread evidence that the type of pine trees (the **gymnosperms**) that dominated the forests at this time were destroyed and replaced by opportunistic ferns. Geological evidence for increased sedimentation during this time suggests that surface plants had died back, leaving soils exposed and subject to higher rates of erosion.

Particularly hard hit were marine organisms that have calcareous shells such as the **foraminifera** (or forams for short; (Figure 15.12)). Over 90% of them became extinct. This suggests ocean acidification, possibly by volcanic gases. The end-Permian oxygen isotope record suggests that temperatures of the ocean rose to 40 °C, caused by a period of intense global warming, perhaps itself caused

by volcanic CO_2. The warming of the oceans would have caused stagnation, which is evidenced by black pyrite-rich shales in the rock record that contain very few fossils, a sign that the oceans became putrid, stagnant and anaerobic (**anoxic**). Carbon isotope values, which vary wildly during this time, point towards a collapse, or numerous collapses, of photosynthesis in the oceans and the injection of carbon into the Earth system (such as through CO_2 in volcanic eruptions).

Changing nitrogen isotope values, seen as a decrease in $\delta^{15}N$, point towards enhanced nitrogen fixation. As nitrogen fixation requires low oxygen levels, this suggests that the oceans became oxygen-poor. Biomarker studies across the boundary suggest an enhancement of the compounds, isorenieratane and isoprenoids. These are compounds associated with green sulfur bacteria (Chapter 5) which are anoxygenic photosynthesisers. They would have proliferated in an environment with higher levels of H_2S that they could use for photosynthesis. Again, this evidence is consistent with stagnant anaerobic water. In general, the extinction evidence suggests that the marine victims of the extinction were organisms with a poor ability to survive rapid changes in temperature, pH, low oxygen availability and water CO_2 content.

All together, the evidence clearly suggests a catastrophe in the oceans and on land, which could have occurred over just a few hundred thousand years. It took five million years before stable ecosystems were established once more. One of the newly emerging groups that benefited from the stable conditions was the dinosaurs, whose ancestors appeared at this time.

15.7.4 End-Triassic Mass Extinction

The loss of up to 50% of species ~201 Ma ago, including a variety of marine organisms such as clams and many molluscs has led to suggestions that ocean **anoxia** could have been one cause of this Triassic–Jurassic extinction event. It appears that the extinction was rapid, possibly occurring in the space of 10 000 years. This would favour a role for volcanism. The extinction coincided with the extrusion of flood basalts accompanying the spreading of Atlantic Ocean. The extinction of reptiles and amphibians on land suggests effects were not localised to the oceans. Volcanism could have caused the release of CO_2, triggering global warming.

15.8 Impacts and Extinction

Asteroid and comet impacts as a mechanism of extinction have been popular. There is something coldly fascinating about this possibility. However, aside from this more popular view of them, they are significant in the history of life because they are an unambiguous example of a perturbation to the Earth system that is not caused by life, but which has dramatically influenced the biosphere in the past. Let's explore their influence a little more.

The evidence for impacts at extinction boundaries is not always strong. An asteroid impact at the end-Permian was once a popular suggestion, but the evidence at the boundary for impact is not convincing. The strongest evidence for impact is at the K-Pg boundary, which as we have seen, comes in the form of various lines of evidence, including an iridium anomaly, tektites, shocked quartz, soot and a potential candidate crater linked to the event.

There is little doubt that impacts are a serious threat to life. The frequency today with which an object capable of causing global-scale perturbation to the biosphere collides with the Earth is estimated to be about once every 100 million years (Figure 15.13). However, events capable of causing local devastation are much more frequent. The type of event that occurred above Tunguska in Siberia in 1908, which flattened over 2000 km² of forest, may occur with a frequency of once every thousand years or more. The blast was equivalent to 30 megatons of TNT,

although no crater was formed. The projectile exploded in the atmosphere as an air burst. Although this event occurred in a largely unoccupied region of the Earth, such an event, if it occurred near a populated area, could annihilate a city and kill many people.

Many space scientists and space agencies consider the threat serious because, although the chances of an extinction-causing event are low, it is just a game of statistics. Once every one hundred million years could mean that three come tomorrow and then none for a long time. Efforts to map asteroids in our Solar System and so be in a position to detect them early have been implemented in asteroid detection programmes such as **Spaceguard**.

Deflecting an asteroid from the Earth would not be a trivial task, but numerous schemes have been proposed including: nudging them away from the Earth using explosives, painting one side white so as to cause radiation to be reflected, giving it a small tendency to veer off course, or firing lasers at one side to vapourise material and cause it to deflect (Figure 15.14). All of these methods depend on catching the threat early so that a very small angular deviation will be translated into a large deflection once it reaches the Earth. This in itself depends upon having good maps of where asteroids are and being able to reliably track them. We also lack good information about the density of the different varieties of these objects, which makes estimating the energy needed to deflect them uncertain.

Even more of a challenge is the diversion of the more unpredictable, more difficult to map, comets. They travel considerably faster than asteroids. Typical velocities are ~15 km s⁻¹ for asteroids and ~50 km s⁻¹ for comets, although comet velocities vary greatly depending on their orbit and they can travel faster than this. Aside from their more rapid movement, they have highly elliptical orbits making them liable to appear with less warning. An Earth-intersecting comet could give us very little time to act.

15.9 Some Questions About Extinctions and Life

Mass extinctions raise some important questions about how changes in the biosphere are driven. We are used to the Darwinian tenet that as environmental conditions change they force a selection pressure on a population of organisms that have variety. The selection pressure removes some variations, leaving other ones alive that go on to produce offspring and so on. However sudden mass extinctions, particularly dramatic ones, such as

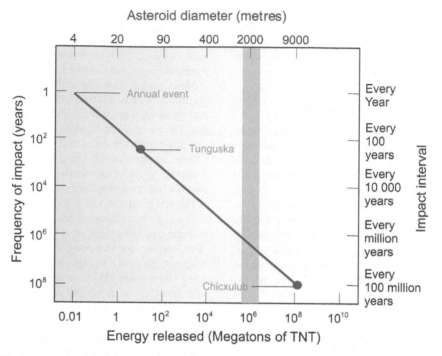

Asteroid diameter (metres)

Energy released (Megatons of TNT)

Figure 15.13   *The frequency and energy of impact events. Energy is given in the commonly used units of equivalent megatons of the explosive, TNT. Also shown are the Tunguska event and the end-Cretaceous impact at Chicxulub. The grey bar represents the region in which the energy of an impact transitions from causing local-scale effects to global-scale perturbation (potentially a mass extinction).*

Figure 15.14   *A concept for deflecting an asteroid on a collision path with the Earth using a fleet of small spacecraft equipped with intense lasers to vapourise material on one side, thus nudging the rock and changing its trajectory by a small angle (Source: NASA).*

those caused by an asteroid impact, hardly provide time for evolution to work in this gradual way. The interval between extinction events is so large and the causes diverse, that by the time the next similar extinction occurs, there are very few of the species, if any, left from the last extinction to be better suited for survival. We might ask whether some extinctions are more a case of survival of the luckiest rather than survival of organisms best adapted to their slowly changing environment.

The study of mass extinctions has raised many questions of interest to biologists, not least the question of what constitutes the characteristics of species that survive these events. There is no hard rule for the characteristics of successful survivors as this will surely depend on the nature of each extinction. Nevertheless there are some patterns. Survivors are generally organisms that have wide geographical distribution. Many of them tend to be organisms that have large population sizes or are capable of foraging, burrowing and otherwise possess traits that

Figure 15.15 Lystrosaurus, *a pig-sized survivor of the end-Permian extinction and member of the so-called 'disaster taxa'* (Source: wikicommons, Dmitry Bogdanov).

allow for survival of rapid changes in environmental conditions and the imposition of environmental stresses. This isn't always true, however. The end-Permian extinction was so severe that some groups that had global distribution, such as the trilobites, went extinct. So-called **disaster taxa** included the end-Permian *Lystrosaurus*, a pig-sized herbivore about 1 m long that dominated the Earth for a short time after the extinction. It has been suggested to have been robust because of its burrowing habitat and small size (Figure 15.15), although its survival could have been luck.

15.10 The Sixth Mass Extinction?

Are we currently in a sixth extinction? There seems to be no doubt that the rate of extinction has been increased by human activity. In 2014, a report (shown in *Further reading* as Pimm *et al.*, 2014) suggested that the rate of extinction is now about 1000 times higher than the background rate of extinction.

We can argue about the specifics of facts and quotes. This is not a specialist book about extinction and, as you can understand, this is a hotly debated and controversial area. However, to get into detailed arguments about individual sets of data would probably be nit-picking at the general broad consensus that seven billion human beings with their industrial activity is having profound and, in geological terms, extraordinarily rapid effects on the Earth's biological diversity.

A human-caused mass extinction raises important points, or questions, that have an astrobiological flavour. Do the rise of intelligence and the spread of a civilisation across a planetary surface usually imply the loss of diversity of other species that share the planet? Does a mass extinction, if you choose to classify the current period, always accompany the rise of intelligence? The current crisis also encourages us to ask whether past extinctions might tell us about the potential effects of human-caused changes. Can we learn anything about extinctions caused by increasing levels of CO_2, for example during the end-Permian extinction, that would provide us with a window for understanding what our own impact might be if we continue to inject CO_2 into the atmosphere?

Quite apart from the obvious and immediate need for us to address the pressing issue of diversity loss and the potential deleterious feedback this may have on our civilisation, the link between the rise of humanity and the other organisms on the planet force us to address broader

---

Debate Point: The sixth extinction?
The current rate of species loss caused by human industrial activity and environmental change has been argued by some to constitute a sixth mass extinction. Do you agree with this assessment? A judgement of this matter depends on a variety of things, including the definition one chooses of mass extinctions, the data one uses to define the background rate of extinction and data collected on the rate of present-day extinction. Although we have records of extinctions caused by humans, including the passenger pigeon, the dodo, the zebra-like quagga and the Tasmanian tiger, these are in some sense anecdotal losses. We have no systematic record of which species have gone extinct in recent years, many of which may not have been recorded. From an astrobiological point of view, measuring the current rate of extinction is an important exercise in developing the methods and metrics for understanding and quantifying the composition of a biosphere. Discuss the evidence for a sixth extinction.

Barnosky, A.D., Matzke, N., Tomiya, S. et al. (2011) Has the Earth's sixth mass extinction already arrived? *Nature* **471**, 51–57.

Kolbert, E. (2014) *The Sixth Extinction: An Unnatural History*. Henry Holt and Co., New York.

questions about the pattern and links between extinctions and life on a planet.

15.11   Conclusions

Throughout the history of the planet, life has been subjected to mass extinctions. We know little about extinction in the Precambrian because of the poor preservation of microorganisms and the difficulty in identifying microbial fossils to species level. Much more detailed information exists for the five major extinctions of multicellular organisms in the Phanerozoic. The cause of these is varied. Volcanism, asteroid impact, biologically-induced changes, astrophysical changes such as supernova explosions and changes in ocean conditions including anoxia have all been suggested as triggers of these events. Periods of extinction have opened up new niches for organisms, but they have not substantially altered the long-term trend to greater species diversity on the planet. The largest of these extinctions, in terms of loss of species, was the end-Permian extinction. The end-Cretaceous extinction was less devastating, however its role in the extinction of the dinosaurs and the strong evidence for the role of an asteroid impact has made it a focus of study. All of these extinctions emphasise that once life becomes established on a planet it faces inevitable disruption. If planetary conditions do not perturb it, then extraterrestrial events, including asteroid and comet impacts, almost certainly will. It seems a reasonable speculation to suggest that if life does exist on other planetary bodies, it too will be subject to extinctions.

Further Reading

Books

Benton, M.J. (2008) *When Life Nearly Died: The Greatest Mass Extinction of All Time*. Thames and Hudson, London.

Kolbert, E. (2014) *The Sixth Extinction: An Unnatural History*. Henry Holt and Company, New York.

Papers

Alvarez, L.W., Alvarez, W., Asaro, F., Michel, H.V. (1980) Extraterrestrial cause for the Cretaceous-Tertiary extinction. *Science* **208**, 1095–1108.

Huey, R.B., Ward, P.D. (2005) Hypoxia, global warming, and terrestrial late Permian extinctions. *Science* **308**, 398–401.

Jablonski, D. (2005) Mass extinctions and macroevolution. *Paleobiology* **31**, 192–210.

Kring, D.A. (2003) Environmental consequences of impact cratering events as a function of ambient conditions on Earth. *Astrobiology* **3**, 133–152.

Payne, J.L., Clapham, M.E. (2012) End-Permian mass extinction in the oceans: an ancient analog for the twenty-first century? *Annual Review of Earth and Planetary Sciences* **40**, 89–111.

Pimm, S.L., Jenkins, C.N., Abell, R., Brooks, T.M., Gittleman, J.L., Joppa, L.N., Raven, P.H., Roberts, C.M., Sexton, J.O. (2014) The biodiversity of species and their rates of extinction, distribution, and protection. *Science* **344**, 1246752

Roberston, D.S., McKenna, M.C., Toon, O.B., Hope, S., Lillegraven, J.A. (2004) Survival in the first hours of the Cenozoic. *GSA Bulletin* **116**, 760–768.

Ryskin, G. (2003) Methane-driven oceanic eruptions and mass extinctions. *Geology* **31**, 741–744.

Scalo, J., Wheeler, J.C. (2002) Astrophysical and astrobiological implications of gamma-ray burst properties. *Astrophysical Journal* **566**, 723–737.

Stigall, A.L. (2012) Speciation collapse and invasive species dynamics during the late Devonian Mass Extinction. *GSA Today* **22**, 4–9.

Wang, S., Marshall, C. (2004) Improved confidence intervals for estimating the position of a mass extinction boundary. *Paleobiology* **30**, 5–18.

16

The Habitability of Planets

Learning Outcomes

➤ Understand the concepts of habitability.
➤ Understand what defines the classical habitable zone and its limitations.
➤ Understand the concept of the galactic habitable zone.
➤ Describe some of the factors that might influence the long-term habitability of a planet including the presence of a Moon, the carbonate–silicate cycle and plate tectonics.
➤ Be able to calculate the effective temperature of a planet.
➤ Understand how the greenhouse effect works and its consequences for habitability.
➤ Understand that the presence of habitable conditions does not imply the presence of life.

16.1 What is Habitability?

Throughout astrobiology texts and papers you will come across the term 'habitable'. Scientists speak of habitable worlds, habitable environments and habitable zones. What do we mean by habitable?

Defining this word exactly is not easy. The word habitat comes from the Latin verb, *habitare* (to live), and it translates as 'it lives'. Bearing in mind the difficulties in defining 'life' (Chapter 1), we could nevertheless define a habitable environment to be 'a location that has the necessary conditions for at least one known organism to

be active'. Active might be taken to mean metabolically active, where activity could include maintenance, growth or reproduction of an organism sustained by the habitat. Some researchers have defined a habitat as a place where an organism can at least survive. Ultimately all organisms need at least one habitat where they can reproduce or their population cannot be sustained.

An important facet of understanding habitability is the concept of the **niche**. Ecologist Eugene Odum (1913–2002) famously made the distinction that a habitat can be considered to be the 'address' of an organism (i.e. its physical location), whereas a niche describes its 'profession'. Ecologist George Hutchinson (1903–1991) considered the niche to be the abstract n-dimensional hypervolume made up of the n different variables (environmental and biological) that define the conditions within which an organism can make a living. For example the niche of a microbe carrying out iron reduction would be the variety of environmental and biotic factors required for this organism to carry out this function in an **anaerobic** environment. From a strict ecological perspective a niche is therefore not a physical space, but a functional definition. However, in the planetary science literature it is common to find the term niche incorrectly used interchangeably with habitat and meant as a purely physical space.

There is much debate and historic discourse in the ecological literature on what these different terms should mean, and a great deal of nuanced discussion. Broadly understood to be the capacity to support the metabolic activity of at least one known organism, habitability so defined provides an adequate place from which to discuss

Astrobiology: Understanding Life in the Universe, First Edition. Charles S. Cockell.
© 2015 John Wiley & Sons, Ltd. Published 2015 by John Wiley & Sons, Ltd.
Companion Website : www.wiley.com/go/cockell/astrobiology.

in the rest of the chapter those conditions required to make a place habitable.

The terms 'habitable' and 'habitability' necessarily are assessments defined, and limited, by our knowledge of life on Earth. It also follows, assuming our knowledge of the limits of life in the Universe is incomplete, that we are likely to overestimate the extent of uninhabitable places and underestimate the extent of habitable spaces in the Universe. As we saw in Chapter 3, people have imagined other biochemistries, such as life using liquid ammonia as a solvent, which would dramatically change the types of putative environments in which life might exist, and therefore the conditions required for habitability. However, as we cannot empirically assess these possibilities easily, nor understand what criteria must be met for habitability in these alternative chemistries, then we usually assume what has been called the 'carbon–water chauvinism' – that life is carbon-based and requires liquid water with its corresponding implications for habitability. If there are no alternative chemistries for life, we might even find that there is no chauvinism and that terrestrial life reflects a universal norm.

Habitability can be assessed at different scales, from the scale of a microorganism about one **micron** in size to the scale of a galaxy many tens of thousands of light years in diameter.

Coming up with a reasonably sensible operative definition of habitability at the scale of a microorganism is not very difficult, as we have already done throughout this book. It is possible to list a series of requirements for an environment to be habitable to a given organism at any given point in time (we might call this 'instantaneous' habitability). As a minimum we need liquid water to act as a solvent together with essential elements (CHNOPS; Chapter 3), an energy source (Chapter 5) and appropriate physical and chemical conditions that allow for growth (Chapter 7). Particular organisms might have other requirements such as iron or metals that can be used as co-factors in enzymes and so on.

The task of defining what is required for an environment, or planet, to be habitable over geological time periods (what we might call 'continuous' planetary habitability) becomes more challenging since it requires us to identify the conditions that allow the requirements for life just discussed to be sustained in environments over long time periods.

16.2   The Habitable Zone

Perhaps one of the most pervasive metrics for assessing habitability on the planetary scale has been the **habitable zone**. The habitable zone is defined as the zone around a star where liquid water is stable at the surface of a planetary body. The phrase, habitable zone, was coined by astrophysicist, Su-Shu Huang (1915–1977) in 1959. The theory of habitable zones was further developed in 1964 by Stephen Dole (1917–2000) in his book, *Habitable Planets for Man*, in which he elaborated the idea of the circumstellar habitable zone as well as various other determinants of planetary habitability, leading him to estimate the number of habitable planets in our galaxy to be about 600 million.

At once we can see some limitations to this definition. Life needs much more than just liquid water. A planet could have liquid water, but be lacking some vital element, such as enough biologically available nitrogen. Therefore not all planets in the habitable zone, even those that host liquid water, can be definitively said to be habitable.

The habitable zone is also limited because planetary bodies need not have surface liquid water, but they could still host sub-surface liquid water that might be habitable. The icy moons of the outer Solar System, Europa and Enceladus, for example, are a case in point as we will explore in Chapter 18. Therefore, just as not all planets in the habitable zone need be habitable, not all habitable planetary bodies need be in classical habitable zones.

Nevertheless, the concept of the habitable zone is useful because it does at least allow us to constrain the search space in the Universe we are interested in to a narrower zone than merely 'everywhere'.

The habitable zone is defined by an inner and outer boundary (Figure 16.1). The inner boundary is set by the distance from the star at which solar luminosity sets in place a **runaway greenhouse effect**, such as that currently seen on Venus. At this distance, the water reservoir is boiled away, it is photodissociated into hydrogen and oxygen, the former being lost to space, the latter recombining with rocks to form oxidised compounds. In our own Solar System the inner limit, as defined by the runaway greenhouse effect, is ~0.85 AU (light green zone in Figure 16.1). However, it is possible that cloud formation might cause water to gather in the upper atmosphere where it is photodissociated, leading to a less extreme 'moist' greenhouse effect long before an extreme runaway greenhouse sets in, but nevertheless a process that ultimately causes water to be lost with its fatal consequences for life. This distance is a surprisingly close 0.97 AU (dark green zone in Figure 16.1), remarkably near to our current position. Thus, the exact position of the habitable zone depends upon the assumptions used.

The outer limit of the habitable zone is set by the distance at which the greenhouse effect is insufficient to generate temperatures to sustain liquid water at the surface.

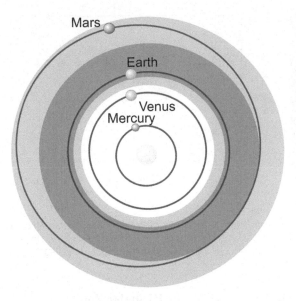

Figure 16.1 *The habitable zone in our Solar System. The dark green zone shows the zone within which liquid water would be stable on a planetary surface using quite conservative estimates and the light green zone less conservative estimates. These limits are explained in the text (Source: Charles Cockell).*

of the planet or where the CO_2 in the atmosphere starts to condense out, again preventing greenhouse warming. Both of these cases produce an outer limit of the habitable zone in our Solar System of 1.7 AU (light green zone in Figure 16.1) and 1.4 AU (dark green zone in Figure 16.1), respectively. These numbers are consistent with what we observe about Mars. Although it may once have hosted long-term surface liquid water (Chapter 17), the planet today is a desert and has an insufficient greenhouse effect or atmospheric pressure to keep the planet warm and sustain surface liquid water, although sub-surface water cannot be discounted. Mars is on the border line of habitability.

It is unsurprising that sat between an inner boundary that is too hot and an outer boundary that is too cold, the habitable zone is sometimes called the **Goldilocks zone**.

Within our own Solar System, Mercury lies well outside the inner boundary of the habitable zone and is too hot to support life. The atmosphere of the planet is so thin anyway (10^{-14} bar or 1 nPa) that there would be no possibility of liquid water. Venus lies just outside the inner edge of the habitable zone. The **runaway greenhouse** effect creates a surface temperature of 467 °C and there is no liquid water. The planetary surface is therefore thought to be completely uninhabitable. There have been some intriguing speculations about the possibility of life in the Venusian cloud deck. At an altitude of ~50 km the temperature is about 40 °C and the pressure is about 1 atm (0.1 MPa), which is biologically clement. However, the clouds contain H_2SO_4 at greater than 80% concentration and so they are unlikely places for complex organic chemistry required for life.

The distance and width of the habitable zone will be influenced by the type of star around which a planet orbits (Figure 16.2). Stars hotter than the Sun (e.g. F stars) will have habitable zones further away. Stars cooler than the Sun, for example M stars, will have habitable zones so close in that they would be within the equivalent orbit of Mercury (0.4 AU). For example the star Proxima Centauri, an M star 4.24 light years away, would require a planet to be about 0.032 AU away to have temperatures similar to the Earth.

16.2.1 Star Types

Star types have secondary impacts on habitability that go beyond merely changing the distance of the habitable zone.

M star habitable zones have particularly specific problems. They have the complication that the habitable zone is sufficiently close to bring planets into the zone where they become tidally locked (Figure 16.2b) – the rotation of the planet is synchronous with the star such that the same face always faces the star. This was thought to be a major showstopper for life because the atmosphere would freeze out on the dark side. However, models have shown that with sufficiently vigorous atmospheric circulation heat can be transported from the star-facing side to the dark side and prevent the freeze-out of the atmosphere. The lack of rotation in low mass planets can contribute to weak magnetic fields which can leave the atmosphere exposed to erosion by the stellar wind (the same phenomenon as the **solar wind** from the Sun) or **coronal mass ejections**.

M stars tend to be prone to flares of UV radiation, such that planets orbiting these stars may be subject to intermittent elevated UV radiation doses, one to three orders of magnitude above background. As we discussed in Chapter 10, however, there are ways in which life, for example protected within rocks, might be able to survive or even grow under these periodic extremes of UV radiation.

An advantage of M stars is that they have long life times. The lives of M stars can exceed trillions of years (Chapter 8), giving any planets orbiting them plenty of time for the evolution of life, even intelligence. As they make up 75% of all stars in the Universe they are also numerous. K stars, a little more massive than M stars, also

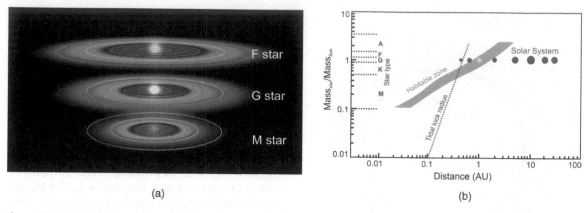

(a)   (b)

Figure 16.2 *Habitable zones around different stars. (a) The general principle is shown that the hotter the star (an F star is hotter than a G star, which is hotter than an M star), the further away the habitable zone. (b) A more detailed plot showing the habitable zone at given distances from different stars. Our Solar System is shown. Redrawn from Kasting et al., (1993) Icarus* **101**, *108–128.*

have long lives (~20 billion years) and make up another 15% of all stars, showing that low mass stars are good candidates for hosting habitable planets from a purely statistical consideration.

Stars more massive than the Sun produce more UV radiation but, again, as we discussed in Chapter 10, this type of radiation can be screened out by a variety of natural substrates such as rocks. Their more problematic feature is their relatively short lifetime. Star types from O to F have lifetimes measured in values of two billion years or less. Given that it took the Earth about four billion years to go from its formation to the evolution of multicellular life and a further ~600 million years to give rise to intelligence, it is possible that even if these stars do host life, their evolution could be truncated, unless planetary conditions allow for accelerated evolutionary transitions in comparison to the Earth.

16.2.2   Continuously Habitable Zone

The habitable zone around a star is augmented in the concept of the **continuously habitable zone**. This is the zone around a star where a planet can sustain liquid water for the planetary lifetime. As a star's luminosity increases as it ages, so the edges of the habitable zone will move outward. Therefore the continuously habitable zone will be a narrower width than the habitable zone and its exact dimensions will depend upon the star. This continuously habitable zone is an important concept as it defines the zone within which liquid water on a planetary surface will be around sufficiently long for life to undergo billions of years of evolution.

16.2.3   The Galactic Habitable Zone

The idea of the habitable zone can potentially be applied on a galactic scale in the concept of the **galactic habitable zone**. Stars too close to the centre of a galaxy are likely to be within regions of dense star clusters and might be subject to close supernova explosions, intense gravitational perturbations and other astrophysical violence with potentially devastating effects on any would-be biosphere over geological timescales. In our own galactic centre there are 20 million stars packed into a volume just over three light years across, each star being, on average, only about 1000 AU from another star. Collisions occur about once every million years and X-rays and gamma rays stream from the centre. Even more extreme is the supermassive **black hole** at the centre, a region in which life-bearing planets and their stars cannot exist. Stars far out in a galaxy tend to be of lower metallicity (they have fewer heavy elements) and so may not be so conducive to hosting the wide diversity of elements useful for building life (Figure 16.3). The zone between these two extremes (between about 15 000 and 38 000 light years from the galactic centre) may be optimal for hosting stars with sufficient concentrations of elements, but also in sufficiently quiet neighbourhoods to allow for the emergence of biospheres over long uninterrupted timespans.

A case in point is our own Solar System. The Solar System is in the inner edge of the Orion Arm of the Milky Way, a spiral-shaped concentration of dust and gas. Our galaxy is about 100 000 light years across and the Sun is about 26 000 light years from the centre (Chapter 8). Several features of this region make it suitable for life. It

Figure 16.3 *The galactic habitable zone shown in green in a schematic of the Milky Way. The location of the Sun with respect to the zone is shown* (Source: NASA).

is not in a globular cluster where star densities might cause high radiation and gravitational disturbance. It is not near an active gamma ray source and it is not near the galactic centre where star densities increase the likelihood of high ionising radiation or astrophysically disruptive events such as supernovae.

16.2.4   The Right Galaxy?

Even our galactic type could be important. Elliptical galaxies, which are near-spherical ellipsoidal collections of stars, have very little dust and gas suggesting that star formation has generally long since ceased. They comprise about 15% of the galaxies in the Virgo Supercluster. Although they could be the site of planets that have ancient ecosystems on them, these galaxies also tend to be comprised of ancient Population II stars, much older than our own Sun (a Population I star). They tend to have low metallicity. Irregular galaxies, by contrast, have much larger amounts of gas and dust than spiral galaxies (which make up about 60% of observable galaxies) and although this makes them good places for star formation, they may be places in which there is a greater probability of life-destroying astrophysical violence. Starburst galaxies and Active Galactic Nuclei (AGN) may also be places with high levels of potentially life threatening activity

and radiation exposures. We know very little about how large scale galactic structure influences the prospects for life.

16.3   Maintaining Temperature Conditions on a Planet Suitable for Water and Life

We can identify the major requirements at any given point in time for an environment to be able to support the metabolic activity, growth and/or reproduction of any given organisms. However, for a planet to be able to support life over giga-year time scales, it must be possible for these requirements to be met continuously. The temperature must be sustained within a range where liquid water can exist, which as we have seen, defines the classical habitable zone. How are temperatures on a planetary surface sustained at values suitable for liquid water?

16.3.1   Effective Temperature and the Greenhouse Effect

The temperature of a planetary environment is critical for the presence of water and life. At the current time we think that life must exist within the boundary of circa −15 to 122 °C (Chapter 7) to be able to reproduce. Note that this range is smaller than the range for liquid water. Salts can depress the liquid water to temperatures well below −20 °C and under high pressure deep sub-surface environments can host liquid water well above 122 °C. However, the range given is for demonstrated reproduction in the laboratory. It may be conservative and further research may expand this range, but it provides a basis with which to examine the habitability of other worlds. How do we calculate the surface temperature of a planet to assess its habitability for life?

16.3.1.1   Calculating the Effective Temperature of a Planet

A planetary body absorbs radiation from its host star and transmits energy as it is a blackbody source of radiation, giving it an overall effective temperature that we can calculate. For many rocky planetary bodies this effective temperature is quite a simple calculation since there is no atmosphere to complicate the absorption and re-radiation of energy.

The energy absorbed by a planetary body, such as the Earth, can be calculated by assuming that the planet (with a radius, r, and distance, a, from the star), absorbs some of the energy being emitted by the star.

First we know that the energy per unit time emitted by a star or its luminosity, L, is given by the Stefan–Boltzmann law (this was introduced in Chapter 8):

$$L = 4\pi R^2 \sigma T^4 \qquad (16.1)$$

Where R is the radius, T is the temperature and σ is the Stefan–Boltzmann constant.

This energy spreads out into space in a sphere, given by an area, at the distance of the planet, of $4\pi a^2$.

Therefore, the power received per unit area, S, at the planet can be given by:

$$S = (4\pi R^2/4\pi a^2)\sigma T^4 \qquad (16.2)$$

$$= (R/a)^2 \sigma T^4 \qquad (16.3)$$

In the case of the Earth, S is known as the **solar constant** and its standard value is usually taken to be 1367 W m^{-2}.

The planet is a disc of cross section πr^2, where r is the radius of the planet, so the energy intercepted by the planet (P_{in}) will be S multiplied by this area:

$$P_{in} = S\pi r^2 (1-A) \qquad (16.4)$$

You'll also notice that we've snuck in another term, $(1 - A)$. This is the fraction of energy actually absorbed and takes into account that some of the radiation will be reflected back into space by the planet's surface or clouds without being absorbed, where the **albedo**, A, is the fraction reflected. For the Earth, the value of the albedo is ~0.37, but varies depending on cloud cover, presence of dust in the atmosphere and surface cover (ice has a higher albedo than dark volcanic rocks). The Moon and Mercury, which have no appreciable atmosphere, have albedo values of 0.113 and 0.138, respectively.

What about the outgoing blackbody energy from the planet? The total blackbody energy being emitted (which does not include that merely reflected by its surface which is in the albedo term) is again given by the Stefan–Boltzmann law. The total energy (analogous to the luminosity measurement for a star) is the energy per unit area multiplied by the surface area of the planet:

$$P_{out} = 4\pi r^2 \sigma T_e^4 \qquad (16.5)$$

where in this case T_e is the blackbody temperature of the planet.

Assuming that the incoming and outgoing energy are in balance then the effective temperature of the planet, T_e is given by the equation:

$$\mathbf{T_e = (S(1-A)/4\sigma)^{1/4}} \qquad (16.6)$$

where S is the solar constant, A is the albedo and σ is the Stefan–Boltzmann constant.

16.3.1.2 The Greenhouse Effect

This last equation gives us the effective temperature of the planet. However, there is a complication. The radiation arriving at the surface of the planet is from the star, which is a much hotter blackbody than the planet. As we saw in Chapter 8, we can calculate the peak wavelength of a blackbody from Wien's law, which states that this peak wavelength (λ_{max}) is given by:

$$\lambda_{max} = 2.898 \times 10^{-3}/T \ (K).$$

As the planet is a colder blackbody than the star, the radiation re-emitted by the planet will be at longer wavelengths than the radiation absorbed (the peak wavelength is shifted from the optical to the infra-red; review Figure 8.11 in Chapter 8). In the absence of an atmosphere this fact would be one of passing academic interest. However, gases such as H_2O, CO_2, CH_4 and NH_3 tend to be transparent at short wavelengths and opaque at longer wavelengths, meaning that they let in short wavelength radiation to the surface of a planet, but tend to block the longer wavelengths from escaping. The result is that the energy is trapped, heating the surface of the planet in the **greenhouse effect** (hence they are called **greenhouse gases**). The absorption bands can be made more effective at trapping gases by the influence of other gases. For example the absorption bands of CO_2 are broadened by the presence of N_2 and O_2, enhancing the greenhouse effect.

On the Earth today, H_2O and CO_2 are the two most important greenhouse gases, accounting for almost all of the greenhouse effect, with further more minor contributions from CH_4, N_2O and O_3. CH_4 is about 20 times more effective as a greenhouse gas than CO_2 because of its stronger absorption of infra-red radiation (7.5–12.0 μm wavelength), although its concentration in the present-day atmosphere is low. In total, the greenhouse effect is about 30 K for the Earth, but it makes all the difference because it means our mean temperature remains above freezing.

Bodies without an atmosphere such as the Moon or Mercury have no greenhouse effect. Some other planets have much more extreme greenhouse conditions than the Earth where temperatures are so high that liquid water cannot persist on a planetary surface. An example of this **runaway greenhouse effect** is to be found with Venus, which has a dense 92 bar atmosphere of 96.5% CO_2. It has a greenhouse effect of ~510 K, taking it from the effective

Debate Point: Atmospheric habitability and airborne biospheres

So far we have made the assumption that we are looking for habitable environments on the surface and sub-surface of planets. However, could atmospheres of planets harbour active biospheres even when the surface is inclement? In a wild, but fascinating paper, Sagan and Saltpeter speculated about ecologies in the atmosphere of Jupiter. They hypothesised the existence of 'floaters' and 'hunters'. The 'floaters' were giant bags of gas like hot air balloons, metabolising with sunlight and molecules in the atmosphere to keep their gas warm. The 'hunters' were squid-like creatures, using jets of gas to propel themselves into the 'floaters' and consume them. Other scientists have discussed the possibility of life in the cloud deck of Venus, even today, where in some regions temperatures and pressures may be clement for life. Do you think these ideas are plausible? How would life maintain itself in an atmosphere? Could you imagine or calculate whether bacteria-sized particles would be able to reproduce fast enough, as some of them drift out of an atmosphere in wind currents, to maintain a viable population within a planetary atmosphere. We know that there are microorganisms in the Earth's atmosphere, but would an atmospheric biota be sustained if the Earth's surface became uninhabitable?

DeLeon-Rodriguez, N., Lathem, T.L., Rodriguez, L.M., Barazesh, J.M., Anderson, B.E., Beyersdorf, A.J., Ziemba, L.D., Bergin, M., Nenes, A., Konstantinidisa, K.T. (2013) Microbiome of the upper troposphere: Species composition and prevalence, effects of tropical storms, and atmospheric implications. *Proceedings of the National Academy of Sciences* **110**, 2575–2580.

Sagan, C., Saltpeter, E.E. (1976) Particles, environments and possible ecologies in the Jovian atmosphere. *Astrophysical Journal Supplement Series* **32**, 737–755.

Schulze-Makuch, D., Grinspoon, D.H., Abbas, O., Irwin, L.N., Bullock, M. (2004) A sulfur-based UV adaptation strategy for putative phototrophic life in the Venusian atmosphere. *Astrobiology* **4**, 11–18.

temperature it would have without an atmosphere of 257 K to a greenhouse-induced temperature of about 740 K, more than 300 K above the known upper temperature limit for life. Venus is uninhabitable. The greenhouse effect matters.

We cannot rule out the possibility that the situation was very different on early Venus. During the Earth's Hadean, the Sun was about 30% less luminous. During this time, Venus may have hosted early oceans since the inner limit of the habitable zone would have been nearer the Sun. We do not know exactly at what time the greenhouse effect would have set in, perhaps a few hundred millions years after the formation of the planet.

As we learned in Chapter 10, it remains a point of discussion about how, when the Sun was less luminous than it is today in the early history of the planet, the Earth remained warm enough for oceans and liquid water. The greenhouse mechanisms and gases available during this time have attracted the interest of astrobiologists seeking to solve the **faint young Sun paradox**.

The greenhouse effect on the Earth is a modern concern to us because, by injecting CO_2 into the atmosphere, humans are thought to be raising the temperature.

A matter that arises from this discussion is whether the characteristic that the Earth has sustained liquid water at its surface through billions of years is just a matter of luck – that the incoming radiation just happens to nicely balance the outgoing radiation to allow for liquid water. It turns out that it is not, as we shall now discuss.

16.3.2 The Carbonate–Silicate Cycle

One significant control on planetary temperature is the carbonate-silicate cycle (Figure 16.4).

As CO_2 in the atmosphere mixes with rainwater, it forms carbonic acid (H_2CO_3), a weak acid. This solution gently erodes silicate rocks (generically illustrated in Figure 16.4 as $CaSiO_3$), thus **weathering** them, causing the release of cations such as Mg^{2+} and Ca^{2+} into rivers and streams.

In water bodies, such as rivers, lakes and oceans, some of these cations bind to bicarbonate ions (HCO_3^-) to form carbonates (shown in Figure 16.4 as $CaCO_3$) by both biological and abiotic processes, which sink to the bottom of the ocean and other water bodies, thus sequestering CO_2 from the atmosphere and burying it into the rock record. The net loss of the CO_2 is not permanent. On the Earth, plate tectonics will eventually cause the subduction of the carbonates, heating them in metamorphism with silicates, releasing the CO_2 back into the atmosphere in volcanic exhalations and eruptions.

Within this cycle there is a feedback process (Figure 16.4b). If CO_2 in the atmosphere rises, then temperatures rise as a result of the greenhouse effect. Higher temperatures increase the rate of chemical reactions.

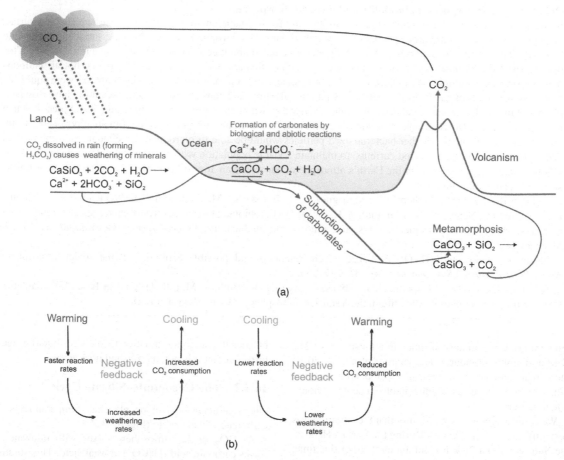

Figure 16.4 *The carbonate–silicate cycle. (a) One principal mechanism by which temperatures on the surface of the Earth are regulated through the feedback control of the greenhouse gas, CO_2. (b) The carbonate–silicate cycle works by a negative feedback process.* (Source: Charles Cockell).

The link between temperature and chemical reaction rates is given by the **Arrhenius relationship**, first elaborated by Swedish physicist Svante Arrhenius (1859–1927). The Arrhenius relationship is a simple equation which gives the relationship between the rate constant of a chemical reaction, k, to the absolute temperature, T. That relationship is:

$$k = Ae^{-E_a/(RT)}$$

where R is the Universal Gas Constant and E_a is the activation energy of the reaction. In general, chemical reaction rates about double for every 10° increase in temperature. Therefore, as temperatures increases, rock weathering reactions increase.

As weathering reaction rates increase, so more CO_2 is consumed in these reactions, reducing its atmospheric concentration (assuming that volcanic production is unchanged) and thus reducing CO_2 greenhouse warming and lowering the global temperature. An increase in CO_2 concentrations will also increase the acidity of solutions, further enhancing weathering rates.

Conversely, a reduction in atmospheric CO_2 concentration or temperatures causes the reverse effect – the rate of weathering reactions decreases, reducing CO_2 consumption and thus allowing it to build up in the atmosphere. The net effect is that the carbonate–silicate cycle can be regarded as an enormous thermostat, maintaining temperatures within a habitable range (Figure 16.4).

Feedback processes such as this are common in climate cycles. In Chapter 14, we discussed how feedback processes were involved in the Earth's entry and exit from a Snowball state.

These long-term changes in carbonate–silicate thermostatic regulation have imposed upon them other more short-term forcings, for example Milankovich cycles, named for Serbian geophysicist, Milutin Milanković (1879–1958). Rotation of the axis of the Earth like a spinning top (precession), changes in **obliquity** and orbital **eccentricity**, introduce 20 000, 40 000 and 100 000 year cycles, respectively, into the climate. Even shorter-term changes, on the order of 10–1000 years are introduced into the climate and temperature system, possibly caused by changes in ocean circulation. Nevertheless, the carbonate–silicate cycle accounts for the long-term maintenance of suitable temperature conditions for liquid water.

16.4   Plate Tectonics

As a corollary to the contribution of the carbonate-silicate cycle to habitability, plate tectonics may be essential for maintaining habitable conditions. It provides the return pathway for buried carbon to be re-released into the atmosphere as CO_2, thus completing the carbonate–silicate cycle.

An important factor in long-term habitability is the maintenance of available energy and nutrients. Once an organism has used up nutrients or its local energy supply, it will no longer be able to be active. Therefore, planets must not only have conditions that provide clement physical conditions for microbial activity over long periods, but there must also be mechanisms to prevent what one might broadly refer to as 'biogeochemical rundown' – it must be possible to recycle required resources for life. In addition to returning carbon, plate tectonics acts to recycle elements such as different redox states of sulfur and iron, maintaining a vigorous circulation of elements used by life and generating widespread and constantly changing geochemical disequilibria which life uses as a source of energy.

Importantly, plates must be lubricated for movement to occur and water is the best substance for making the plates deformable enough to move and subduct. The presence of water, one fundamental requirement for habitability, is therefore linked to plate tectonics. Water plays other important roles. The weathering of rocks produces hydrous minerals. When these are subducted

at plate boundaries, they are melted and tend to form silica-rich minerals such as granite and rhyolite. High silica rocks (such as granites) are less dense than the basalts from which the oceanic crust is made and so they float to the top, forming the continents. You can think of the continents as fluffy silica-rich froth that has solidified. These continents provided the setting for the emergence of intelligent land life. Water, not only required as a solvent for life, may be inextricably linked to continent formation through plate tectonics.

The story with plate tectonics is not merely limited to near-surface processes. It may also play a fundamental role in maintaining the right temperature conditions in the interior for mantle convection to occur. Water taken down into the mantle in plate tectonics reduces the **viscosity** of the mantle, enhancing its circulation. This circulation in turn may influence heat exchange with the outer liquid core that produces the magnetic dynamo responsible for the magnetic field. The magnetic field (**magnetosphere**) protects the atmosphere from the **solar wind** (predominantly made of electrons and protons with energies between 1.5 and 10 keV). As particles in the solar wind are charged they cannot cross magnetic field lines and are channelled around the Earth or trapped within the Van Allen Belts, radiation belts which extend from 1000 to 60 000 km above the Earth. The magnetosphere thus prevents the atmosphere from being sputtered away, with obvious consequences for the maintenance of a life-supporting atmosphere.

It therefore seems that plate tectonics, liquid water and the magnetic field are all intimately linked into a single system. Planets that are too small and cool rapidly after formation will not sustain plate tectonics and will rapidly lose their magnetic dynamos. Mars is such an example. Thus, planetary size is an important factor in plate tectonics and thus planetary habitability. In the case of Venus, although the planet is larger, it is too near the Sun. The runaway greenhouse effect has caused the loss of the oceans. The lack of water makes the crust solid and not as ductile as the Earth's crust to allow for plate tectonics. The lack of water stymies the production of silica-rich rocks to form continents. On Venus we see how the position in the habitable zone affects water availability and thus the regime for plate tectonics with its implications for long-term habitability. Although one should be careful about Earth-centric assessments of habitability, there is a remarkable number of interlinked factors in plate tectonics that suggest that over very long geological time periods plate tectonics plays a role in creating habitable conditions on the surface of a planet.

Debate Point: The factors required for habitability

What factors are required for a planet to be habitable? This has been a long-term debate and there is no clear-cut answer. However, the answers are essential for assessing other planetary locations as supportive of life and, closer to home, understanding what factors influenced the success of different types of life, from prokaryotic to multicellular life, in the past history of the Earth. Draw up a list of conditions that are required for habitability at the micron scale and those conditions that might be required to support these conditions over geological time scales. Try ranking these conditions in terms of their importance, identifying those for which we have unequivocal evidence of their role in sustaining habitable conditions over prolonged time periods and those for which we do not. Identify factors that may not have been discussed in this chapter.

Javaux, E.J. and Dehant, V. (2010) Habitability: from stars to cells. *Astronomy and Astrophysics Reviews* **18**, 383–416.

Lammer, H., Bredehoft, J.H., Coustenis, A., Khodachenko, M.L., Kaltenegger, L., Grasset, O., Prieur, D., Raulin, F., Ehrenfreund, P., Yamauchi, M., Wahlund, J.E., Griessmeier, J.M., Stangl, G., Cockell, C.S., Kulikov, Y.N., Grenfell, J.L., Rauer, H. (2009) What makes a planet habitable? *Astronomy and Astrophysics Review* **17**, 181–249.

16.5   Do We Need a Moon?

There has been a great deal of speculation that the Moon plays a fundamental role in habitability. Early models suggested that it plays a role in stabilising the **obliquity** of the Earth (Figure 16.5) but that, over long time periods (hundreds of millions of years), without a moon, the Earth's obliquity would vary chaotically, leading to tilts up to 50°. More recent models suggest instead that a moonless Earth, although exhibiting greater obliquity changes, would still maintain obliquity variations within a 20–25° range for hundreds of millions of years. Regardless, by stabilising

obliquity to some extent, the Moon might influence variations in terrestrial climate. Thus, the argument goes that, without a moon, Earth's climate would be unstable over geological time periods.

Certainly the models suggest that the Moon does exert a stabilising influence on the tilt of the planet and, by inference, on its climatic state. However, caution should probably be exercised in the biological interpretation. One might equally argue that if there was no Moon, the varying obliquity would select for very generalist organisms capable of dealing with geologically frequent climatic change. This could lead to groups of organisms better able to cope

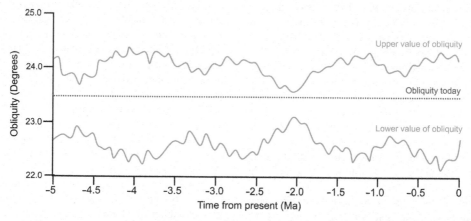

Figure 16.5 *Moon-induced stabilisation of the Earth's obliquity showing the upper and lower value of the obliquity for the past five million years. Note that the obliquity varies only from about 22.0° to 24.5°. The red line is the value today.* [Source: Adapted from Berger (1976)].

with sudden and catastrophic climate change, such as freezing or warming episodes induced by impact events. We could just as well speculate that the presence of a Moon led to the evolution of extinction-prone specialist organisms.

The arguments about the Moon highlight the more general point that one should be careful about anthropocentric assessments of habitability. There are clearly things that organisms do need to be able to inhabit a planet (such as an energy source – without this they cannot grow or reproduce). These requirements are irrefutable. However, there are other factors that might well modify the characteristics of habitable conditions, their long-term stability and the course of evolution, but may not in themselves make a categorical difference to whether a planet is habitable or not. The presence of a Moon might be such a factor.

16.6 Surface Liquid Water, Habitability and Intelligence

Many of the factors for long-term habitability that we have discussed, such as plate tectonics, a magnetic field and the formation of continents, apply to planetary bodies with liquid water on their surface. However, habitable conditions might exist in sub-surface oceans within tidally distorted icy moons or even within planets on which the greenhouse effect is not sufficient to maintain liquid water, but radioactive heating is sufficient to maintain liquid water.

The distinction between these two types of worlds from an astrobiological point of view may well be categorical. In the case of worlds with liquid water on their surfaces, the water may occur spatially co-terminus with the presence of light, thus allowing for photosynthesis. If we assume that the emergence of multicellularity and intelligence is linked to the presence of oxygen from photosynthesis (Chapter 14), then these worlds may be the only type capable of producing the conditions for planetary habitability leading to intelligence. By contrast, worlds with water only in their interior may have habitable conditions for a range of metabolisms, but not conditions for photosynthesis. Therefore, although the planet could retain continuous planetary habitability for certain anaerobic metabolisms in its sub-surface, the surface becomes uninhabitable for phototrophs. This truncates the conditions for habitability required for the large-scale increase in atmospheric oxygen and thus multicellular life and intelligence.

16.7 Uninhabited Habitats: Habitats Need Not Always Contain Life

Normally, when we talk about a planet being habitable, we make the implicit assumption that it hosts life. This is a paradigm rooted in the nature of life on the Earth. On the present-day Earth, photosynthesis generates large quantities of organic carbon and oxygen, both of which are distributed globally in the hydrological cycle and the atmosphere. These two products of photosynthesis can be readily combined for aerobic respiration, so that a newly formed habitat, for example one formed as a result of the deposition of new lava flows, or by the ponding of water in a depression from melting ice, becomes rapidly colonised by active organisms. Oxidised compounds, such as sulfate and nitrate, whose abundance is a consequence of the Earth's oxidised atmosphere and crust, also provide electron acceptors for the anaerobic oxidation of photosynthetically-produced organic matter. These conditions mean that almost all habitats are colonised. In other words, in studies of terrestrial biology there has always been an assumption, almost a paradigm, of the coupling between habitability and the presence of life.

This paradigm might be more obviously invalid on other planetary bodies. For example, lifeless planets might harbour locations that have all the requirements for certain organisms at small scales, but the lack of an origin of life or the interplanetary transfer of life to the planet from a nearby life-bearing planet (assuming this process is even possible; this will be discussed more in Chapter 17), could result in uninhabited, but habitable spaces. These spaces could include habitats that are completely devoid of organics or habitats that have abiotic organics, but no life. On sparsely inhabited planets with a weak hydrological cycle, a lack of connectivity between habitats could result in habitats in which there is no active life. We can call these places **uninhabited habitats**, and we could define such a habitat as 'an environment capable of supporting the activity of at least one known organism, but containing no such organism'.

An environment can be demonstrated to be an uninhabited habitat by a number of means. It could be proven directly by showing that the environment will support the activity of an organism inoculated (potentially irreversibly) into that environment or by removing physical material from the environment and showing in a laboratory that it will support the activity of certain organisms

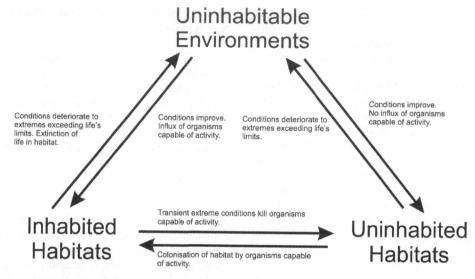

Figure 16.6 *Environments in the Universe can be split into three types. In many places on and just beneath a planetary surface, they can be interchangeable. Note that in the case of an uninhabitable environment making the transition to an inhabited habitat the appearance of habitable conditions must occur contemporaneously with the arrival of organisms or the environment transitions through being an uninhabited habitat* (Source: Charles Cockell).

under the same environmental conditions. It could be shown indirectly, by a comprehensive geochemical and environmental analysis which demonstrates that the environment has the necessary conditions to support the activity of at least one known organism.

From a biological perspective, all environments in the known Universe can therefore be split into one of three types (Figure 16.6).

Given the extent of biologically deleterious inter-stellar space with its extreme conditions, of the three environments shown in Figure 16.6, **uninhabitable environments** ('an environment in which no known organism can be active') are the most common type of environment in the Universe. The second two environment types are both types of habitat – inhabited and uninhabited habitats. In the former there is life active or capable of activity in the environment, in the latter there is not. In some locations in the Universe, particularly on and in planetary crusts, these three environments can be interchangeable as local conditions alter (Figure 16.6). The arrows in Figure 16.6 are applied to a fixed point in or on a planetary body and represent the dynamic nature of environments.

The significance of uninhabited habitats is that they would provide us with control environments to inves-tigate what influence biology has on the geology and

geochemistry of planets. When we investigate the early Earth, for instance, and its isotopic evidence for life, we do not have lifeless control environments with which to understand what signatures are unequivocally caused by life. If uninhabited habitats are to be found elsewhere, for example, on Mars, then they could provide us with insights into the geochemistry of a planet without the confounding influence of life.

16.8   Worlds More Habitable than the Earth?

We take the Earth as our benchmark for habitability, but are there planets in the Universe in which conditions for biology might even be better? In Chapter 14 we opened a discussion about whether, on a smaller planet, the rise in atmospheric oxygen concentrations could occur sooner than on the Earth. There are other ways in which we could imagine improved conditions, for example, a more vigorous hydrological cycle might make more nutrients available for life.

Yet again, we must be cautious. The phrase, 'more habitable', in particular, may not be specific enough to be useful. Do we mean a greater biomass sustained, or a larger diversity of species, or something else? To make a comparison between the Earth and another planet hosting a biosphere we would have to be clear about exactly what aspects of the biosphere we were comparing. We could plausibly argue that a planet is either habitable to life or it isn't. If this is a binary decision, then the follow-up question is not 'are some planets more habitable?' but a more specific question, such as 'are the planet's habitable conditions such that it can sustain a greater biomass per unit surface area?'.

16.9 The Anthropic Principle

The proposed requirements for habitability and the conditions for it to be sustained over long periods fall into the purview of one important principle in cosmology and astrobiology – the anthropic principle.

The argument simply states that the characteristics of the Universe must be compatible with life for us to be here to observe it in the first place. For example, the physical laws of the Universe must be such that planets can come into existence and support life, or else we could

not observe this fact. Even tiny changes in the physical constants that allow for stable atomic nuclei binding, for instance, would lead to a Universe where the matter that we are familiar with could not exist. Is this a strange lucky coincidence or is there some deeper structure that makes the Universe well 'tuned' for life? This question can become highly philosophical and it has been severely criticised for its lack of testability or falsifiability. However, it does bear on the more empirical question of whether planet Earth is extremely rare, the one location where all the potentially many requirements for habitability have been met, or whether the requirements for habitability are sufficiently relaxed in the Universe for many planets in the Universe to host life, even intelligence. How stringent are the physical laws and the resulting nature of the physical Universe with respect to the origin and emergence of life?

16.10 The Fate of the Earth

Despite having a thermostatic control on temperature through the carbonate–silicate cycle, this system cannot indefinitely prevent the inevitable. Eventually the increase

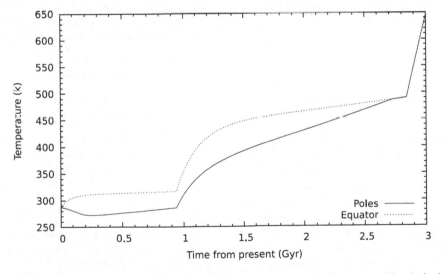

Figure 16.7 *Change in global mean temperatures over time with increasing solar luminosity. The dashed line represents equatorial temperature and the solid line represents polar temperature. After about 1 Gyr a moist greenhouse begins when temperatures reach 330 K, allowing the water vapour content of the stratosphere to increase rapidly. When temperatures reach approximately 420 K life would likely no longer be able to survive. A runaway greenhouse regime begins after approximately 2.8 Gyr. [Source: O'Malley-James et al. (2013)].*

in solar luminosity as the Sun begins its journey to Red Giant status (Chapter 8) will cause the Earth to enter into a moist greenhouse state where the oceans begin to evaporate in about one billion years from now (Figure 16.7). This may occur when the Sun's luminosity is 10% greater than today. During this state, which is a less extreme version of a **runaway greenhouse effect**, the water vapour will collect in the stratosphere and be **photolytically** split, with the hydrogen escaping into space and the oxygen remaining in the atmosphere. Microbial life might persist in localised habitats up to 2.8 Ga from today. Beyond this state, the planet will become Venus-like as it enters a runaway greenhouse effect and the oceans are completely boiled away. The last vestiges of life on Earth will die.

The biosphere will, in a crude way, reverse its emergence. Animals are expected to die off first as temperatures exceed their tolerance. Plants will follow. The last ecosystems will be microbial, perhaps comprising of organisms found in high temperature, saline habitats and located in high altitude environments, caves or other localised environments where liquid water persists. You might like to consider the fate of intelligence on our planet if it still exists. Can the technology exist for a species to escape a planet permanently that is shortly destined to be engulfed by a runaway greenhouse effect?

16.11   Conclusions

Habitability is a broad, and often ill-defined, term, but is generally used by astrobiologists to define places that can support the maintenance, growth and reproduction of organisms. We can define conditions at the small scale required at any point in time for a place to be habitable for given organisms. They include the requirement for an energy source, CHNOPS elements, liquid water and appropriate physical and chemical conditions. Habitability is necessarily a concept circumscribed by what we know about life on the Earth and it must always be defined with reference to particular organisms. An environment habitable to a one particular type of organism can be uninhabitable for another. More difficult to identify are the factors that over geological timescales allow habitable conditions to be sustained. Plate tectonics and the carbonate–silicate cycle are just two processes that on the Earth are thought to contribute to long-term habitable conditions on the surface. When we want to expand these concepts to a more universal scale then we need to constrain the

sorts of environments where habitable conditions might be able to persist. Concepts such as the habitable zone prove to be useful ways to do this, but they must be used with care. We may find habitable bodies outside the classical habitable zone, such as in icy moons. Being within the habitable zone does not necessarily imply habitability. Habitable conditions do not necessarily imply the presence of life, even on an inhabited planet.

Further Reading

Books

Langmuir, C.H., Broecker, W. (2012) *How to Build a Habitable World*. Princeton University Press, Princeton.

Kasting, J. (2010) *How to Find a Habitable Planet*. Princeton University Press, Princeton.

Ward, P.D., Brownlee, D. (2004) *Rare Earth: Why Complex Life is Uncommon in the Universe. Copernicus*. Springer, New York.

Papers

Berner, R.A., Lasaga, A.C., Garrels, R.M. (1983) The carbonate–silicate geochemical cycle and its effect on atmospheric carbon-dioxide over the past 100 million years. *American Journal of Science* **283**, 641–683.

Campbell, I.H., Taylor, S.R. (1983) No water, no granites – no oceans, no continents. *Geophysical Research Letters* **10**, 1061–1064.

Cockell, C.S., Balme, M., Bridges, J.C., Davila, A., Schwenzer, S.P. (2012) Uninhabited habitats on Mars. *Icarus* **217**, 184–193.

Hall, L.S., Krausman, P.R., Morrison, M.L. (1997) The habitat concept and a plea for standard terminology. *Wildlife Society Bulletin* **25**, 173–182.

Heller, R., Armstrong, J. (2014) Superhabitable worlds. *Astrobiology* **14**, 50–66.

Hoehler, T.M. (2007) An energy balance concept for habitability. *Astrobiology* **7**, 824–838.

Jakosky, B.M., Shock, E.L. (1998) The biological potential of Mars, the early Earth and Europa. *Journal of Geophysical Research* **103**: 19359–19364.

Kasting, J.F., Catling, D. (2003) Evolution of a habitable planet. *Annual Review of Astronomy and Astrophysics* **41**, 429–463.

Kasting, J.F., Whitmire, D.P., Reynolds, R.T. (1993) Habitable zones around main sequence stars. *Icarus* **101**, 108–128.

Knoll, A.H. (2011) The multiple origins of complex multicellularity. *Annual Review of Earth and Planetary Sciences* **39**, 217–239.

Lineweaver, C.H., Fenner, Y., Gibson, B.K. (2004) The galactic habitable zone and the age distribution of complex life in the Milky Way. *Science* **303**, 59–62.

Nisbet, E., Zahnle, K., Gerasimov, M.V., Helbert, J., Jaumann, R., Hofmann, B.A., Benzerera, K., Westall, F. (2007) Creating habitable zones, at all scales, from planets to mud-micro-habitats, on Earth and on Mars. *Space Science Reviews* **129**, 79–121.

O'Malley-James, J.T., Greaves, J.S., Raven, J.S., Cockell, C.S. (2013) Swansong biospheres: refuges for life and novel microbial biospheres on terrestrial planets near the end of their habitable lifetimes. *International Journal of Astrobiology* **12**, 99–112.

Sleep, N.H., Zoback, M.D. (2007) Did earthquakes keep the early Earth habitable? *Astrobiology* **7**, 1023–1032.

Stoker, C.R., Zent, A., Catling, D.C., Douglas, S., Marshall, J.R., Archer, D., Clark, B., Kouvanes, S.P., Lemmon, M.T., Quinn, R., Renno, N., Smith, P.H., Young, S.M.M. (2010) Habitability of the Phoenix landing site. *Journal of Geophysical Research* **115**, E00E20.

17

The Astrobiology of Mars

Learning Outcomes

➤ Understand some of the major geological stages of Martian history.

➤ Understand the history of water on Mars.

➤ Understand how to calculate the root mean square speed of gases in an atmosphere, the factors that determine the escape of gases from a planetary atmosphere such as Mars and some factors that caused Mars to lose its atmosphere.

➤ Understand what we know about the distribution of biologically essential elements on Mars.

➤ Understand past attempts to ascertain whether there is life on Mars, including the Viking missions and the study of meteorites.

➤ Know why planetary island biogeography is important to astrobiology.

17.1 Mars and Astrobiology

In this chapter we take the first detailed look at another planetary body other than the Earth. You'll find, as we investigate Mars, that a whole string of ideas that we have been investigating in previous chapters come together either explicitly or embedded within a range of facts and ideas. This is the chapter in which you will finally see why astrobiology requires a grasp of many areas of planetary sciences, earth sciences, astronomy – and when we are talking about habitability, biology. As you read this chapter consciously think about the concepts that underlie statements, facts and theories about Mars and refer back to previous chapters if you need to. You'll also realise that it is very difficult to do a comprehensive job of explaining everything in a book like this. Take the opportunity to find other sources, read scientific papers and look on the internet for more information to deepen your knowledge of the concepts and ideas provided in this textbook.

We have a growing amount of knowledge about Mars, thanks to many orbiter and surface missions. These missions, together with evidence from meteorites, have provided us with a comprehensive understanding of one of the most astrobiologically interesting planetary bodies in the Solar System.

Named after the God of War by the Romans, on account of its red surface, Mars, which has a radius (3397 km) about half that of the Earth, has always caught the interest and imagination of scientists. Its colour, caused by iron oxides, has often caused it to be referred to as the Red Planet. As our neighbouring world, it has attracted a variety of strange and obsessed characters and ideas, from Percival Lowell (1855–1916) and his books about the canals of Mars – channels carved by intelligent Martians to take water to their beleaguered, desiccated cities – to strange claims of monuments, faces and pyramids on Mars that turn out to be nothing more than low resolution aberrations in orbital photographs of Martian hills. The history of the interest in Martian civilisations is entertaining and a serious lesson from the past about allowing optimism to get the better of data. We won't pursue this history any further here, although it is well worth reading about.

Astrobiology: Understanding Life in the Universe, First Edition. Charles S. Cockell.
© 2015 John Wiley & Sons, Ltd. Published 2015 by John Wiley & Sons, Ltd.
Companion Website : www.wiley.com/go/cockell/astrobiology.

There are many ways to approach the study of Mars. We cannot hope to cover everything that is known about the geology, geochemistry and potential habitability of this planet. As this is a textbook about astrobiology, then I have no hesitation in focusing on those factors about Mars that are important for habitability.

17.2   Martian History: A Very Brief Summary

Mars is geologically distinctive at the large scale. The southern hemisphere is made of ancient cratered terrain and the northern hemisphere is flat, more recent (less impacted) terrain that is at lower elevation. This distinctive geology is called the **crustal dichotomy**. Its origins are not known. Hypotheses include a role for single or multiple large impacts that created the depressed lower elevation of the northern hemisphere. Alternatively, the separation was the result of inhomogeneous mantle convection in an early plate tectonic regime.

The Martian surface is punctuated by the deep Hellas and Argyre Basins in the southern hemisphere, formed by ancient impacts, and the giant Tharsis bulge in the north, which boasts Olympus Mons, an enormous **shield volcano** two and a half times the height of Mount Everest. A rift, formed when the volcanic bulge split, is the Valles

Marineris, a canyon system over 4000 km long, about 200 km wide and up to 7 km deep. In comparison, the Earth's Grand Canyon is ~500 km long and ~2 km deep. The Elysium is another distinctive volcanic region.

Like the Earth, the history of Mars has been split into distinctive geological eons. The eons broadly reflect large scale environmental changes on Mars (Figure 17.1) as we shall shortly explore. Significantly, the eon that overlaps with the emergence of life on the Earth, the **Noachian** (4.1–3.7 Ga ago), which follows the Pre-Noachian (before 4.1 Ga ago) is a time when there is evidence for sustained bodies of liquid water on Mars. We will explore the evidence for this shortly. This is a time associated with the formation of clays in water-rich environments. Often the Pre-Noachian and Noachian are lumped together into the Noachian and in this chapter I will use this word to refer to the time before 3.7 Ga ago.

Following the Noachian, water begins to dissipate and freeze and the low abundances of water, mixed with sulfur gases, lead Mars into an age, the **Hesperian** (3.7–3.0 Ga ago), characterised by acidic conditions and the formation of sulfate salts, such as iron sulfate. During this time there was a great deal of volcanism, with activity at Tharsis and Elysium. Eventually, even this water freezes or evaporates and Mars transitions into a desiccated, barren world during the **Amazonian** (3.0 Ga ago to the present day), when volcanism also declined, a condition

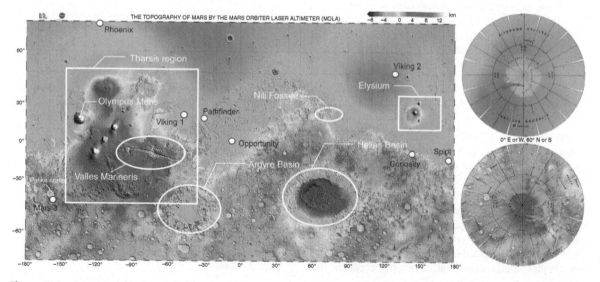

Figure 17.1 *A topographic map of Mars obtained with the Mars Orbiter Laser Altimeter (MOLA) onboard the Mars Global Surveyor (the poles are shown on the right). The crustal dichotomy is clear both in the topography and degree of cratering. Some of the major features discussed in the text are marked (in white) and the landing sites of spacecraft (in black; Source for main image: NASA).*

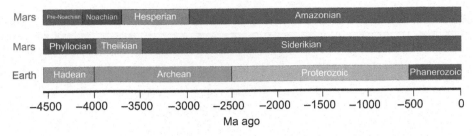

Figure 17.2　*Two geological time scales for Mars compared with the Earth.*

that dominates its surface for three billion years and fashions the desert world with which we are all familiar. The distinctive changes in environmental conditions have led to suggestions of another geological timescale (Figure 17.2) based on changes in planetary-scale geochemistry that reflect the change from a predominantly clay (**phyllosilicate**)-forming eon (the Phyllocian) into a very acidic sulfur-dominated eon (Theiikian) and then into the modern desert world (Siderikian).

We can therefore view the general history of Mars as one in which there has been a deterioration in conditions, at least with respect to water availability and therefore the biological potential of the planet.

17.3　The Deterioration of Mars

An obvious question is how a planet, that on the face of it seems quite similar to the Earth in its early history (it hosted large quantities of liquid water) became so barren.

The answer to this question is by no means simple, but one large scale factor that we can identify as being a contributor is the small size of Mars. Mars cooled quicker than the Earth because of its smaller radius. As a result, the dynamo that generates the magnetic field, and on the Earth protects our atmosphere from being **sputtered** away by the **solar wind**, shut down. Magnetic stripes (Figure 17.3) in rocks of the southern hemisphere of Mars have been suggested as evidence for an early tectonic regime where rocks were being exuded from the surface retaining the magnetic field of this early history.

The sputtering away of the atmosphere cannot entirely account for all of the lost atmosphere. Two other processes made a very significant contribution – hydrodynamic escape and impact erosion.

Hydrodynamic escape is best understood by considering the behaviour of gases.

The escape velocity, or the minimum velocity required for an object to escape the gravitational pull of a planet, is given by:

$$v_{esc} = \sqrt{2GM/R} \qquad (17.1)$$

where G is the Gravitational constant, M is the mass and R is the radius.

This equation gives an escape velocity for Mars of 5.03 km s^{-1} (compared to the Earth's of 11.19 km s^{-1}).

The escape velocity influences the extent to which gases can escape. To understand this it is necessary to know that the distribution of speeds of molecules of a gas conforms to a Maxwell–Boltzmann distribution (Figure 17.4) and that as the temperature of a gas increases, so there will be a greater number of gas particles with a higher speed. The speed distribution means that there is a 'tail' of gas molecules that have speeds much greater than the average speed.

As a general rule of thumb, if a gas has a mean speed about a sixth of the escape velocity, then there will be a significant tail of molecules that are travelling fast enough to escape the planetary atmosphere into space. Thus, the heating of the upper atmosphere of Mars, in its past, would have had a higher chance of driving off volatiles compared to a higher mass planet, all other things being equal, because of Mars's lower escape velocity.

Furthermore, lighter gases, such as hydrogen (produced from the light driven break-down, or **photolysis**, of water) will escape more effectively than heavier molecules such as O_2. This is schematically shown in Figure 17.4.

This relationship can be quantitatively expressed. The root mean square speed of a gas particle (v_{rms}), defined as the square root of the average speed-squared of the molecules in a gas – a common means of expressing speed of gases – is inversely proportional to its molecular mass and given by:

$$v_{rms} = \sqrt{3kT/m} \qquad (17.2)$$

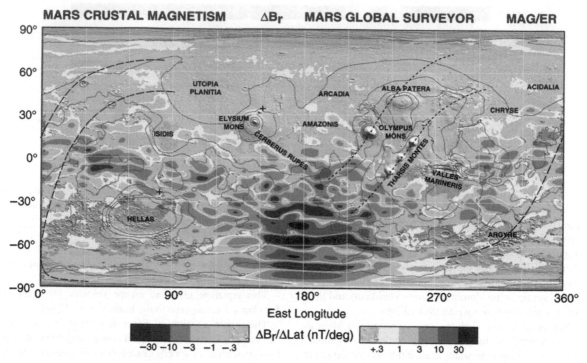

Figure 17.3 *Magnetic stripes in rocks in the ancient cratered surface of Mars give tantalising evidence of the possible record of a changing early magnetic field in rocks being extruded onto the surface of early Mars. Data was collected by the Mars Global Surveyor. Units in nanoTesla/degrees are given underneath the image* (Source: Mars Global Surveyor, NASA/JPL).

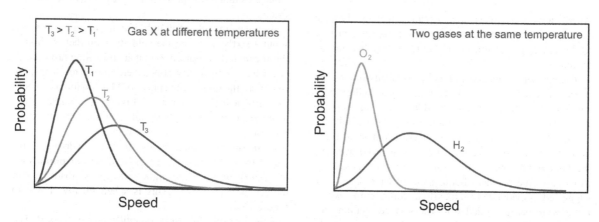

Figure 17.4 *A schematic of the Maxwell–Boltzmann distribution of speeds in gas atoms or molecules shown according to a probability distribution (the area under each curve equals unity). On the left is shown the effect of temperature on a given gas. The higher the temperature the more the speed distribution is skewed to higher speeds. On the right is shown how molecular mass effects the speed distribution for different gases. The lighter the gas, the higher the average speed at a given temperature. The molecular mass of hydrogen is 2, compared to oxygen, which is 32* (Source: Charles Cockell).

where k is the Boltzmann constant, T is the temperature and m is the mass of one molecule of the gas.

The equation may alternatively be written as:

$$v_{rms} = \sqrt{3RT/M_m} \qquad (17.3)$$

where R is the Universal gas constant, and M_m is the mass of one **mole** of the gas.

Although light gases such as hydrogen will escape more easily, as it escapes into space it tends to 'drag' other gases along with it as it collides with heavier atoms, leading to a large scale depletion of the atmospheric mass. The process is called **hydrodynamic escape**.

Evidence for hydrodynamic escape is to be found by measuring the deuterium/hydrogen ratio in the Martian atmosphere. The ratio is about five times higher than that found in terrestrial seawater and suggests that the lighter hydrogen has preferentially escaped compared to the heavier deuterium. Other elements, such as nitrogen and argon, which are depleted in their lighter isotopes, lend further support to this idea.

In addition, the period of early intense bombardment (Chapter 10) would have lent a helping hand. Giant impacts would have heated atmospheric gases, aiding their escape from the atmosphere compared to a more massive planet. It is thought that this **impact erosion** may have caused about 90% of the loss of the atmosphere, with the remaining loss accounted for by the other processes described earlier.

It is also clear that Mars never developed a fully functioning tectonic regime. Carbon dioxide locked up as carbonates and buried underground was never returned to the atmosphere – the carbonate–silicate cycle (Chapter 16) did not develop, possibly contributing to atmospheric loss.

All of these factors meant that the atmosphere thinned, water froze and eventually the atmospheric pressure reached values near the triple point, precluding the possibility of surface liquid water. Today, the average surface atmospheric pressure is almost exactly on the triple point (6 mb; Chapter 2), meaning that water ice, when heated, sublimes rather than forming standing bodies of liquid water. Even in the deepest part of Mars, the Hellas Basin, the pressure only reaches 11.6 mb, whereas at the highest point, the summit Olympus Mons, the pressure is 0.3 mb.

These atmospheric characteristics have a very profound influence on surface temperatures by severely limiting the potential for greenhouse warming. The average temperature of Mars today is −55 °C. Air temperatures at the poles can drop to −153 °C. However, they can rise at the equator to 20 °C. Soil temperatures were found to transiently reach 27 °C at the Viking lander sites.

The broad geological changes since the formation of Mars have had, superposed upon them, more frequent orbitally-driven changes. We saw in Chapter 16 that the presence of the Moon stabilises the obliquity of the Earth. This is not the case for Mars. Its obliquity is subject to cyclical shifts every 100 000 years with longer-term changes every million and 10 million years caused by rotational and orbital dynamics. The planet's obliquity varies from 5 to nearly 50°, with a high obliquity period about six million years ago. Since periods of high obliquity will tend to deliver heat to the poles when they are facing the Sun, thus vapourising the ice caps more effectively and potentially increasing atmospheric pressure and water content, these oscillations in obliquity are important for understanding the past shifting patterns of water on Mars.

17.4 Missions to Mars

Much of what we know about Mars has been provided by the large number of missions since the 1960s. They have sequentially added to our understanding of the past history of water and the geochemical environment as it relates to habitability. In addition to space missions, we have learnt a lot from Martian meteorites (and we will see how this is possible later in the chapter). Here we will briefly review these missions.

The early exploration of Mars involved the Mariner spacecraft. There were three fly-by missions by Mariners 4, 6 and 7 and one orbiter mission (Mariner 9). These missions revealed a barren landscape covered in impact craters, not unlike the Moon. The surface of Mars was obscured from the view of Mariner 9 by a planetary-scale storm, giving scientists the first glimpse of these enormous storms that can start locally and enshroud the entire planet, lasting many months. These missions also revealed for the first time the great Tharsis volcanoes (Figure 17.5).

In 1976 the Viking Orbiters and landers further advanced our knowledge of the ancient surface of Mars. They revealed valley networks, tear-drop shaped islands and other indications of water. The two landers carried the first biology experiments sent to another world to seek signs of life in the Martian soils. We will discuss their results in more detail later in this chapter.

From 1997 until 2006, the Mars Global Surveyor marked the return of the United States to Mars exploration. It was equipped with instrumentation to provide detailed images of Mars, measure its magnetism, surface composition and atmospheric temperatures. Its Mars Orbiter Laser Altimeter (MOLA) provided stunning topographical maps of Mars (Figure 17.1) and the mission

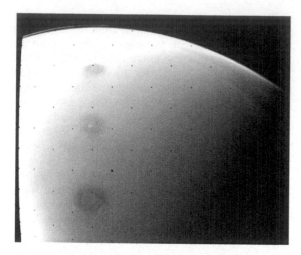

Figure 17.5 *Early Mariner 9 image of Mars taken in 1971. The towering peaks of the Tharsis volcanoes begin to appear through a global dust storm (Source: NASA).*

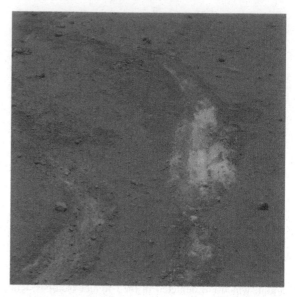

Figure 17.6 *Pale silica deposits on Mars revealed by the wheels of the Spirit rover. The groove is ~20 cm across. The deposits are evidence for water–rock interactions (Source: NASA).*

took many images that would later be used to decide the landing sites for Phoenix and Curiosity (the Mars Science Laboratory).

Mars Pathfinder was a technology demonstration mission in which a static lander was combined with a small robotic rover (Sojourner) that could analyse nearby rocks seen by the lander. It landed in 1997 in the Ares Vallis, in a region called Chryse Planitia. It had X-ray spectrometers and cameras which provided chemical analysis of rocks near the landing site. This mission paved the way for future rovers by demonstrating landing technologies including an airbag landing. The Sojouner was followed by a new robotic orbiter, Mars Odyssey, which began mapping the surface of Mars in 2002. It was equipped with instruments to measure surface temperatures, and a gamma ray spectrometer to analyse surface minerals.

Following the success of the Sojourner rover and the scientific questions that arose from data sent back from Mars Global Surveyor, two identical rovers, the Mars Exploration Rovers (MER), 'Spirit' and 'Opportunity', were sent to two different locations to explore the geology of Mars, beginning in 2004. To achieve this, they were equipped with instruments designed to analyse the fine-scale textures and chemical composition of rocks and soils including a microscopic imager, a miniature thermal emission spectrometer (Mini-TES), X-ray spectrometry (APXS) and a rock abrasion tool (RAT) that allowed for investigations just below the surface of rocks. Opportunity was sent to Meridiani Planum to investigate hematite-rich

sedimentary outcrops that formed through the result of acidic fluids leaching through the rocks, leaving this iron oxide mineral (hematite has the formula, Fe_2O_3). Spirit was sent to Gusev Crater, and explored the nearby feature 'Home Plate'. This area was found to be volcanic, with evidence for hydrothermal activity in the form of 90% pure silica deposits suggesting an ancient hydrothermal system (Figure 17.6). Both these findings suggested liquid water environments that might once have been habitable. The rovers significantly outlasted their 90 **sol** nominal mission and were operating ten Earth years after their arrival.

Alongside the MER rovers came the European Space Agency's Mars Express Orbiter Mission, which flew instruments to map the subsurface (MARSIS; Sub-Surface Sounding Radar Altimeter) and map minerals on the surface (OMEGA; Visible and Infrared Mineralogical Mapping Spectrometer). It also had a high resolution camera that gathered spectacular images used by geologists – the High Resolution Stereo Camera (HRSC). The craft also had a landing element, the Beagle 2 lander, which successfully landed on Mars, but did not fully deploy. It would have carried out surface geology and astrobiology experiments including the search for carbon.

A few years after Spirit and Opportunity, another orbiter (the Mars Reconnaissance Orbiter) was sent to

Figure 17.7 *False colour image of the edge of Jezero crater (49 km in diameter) in the northern hemisphere of Mars (at 18.4° N 282.4° W) in the Nili Fossae region showing delta-like features near the end of a putative ancient water channel (coming in from the left). Phyllosilicate-bearing materials are green, olivine-bearing materials are yellow, low-calcium pyroxene-bearing materials are blue and purple–brown surfaces have no distinct spectral features. The image shows an area approximately 10 km across* (Source: NASA/JPL/JHUAPL/MSSS/Brown University).

Mars equipped with instruments to image the surface at high resolution. These included HiRISE (High Resolution Imaging Science Experiment), which could resolve images ~1 m in size, CTX – a context camera to provide three dimensional images of the Martian surface, and CRISM (Compact Reconnaissance Imaging Spectrometer for Mars) – a spectrometer to image the surface at wavelengths spanning the visible to infrared, revealing the fine-scale mineralogy of Mars. CRISM in particular found the first extensive deposits of 'hydrated minerals' – minerals including clays (phyllosilicates), sulfates (e.g. gypsum) and carbonates that could only form in the presence of liquid water. The identification of clay minerals hinted at once clement environments on Mars with neutral pH and sustained liquid water. This was particularly important when minerals such as clays and carbonates coincided with geomorphological features that were indicative of plentiful water, such as channel and delta deposits in Jezero Crater (Figure 17.7).

In 2008, the NASA Phoenix lander was sent to the north polar region of Mars to study water-ice deposits.

It landed at 68.218°N 234.251°E in the Vastitas Borealis (Figure 17.8). Perhaps one of the most startling discoveries of the lander was near-surface ice, which was imaged by the lander cameras and was also observed in regions where the rocket exhaust had blasted away the surface dust. The Phoenix lander had a wet-chemistry laboratory on board, which showed the presence of perchlorates (compounds containing the perchlorate ion; ClO_4^-) within the Martian soil. This was an unexpected finding. The presence of perchlorates may provide an explanation of the lack of detection of organics in earlier missions, such as Viking. Perchlorates would have oxidised organic molecules when heated to high temperatures, such as during pyrolysis (the heating step) during in the **mass spectrometry** analysis of Mars soils by the Viking landers.

The wet chemistry laboratory showed the surface soil was moderately alkaline (between pH 8 and 9). Magnesium, potassium, sodium and chloride ions were detected and the salinity was low, suggesting a very benign environment from the point of view of life.

Figure 17.8   *Panorama view around the Phoenix lander (Source: NASA).*

The most recent rover to explore the Martian surface was the Curiosity rover. This rover was the most advanced ever sent to Mars, with instruments to assess the past habitability of Mars, landing in the 150-km diameter Gale Crater. These instruments include the SAM (Sample Analysis at Mars) instrument, to investigate the soils for organics and analyse gases, and CheMin (Chemistry and Mineralogy) to examine mineral composition by X-ray Diffraction and X-ray Fluorescence. The rover discovered clement environments that might once have been habitable and it returned new detailed information about surface mineralogy.

17.5   Mars and Life

Assessing the habitability of Mars has been an objective of scientists for a long time, but has recently become a sustained focus in the light of data being returned from the planet and growing knowledge about life in extreme environments. This endeavour is important for elucidating the history of Mars and providing a comparative data point for understanding the geological and biological history of the Earth. By investigating the habitability trajectories possible for rocky planets, we can gain information with which to assess the habitability of Earth-sized extrasolar planets, which will be discussed in Chapter 19.

17.5.1   Liquid Water and Mars

Liquid water is the essential solvent for life (Chapter 3) and its history on Mars underpins our understanding of the habitability of the planet. Generally speaking, Mars has had a history of declining liquid water availability over time. In the earliest history of the planet there is a great deal of evidence for liquid water.

The presence of water was long suspected on Mars, even during the early Mariner and Viking missions when sinuous channels observed from orbit and teardrop-shaped islands suggested the flow of a liquid on the surface. It would take mineral mapping to confirm these suspicions. The presence of liquid water on ancient Mars is supported by observations of clay compounds produced when water reacts with volcanic rocks – the phyllosilicates. **Phyllosilicates** are found in many Noachian terrains. In some regions magnesium carbonates are found associated with the clays, also evidence of liquid water. Although carbonates form part of the globally ubiquitous Martian dust (~5% or less by mass of its content), bedrocks of carbonate are strong evidence for sustained liquid water in which they would have been deposited. Phyllosilicates are also observed associated with ancient Noachian crust exposed by impact craters (hence the proposed alternative name for this eon – the Phyllocian). For example, the Nili Fossae region of Mars (Figure 17.9) exhibits an extraordinary range of minerals and is comprised of distinct units. Information on this region has been obtained using the Mars Reconnaissance Orbiter (MRO) CRISM and the HiRISE instruments.

One unit in the Nili Fossae has iron and magnesium-rich clays which is made of metre- to kilometre-sized blocks of altered and unaltered rock and is inferred to represent the ancient crust of Mars, torn up in subsequent impact events. These Fe/Mg-**smectite** clays are the most abundant clays on Mars and are thought to have been formed by the interactions of water with volcanic rock.

Asteroid and comet impact craters provide information on the composition of the Martian subsurface and its geochemical characteristics in the ancient past. The so-called **central peaks** of craters formed when the material in the middle of the crater rebounds, show indications of hydrated phases and hydrothermal alteration products during water circulation. The central uplift of the equatorial Leighton Crater has excavated material from 6 km depth in the Martian subsurface and contains carbonates.

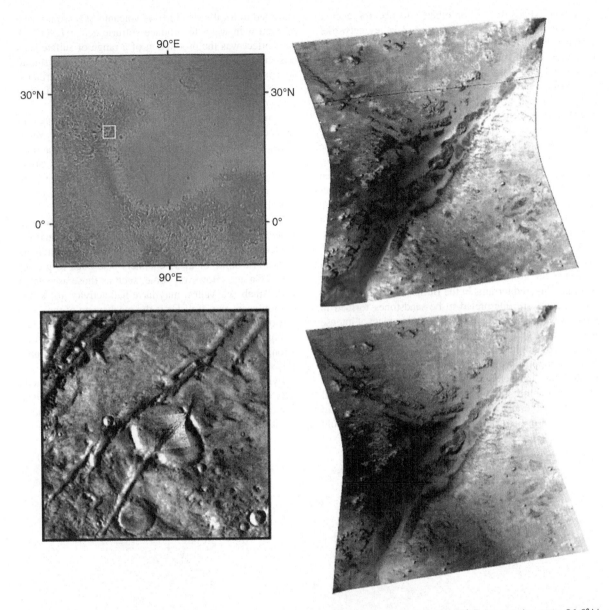

Figure 17.9 *Geological diversity in the Nili Fossae, Mars. Image of part of a fracture in the Nili Fossae region near 21.9°N, 78.2°E; Compact Reconnaissance Imaging Spectrometer for Mars (CRISM). The top left shows the Isidis basin image (small square) superposed on a Mars Orbiter Laser Altimeter (red higher elevations, blue lower). The fracture shown, which is 11 km at its narrowest point (top right) is overlain on a Viking orbiter digital image (lower left) to show topography. The top right image also shows the region in the infrared channels, false coloured. Bright green are phyllosilicates, yellow-brown are olivines, purple are pyroxenes. The CRISM data is superposed on High-Resolution Imaging Science Experiment (HiRISE; lower right) showing that phyllosilicates are in small eroded outcrops of rock and olivines in sand dunes. (Source: NASA/JPL/JHUAPL/Brown University).*

A non-exhaustive list of other evidence for ancient water on Mars includes the presence of valley networks, which are collections of long sinuous channels carved by water, controversial evidence of a possible northern ocean and sediments in ancient lakes, particularly those that formed in impact craters.

Like the Earth, early Mars too confronts us with a **Faint young Sun paradox**. How was the liquid water sustained on the surface when the Sun was less luminous? Various proposals have been advanced, from a greenhouse effect caused by CO_2 ice clouds to high concentrations, or episodic production, of volcanic SO_2. It has also been proposed that much of the water could have been under glacial ice sheets rather than directly exposed at the surface, invoking ideas of a cold, wet Mars rather than a warm, wet Mars.

As the Noachian transitioned into the Hesperian, water bodies became less abundant, but nevertheless there is still evidence for groundwater activity during this time. For example, layered terrains in the Burns Formation, Meridiani Planum are interpreted to be sandstones formed in shallow water systems and have been the subject of intense geochemical and geological discussion. The widespread presence of sulfate salts and observations of hematite concretions support a model of water in the ground seeping up and altering the surface, leaving iron oxides and sulfate salts behind. During the Hesperian, the hydrological cycle was generally characterised by low water-rock ratio interactions in the near-surface environment. This would

have led to localised acidity as volcanic SO_2 would have reacted with water to produce sulfuric acid (H_2SO_4). A side effect was the production of a range of sulfate salts associated with acidic environments such as iron sulfate and **jarosite**. This accounts for the proposed alternative name for this eon (Figure 17.2) – the Theiikian, derived from the Greek for sulfur (*theion*).

Catastrophic outflow channels provide particularly compelling evidence for subsurface and surface water since the Noachian. These features begin from a fracture or region of chaotic terrain and consist of broad depressions tens to thousands of kilometres long with streamlined islands and deposits around craters along their beds (Figure 17.10).

The mechanisms by which they might have been formed include the release of groundwater from frozen permafrost melted by impact events, earthquakes or volcanic magma intrusions. Some of these channels date back to greater than 3 Ga ago. However, some, such as those associated with Athabasca Valles, may have had activity just a few million years ago. If water is the mechanism of their formation, then outflow channels would suggest the presence of liquid water in the subsurface from the early Hesperian through to the geologically very recent past in catastrophic episodes.

Evidence for short-duration water–rock interactions after the Noachian is also found in Martian meteorites. For example, alteration textures in the Nakhlite meteorites in which minerals and glass have been altered

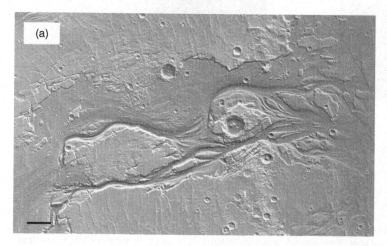
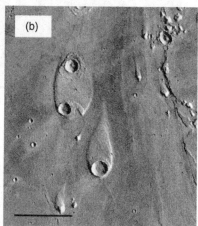

Figure 17.10 *Catastrophic outflow channels on Mars. (a) Kasei Vallis (25° N, 300° E), the largest outflow channel on the planet, emerges from a shallow north–south canyon to the west. Scale bar 100 km. (b) Streamlined islands near the mouth of the Ares Vallis outflow channel in Chryse Planitia. This image shows two teardrop-shaped scarps with heights of 400 m (upper island) and 600 m (lower island) formed by the erosive power of the flood that passed through this region early in Martian history. The lower crater is 10 km in diameter (Source: NASA).*

to clays record aqueous interactions in the Amazonian ~600 Myr ago.

After the Noachian and throughout the Hesperian, Mars's declining heat flow as it cooled down is hypothesised to have led to freezing of most of the water in the surface environment and at gradually increasing depth.

Mars today hosts a large body of frozen water. The near-surface (to a depth of tens of centimetres) of Mars harbours water ice deposits that vary from 2% weight at the equator to pure ice at the polar regions mixed with surface volcanic **regolith**. These ice deposits have been detected with the Gamma Ray Spectrometer on the Mars Odyssey spacecraft.

Gamma Ray Spectrometry is an ingenious method. When elements are bombarded with cosmic rays from the Sun and interstellar space, they tend to release neutrons, which themselves collide with other nuclei causing them to rise to a higher energy state. They revert back to a lower energy state by emitting gamma rays. The energy of these gamma rays can be measured in orbit and the resulting spectrum of energy can be used to determine the concentrations of about 20 different elements in a planetary crust.

Hydrogen atoms (for example in water) are particularly good at absorbing the energy from neutrons as they have a similar mass. Thus, gamma ray spectroscopy can be used to detect the tell-tale absorption caused by water and so measure water ice concentrations within the top meter or less of Mars (Figure 17.11). The Gamma Ray Spectrometer measurements were confirmed with the direct detection of near-surface ice at the Phoenix landing site.

Some modelling estimates have put the **cryosphere** depth (the depth of frozen water) to be ~2.5 km at the equator to ~6.5 km at the poles. However, downward revisions of the geothermal gradient of Mars in other models suggest that the depth could be up to two to three times greater. Salt solutions (such as the presence of perchlorates or NaCl) would lessen these depths by depressing the freezing point. The Martian cryosphere is estimated to contain an equivalent global layer of water of ~35 m.

What about water in the deep subsurface of Mars? Direct observations of the present-day Martian deep subsurface were made using the Mars Advanced Radar for Subsurface and Ionospheric Sounding (MARSIS) instrument on the Mars Express spacecraft, which has a theoretical penetration depth of ~5 km. The instrument successfully provided information on the subsurface of the north polar deposits on Mars down to ~1.8 km, confirming that the deposits are mainly composed of pure

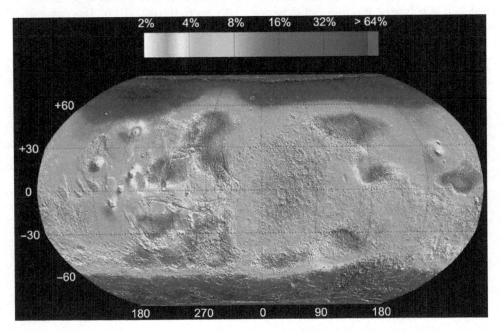

Figure 17.11 *Hydrogen in the surface of Mars detected by Gamma Ray Spectroscopy, interpreted as the lower limit of water in the near surface of Mars (mass fraction shown in the colour scale at the top of the image;* Source: NASA/JPL-Caltech/Los Alamos National Laboratory).

water ice. The Shallow Radar (SHARAD) instrument on Mars Reconnaissance Orbiter (MRO) was used to study subsurface structures in lobate debris aprons (LDAs), which are broad, lobate features that extend up to 20 km away from steep slopes in equatorial regions of Mars. These observations suggest that there are buried glaciers on Mars. The data from the SHARAD instrument suggests that up to ~28 000 km³ of water ice might be sequestered in LDAs in the Hellas Basin region of Mars alone, equivalent to a global water layer ~20 cm thick. These buried glaciers are testament to climatic changes on Mars over timescales of several million years. However, neither of the radar instruments detected subsurface bodies of *liquid* water.

In general, the water ice on Mars follows its predicted depth of stability under current climatic conditions, consistent with the idea that the subsurface ice conditions of Mars follow orbitally-driven climate cycles, with local heterogeneities reflecting differences in topography and material type and preservation of icy deposits from previous epochs.

Quite apart from massive ice deposits at the poles, with an estimated volume of 1.2–1.7×10^6 km³, and buried ice just discussed, there is a large literature on other glacial and **periglacial** features on Mars. Evidence for subsurface ice includes distinctive ordered features in the ground such as polygonal structures (which form when ground freezes and thaws), parallel sorted stone stripes (which also form from freezing and thawing of the ground), gullies and ice-sublimation related features among others. Pingos, which are mounds produced by liquid water being injected into the subsurface that then freezes and expands, causing upheaval, have been suggested. They are of particular interest as their formation mechanism involves bulk liquid water movement. Observations from orbit include recently excavated small craters which reveal water ice. The ice is observed to sublimate away after several months (Figure 17.12).

Liquid water at the surface of Mars today is rendered unstable, partly because much of the surface is at the triple point and partly because the low humidity means that when liquid water is formed, it will rapidly evaporate, even if it does not boil. However, evidence has been suggested for near-surface present-day liquid water. Gullies, which have characteristic alcoves located on a steep slope with a sinuous channel leading down to an apron of deposited material, have been proposed as evidence of present-day liquid water. However, gullies could also be formed by CO_2-induced processes such as the fluidisation of the regolith, caused by the sublimation of CO_2.

Other evidence for present-day water has been suggested. Dark slope streaks which are observed on Mars and change seasonally could be present-day near-surface salty water (Figure 17.13). They are called Recurring Slope Lineae (RSL).

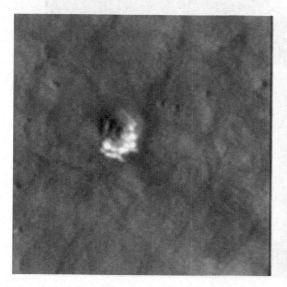 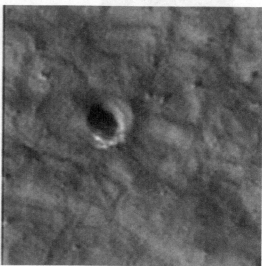

Figure 17.12 *Recent small craters on Mars reveal sub-surface ice which is observed to sublime away after a few months. This is a fresh, 6 m wide, 1.33 m deep crater on Mars photographed on 18 October 2008, and again on 14 January 2009, by the Mars Reconnaissance Orbiter's HiRISE camera (Source: NASA).*

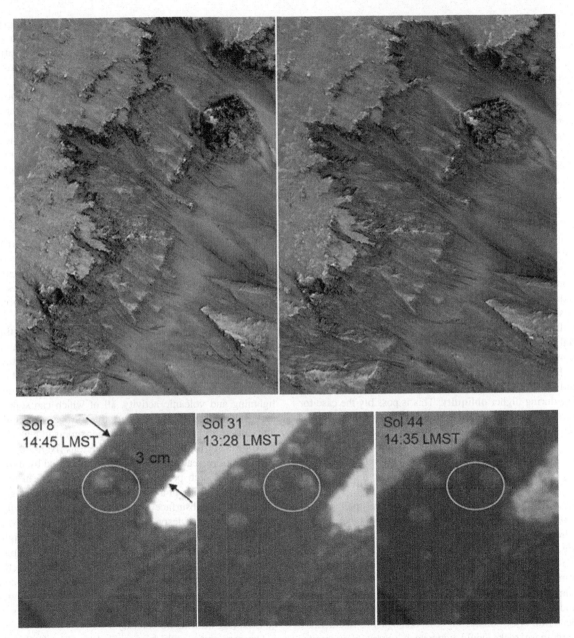

Figure 17.13 *Support for the stability of brines on present-day Mars? Top: Palikir Crater, which is inside the much larger Newton crater (41.6°S, 202.3°E), contains thousands of individual flows called Recurring Slope Lineae, or RSL. In the Southern middle latitudes, RSL form and grow every summer in certain places, fading in late summer and autumn (an example is shown with a red arrow). This image shows a comparison of RSL from 2011 to 2013 over a small piece of Palikir Crater's steep northwest-facing slopes. The new image shows RSL are slightly more extensive and longer than at nearly the same time the year before (Source: NASA/JPL-Caltech/Univ. of Arizona). Below: Putative drops of brine on the legs of the Phoenix lander that change positions over many days (LMST is Local Mean Solar Time; Source: NASA).*

Interestingly, droplets of putative salt solutions that were observed on the legs of the Phoenix lander have been suggested to be evidence for the stability of brines on Mars, perhaps formed from perchlorates and sulfates in the soils (Figure 17.13).

On present-day Mars, groundwater could exist deep underground where heating from radioactive minerals combined with deep subsurface (**lithostatic**) pressures would allow liquid water to exist above the freezing point. Impact events would be another mechanism by which the present-day permafrost could be melted and disrupted to create a link between the subsurface and surface.

Liquid water could exist today in the form of thin water films on soil grains. Dielectric measurements of soils at the Phoenix landing site suggested the presence of liquid water, but in very thin layers of one to three monolayers thickness. Whether microorganisms could ever access thin layers of water or use it as a solvent, since it is tightly bound to the grains, is not known. However, a serious limitation of this water is that if it remains static (as is suggested for the Phoenix lander site), the habitat it creates will geochemically run down, becoming depleted in essential nutrients, making this water a poor environment for the long-term activity of life (Chapter 16).

In the more recent geological history of Mars, bulk liquid water might have become available as a result of warming during higher **obliquity**. This is possibly the case for ices at the Phoenix landing site where, during the last 5 Ma, obliquity increases up to 50° would generate surface temperatures in excess of 0 °C up to 100 days per year.

17.5.2   Basic Elements for Life on Mars

Life requires six basic elements to construct macromolecules (C, H, N, O, P, S; Chapter 3). Let's look at where these might come from on Mars. Carbon atoms are likely to have been, and continue to be, present in the surface and subsurface of Mars as a consequence of atmospheric exchange (in the present day atmosphere the composition is 95.32% CO_2 and includes 800 ppm CO) and could be acquired by life through autotrophy. The detection of carbonates suggests that aqueous interactions with these rocks could generate a source of inorganic autotrophically available carbon throughout Martian history as bicarbonate ions. The concentration of organic carbon on Mars, a potential source of carbon for heterotrophs, in different regions and depths on Mars, is unknown. The in-fall of **carbonaceous chondrites** and other organic carbon-bearing material is expected, but

organics are likely to be destroyed by **reactive oxygen species**, ultraviolet radiation and **ionising radiation** in the near-surface environment. These factors will greatly influence its preservation and concentrations in different regions and depths on Mars.

Hydrogen atoms are available from water throughout the Martian depth profile, which could be split **radiolytically** (by radioactivity) in the subsurface to produce hydrogen. Hydrogen could also be generated in chemical reactions. The presence of serpentine in impact craters suggests the possibility of hydrogen production through **serpentinisation** reactions (the reaction sequence was shown in Chapter 12), particularly when water flow was more extensive in the Noachian.

Nitrogen gas is present in the modern atmosphere at 2.7%. Fixed nitrogen compounds have been reported in Martian meteorites and confirmed on the surface of Mars. They have been predicted to include nitrate and ammonium based on terrestrial analogues. To be used in biological systems, nitrogen must be in a fixed form. One potential pathway is biological fixation (Chapter 5), which was shown to be possible at a nitrogen **partial pressure** (pN_2) of 5 mb, but not below 1 mb, suggesting that this pathway was plausible in a denser early Martian atmosphere, but unlikely today. Nitrogen fixation on Mars could occur by abiotic processes, including impact events, lightning and volcanic activity, all of which can supply sufficient energy to break apart the nitrogen molecule and cause it to recombine with oxygen atoms to generate fixed forms of nitrogen such as nitrate ions (NO_3^{2-}). The concentrations reached and the depths achieved by nitrogen fixed in such processes throughout Martian history are unknown. Without a continuous flow of fixed nitrogen into the deep subsurface of Mars, particularly following the cessation of widespread surface hydrological activity on Mars in the Noachian, nitrogen might be, and might have been, one of the limiting factors for life. Despite the detection of fixed nitrogen in meteorites and directly on the surface of Mars, determining the distribution and form of fixed nitrogen in the Martian crust, past and present, remains one of the most important challenges in constraining Martian habitability.

Oxygen atoms could be provided by CO_2, H_2O, sulfates, perchlorates, ferric oxides and **reactive oxygen species**. Oxygen atoms are bound to many of the biologically accessible compounds discussed here in association with other elements (C, H, N, P, S).

Phosphate has been reported in Martian meteorites and on the surface of Mars in a number of missions.

For example, the Mars Exploration Rovers found rocks containing **apatite** (a group of phosphate minerals) at between 0.1 and 2.4% weight. Some rocks with phosphorus abundances of over 5% were observed in Gusev crater by the Spirit rover. Phosphorus was observed in alkaline basalts studied in Gale Crater by the Curiosity Rover at <1% weight abundances.

Sulfur has been detected on Mars in meteorites and on the surface of Mars in the form of sulfate salts including gypsum, ferric sulfates, jarosite and other S-bearing species in different oxidation states, including sulfides. The extent of these compounds in the subsurface is not known, but the dominance of the sulfur cycle on Mars suggests that sulfur species would have been distributed from the mantle to the surface throughout Martian history, potentially including sulfur in microbially accessible gaseous phases such as H_2S and SO_2.

In summary, we can see that there are readily observable sources of all of the major elements required for life on Mars, apart from nitrogen, for which we currently do not know the concentration of its fixed forms in different regions and depths on Mars.

Figure 17.14 *Rocks on Mars have many of the trace elements required for life. This is a vintage image from the Viking 2 landing site with a landing leg in the bottom right and a protective canister from one its instruments on the surface. The landing pad diameter is approximately 30 cm (Source: NASA).*

17.5.3 Trace Elements for Life on Mars

The presence of widespread **ultramafic** and **mafic** rocks on Mars and their alteration products show that a range of major and trace elements required by life are available, in the same way as are igneous rocks and their alteration products on the Earth. Fe is abundant in the ferrous state in olivines and in the ferric state in a variety of materials from clays to ferric oxides (crystalline and amorphous) and sulfates measured both from orbit and on the surface. Mg and Ca are present in materials such as clays and pyroxenes, and K and Na in materials such as muscovite, illite, plagioclases and K-feldspar. As would be expected for igneous rocks, other biological trace elements including Mn, Cr, Ni, Zn have been observed. There is no major or trace element used by biota on the Earth and accessible from igneous rocks or their alteration products that is obviously lacking in Martian rocks (Figure 17.14).

There is likely to be a depth and spatial dependence in which the form of these elements is found. In the Noachian deep subsurface (and possibly in the deep subsurface today), in confined closed aqueous systems, water may have been heavily enriched in Ca, Mg and Fe. In the near-surface environments, both in the present-day and the past, many of these elements would be partitioned in brines, oxides, sulfates and a variety of minerals with implications for the co-existence of biologically required suites of elements in any given location at microscopic scales.

17.5.4 Energy and Redox Couples for Life on Mars

Before you read this section, you might like to review Chapter 5.

A variety of microbial energy sources can be assessed for Mars (Table 17.1). Photosynthesis would be a plausible mode of metabolism if surface water was available. There are depths on the order of millimetres or less in the near-surface where the ultraviolet (UV) **biologically effective irradiances** are no worse than on Earth today, but where photosynthetically active radiation (PAR) is sufficient for phototrophy, such as anoxygenic photosynthesis using ferrous iron or reduced sulfur species. The lack of liquid water on the surface today precludes a productive surface photosynthetic biosphere. As for subsurface life on Earth, photosynthesis is eliminated at depth.

Chemoautotrophic redox couples are an alternative energy source. Ferric and sulfate ions as electron acceptors, both detected on Mars, can be reduced with hydrogen

---

Focus: Astrobiologists. Ralf Moeller

What do you study?

In my research I study the survivability and robustness of bacterial spores towards extreme terrestrial and extraterrestrial conditions. I am focused on investigating the molecular mechanisms involved in the unique resistance of spores. For my research I am testing spores of different strains of *Bacillus subtilis* towards simulated space and planetary conditions (e.g., Mars).

Where do you work?

I work at the German Aerospace Center, Institute of Aerospace Medicine, Radiation Biology, Research Group Astrobiology in Cologne (Köln), Germany. I studied Biology with a focus on Microbiology, Biotechnology and Genetics at the Technical University of Braunschweig, Germany. I received my PhD in Microbiology in 2007. I was a visiting scientist at the NASA's Kennedy Space Center (USA), National Institute of Radiological Sciences in Chiba (Japan), and the Max Planck Institute for Infection Biology, Berlin (Germany).

What was your first degree subject?

My first degree was a Diploma in Biology, received in 2003 from the Technical University of Braunschweig, Germany.

What science questions do you address?

The following scientific questions I like to address in my research: (i) Is there past or present (microbial) life on other planetary bodies (e.g., Mars), and if so, what would such life look like; (ii) What are the mechanisms responsible for the unique microbial resistance towards extraterrestrial conditions; and (iii) How will (terrestrial) microorganisms evolve under non-Earth-like conditions?

What interests you about astrobiology?

My interest in astrobiological research relates to the response and ecology of different microbial species towards space conditions; the impact and prevention of microbial contamination with regard to human and robotic spaceflight; and the evolution of life outside Earth.

(suggested from the presence of **olivine**, **serpentine** and other substrates or products of hydrogen-evolving mineral weathering such as **serpentinisation**). On the Earth, hydrogen can act as the electron donor in the subsurface for microbial redox reactions with sulfate and ferric iron.

Large resources of ferrous (Fe^{2+})-bearing minerals such as olivines are available for chemoautotrophic iron oxidation. However, Mars probably lacks suitable electron acceptors. Oxygen in the atmosphere (0.14%) is at insufficient concentrations today for aerobic iron oxidation, although localised oxygen production on Mars, produced by abiotic pathways during an early oxidised past, is not ruled out. Anaerobic ferrous iron oxidation linked to nitrate could be possible since nitrate is known to exist on the Martian surface. However, the depth and surface distribution of nitrates are not known. Although perchlorates can be used as an electron acceptor for iron oxidation, to date this couple has not been shown to conserve energy for growth.

The presence of reduced sulfur species such as sulfides, found in Martian meteorites and on the surface of Mars, suggest the possibility of sulfur species oxidation. However, anaerobic conditions prevent chemoautotrophic sulfur species oxidation using oxygen as the terminal electron acceptor. Sulfur can be oxidised using ferric iron as the electron acceptor. This reaction occurs in acidic conditions and sulfur has been tentatively identified on Mars (it would be expected to exist from volcanic activity given the widespread presence of sulfur compounds).

Other chemoautotrophic redox couples could include **methanogenesis** and **acetogenesis**, both using CO_2 from the atmosphere or from dissolved carbonates as the electron acceptor and H_2 from serpentinisation reactions, as observed in the subsurface of the Earth, as the electron donor. Methane itself can be oxidised by microorganisms as a source of energy and could be produced abiotically. Serpentinised ultra-mafic rocks are known to host thriving microbial communities in the subsurface of the Earth and

Table 17.1 Examples of potential redox couples for life on past and present day Mars. Reactions are selected from anoxic redox couples. The use of O_2 as an electron acceptor for microaerophilic reactions such as hydrogen oxidation on past or present-day Mars is not explicitly ruled out. The table does not include many redox reactions that use different oxidation states of nitrogen (e.g. anaerobic ammonium oxidation with nitrite) since the fixed state of nitrogen in the Martian crust is not known. Note that redox couples involving the oxidation and reduction of iron can be performed with other variable valence cations (e.g. Mn, U) which could be present in varying oxidation states in the Martian crust.

Electron donor	Electron acceptor	Name	Comment
Photosynthesis			
Fe^{2+}		Photoferrotrophy	Depends on clement surface conditions. Unlikely since the Noachian.
S/S^{2-}		Anoxygenic sulfur photosynthesis	Depends on clement surface conditions. Unlikely since the Noachian. Sulfide suggested in Curiosity data. Sulfur suggested at Gusev crater.
Organics		Photoheterotrophy	The distribution of organic carbon in different regions and different depths is not known.
H_2O		Oxygenic photosynthesis	Cannot be discounted on early Mars, but no atmospheric evidence for this reaction on present-day Mars.
Chemoautotrophy			
Fe^{2+}	NO_3^{2-}	Anaerobic iron oxidation	Distribution of NO_3^{2-} on Mars not known although fixed nitrogen is inferred.
Fe^{2+}	Perchlorates	Anaerobic iron oxidation	Perchlorate can be used to oxidise iron, but not shown to be used for growth in organisms. It is included to highlight the need for investigation of perchlorate containing redox couples.
H_2	CO_2	Methanogenesis, acetogenesis	Hydrogen inferred from presence of olivine and serpentine – substrates and products for H_2-evolving water–rock reactions.
H_2	Fe^{3+}	Iron reduction	As above for hydrogen.
H_2	SO_4^{2-}	Sulfate reduction	As above for hydrogen.
H_2	Oxidised nitrogen species		Distribution of oxidised nitrogen species on Mars not known.
S	NO_3^{2-}	Sulfur oxidation	Sulfur suggested at Gusev crater.
S	Fe^{3+}	Anaerobic sulfur oxidation	Occurs in acidic conditions
CO	NO_3^{2-}	Anaerobic carboxydotrophy	Carbon monoxide in atmosphere.
Chemoheterotrophy			
Organics	Fe^{3+}	Iron reduction	The distribution of organic carbon in different regions and different depths is not known.
Organics/CH_4	SO_4^{2-}	Sulfate reduction	As above for organics. Methane could be oxidised in a **syntrophy** with sulfate-reducing organisms.
Organics/CH_4	NO_3^{2-}	Nitrate reduction	As above for organics. Methane could be oxidised by organisms that also have genes for nitrate reduction (known on Earth).
Organics	Perchlorate	Perchlorate reduction	As above for organics.
Fermentation (disproportionation)			
Organics	Organics	Fermentation	As above for organics.

could provide analogies to potential water-rock-microbial interactions for the Martian subsurface.

The presence of present-day life in the subsurface of Mars remains an open question. Transient peaks of methane up to 7.2 parts per billion (volume) have fuelled continued debate about the possibility of subsurface life on Mars. It could be argued that some other observations do not suggest a deep biosphere. The presence of CO in the atmosphere, which can be used as an electron donor in anaerobic carbon monoxide oxidation (**carboxydotrophy**), has been suggested to show the lack of a significant biological sink. Although Martian sources and sinks of CH_4 and CO are not fully understood, the data provide plenty of scope for debate about a present-day active chemoautotrophic biosphere on Mars.

Chemoheterotrophy could provide energy for growth, past and present. Iron and sulfate reduction can be accomplished with organic electron donors.

There is a wide diversity of other, more unusual, redox couples involving alternative anions or cations including arsenite oxidation, uranium reduction and many others that can be used by microorganisms to conserve energy for growth. The study of more geochemically heterogeneous sites on Mars and a greater investigation of localised distributions of a variety of anions and cations will allow alternative redox couples to be assessed.

The accessibility of theoretically available energy for life on Mars would have tracked liquid water availability. As much of the surface water disappeared in the late Noachian, photosynthesis as a theoretical metabolism would have been eliminated. As a greater proportion of the surface and near-surface environment became desiccated and frozen, so near-surface exogenous organics from meteoritic material would be unavailable as a source of energy. Ultimately, the theoretical energy sources would be restricted primarily to chemoautotrophic deep subsurface redox couples and chemoheterotrophy where organics could be supplied from dead organisms (necromass).

Although there are potentially a wide variety of energy sources, a current assessment of energy availability on Mars, past and present, based only on unambiguous detection of half-reactions (Table 17.1), shows that almost all microbial metabolisms lack definitive detection of a complete redox couple with the conclusion that many Martian environments were and are extremely energy limited. Redox reactions for which available electron donors have been detected tend to lack definitively detected or abundant electron acceptors. Those with confirmed electron acceptors lack abundant electron donors. This conclusion of extreme energy limitation can only be removed with the direct detection, determination of the concentration and biological accessibility of a greater variety of electron donors and acceptors in recent and ancient terrains.

17.5.5 Physical Limits to Life: Radiation

Mars lacks any significant amount of ozone or other gaseous absorbers in the atmosphere that can absorb ultraviolet (UV) radiation above 200 nm. Like the early Earth (Chapter 10), the surface of Mars therefore experiences UV radiation down to 200 nm (CO_2 absorbs most of the radiation below 200 nm). However, UV radiation is rapidly attenuated in the subsurface, so although the surface flux generates biologically-effective DNA damage about three orders of magnitude higher than on the surface of the Earth, within a depth of a just few tens of microns to millimetres, depending on soil particle size, UV radiation is extinguished. Today, the surface of Mars is uninhabitable because of desiccation. However, UV radiation would not, in itself, create uninhabitable conditions if other conditions for life were met.

Ionising radiations of solar energetic particles (SEP) and galactic cosmic rays (GCR) are more penetrating. The total dose of ionising radiation experienced on the Martian surface has been measured as 76 mGray (mGy) year^{-1}, much lower than the fluxes that can be tolerated by radioresistant organisms such as *Deinococcus radiodurans*, which can withstand doses in excess of 5 kGy without appreciable loss of viability (Chapter 7). However, inactivity would result in accumulated damage such that at 2 m depth in the Martian crust, a *D. radiodurans* population was estimated to suffer an approximately six-order of magnitude reduction in viability after 450 000 years. For a deep subsurface biota at just a few metres depth or greater on Mars, particularly one that is active and can repair damage in an environment where liquid water is available, radiation would not render the subsurface uninhabitable.

17.5.6 Physical Limits to Life: pH

In a variety of Martian settings, pH ranges are within the boundaries for life. The pH of the Martian near subsurface was measured at the Phoenix lander site. It was found to be slightly alkaline, pH 7.7–7.9, and carbonate-buffered. The pH at Yellowknife Bay, Gale Crater was also inferred to be neutral from Curiosity data. This pH range is benign for organisms. Although we have no direct measurements of pH in the deep subsurface, past or present, reactions of fluids with mafic and ultramafic rocks control solution chemistry. It would be expected, as on the Earth, that fluids would be anoxic and alkaline or ultra-basic (pH > 10).

Although many environments in the ancient history of Mars may have been neutral to alkaline, the presence of certain sulfate minerals on the surface of Mars suggests locally acidic conditions. Sulfates are found as Hesperian layered sulfates, polar deposits, sediments in craters, within the globally ubiquitous Martian dust (which may contain 5–10% sulfates) and as sulfate veins within rocks. Although many of these sulfates are gypsum ($CaSO_4$), which is not associated with acidic conditions, jarosite and iron sulfates suggest acidic environments. These minerals suggest a period of acidic weathering (pH 2–5) in the Hesperian, during which exhalations of SO_2 from Martian volcanic activity would have produced acidic conditions, which then weathered Martian basalts to produce secondary sulfate minerals in low water–rock ratio interactions. These sulfate salts are also testament to the fact that the surface and subsurface geochemical cycles of Mars have been strongly influenced by the sulfur cycle, as compared to the Earth, where the carbon cycle generally dominates. Acidic environments on the Earth are not uninhabitable, but they do restrict the range of organisms capable of active growth. The restriction of water availability and the presence of acidic conditions during the Hesperian would attest to more widespread inclement environments in the surface and possibly in the subsurface. However, the lowest pH values predicted for Mars would not in themselves make environments uninhabitable. Local differences in pH would merely change the suitability of putative habitats for particular organisms.

17.5.7 Physical Limits to Life: Salts

Brines can constrain the boundaries of active life by influencing **water activity** and other parameters such as **chaotropicity** (the degree of disorder induced in macro-molecules). Extremely low water activities (Chapter 7) and high chaotropicities can be generated by brines such as chlorides ($CaCl_2$) and mixed sulfate brines.

On present-day Mars, seasonally Recurrent Slope Lineae (RSL) could be formed from concentrated briny solutions. Some Martian brines are calculated to have water activities below those required for life and would not be habitable environments. Thus, as the hydrological environment of Mars transitioned from the Noachian into the Hesperian and salt saturated solutions became prevalent in groundwater environments, some of these briny solutions could have rendered localised environments uninhabitable.

17.5.8 Habitat Space for Microbes on Mars

The porosity of the subsurface of Mars may be greater than the Earth at comparable depths and lithologies because of the lower gravity (0.38 of Earth's gravity) on the planet. As on the Earth, porosity in any location will be controlled by factors including secondary mineral infilling, sediment deposition, local rock pressure environments and other geological processes, but fundamentally there is no reason why the Martian subsurface environment should not be accessible to life. The permeability of subsurface environments will affect the connectivity of environments.

We do not have direct measurements of temperature and pressure profiles into the deep subsurface of Mars. As for the Earth, deep habitability is likely to be constrained when the temperature set by the **geothermal gradient** exceeds the upper temperature limit for life. Geothermal gradients of between ~10 and ~20 °C km^{-1} imply a lower depth of habitable temperatures on the order of ~6–15 km depth.

17.6 Trajectories of Martian Habitability

Supported by this previous synthesis of environmental conditions that would have influenced conditions available to support the activity of organisms, it is possible to think about trajectories of the habitability of Mars that are consistent with these data (Figure 17.15).

All trajectories of Martian habitability begin with the formation of Mars. From early planetesimals an uninhabitable planet formed, much like the Earth (Chapter 8). As water condensed and the environment cooled, the planet was at a branch point in its long-term trajectory of biological conditions. In one set of trajectories, the planet is defined by its condition as uninhabited (neither an origin of life occurs, nor does life transfer to the planet from the Earth in meteoritic matter). In the second set of trajectories, the planet is defined by the establishment of life, an event that changes the use of habitable conditions and through feedback effects, would itself change the habitability of environments.

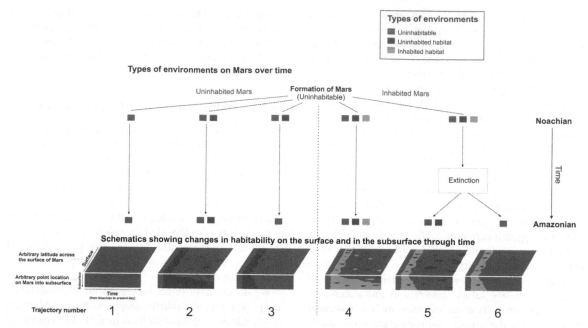

Figure 17.15   *Martian habitability trajectories. Different trajectories of the habitability of Mars through time, beginning with the branch point of an uninhabited and inhabited Mars. In the lower part of the diagram are shown schematics illustrating the trajectories. The 'surface' section represents the surface of Mars. The 'sub-surface point' is an arbitrary location through the Martian crust to an arbitrary depth of several kilometers. Experimental investigations of Mars will allow for the determination of which trajectory applied to Mars and the relative abundance of the different environments (Source: Charles Cockell).*

17.6.1   Trajectories for an Uninhabited Mars

It is not known exactly the conditions that are required to transform prebiotic chemicals into self-replicating life forms. It is possible to consider various environments and conditions in which an origin of life might have occurred (Chapters 10 and 11) and how prevalent these might have been on early Mars. However, at the current time it is not possible to quantify the probability of life's origin on Mars and we do not know whether it did originate.

These factors, taken together, mean that an uninhabited Mars remains a plausible condition for the planet throughout its history and must be taken as one early branch point in systematically identifying habitability trajectories.

There are three possible trajectories of Martian habitability for an uninhabited Mars. In Trajectory 1 (Figure 17.15), Mars might always have been uninhabitable. Even if all the requirements for life are met, but never all together in one place at the scale of microorganisms (micron to sub-micron scales), then environments would

be uninhabitable. This scenario seems less plausible for early Mars compared to the present day, since water flow might be expected to solubilise many elements and nutrients and generate environments in which a diversity of chemical species co-exist at small spatial scales, including the elements required for life. As conditions deteriorated during the Hesperian and sustained hydrological processes were terminated, a greater number of conditions leading to uninhabitable environments, for example lack of liquid water, low water activity in extreme brines and acidity, would have been realised and co-localised in a greater number of environments, particularly on the surface. In combination, they could have ensured that many environments remained uninhabitable.

Mars might have always lacked a fundamental requirement for life at sufficient concentrations. Although fixed nitrogen has been detected in Martian meteorites and on the surface of Mars, if these species had not been produced at sufficient concentrations over large scales or

Facilities Focus: Mars Simulation Chambers

It is very expensive to get to Mars. Another approach to investigating conditions on the Red Planet is to re-create them on the Earth. Mars Simulation Chambers (Figure 17.16) can be used to simulate conditions on Mars, past and present. Most Mars chambers are some sort of vacuum chamber into which gases with the atmospheric composition of Mars, either in the present-day or inferred for its past, are injected. The chamber is connected to a computer-controlled vacuum pump which allows the pressure to be regulated to the desired value (for example ~6 mb for present-day Mars). The gas is either pre-mixed or bottles of the different component gases to be found in the Martian atmosphere are mixed as they go into the chamber using mass-flow controllers. The Mars chamber may contain a cold finger or some type of circulated cooling which allows the low temperatures to be generated. Liquid nitrogen can be used as a coolant. Using a quartz window or fiber-optics, the surface light environment can even be created. Xenonarc lamps can be used to generate a solar-like spectrum containing the high UV radiation on the Martian surface. Computer control can be used to generate diurnal or seasonal temperature and solar radiation variations equivalent to those observed by spacecraft. The design of a Mars chamber will depend on the planned studies. For the examination of organics on Mars it is essential to have a chamber that is very clean and for which pieces of the chamber can be sterilised and cleaned of organic contamination. For microbiology experiments, either the samples are enclosed within flasks or dishes that have a gas-permeable, but microbe impermeable, seal to minimise the chances of contamination from within the chamber, or air going into the chamber must be filtered to reduce the chances of contamination.

Figure 17.16 A Mars simulation chamber used to re-create conditions in the laboratory equivalent to those on the surface of Mars, past or present (Source: Spanish Centre for Astrobiology).

co-localised at micron scales with other requirements for life, then this element could have been limiting to life.

However, studies by the Curiosity rover do not suggest globally uninhabitable conditions. In fact, the rover team published a paper in early 2014 suggesting a clement early Hesperian aqueous environment at Yellowknife Bay, Gale Crater with the presence of essential elements and diverse redox states of elements that could have supported life. If this conclusion is correct, then it may already be possible to discount this trajectory.

Another trajectory of an uninhabited Mars is one involving habitable conditions, despite the lack of life throughout the planetary history (Trajectory 2, Figure 17.15). During the Noachian, when liquid water was more widespread, **uninhabited habitats** could have been abundant. This trajectory supposes that the circulation of ancient water on Mars generated environments where dissolved elements and gases provided vital elements (CHNOPS), nutrients and energetic redox couples co-located at microscopic scales with clement pH conditions and water activities. As surface conditions deteriorated during the Hesperian, uninhabited habitats would become more confined and localised as an increasing number of surface environments became uninhabitable at all scales on account of desiccation, long-term irradiation, acidity, the presence of low water activity brines and other environmentally deleterious conditions.

Up until the present-day, uninhabited habitats would become confined to subsurface and near-surface environments where liquid water became transiently available, for example, in impact-induced melting of permafrost at macroscopic scales, layers of water within soil grains at microscopic scales and production of local liquid water during obliquity changes. They could exist if the geothermal gradient allows for deep aquifers or if habitable liquid water is transiently sustained in near-surface environments such as dilute brines. Thus, this trajectory is characterised by a changing relative abundance of uninhabited habitats and uninhabitable environments through time. The trajectory assumes that locations on Mars have harboured habitable conditions throughout its history.

Yet another trajectory (Trajectory 3, Figure 17.15) is a more extreme version of the previous one. In this scenario, uninhabited habitats existed during the early history of Mars, but deteriorating conditions eventually led to their constriction and localisation until they completely disappeared, even at microscopic scales, rendering the entire planet uninhabitable, a condition that remains to the present day. A hypothetically plausible scenario

would be the desiccation of the surface, the freezing of the subsurface as the cryosphere extended in depth and, even if deep aquifers persisted to the present day, a geochemical run-down of the availability of nutrients caused by insufficient turn-over. It would also require that the formation of transient liquid water in recent times occurred in locations where conditions rendered it uninhabitable, for example, where brine concentrations and water activities of aqueous solutions were inappropriate for growth or where there was insufficient energy, nutrients and other basic elements co-localised at the scale of microorganisms. Thus, this trajectory posits a well-defined biphasic history for Mars where it was once habitable but is now uninhabitable.

17.6.2 Trajectories for an Inhabited Mars

A second set of trajectories is realised if Mars became inhabited, either by an indigenous origin of life or a transfer of life from Earth to Mars. There are three plausible trajectories that can be identified for an inhabited Mars.

In one trajectory (Trajectory 4, Figure 17.15), uninhabitable environments (in places such as extreme desiccated surface environments), uninhabited habitats (newly formed habitats disconnected from inhabited regions, even if only transiently) and inhabited habitats that contain Martian life all exist and existed on Mars.

Similarly to other trajectories, the geologic evidence observed on Mars suggests that the relative abundance of these three environments would have changed over time. As surface conditions deteriorated and liquid water ceased to be as abundant, one would expect the diversity and area of inhabited habitats to decrease from the Noachian through to the Amazonian. As surface environments became desiccated, more acidic and evaporative brines became more extreme, the area of surface uninhabitable environments would have increased through time. The diversity and area of uninhabited habitats would also have concomitantly changed. As the hydrological cycle reduced in scope, so more habitats would have become isolated and separated from inhabited habitats, potentially increasing the number of uninhabited habitats.

Another trajectory (Trajectory 5, Figure 17.15) posits the existence of life on early Mars, but as hydrological conditions deteriorated and geochemical turnover become less efficient, life was eventually constrained to such small pockets of existence that it become functionally extinct and eventually a total extinction occurred. At this point Mars transitioned into a planet harbouring only uninhabitable environments and uninhabited habitats. The extinction event would not have precluded new habitable places becoming available, for example from obliquity-driven liquid water formation or in impact-induced hydrothermal systems, but a lack of connectivity and sufficient water flow prevented their colonisation from the last remaining vestiges of life until eventually, when life became extinct, there was no life to occupy uninhabited habitats that persist, or are transiently produced, to this day.

Finally, yet another more extreme trajectory (Trajectory 6, Figure 17.15) posits the existence of life in the early history of Mars, but as conditions deteriorated it became extinct and conditions became so extreme that all habitable conditions disappeared. At this point, Mars transitioned into a planet harbouring only uninhabitable environments. This transition could have occurred through a phase where isolated uninhabited habitats existed for a period of time, that is Mars hosted inhabited and uninhabited habitats until conditions became so extreme that even uninhabited habitats ceased to exist. This trajectory requires that the combined environmental, chemical and physical conditions on Mars eventually placed all environments outside the capabilities of all known microorganisms to sustain activity.

There are other trajectories that can be suggested. Examples include an inhabited Mars on which life becomes extinct and then re-originates (or is transferred from Earth) at some later time. Combinations of the major trajectories discussed above are possible. However, from a position of parsimony, here just six major trajectories have been discussed based on what we currently know about the history of the Martian environment.

The habitability of Mars and its biological condition has been of long-term interest to scientists. The clarification of trajectories of habitability has implications for the interpretation of data returned from missions to Mars examining habitability on the planet. With sufficient research and missions, it is possible to determine which of these proposed trajectories the planet Mars took and quantify the relative abundance of its component environments. This approach to identifying habitability trajectories can be applied to other planetary bodies.

17.7 The Viking Programme and the Search for Life

The NASA Viking missions were motivated by the possibility of life being present on Mars. The mission had two landers that arrived on the surface of Mars in 1976, Viking I and II (Figure 17.17) and two orbiters. The Viking landers carried out other scientific objectives, including the first surface measurements of the composition of the Martian atmosphere and surface rocks. However the principal objective was to establish whether or not there was life on Mars.

The Viking landers had four experiments to look for evidence of active microbial metabolism. Let's look at these experiments. Although they are now rather old, they are a very good example of applying the scientific method on other planets and the importance of controls.

17.7.1 GCMS Analysis

GCMS (Gas Chromatography Mass Spectrometry) was used to analyse the components of Martian soil, and particularly those components that were released as the soil was heated to different temperatures. It could measure molecules present at a level of a few parts per billion. The GCMS measured no significant amount of organic molecules in the Martian soil. The only organic chemicals identified when the Viking landers heated samples of Martian soil were chloromethane and dichloromethane, compounds interpreted at the time as likely contaminants from cleaning fluids.

That this experiment was negative was difficult to explain since organic material is continually being delivered to Mars through meteoritic in-fall including organics in **carbonaceous chondrites** (Chapter 11). An explanation for this observation wasn't provided until the 2008 Phoenix mission when the lander found perchlorate salts in the Martian surface. A recent interpretation is that the perchlorate, when heated with indigenous organics from meteorites during pyrolysis, produced the chlorinated compounds that were observed at very low concentration.

17.7.2 Gas Exchange Experiment

The Gas Exchange (GEX) experiment looked for gases given off by an incubated soil sample by first replacing the Martian atmosphere with the inert gas, helium. It applied a liquid mixture of organic and inorganic nutrients, first with nutrients added, then with water added as a control. After incubating for 12 days, the instrument sampled the atmosphere of the chamber and used a gas chromatograph to measure the concentrations of several gases, including oxygen, carbon dioxide, nitrogen, hydrogen, and methane. The scientists hypothesised that metabolising organisms

Figure 17.17 *One of the Viking Landers showing its scoop in operation in a model Martian setting* (Source: NASA).

would produce one of these gases. The result of the experiment was the production of oxygen. However, the gas was also produced in controls heated to 145 °C, which was not consistent with microorganisms, but could be explained by the presence of reactive compounds in the soil.

17.7.3 Labelled Release Experiment

In the Labelled Release (LR) experiment, a sample of Martian soil was inoculated with a 1-ml drop of very dilute nutrient solution. The nutrients, which comprised seven organics (formate, glycolate, glycine, D-alanine, L-alanine, D-lactate and L-lactate) were tagged with radioactive ^{14}C. After incubation for 10 days, the air above the soil was monitored for the evolution of radioactive $^{14}CO_2$ gas as evidence that microorganisms in the soil had metabolised one or more of the nutrients. The result was the production of radioactive gases. The experiment was done by both Viking landers, the first using a sample from the surface exposed to sunlight and the second probe taking the sample from underneath a rock. Both experiments gave positive results. The control samples heated to 160 °C did not give the same result, which was consistent with biology. However, subsequent injections into the experimental chambers a week later did not give the same reaction, which would not be expected for biology (with

new nutrients, we would expect living things to begin to grow again). Under cold and dry conditions and in a carbon dioxide atmosphere, ultraviolet radiation can cause carbon dioxide to react with soils to produce various oxidised compounds, including highly reactive superoxides such as hydrogen peroxide which could have reacted with the organics to produce the gases. Superoxide chemistry also accounts for the fact that the addition of yet more nutrients did not elicit a reaction since the chemically active component in the soil would have been exhausted.

17.7.4 Pyrolytic Release Experiment

Light, water, and an atmosphere of carbon monoxide (CO) and carbon dioxide (CO_2), simulating that on Mars, were introduced into the experimental chamber with a 0.25 cm^3 soil sample. The carbon-bearing gases were made with ^{14}C. If there were photosynthetic organisms, they would incorporate some of the carbon as biomass. After 120 h of incubation using a xenon arc lamp to provide light, the experiment removed the gases, baked the remaining soil at 625 °C and collected the products in a device which counted radioactivity. If any of the ^{14}C had been converted to biomass, it would be vaporised during heating and the radioactivity counter would detect it as evidence for life. Controls using heated soil were also investigated. In both the experimental and heated control samples, gases were released from apparently small amounts of fixed carbon. This was not consistent with microorganisms.

Overall, the results from Viking are thought to be explained by active chemical components in the Martian soils, not biology.

17.7.5 Viking: A Lesson in Science

The Viking experiments are an important lesson in science. If you go back through the foregoing and think about the results, you will realise that all of the experiments, on their own, suggest life. However, it was the heat-sterilised controls that suggested a chemical explanation by displaying activity when we would not expect organisms to be alive. Combined with the GCMS data that did not find substantial organics, the chemical explanation is the strongest one. Controls are a very important part of a good scientific experiment. The Viking experiments, although they did not find life, were well planned.

17.8 Martian Meteorites

Throughout the lifetime of the Solar System, rocks have been ejected from the surface of Mars in asteroid and comet impacts and these wandering Solar System materials eventually land on Earth as meteorites. The number of Martian meteorites identified is small. There are just over 130, although this number continues to grow. Their very different composition from Earth rocks and other meteorites marked them out from the early days of meteorite hunting. The meteorites contain pockets of gas trapped in glasses when the rock was melted during ejection from the surface of Mars. These pockets of gas have essentially sampled the atmosphere of Mars, one of the lines of evidence that demonstrates they are from Mars. The gases have elemental and isotopic compositions that are similar to atmospheric analyses on Mars (Figure 17.18). Furthermore, the mineral composition of the rocks themselves is consistent with a Martian origin.

The Martian meteorites are divided into three groups, the shergottites, nakhlites and chassignites (or SNC meteorites, the general name given to Martian meteorites). There are also two 'grouplets' called the orthopyroxenite and basaltic breccias.

About three-quarters of all Martian meteorites are shergottites, named after the Shergotty meteorite, which fell at Sherghati, India in 1865. The formation ages for many shergottites are mostly about 150–575 Ma ago, in other

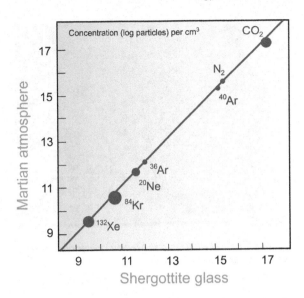

Figure 17.18 *The abundance of gases found trapped in the glasses of the Martian shergottite meteorite EET A79001 compared with the Martian atmosphere (as determined from spacecraft) suggests that samples of Martian atmosphere was trapped in melted glass when the meteorite was ejected from the Martian surface.*

words very young. The age dating of Martian meteorites is obviously an important topic because these rocks tell us about the geochemistry and environment of one planet of enormous interest to astrobiologists. Accurate assessments of the timing of surface processes, such as the presence of water, are crucial.

The nakhlites are named after the Nakhla meteorite, which fell in El-Nakhla, Alexandria, Egypt in 1911 and had an estimated weight of 10 kg. They are all igneous rocks formed from basaltic magma about 1.3 billion years ago. They contain **olivine** crystals. Their ages suggest that they formed in the large volcanic regions of Mars. It has been shown that the nakhlites were infiltrated with water around 620 million years ago and that they were launched from Mars in impact events around 11 million years ago. Most of the collected meteorites fell to the Earth during the last 10 000 years.

The chassignites are named after the Chassigny meteorite, which fell at Chassigny, Haute-Marne, France in 1815.

Perhaps the most famous Martian meteorite is ALH84001 (Figure 17.19). The rather cumbersome name does have a logic. The 'ALH' refers to the Allan Hills region of Antarctica where it was found; the '84'

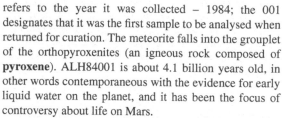

Figure 17.19 ALH84001, a Martian meteorite (Source: NASA/JSC/Stanford University).

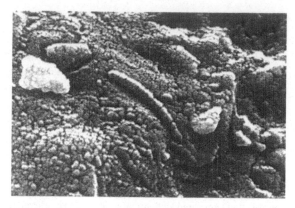

Figure 17.20 Putative microfossils in ALH84001. The putative microfossil shape is ~400 nm long. These images are highly suggestive and caught the imagination of the media, but morphology alone is not compelling evidence for life.

refers to the year it was collected – 1984; the 001 designates that it was the first sample to be analysed when returned for curation. The meteorite falls into the grouplet of the orthopyroxenites (an igneous rock composed of **pyroxene**). ALH84001 is about 4.1 billion years old, in other words contemporaneous with the evidence for early liquid water on the planet, and it has been the focus of controversy about life on Mars.

In 1996 a bold claim was made in the journal, *Science*, by a group led by David McKay (1936–2013) at the NASA Johnson Space Center, stating that evidence for primitive Martian life had been found in ALH84001. Many lines of evidence were presented to prove the case for Martian life. An examination of the meteorite showed that: (i) it contained tiny carbonate globules which were interpreted as being formed at low temperatures conducive to life, (ii) it contained polycyclic aromatic hydrocarbons (*PAHs*; Chapter 9) associated with the carbonate globules, (iii) it contained shapes that resembled microfossils of primitive bacteria about 300 nm long (Figure 17.20) and (iv) it contained **magnetite** mineral phases of ~40–60 nm in size, consistent with by-products of bacterial activity and particularly **magnetotactic bacteria**. These organisms are known to produce chains of iron oxide minerals with which they can orient themselves along the Earth's magnetic fields as they search for the optimum oxygen concentration. The magnetite crystals were claimed to be the products of biological activity. The original claim was that each one of these lines of evidence could be separately accounted for by non-biological processes, but together they constitute evidence for life.

Today this evidence remains controversial. There has been work claiming that each one of these lines of evidence can be produced by non-biological processes (see Chapter 12 for a discussion on criteria for determining whether ancient preserved material can be assigned as biological). The carbonates have been claimed to be produced at high temperatures, between 200 and 500 °C, much higher than the currently known upper temperature limit for life. The microfossils may be shapes produced by non-biological means. They are very small, at the limit or smaller than what is thought to be the minimal size of cells (Chapter 4). Many of them could not include more than a few ribosomes, at least using terrestrial cell size as a reference. The PAHs found in the meteorite, although associated with the carbonate globules, are known to have potential non-biological origins (Chapter 9). The magnetic crystals have remained the most contentious point of debate. The discussion has revolved around arguments about whether their shapes are distinctively biological or whether the crystals could be produced by non-biological processes. Nevertheless, the work on the meteorite stimulated many discussions of relevance to astrobiology, including the minimum size of cells and the criteria for assigning biogenicity of putatively preserved remains of life. The debate remains open concerning whether Martian meteorites contain any evidence for biogenic activity.

Debate Point: Science and astrobiology

Astrobiology seeks to address a hypothesis that there is life beyond the Earth. The null hypothesis is that there is no life beyond the Earth. Good scientists care about neither outcome; they merely wish to objectively find out which one is correct. However, it is very easy to get caught up in a 'quest' for life whereby the objective is to prove its existence. After all, the payback from discovery would be enormous for individuals and organisations. The planet Mars provides particularly excellent examples of the 'desire' to find life. Percival Lowell became convinced of his canals to the point that he left the realms of finding ways to test the hypothesis of whether they were artificial, and instead began to focus on their nature and characteristics. The study of ALH84001 was a robust investigation, but the images of microfossils fuelled the 'hope' of alien life that, at least in the public domain, rapidly left the realms of objective analysis. Even President Clinton felt compelled to laud the discovery on the White House lawn. What lessons are there from these episodes for astrobiologists? How do astrobiologists ensure that the enticing and incredibly compelling question of life beyond Earth remains objectively tested? You might consider setting up teams to debate the evidence for microscopic life on Mars and assess the scientific evidence for life on Mars today.

17.9   Mars Analogue Environments

A Mars analogue environment is an environment on the Earth that hosts conditions comparable to either past or present environments on Mars and in some way allows us to understand the geophysics, geochemistry or the potential habitability of the planet. There is no environment on the Earth that is truly identical to Mars (the surface and even sub-surface of the Earth has been influenced by over 2.4 Ga of atmospheric oxygen; Chapter 14). However, some environments have characteristics that could be useful for the study of Martian environments.

Mars analogue environments have been used by astrobiologists for a number of purposes, for example, to study geological processes and investigate how water–rock interactions might have influenced habitability. They are used as locations from which to isolate microorganisms to investigate how they adapt to extreme environmental conditions of relevance to Mars. Analogue environments are also used to test instrumentation for use in planetary exploration.

Examples of Mars analogue environments include:

1. Acidic environments, which provide insights into the geochemistry and habitability of Hesperian acidic terrains. An example is the acidic Rio Tinto River, Spain.
2. Permafrost, which is used to understand possible processes in Martian permafrost. Examples include the Canadian High Arctic (e.g. Axel Heiberg Island, Devon Island) and Greenland.
3. Dry regions used to understand the habitability of extreme desiccated environments relevant particularly to present-day Mars. Examples are the Dry Valleys of Antarctica and the Atacama Desert, Chile.
4. Volcanic environments used to study ice–volcano interactions and the habitability of igneous rocks and volcanic environments. An example is the many terrains of Iceland (Figure 17.21).
5. Caves, used to investigate near-surface processes and the habitability of enclosed subsurface regions. An example would be lava tubes and caves on the island of Lanzarote, Spain.
6. Underground laboratories used to study deep subsurface processes that could be of relevance to Mars (Chapter 7).

17.10   Panspermia – Transfer of Life Between Planets?

The similarity of conditions on the early Earth and Mars has led to inevitable questions about whether they could have shared life in their early history. The possibility that life could be transferred between planets is sometimes called **panspermia** (from the Greek, *pan*, meaning 'all', and *sperma*, meaning 'seed'). The word is egregious because its rough translation, meaning 'life everywhere', pre-supposes the outcome of the very hypothesis one is supposed to be objectively testing. I prefer to merely think of it as planetary island biogeography – an extrapolation of **island biogeography**, a field of ecology which

Figure 17.21   *Iceland is an environment with volcanic rock–ice interactions which provides an analogue location for understanding the past geochemistry and habitability of Mars. This image shows lava flows in the south-east of the country.*

considers how life can be transported between habitats and the factors that limit dispersal. On the Earth, some organisms, such as microbes and small invertebrates, can be transferred between continents, although large organisms, such as some mammals, are frustrated by obstacles such as mountain ranges. A logical question progressing from these observations is: Can organisms be transferred between planets (Figure 17.22) and what physical barriers or **dispersal filters**, might prevent this? Are planets biogeographical islands?

The best way to approach the question of the transfer of life between planets is to consider the dispersal filters acting on organisms travelling between planets.

At the moment we do not know if there was, or is, life on Mars. We know there is an exchange of material between both planets (evidenced by the collection of Martian meteorites). This exchange suggests that the question of the transfer of life is important. If we never find life on Mars, but the planet hosted habitable conditions, does that mean that dispersal filters prevented a transfer from occurring? If we find life, is it related to life

on Earth? If it is, maybe Earth and Mars life are related through interplanetary transfer. Maybe life evolved on Mars and we are all Martians! If Martian life is not related to Earth life, why wasn't Mars inoculated with Earth life? Therefore, whatever the outcome of the search for life on Mars, the question of whether planets are biogeographical islands will be part of the explanation of the data.

Let's look at some of the data in more detail and consider the process of the transfer of life between planets.

17.10.1   Ejection from a Planet

The dispersal filter acting on organisms during the ejection of rocky material from a planetary surface involves acceleration, shock and high temperatures associated with the impact that causes the launch of material to escape velocity.

The effect of acceleration during ejection has been investigated with both *Bacillus subtilis* and *Deinococcus radiodurans* in experiments where the organisms were fired from a rifle into a hard target to study their

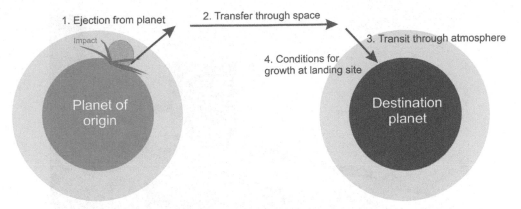

Figure 17.22   *Are planets biogeographical islands? A schematic showing the dispersal filters that life must survive to be transferred from one planet to another. The thickness of the atmosphere is exaggerated* (Source: Charles Cockell).

survival. Survival values of 40–100% after accelerations up to 4.5×10^6 m s^{-2} suggested that acceleration is not a strong dispersal filter. *Bacillus subtilis* spores, isolated from Sonoran desert (USA) basalt, similarly survived accelerations up to 1.5×10^{10} m s^{-2}. *Bacillus subtilis* is a spore-forming organism (Chapter 7) and therefore might be resistant to jerk and acceleration on account of the tough spore coat.

Planetary ejection also requires survival of shock pressures. We can get empirical knowledge of shock pressures experienced by ejected rocks by looking at Martian meteorites, which show shock pressures from approximately 5 to 55 gigapascal (GPa). Shock experiments using **gas gun** (Figure 17.23) and **plate-flyer** apparatus carried out by numerous researchers show clearly that survival depends both on the nature of the cell and its local microenvironment. Spores of *Bacillus subtilis* show high resistance to shock, with survival of shock pressures greater than 50 GPa. In contrast, vegetative cells are susceptible to disruption by low shock pressures. *Chroococcidiopsis* sp., a photosynthetic cyanobacterium introduced in Chapter 7 as a natural denizen of rocks, was killed at shock pressures greater than 10 GPa. This organism was chosen because it naturally inhabits the interior of rocks. The maximum temperatures experienced during these intense shocks will depend upon the nature of the propagation of the shock wave, the thickness of the biological material and the characteristics of the organisms.

The experiments show that even for desiccated vegetative cells, shock pressures at the lower end required for escape velocity from Mars (and Mars-like planets) can be survived, although in the case of *Chroococcidiopsis* sp., its threshold of survival was at the lower end of shock

pressures for rocks to reach escape velocity from Mars. The escape velocity of the Earth (11.19 km s^{-1}) is approximately twice that of Mars (5.03 km s^{-1}) implying higher minimum shock pressures required to reach escape velocity. The presence of spore-forming states will increase the range of shock pressures under which organisms can survive the dispersal filter, increasing the chances of their survival of planetary ejection. The primary molecular basis of this dispersal filter is therefore exceeding the mechanical strength of cell walls/membranes. A contribution of the thermal inactivation of cellular macromolecules might also be important.

17.10.2   Interplanetary Transfer Phase

Organisms must also survive the journey through space. During the interplanetary transit phase organisms near the surface of the rocks are potentially subject to high UV radiation. Interplanetary UV radiation has been considered to be a significant detrimental factor for organisms being transferred between planets. The flux of UV radiation penetrating to an organism will depend upon the distance from the star that the rock is located, its orientation, the thickness and UV absorption and scattering properties of the material covering the organisms. Without the protective effects of atmospheric absorption, the surface of rocks in interplanetary space will be exposed to the full UV wavelength range emitted by the star.

A diversity of published work shows that layers of different mineral types can significantly reduce UV radiation. One study showed that *Bacillus subtilis* spores mixed within clay, sandstones or Martian meteorite (Zagami) were protected against UV radiation, although thin layers

Figure 17.23 *A light gas gun used to simulate asteroid impact* (Source: NASA).

were ineffective, possibly due to UV penetration through cracks or the formation of toxic photochemical products.

Another detrimental factor is cosmic (**ionising**) radiation, which is more penetrating than UV radiation. Models using experimental data on the survival on *Bacillus subtilis* spores to heavy ion doses suggest that, behind 1 m of rock, spores could survive for about one million years. This survival period drops to about 300 000 years behind 10 cm of shielding due to secondary radiation generated from cosmic radiation impinging on the surface of the rock.

Despite these strong filters, evidence suggests that organisms can survive other dispersal filters associated with the conditions in space. A whole range of organisms have survived in space conditions. *Bacillus subtilis* spores were shown to survive for six years in Low Earth Orbit. Lichens, cyanobacteria, fungi and a range of bacteria have all been exposed on the European Space Agency's **EXPOSE facility** outside the International Space Station (Figure 17.24) and can survive the desiccation and extreme vacuum of space for years, albeit generally much shorter than transit times between planets (on the order of thousands to millions of years).

17.10.3   Arriving at the Destination Planet

Both the launch of a rock and its landing on the destination planetary body require that it pass through the dispersal

filter of transit through the origin and destination planetary atmospheres. As with planetary ejection, this has proven to be experimentally tractable using sounding rockets and orbital vehicles.

Bacillus subtilis spores inoculated onto **granite** slices were subjected to hypervelocity atmospheric transit by launch to ∼120 km altitude on a rocket. Spores were shown to have survived on the sides of the rock, but not on the forward-facing surface, which was subjected to a maximum temperature of 145 °C.

In the context of the **STONE experiment**, an experimental campaign designed to investigate the survival of artificial meteorites during atmospheric entry, the effects of atmospheric entry on photosynthetic organisms was also studied (Figure 17.25). Photosynthetic organisms must inhabit a rock at a depth where light levels are sufficient for photosynthesis. In a diversity of rock types normally conducive to the growth of photosynthetic organisms, the depth required for photosynthesis corresponds to approximately 5 mm or less. *Chroococcidiopsis* sp. was inoculated into some rocks that had been naturally shocked by asteroid and comet impact. The organisms were inoculated into the rock to the depth at which phototrophs are found growing in the material in the natural habitat. The rock was fixed into the heat shield of a Russian FOTON space capsule and launched to orbit. After orbiting the Earth for 16 days the sample

Figure 17.24 *The EXPOSE facility. The apparatus is attached to the outside of the International Space Station. Each small hole within the container can fit a sample of biological material within it. Above it are filters that can be chosen to allow certain wavelengths of solar radiation to penetrate* (Source: ESA).

Figure 17.25 *The Foton spacecraft returns to Earth in the steppes of Kazakhstan. Embedded within the heat shield (the small white circles on the left of the spacecraft) are artificial meteorites to study how rocks respond to atmospheric entry and whether organisms could survive.*

was de-orbited and retrieved in Kazakhstan. Neither the organisms nor their biomolecules survived providing a simple, although empirical demonstration of the effectiveness of atmospheric transit as a dispersal filter in the transfer of photosynthetic organisms that need to live near the surface of rocks to grow.

Despite these results, it is well known that the interior of meteorites can remain cool during atmospheric transit. Although the temperatures reached on the outside of the rock are high, transit times when traveling at $\sim 10\,\text{km s}^{-1}$ are on the order of tens of seconds, such that the heat does not have time to penetrate into the interior. For sufficiently sized rocks, any putative organisms in the interior of rock would not necessarily be affected by high temperatures.

In summary, at the current time there is strong evidence that some microbes can survive the conditions required for ejection from a planetary surface and that the interior of rocks remains cool enough for organisms deep within rock to potentially survive atmospheric entry. Although some rocks may be transferred between planets in short time periods, a major dispersal filter to organisms is the usually long journey through interplanetary space.

Some researchers have raised the question of the transfer of life between stars. Interstellar space is likely to act as an effective dispersal filter to organisms. Quite apart from the low probability of ejected rocks intersecting with a destination planet, the integrated exposure to cosmic radiation over long durations (many millions of years) will likely be sterilising.

The question of whether planets are biogeographical islands remains a fascinating question in astrobiology. Although there is no evidence that it has occurred in our Solar System, it nevertheless drives us to ask questions about the limits of microbial survival and draws us into extrapolating ecological questions beyond the home world. It will eventually be needed to explain why Mars is inhabited or lifeless.

17.11 Conclusions

Mars is the most Earth-like planet in the Solar System, yet its history has been very different to Earth. From the Noachian to the present day, the planet has experienced three general phases of geological evolution, each one sequentially more inimical to life. The planet has most of the requirements for life, but the desiccated and irradiated surface suggests that if there is any life there today it must be in the subsurface. Robotic missions to

Mars have significantly advanced our understanding of the planet and its past. Viking was the first mission to attempt to search for life on Mars. It remains an open question as to whether Mars was or is inhabited and the extent of habitable conditions. Part of understanding the history of Martian habitability and its biological status is encompassed by the question of whether planets are biogeographical islands.

Further Reading

Books

Pyle, R., Manning, R. (2012) *Destination Mars: New Explorations of the Red Planet*. Prometheus Books, New York.

Vogt, G.L. (2009) *Landscapes of Mars: A Visual Tour*. Springer, New York.

Papers

Benner, S.A., Devine, K.G., Matveeva, L.N., Powell, D.H. (2000) The missing organic molecules on Mars. *Proceedings of the National Academy of Sciences* **97**, 2425–2430.

Bibring, J.-P., Langevin, Y., Mustard, J.F., Poulet, F., Arvidson, R., Gendrin, A., Gondet, B., Mangold, N., Pinet, P., Forget, F., the OMEGA team (2006) Global mineralogical and aqueous Mars history derived from OMEGA/Mars Express data. *Science* **312**, 400–404.

Boston, P.J,, Ivanov, M.V., McKay, C.P. (1992) On the possibility of chemosynthetic ecosystems in subsurface habitats on Mars. *Icarus* **95**, 300–308.

Burr, D.M., Grier, J.A., McEwen, A.S., Keszthelyi, L.P. (2002) Repeated aqueous flooding from the Cerberus Fossae: evidence for very recently extant, deep groundwater on Mars. *Icarus* **159**, 53–73.

Cockell, C.S. (2014) Trajectories of Martian habitability. *Astrobiology* **14**, 182–203.

Cousins, C.R., Crawford, I.A., Carrivick, J.L., Gunn, M., Harris, J., Kee, T.P., Karlsson, M., Carmody, L., Cockell, C.S., Herschy, B., Joy, K.H. (2013) Glaciovolcanic hydrothermal environments in Iceland and implications for their detection on Mars. *Journal of Volcanology and Geothermal Research* **256**, 61–77.

Ehlmann, B.L., Mustard, J.F., Murchie, S.L., Poulet, F., Bishop, J.L., Brown, A.J., Calvin, W.M., Clark, R.N., DesMarais, D.J., Milliken, R.E., Roach, L.H., Roush, T.L., Swayze, G.A., Wray, J.J. (2008) Orbital detection of carbonate-bearing rocks on Mars. *Science* **322**, 1828–1832.

Grotzinger, J.P., the MSL Science Team (2013) A habitable fluvio-lacustrine environment at Yellowknife Bay, Gale Crater, Mars. *Science*. doi: 101126/science1242777.

Hecht, M.H., Kounaves, S.P., Quinn, R.C., West, S.J., Young, S.M., Ming, D.W., Catling, D.C., Clark, B.C., Boynton, W.V., Hoffman, J., Deflores, L.P., Gospodinova, K., Kapit, J., Smith, P.H. (2009) Detection of perchlorate and the soluble chemistry of martian soil at the Phoenix lander site. *Science* **324**, 64–67.

Horneck, G., Rettberg, P., Reitz, G., Wehner, J., Eschweiler, U., Strauch, K., Panitz, C., Starke, V., Baumstark-Khan, C. (2001) Protection of bacterial spores in space, a contribution to the discussion on panspermia. *Origins of Life and Evolution of Biospheres* **31**, 527–547.

Jakosky, B.M., Nealson, K.H., Bakermans, C., Ley, R.E., Mellon, M.T. (2003) Subfreezing activity of microorganisms and the potential habitability of Mars' polar regions. *Astrobiology* **3**, 343–350.

Malin, M.C., Edgett, K.S. (2000) Evidence for recent groundwater seepage and run-off on Mars. *Science* **288**, 2330–2335.

Mancinelli, R.L., Banin, A. (2003) Where is the nitrogen on Mars? *International Journal of Astrobiology* **2**, 217–225.

McEwan, A.S., Ojha, L., Dundas, C.M., Mattson, S.S., Bryne, S., Wray, J.J., Cull, S.C., Murchie, S.I., Thomas, N., Gulick, V.C. (2011) Seasonal flows on warm Martian slopes. *Science* **333**, 740–743.

McKay, D.S., Gibson, E.K., Thomas-Keprta, K.L., Vali, H., Romanek, C.S., Clemett, S.J., Chillier, X.D.F., Maechling, C.R., Zare, R.N. (1996) Search for past life on Mars: Possible relic biogenic activity in Martian meteorite ALH84001. *Science* **273**, 924–930.

Michalski, J.R., Cuadros, J., Niles, P.B., Parnell, J., Rogers, A.D., Wright, S.P. (2013) Ground water activity on Mars and implications for a deep biosphere. *Nature Geoscience* **6**, 133–138.

Mileikowsky, C., Cucinotta, F., Wilson, J.W., Gladman, B., Horneck, G., Lindgren, L., Melosh, H.J., Rickman, H., Valtonen, M.J., Zheng, J.Q. (2000) Natural transfer of viable microbes in space. Part I: From Mars to Earth and Earth to Mars. *Icarus* **145**, 391–427.

Poulet, F., Bibring, J.-P., Mustard, J.F. *et al.* (2005) Phyllosilicates on Mars and implications for early Martian climate. *Nature* **438**, 623–627.

Schulze-Makuch, D., Irwin, L.N., Lipps, J.H., LeMone, D., Dohm, J.M., Fairén, A.G. (2005) Scenarios for the evolution of life on Mars. *Journal of Geophysical Research* **110**: E12S23.

Weiss, B.P., Yung, Y.L., Nealson, K.H. (2000) Atmospheric energy for subsurface life on Mars? *Proceedings of the National Academy of Sciences* **97**, 1395–1399.

Westall, F., Loizeau, D., Foucher, F., Bost, N., Betrand, M., Vago, J., Kminek, G. (2013) Habitability on Mars from a microbial point of view. *Astrobiology* **13**, 887–897.

18

The Moons of Giant Planets

Learning Outcomes

➤ Understand the characteristics of some of the moons of the giant gas planets in the Solar System.

➤ Understand the evidence for water bodies under the surfaces of some of the icy moons, with special reference to Europa, Ganymede and Enceladus.

➤ Understand some of the geology and characteristics of the surface, atmosphere and carbon cycle on Titan.

➤ Describe some of the ideas about how the conditions for habitability might be met within the interior of moons.

➤ Understand the motivations behind planetary protection and some of the procedures that are used.

18.1 The Astrobiology of Moons

Several decades ago the idea of investigating the outer planet moons for habitable conditions would have been regarded as a science fiction absurdity. As we saw in Chapter 16, the concept of the habitable zone was well established. The paradigm that life would be associated with planets with liquid water on their surfaces, a little like the Earth, was entrenched. It would not have occurred to planetary scientists that small moons in the freezing wastes of the outer Solar System, orbiting the giant and inhospitable gas planets, could be locations of interest to astrobiologists.

Of course, we do not know if moons harbour life, but the detection of water and other requirements for habitability, as we shall soon see, has radically altered our view of these worlds. At the very least, even if they do not host life, they have the potential to reveal important information about the processing of carbon in the Solar System and to provide new ideas about the production of molecules of prebiotic significance. They will certainly tell us about the limits and distribution of habitable conditions in the Universe.

Beyond the orbit of Mars are the giant gas planets and their moons. Many of these moons are small rocky objects such as the 100-km long potato-shaped inner moon of Saturn, Pandora (Figure 18.1). Pandora is representative of the huge array of small rocky moons, either captured by the planets or formed in the same protoplanetary material. Some of these objects are dark and may be carbon-rich. Jupiter and Saturn have over 50 of these small moons, Uranus over 20. All of these little worlds are of interest as we attempt to understand how our Solar System formed, but few of them have special astrobiological interest. The exception is the class of bodies that are large enough to have formed large spherical objects, generally more than ~200 km in diameter. These objects were large enough to hold on to the original **volatiles** endowed to them during the formation of the Solar System. These icy and volatile-rich moons have thrown up some remarkable surprises. In general, as we move further out into the Solar System, we find moons with volatiles that have lower freezing temperatures. In the Jovian system, water ice dominates, but at Triton, the moon of Neptune, there are ices of nitrogen, water, methane and carbon monoxide.

Astrobiology: Understanding Life in the Universe, First Edition. Charles S. Cockell.
© 2015 John Wiley & Sons, Ltd. Published 2015 by John Wiley & Sons, Ltd.
Companion Website : www.wiley.com/go/cockell/astrobiology.

Figure 18.1 *Saturn's rocky moon, Pandora, about 100 km in length, representative of many of the small rocky moons that orbit gas planets* (Source: NASA).

Let's visit some of the candidate bodies of interest to astrobiologists.

18.2   The Moons of Jupiter: Europa

The moons of Jupiter (the Jovian system) have become an intense focus for astrobiology, largely on account of the proposed ocean under the ice crust of Europa. Europa was discovered in 1601 by Galileo Galilei (1564–1642; Figure 18.2). It is the smallest of Jupiter's 'Galilean satellites' (Io, Europa, Ganymede and Callisto).

Most of what we know about Europa is from fly-by missions. In 1973 and 1974, Pioneer 10 and 11 took close-up images. In 1979, Voyager 1 and 2 took the first detailed images of the surface. The highly reflective icy surface prompted discussion on the possibility of a sub-surface liquid ocean. In 1995, the Galileo probe started to provide eight years of the closest and most detailed fly-bys to date, including studies of the magnetic field, the density of the moon and its surface properties.

Europa is slightly smaller than our own moon (3100 km in diameter compared to 3475 km for our Moon) and it orbits Jupiter in three and a half days. Europa is **tidally locked** to Jupiter so that one hemisphere is constantly facing the planet. The average density of the moon is estimated to be 2.94 g cm^{-3} suggesting an interior of the moon that is mostly rocky (silicates), possibly with a metallic core, but overlain by lighter material including a water ice crust and a sub-surface ocean.

Figure 18.2 *Galileo Galilei, discoverer of four moons orbiting Jupiter.*

The Jovian environment is an extreme one. There is a very large radiation field. Jupiter's magnetic field or **magnetosphere** extends up to seven million kilometres in the Sun's direction and almost to the orbit of Saturn in the opposite direction and it traps high-speed charged particles. The radiation doses at the surface of the Galilean moons are broadly correlated with the distance from Jupiter so that Io experiences the most extreme radiation, followed by Europa, Ganymede and finally, Callisto.

Jupiter's mass attracts many comets and other bodies on collision courses. This means that organic material is probably being delivered quite regularly to its moons in collisions of these objects.

An image of Europa taken from afar reveals one obvious feature (Figure 18.3). It is very flat with few impact craters. This suggests that the surface is relatively young geologically and is being renewed by geological activity – perhaps every 10 million years. It is also very reflective with an **albedo** of 0.64, suggesting an icy surface. It is difficult to age-date the surface from cratering because the cratering record and rate, which is better known for our own moon,

 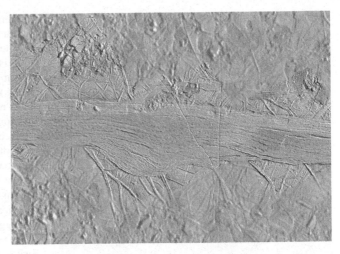

Figure 18.3 *Left: Europa, showing its characteristics lines across the surface. The moon is shown in approximate natural colours. The prominent crater in the lower right is Pwyll and the darker regions are areas where Europa's water ice surface has a higher mineral content. This image was taken on 7 September 1996 by the Galileo spacecraft. Right: Agenor Linea, a bright line feature across the surface of Europa. The image covers an area approximately 130 × 95 km (Source: NASA).*

is not the same in the outer Solar System. When cratering rates are compared on different rocky moons of the outer Solar System it is clear that the cratering rate throughout Solar System history has varied from one location to another, particularly in the outer Solar System. Nevertheless, the lack of craters on Europa is sufficient evidence to suggest a young surface.

The surface of the icy moons is unsurprisingly cold. Europa has a mean surface temperature of 103 K, but in some regions it drops as low as 50 K. Ganymede and Callisto have mean surface temperatures of ~110 and ~120 K, respectively. The different values are caused by the different albedos. Europa is more reflective than Ganymede and Callisto.

Spectroscopic observations of Europa from Galileo show many of the features that one might expect from water ice bombarded with ions caught up in Jupiter's intense magnetic field or from UV irradiation. These include oxygen, ozone and hydrogen peroxide (H_2O_2), all products of the breakdown of water and the production of **oxygen free radicals**. In addition to these constituents, the surface also shows evidence for sulfate salts (including $MgSO_4 \cdot 7H_2O$, which is epsomite) and other sulfur compounds such as SO_2. Some of these salts have been proposed to be components of the Europan ocean that have been expelled onto the surface.

Europa has a set of well-defined features that characterise its surface geology. Particularly prominent are the *lineae* (lines) and the *lenticulae* (spots – named after the Latin for freckles). The vertical range of the moon's surface is limited to a few hundred metres, which is again consistent with a generally geologically young and subdued surface.

The *lineae* are intriguing features comprised of bands of different colours, some of which are dark and some very light and they have distinctive grooves running down their centres. Light bands often occur within dark bands. You can see an example of these bands in Figure 18.3. One hypothesis to explain the dark bands is that they are the fractures along which the ice contorts and buckles as the moon experienced tidal forces, with the ice flexing open and closed as the moon orbits Jupiter. The dark bands also become brighter with age, which may be caused by interactions with radiation as new material is exposed and processed, changing its composition. Some scientists have drawn parallels with plate tectonics on the Earth, particularly as in analogy to plate tectonics on Earth, one can often line up features on either side of these fractures that have obviously been shifted with respect to one another. It also appears that some fractures have moved parallel to one another in analogy to transform boundaries on the Earth (Chapter 13).

The *lineae* can be relative age dated by examining the way in which they cut across each other (Figures 18.3 and 18.4). These observations are based on the principles of relative age dating discussed in Chapter 13.

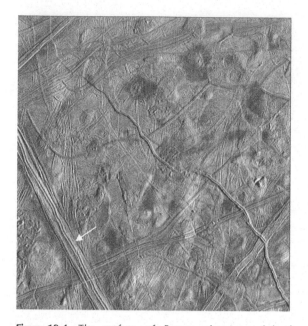

Figure 18.4 *The surface of Europa showing subdued regions and domes. This area covers about 140 × 130 km and is centered at 12.3°N, 268°W. The grooves are* lineae, *the dark splotches,* lenticulae. *The relatively recent nature of the surface is apparent. For example, the 8-km wide set of ridges running diagonally across the lower left corner (white arrow) is younger than most of the terrain seen in this picture. It runs over a narrow band that cuts across from the bottom left to the middle right. In turn, this band is also cut by a narrow 2-km wide double ridge running from the lower right to upper left corner of the picture. Also visible are numerous clusters of hills and low domes as large as 9 km across, many with associated dark patches of non-ice material. The ridges, hills and domes are considered to be ice-rich material derived from the sub-surface.* (Source: NASA/Arizona State University).

Some lineaments have cycloidal (arc-like) geometries, which are consistent with tidal stressing in the ice layer of the moon.

It is thought that the darker patches, or *lenticulae* (Figure 18.4) correspond to regions of younger warm ice rising to the surface and they could represent evidence of local melting. An alternative model is that such features are formed from warm, but solid, ice moving near the surface.

Perhaps one of the most remarkable features on the Europan surface is chaos terrain. These are regions where ice is presumed to have broken up, potentially within or over an ocean. An example is Conamara Chaos (Figure 18.5). The movement of ice within the ice cover is suggested by the observation of fractures that appear to line up, yet are broken apart, in a manner that reminds one of the movements of icebergs, although one should be careful about drawing exact terrestrial analogies.

Rafts rise up to 100 m above the surrounding material. The analogy with icebergs should be viewed with caution since it assumes that the ice is merely floating in liquid, which may not be the case. They might be moving within an ice layer that is tidally heated and flexed and not necessarily melted to liquid water, although the terrain might have begun in a liquid state.

18.2.1   A Sub-Surface Ocean?

Very sensitive gravity and magnetic measurements of Europa allow us to infer the presence of a sub-surface ocean, consistent with the surface geological data presented earlier. The most compelling evidence is the measurement of a small magnetic moment around Europa (a perturbation of Jupiter's magnetic field), which must be induced by a conducting material within Europa interacting with Jupiter's magnetic field. The most plausible

---

Debate Point: Plate tectonics on Europa
Many of the features of Europa, particularly its surface fractures, exhibit similarities to plate tectonics on the Earth. However, the materials involved, ice in the case of Europa and rock in the case of the Earth, are very different. Furthermore, the processes underlying the geological activity are different – plate tectonics is driven by rock density and subduction on the Earth and would be driven by tidal heating on Europa. Despite these obvious differences, there are clear similarities. Review the discussion on plate tectonics in Chapter 13 and discuss what you know about Europa, drawing on other sources if necessary, to compare surface processes on Europa and Earth. What similarities on Europa are there with plate tectonics on the Earth and how are they different? How would you go about testing these ideas using observations and missions?

Kattenhorn, S.A., Prockter, L.M. (2014) Evidence for subduction in the ice shell of Europa. *Nature Geoscience*, doi 10.1038/ngeo2245.

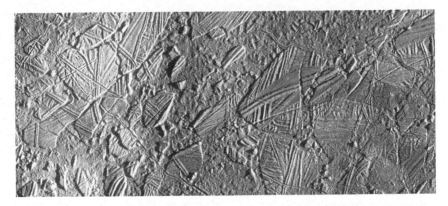

Figure 18.5 *Conamara Chaos, a region of broken chaotic terrain on Europa. The image, centred at 9° N, 274° W covers an area approximately 70 × 30 km (44 × 19 miles), and combines data taken by the Solid State Imaging (CCD) system on NASA's Galileo spacecraft during three of its orbits through the Jovian system. The white and blue colours are thought to be areas that have been blanketed by a fine dust of ice particles ejected at the time of formation of the large, 26 km (16 mile) diameter crater, Pwyll, about 1000 km (621 miles) to the south. The unblanketed surface has a reddish brown colour that may be caused by mineral contaminants carried and spread by water vapour released from below the crust when it was disrupted* (Source: NASA).

candidate for such a material is a salty ocean. The magnetic data cannot easily be explained by merely pockets of water. Geophysical data suggest an ice layer could be up to 30 km thick with the sub-surface ocean potentially 100 km deep. If this figure is accurate this would be equivalent to about twice the volume of all the Earth's oceans combined.

The thickness of the ice crust and its behaviour can be understood better by examining craters formed on Europa. Although there are not many of them, those that do exist have well-defined central peaks (rebound peaks) in their centres. Models suggest that the ice would have to be at least 5 km thick to be able to support these ice peaks.

Observations using the Hubble Space Telescope in 2014 suggested the presence of plumes of water emanating from the surface of the moon. The presence of oxygen was suggested by UV **Lyman-α** emissions at 130.44 nm, observed at least 200 km from the surface. One interpretation of these data was that electron bombardment of water being outgassed from a sub-surface ocean is breaking apart water into its constituent atoms and generating these emissions. The extent to which Europa is ejecting material into space will require further investigations.

An obvious question arises – how is the liquid water beneath the surface of Europa maintained (Figure 18.6). Why doesn't it freeze? The heat to keep the water liquid cannot come from radioactive elements within the rocky core similarly to the Earth's radioactive heating because there is not enough rock to produce the required heat.

Figure 18.6 *A subsurface ocean in Europa? A hypothetical schematic showing the ocean covered by ice. A more solid processed ice layer is shown on top in brown. Also shown are possible links that could exist between the ocean, the ice layers and the surface* (Source: NASA).

For Europa's ocean, the answer is that the heat is produced by tidal distortions from the massive gravitational pull of Jupiter. Europa experiences a small, but significant amount of tidal heating due to its non-circular orbit. Tidal forces on Europa's surface are estimated to distort the moon by ~50 m per orbital period.

The tidal forces generated by gravity experienced by the Jovian moons, like any planetary body, are inversely proportional to the square of the distance. The gravitational force between two objects is given by Newton's Universal Law of Gravitation:

$$F = GM_1M_2/r^2 \qquad (18.1)$$

where r is the distance between the objects, M_1 and M_2 are the masses and G is the Gravitational Constant.

Newton's Law of Gravitation explains simply why Ganymede and Callisto, which are further from Jupiter, feel significantly less tidal heating than Europa. Io, which is closer, experiences the highest tidal forces, accounting for its very active volcanism.

The tidal forces have the effect of causing the moons to become **tidally locked**. If they did not, the bulge caused by the tidal forces would have to rotate with the moon, generating huge temperatures as the rocks were contorted. The most energetically favourable solution to this is to settle into a tidally locked state.

There is another effect at work. Tidal forces tend not only to tidally lock planets, but also to circularise the orbit. For a non-circular orbit, Kepler's Laws (Chapter 8) tell us that the body will move at different speeds at different parts of the orbit (faster when it is closer to the object around which it orbits, slower when further away). This would cause the tidal bulge to move side to side as this orbital speed changed, again generating huge energies. The lowest energy state is for the orbit to circularise, allowing for complete tidal locking. However, the Jovian moons never quite achieve a circular orbit. The gravitational forces of all the Jovian moons influence each other, leading to orbital resonances (interactions between the orbital dynamics of the different moons). Ganymede orbits Jupiter for every two orbits of Europa and Europa orbits for every two orbits of Io. The effect is to cause the moons to develop a small eccentricity in their orbit. Europa tends to rock sideways, back and forwards, as its own rotation is never quite in alignment with its changing orbital speed. This small misalignment generates heat within the moon that drives geological processes.

At the cold surface temperature of the moon, ice mechanically behaves more like concrete than the ice that you and I are familiar with. It tends to fracture, accounting for the cracked surface of the moon.

For Europa the energy input supplied by tidal heating is estimated to be something on the order of 8.7×10^{12} W compared to Io which is 8900×10^{12} W, about three orders of magnitude more. A very crude estimate puts the effective average energy input per unit surface area on Europa to be ~ 0.26 W m^{-2}. For comparison, this is about two orders of magnitude lower than the net flux reaching the surface of the Earth, although local energy input around geological active regions could be much higher.

The possibility of sub-surface liquid water has made Europa one of the main targets in the search for extraterrestrial life. As we have so few constraints on the composition of the ocean or the nature of its interactions with a rocky core, it is difficult to make reliable speculations about whether it could support life. The ocean could contain $MgSO_4$ or $NaCl$ or mixtures of different salts.

If Europa is suitable for life one obvious question would be: what are the potential energy sources? You might like to consider the plausible redox couples based on what you learned in Chapter 5. Methanogenesis would be one possibility, but if the interior was very reducing, most of the carbon would be outgassed as CH_4, not CO_2, making the CO_2 available for methanogens very limited (although CH_4 is itself an electron donor and could be oxidised with O_2, produced on the surface of Europa by the irradiation of water). Either hydrogen produced from the core or organics delivered by meteoritic bombardment and mixed into the ocean from the surface or delivered directly into it by sufficiently energetic impacts might be potential electron donors for iron or sulfate reduction. Oxidised iron would come from the rocky core or from meteoritic bombardment. Sulfates have already been reported to be on the surface of the moon. The ice thickness almost certainly precludes oceanic photosynthesis.

Even if energy sources existed, we do not know what physical stresses life would be exposed to. The lowest reliably observed temperature for the replication of terrestrial life is about $-15\,°C$ (Chapter 7). The Europan ocean might be considerably colder. Without knowledge of the salt concentrations and compositions and their potential to depress the freezing point, it is difficult to accurately predict the ocean temperature. However, tidal heating within the rocky core might produce local fluids with temperatures within the bounds for life.

We can estimate the pressure in the oceans. The pressure under a column of liquid is given by ρgd. We could assume an ocean density (ρ) given by 1100 kg m^{-3} as a very rough estimate on Earth and on Europa (although we don't know the composition of Europa's oceans). The gravitational acceleration, g, is 9.8 m s^{-2} on Earth and 1.31 m s^{-2} on Europa. The depth of the bottom of the Europan ocean might be ~130 km (compared to the Earth's deepest ocean trench, the Mariana Trench at ~11 km, where life is known to grow). Then we can see that the pressure at the bottom of the Europan ocean is about 1.6 times that of the deepest ocean habitats on Earth. This is not considerably greater. Pressure in itself is unlikely to prevent life.

Radiation would be unlikely to be a problem for life. Although the surface of Europa suffers intense bombardment, predominantly by electrons, at an ice depth of just a meter, the doses would be expected to drop to values found in the near sub-surface of the Earth. The dose at the surface of Europa has been estimated at ~5.4 Sv day^{-1}, a dose sufficient to kill a human, which requires about 4.5 Sv.

Even with the most optimistic assessments of the Europan ocean environment, the lack of a photosynthetically productive biosphere would make its ocean orders of magnitude less productive than the Earth's biosphere. The most optimistic estimates give values on the order of 10^{11}–10^{15} g of steady-state biomass, about 1 cell cm^{-3} (compared to the Earth which is ~10^{18} g). You might like to consider the implications of this number for methods used in attempting to detect life in Europa.

18.3 The Moons of Jupiter: Ganymede and Callisto

Ganymede, another Galilean satellite of Jupiter, is the largest moon in the Solar System with a diameter of 5268 km. Like Europa, the measurement of an induced magnetic moment suggests a sub-surface ocean.

Unlike Europa, models of the density of the moon suggest an icy ocean may be sandwiched between the surface ice sheet and the ice in the deep interior. There may even be multiple water layers within the moon. The reasons for this sandwich structure can be understood by considering the phase diagram of water (Chapter 2 and Figure 18.7). As one descends into a planetary interior, the higher temperature will tend to melt ice. However, the extremely high pressures in a planetary interior will tend to form ice. Furthermore, increases in salt content will depress the freezing point. These interacting factors can allow for alternating water/ice layers.

The implications of this sandwich configuration could be important for the internal ocean as a habitable environment. In the case of Europa, circulation through a rocky

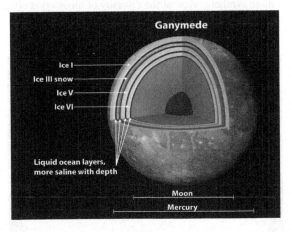
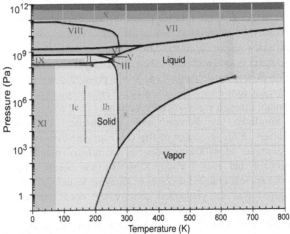

Figure 18.7 Left: A possible complex structure for Ganymede's internal structure, showing different high-pressure forms of ice (including ices I, V and VI) and liquid water sandwiched between them. The size of the Moon and Mercury is shown for comparison. Right: A phase diagram of water, showing how at sufficiently high pressures within a planetary body novel forms of solid ice (shown in blue and designated in Roman numerals) can exist. The phase diagram on the right shows the temperature and pressure conditions for a range of ice phases, including those found in Ganymede. Typical ice on the Earth is Ice Ih (Source: NASA).

Figure 18.8   *Callisto* (Source: NASA).

interior would imply that the ocean water would contain cations and anions entrained from the silicates, if such circulation occurs in the first place. One might hypothesise that the interaction of liquid water with a rocky interior would therefore improve the possibility of the presence of bioessential elements as well as elements and compounds for redox reactions. In the case of Ganymede, if the liquid water is sandwiched between two ice layers, then apart from exogenous delivery, there is little possibility for geochemical turn-over within the water or enhancement of ion content through water–rock interactions.

Similarly to Europa and Ganymede, Callisto (Figure 18.8), the outer of the Galilean satellites (with a diameter of 4800 km), also has an induced magnetic moment and may harbour a sub-surface ocean. The more ancient cratered surfaces of Ganymede and Callisto suggest that the geological activity and ocean–surface interaction is much less intensive than for Europa, which would be consistent with the smaller tidal forces that they have experienced as they are further from Jupiter. In both cases, liquid water may be more than 200 km below the surface.

The differences between the characteristics of Jupiter's moons have not gone unnoticed by ambitious individuals planning for humanity's long distance forays beyond the Earth. The relatively low surface radiation flux experienced on Callisto and its geologically stable surface have

made it a suggested location for a human outpost in the outer Solar System.

18.4   The Moons of Jupiter: Io

Io is a sulfur-rich silicate moon (Figure 18.9) slightly larger (3642 km) than our Moon. It has active volcanism, hosting over 100 volcanic mountains, some higher than Mount Everest. The volcanoes, which are erupting molten silicate material, also spew sulfur and SO_2 up to 500 km off the surface of the moon, aided by its 2.56 km s^{-1} escape velocity, just over four times less than the Earth. This material condenses on the surface of the Moon and is processed by radiation, resulting in its extraordinary colours. Nearer to Jupiter than Europa, the intense tidal forces it experiences, which may create a 100 m tidal bulge, account for its activity. It is unlikely as a target for looking at prebiotic chemistry or life because there is no evidence for liquid water. Nevertheless, as a target of geological exploration the moon, its sulfur chemistry and active volcanism make it an important place with which to understand the fate of primordial sulfur and sulfur chemistry more generally in planetary bodies. It also provides an excellent comparison to the icy moons of interest to astrobiologists in the sense that it completes a more synthetic view of the diversity of moons that orbit gas giants and the influence of different tidal regimes.

Figure 18.9   *Io* (Source: NASA).

It underscores the drastic consequences of large tidal forces for the stability and persistence of liquid water.

18.5   The Moons of Saturn: Enceladus

We now move outwards in the Solar System to Saturn. In orbit around this gas giant, like Jupiter, we find remarkable planetary bodies of interest to astrobiologists.

Saturn's moon, Enceladus, has attracted a great deal of attention. Discovered in 1789 by William Herschel, it has a water ice surface. It is close to Saturn and its rings, which makes it difficult to view from Earth. The moon is embedded in the densest part of Saturn's E-ring and the material ejected from the moon is thought to play an important role in feeding the E-ring of Saturn. Perhaps one of the most distinguishing features of Enceladus is that it is a small moon with a diameter of 504 km. As pointed out at the beginning of this chapter, this is amazing because no-one would have imagined that such a small object, so incapable of supporting liquid water on its surface, could have any relevance to astrobiologists.

In 1980, Voyager 1 and 2 provided the first images and information on the nature of Enceladus (Figure 18.10). Voyager 1 saw a highly reflective surface with few impact craters, not unlike Europa. These data, as for Europa, suggest a geologically young surface. Voyager 2 discovered a heavily cratered mid to high northern latitude and a lightly cratered region closer to the equator.

On the surface of Enceladus, different levels of cratering are found in different regions. All the craters that do exist show signs of deformation and degradation indicative, like Europa, of a geologically young and active surface. Some of the most striking features on the moon are the

Figure 18.10   *The reflective surface of Enceladus* (Source: NASA).

'tiger stripes' in the south polar region which are much younger than 500 000 years old (Figure 18.11). These are ridges over 100 km long and about 2 km wide.

18.5.1   The Plumes of Enceladus

During the Cassini mission, jets of water and gas were observed erupting from the south pole of the moon in

Focus: Astrobiologists. Hunter Waite.

What do you study? I study the composition of the satellites of Jupiter and Saturn with the intent of understanding Solar System formation and the emergence of life within it.

Where do you work? I work at the Space Science and Engineering Division of the Southwest Research Institute, a non-profit research organisation in San Antonio, Texas.

What was your first degree subject? Physics, Chemistry and Mathematics.

What science questions do you address? The questions I am most interested in are: (i) How did the Solar System form; (ii) How did life first begin on Earth; (iii) Is there life elsewhere in the Solar System; (iv) If so, is it similar to life on Earth; and (v) Did other stellar systems follow a course of evolution similar to our own?

What excites you about astrobiology? Astrobiology brings together many different disciplines: physics, chemistry, biology and geology to name a few. Putting all of this information together to explain new and novel planetary bodies is a special privilege I enjoy.

Figure 18.11   *The tiger stripes of Enceladus showing temperatures in their vicinity (in K) as measured by Cassini's composite infrared spectrometer. The data show evidence of heating in the vicinity of the tiger stripes* (Source: NASA).

the vicinity of the tiger stripes (Figure 18.12). Cassini was not designed as a life detection mission, but it did have spectrometers capable of examining gases and carbon compounds. A revision to the mission profile allowed for a fly-by through the plumes, which were found to contain water, methane, carbon dioxide, propane, ethane, acetylene and complex organics. The presence of NaCl at ∼0.5% and $NaHCO_3$ at ∼0.2–0.5% suggest that the plumes represent water from a mildly alkaline (pH 8.5–9.0) water body. The concentrations of carbon-containing components of Encedalus have some similarities to comets, suggesting a primordial origin.

The plumes raise obvious and important questions in astrobiology. The presence of water and organic compounds suggest several possibilities: (i) that primordial organics are being processed in the moon and ejected into space, (ii) that the interior of the moon is a giant organic chemistry reactor, which raises questions about whether the synthesis of prebiotic molecules is occurring within it, (iii) that it might be a habitat for life.

One obvious way to assess the astrobiological potential of the plumes of Enceladus would be to implement a sample return mission to collect material from the plumes and return it to Earth.

The origin of the plumes is uncertain, but they are thought to be material escaping through surface cracks from an internal salty ocean or lake. Figure 18.13 shows a schematic of a model for the tidal heating of ice and the formation of pressurised internal water bodies within the moon. Gravity measurements made by the Cassini spacecraft suggest that there could be a sub-surface ocean ∼15 km thick overlain on a rocky core and covered by ∼30 km of ice.

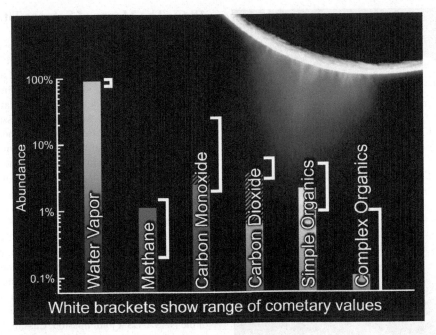

Figure 18.12 *Plumes of water erupted from Enceladus (inset), their composition and abundance. The white bars show the range of cometary values (Source: NASA).*

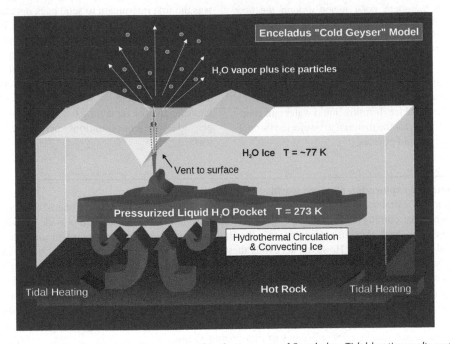

Figure 18.13 *One model for liquid water formation within the ice crust of Enceladus. Tidal heating melts water, generating internal pressurised water that is ejected from the moon in fractures (the cold geyser model; Source: NASA).*

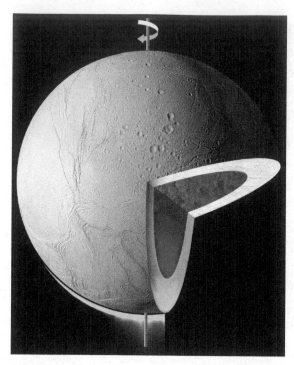

Figure 18.14 *Tidal heating is likely to be responsible for local heating in the south polar region around the tiger stripes and the higher temperatures measured in that region (Figure 18.11) depicted by the red and yellow (within the ice) in this internal schematic (image: NASA).*

There is evidence for heating in the south polar region of Enceladus (Figure 18.14), which may provide the energy needed for melting the ice, forming liquid water and creating the pressurisation for the material to be ejected through the tiger stripes in 'cold geysers'.

18.6 The Moons of Saturn: Titan

Saturn's largest moon is Titan, the second largest moon in the Solar System after Ganymede (with a diameter of 5152 km, 0.40 times that of the Earth). It is even larger than Mercury (with a diameter 0.38 times that of the Earth). Titan is the only moon in the Solar System with a substantial atmosphere, and a surface pressure much larger than Earth. On the surface it has an atmospheric density about four times that of the Earth (10^{20} molecules cm^{-3}). Like the Galilean satellites, and for the same reasons, tidal forces have made it tidally locked – it synchronously rotates around Saturn.

The moon was discovered by Dutch astronomer and physicist Christian Huygens (1629–1695), and in the 1940s astronomer Gerard Kuiper (1904–1975) used absorption spectroscopy (Chapter 2) to show that it had an atmosphere containing methane. Titan was first observed in detail in 1980 as a yellow-orange ball by Voyager 1 (Figure 18.15) and most of the details that we now have of Titan were revealed by NASA's Cassini spacecraft (part of the NASA-ESA joint mission, Cassini–Huygens) when it arrived in the Saturn system in 2004 after a seven-year voyage from the Earth.

The Radio Science System (RSS) on the Voyager craft was the first instrument to send back detailed information about the atmosphere. Unaffected by the haze layer as it operated in the radio region, it was able to show that the temperature at the surface was 94 K and that the atmospheric pressure at the surface was 1.6 bar.

In the stratosphere of Titan, the composition of Titan's atmosphere is roughly 98.4% nitrogen, 1.4% methane and the remaining 0.2% accounted for by hydrogen. Methane concentrations increase to 4.9% near the surface.

Debate Point: The exploration of ocean worlds

The exploration of extraterrestrial oceans beneath the ice covers of planetary bodies is one of the most impressive future frontiers in astrobiology. Even if these oceans turn out to be uninhabited, they will provide us with information about the geochemistry of aqueous environments in the Universe and what might be lacking in such environments for the origin and emergence of life. However, accessing these oceans will be immensely difficult. Assuming that the ice thickness on some of these bodies could be between approximately 10 and 100 km, discuss what sort of methods and technologies might be used to access these environments. There is evidence that in some locations, such as Europa and Enceladus, water is escaping to the surface. Although the task of collecting such samples is less ambitious than direct entry into the oceans, discuss what data might be gathered from plumes and surface features compared to directly acquiring a sample from the ocean. List and discuss the processes that might have changed surface samples and would compromise the search for organics or life in surface or plume samples.

McKay, C.P., Porco, C.C., Altheide, T., Davis, W.L., Kral, T.A. (2008) The possible origin and persistence of life on Enceladus and detection of biomarkers in the plume. *Astrobiology* **8**, 909–919.

Focus: Astrobiologists: Ralph Lorenz

What do you study? I have very wide interests, but the interaction between planetary atmospheres and surfaces is an overarching theme, from seas of hydrocarbons on Titan and dust devils on Mars to sand dunes on Venus. My work includes studying analogues in the field on Earth, modelling and data analysis, but especially the design and operation of instruments and spacecraft for planetary exploration.

Where do you work? At the Johns Hopkins University Applied Physics Laboratory, in Maryland (near Washington, D.C., USA). It is a laboratory that does work for NASA, the United States Navy and other government agencies and has built many spacecraft and instruments.

What was your first degree subject? Aerospace Systems Engineering, at Southampton University, in the United Kingdom. I then worked for a year for the European Space Agency before doing a PhD in Physics and moving more towards science.

What science questions do you address? Whatever seems most interesting at the time! The nature of planetary exploration by spacecraft is that every few years an entirely new window opens up – whether it an instrument to make the first *in situ* measurements of surface chemistry at Titan from the Huygens probe (which I played a part in constructing), to new datalogging instruments that let us study dust devils in Earth's deserts, to global radar maps of Venus. These new capabilities and datasets often stimulate novel analysis approaches, or laboratory investigations or a re-examination of old datasets.

What excites you about astrobiology? Astrobiology touches on a huge range of scientific disciplines, all of them interesting.

A lot of my work revolves around formulating new missions – trying to work out exactly what we should measure and how. While often centered around geophysics questions, sometimes this touches on astrobiology directly (such as what processes of chemical synthesis might be occurring in Titan's hydrocarbon seas and how we would detect evidence of prebiotic chemical processes), sometimes indirectly via issues like planetary protection – assessing whether a Mars lander needs to be sterilised.

Figure 18.15 *Titan in the visible wavelength range imaged by the Cassini spacecraft showing the homogenous haze that enshrouds the moon. (Source: NASA).*

The moon is conspicuous from the Earth in having a dense orange-coloured hydrocarbon smog at high altitudes, containing a rich variety of organic molecules. The hydrocarbon smog is generated by ionisation of molecules high in the atmosphere, primarily by UV radiation from the Sun. Ionised molecules take part in reactions with other ionised and neutral molecules, producing carbon compounds including ethane and propane which eventually rain out of the atmosphere and accumulate on the surface. Some of these hydrocarbons are thought to be complex molecules of a variety of structures, generically referred to as **tholins**, which account for the orange-brown colour of the haze. The Cassini spacecraft used an infrared spectrometer to follow up on pioneering infrared measurements on the Voyager spacecraft, which first detected organics. As well as confirming the earlier findings, it also detected benzene (C_6H_6), hydrogen cyanide (HCN), ethyne (acetylene) (C_2H_2) and other molecules.

The atmospheric chemistry of Titan is driven by a series of **radicals** produced by UV irradiation of the upper atmosphere and these all principally start with methane. Methane, CH_4, can be **photolysed** by UV irradiation to produce CH_2 and H_2, which themselves can react together to produce the methyl radical, CH_3. Two methyl radicals can react to produce ethane (C_2H_6) which rains out of the atmosphere into Titan's lakes.

Reactions between methane and the CH radical can produce ethene (C_2H_4) and ethyne (C_2H_2). Ethyne can react with sunlight in the UV range to produce radicals that recombine to form long-chain hydrocarbons responsible for the brown-orange haze. More complex chemistry occurs when nitrogen gas, which makes up most of the atmosphere of Titan, photolyses and produces N radicals that can react with some of the radicals we have discussed to produce nitrile compounds (which contain CN bonds).

When the Cassini spacecraft arrived in the Saturnian system it carried with it a small probe, Huygens, which made a successful descent through the cloud layer to the surface in 2005. As the Huygens probe descended through the atmosphere, not only did it send back valuable information about the atmosphere of Titan, it also studied the hydrocarbons with its Aerosol Capture and Pyrolyser Instrument (ACP) that fed the products into a Gas Chromatograph/Mass Spectrometer (GCMS). These instruments confirmed that the atmosphere is filled with particles made up of a core of complex hydrocarbons surrounded by a layer of ices and volatiles.

The production of these various compounds is calculated to be enormous. Predictions suggest that over the lifetime of the Solar System, if one was to assume that the processes occurring today have occurred throughout the moon's history, there would be an equivalent layer of ethane 600 m thick and of hydrogen cyanide 20 m thick! However, we don't actually know how constant the processes observed today have been during the history of the Solar System.

Imaging of Titan by the Cassini spacecraft at 928 nm in the infra-red also allowed scientists to peer through the haze which obscures the surface in the visible regions and for the first time to visualise the surface features.

The images showed a rugged terrain similar to that of the Earth, containing dunes, craters, rivers and lakes (Figure 18.16). The lakes are thought to be a mixture of methane and ethane and are mainly localised to the

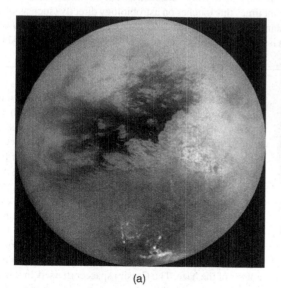

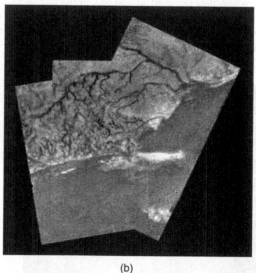

(a)   (b)

Figure 18.16   (a) Surface features of Titan. On the left a mosaic of nine processed images acquired during Cassini's first very close fly-by on 26 October 2004. The view is centred on 15°S, 156°W. The images that comprise the mosaic have been processed to reduce the effects of the atmosphere and to sharpen surface features. Surface features are best seen near the center of the disc, where the spacecraft is looking directly downwards; the contrast becomes progressively lower and surface features become fuzzier towards the outside, where the spacecraft is peering through haze (Source: NASA). (b) An image taken by the Huygens lander as it descended through the atmosphere showing methane rivers and a shoreline (Source: ESA/NASA/JPL/ University of Arizona).

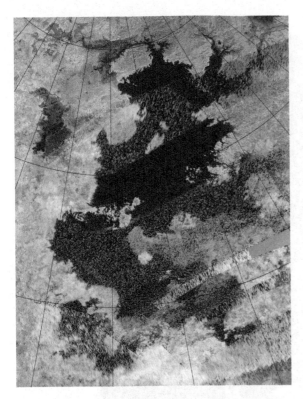

Figure 18.17 *Kraken Mare, a large northern lake on Titan shown in blue (false colour; Source: NASA).*

Figure 18.18 *Titan's surface, as imaged by the Huygens probe. The large boulders in the foreground are water ice (Source: ESA/NASA).*

northern hemisphere. There are many small lakes, but immense ones as well. Kraken Mare (Figure 18.17) covers an area of about 400 000 km², about the same area as California.

Unlike the Earth, methane occupies the role played by water and it is responsible for carving valleys and rivers into the surface as well as replenishing the atmospheric methane lost through reactions driven by **photodissociation**, methane which would otherwise be depleted in just a few tens of millions of years. Water ice, by contrast, behaves like silicate rocks do on the Earth and forms hard boulders, some of which you can see in the surface image of Titan obtained by Huygens (Figure 18.18). Their rounded appearance could have been caused by the sculpting of water ice by methane. This strange methane-driven geology is made possible by the cold surface temperatures.

Cassini fly-bys have also shown that Titan is very much an active world over long time periods. The level of Titan's largest lake in the southern hemisphere, Ontario Lacus (Figure 18.19), was observed to have fallen by a

meter every year over four years, because of enhanced evaporation caused by eccentricity in Saturn's orbit, which brings the moon closer to the Sun during the southern summer. Remarkably, this is evidence of orbitally-forced climate change on the moon, which dramatically influences surface geological (and presumably geochemical) processes.

The surface of the moon is home to vast dunes that slowly ripple across the surface in the wind. These dunes are not thought to be made of sand (quartz) as on our own world. Instead, the dunes are made of hydrocarbons, thick deposits of tholins that have been produced on the surface of the moon or have dropped out of the atmosphere from the organic processing in the high atmosphere. They cover about 30% of the equatorial region and are up to 100 m high.

The dynamic processes occurring on Titan and the general scheme for the processing of methane and hydrocarbons is summarised in Figure 18.20.

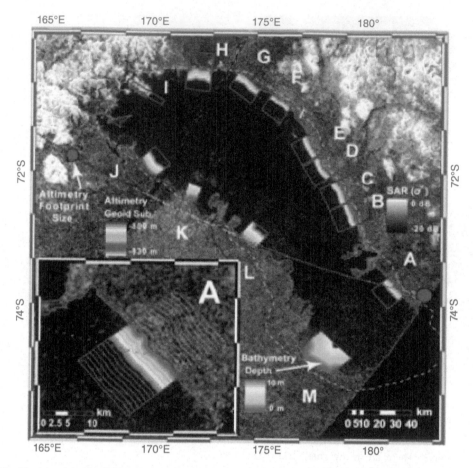

Figure 18.19 *Changing lake levels (imaged in red blocks and shown as coloured stripes) on Ontario Lacus, Titan, bear witness to an active liquid cycle on the surface of Titan and the presence of climate changes* (Source: Radar Science Team, NASA/JPL/Caltech).

The strange similarities of Titan and Earth, at least in terms of liquid cycles that produce rivers and lakes that are subjected to climate change are also reflected in the atmospheric composition. Titan's atmosphere, although about 200 K colder than the Earth's, contains equivalent layers to the Earth's atmosphere: the troposphere, stratosphere, mesosphere and thermosphere. As a result of Titan's weaker gravity, these layers are stretched vertically (Figure 18.21). While the temperature of the Earth's stratosphere increases with height because of UV absorption in the ozone layer, the increase in the temperature of Titan's upper atmosphere is due to absorption of sunlight by the hydrocarbon haze.

Titan may have other astrobiological surprises beneath its surface. Variations in the rate at which Titan spins

about its axis of rotation have shown that surface features on Titan can become displaced by ~30 km as the moon orbits Saturn, out of sync with what would be expected from tidally locked rotation around the planet. Cassini's measurement of a small but significant asynchronicity in Titan's rotation is explained by the separation of the crust from the deeper interior by a liquid layer. You can think of the surface as bobbling around irregularly, decoupled from the tidally-checked interior. These observations indicate that an ocean of liquid might lie below a thick crust of water ice.

Models of Titan's interior (Figure 18.22), including thermal evolution, predict that the satellite may have an ice crust between 50- and 150-km thick containing **clathrates**, lying over a liquid water ocean a couple of

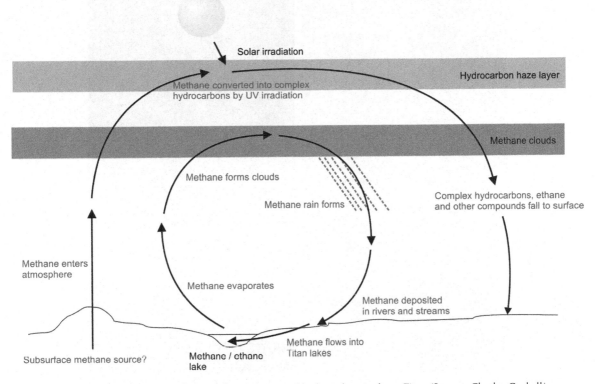

Solar irradiation

Hydrocarbon haze layer

Methane converted into complex
hydrocarbons by UV irradiation

Methane clouds

Methane forms clouds

Methane rain forms

Complex hydrocarbons, ethane
and other compounds fall to surface

Methane enters
atmosphere

Methane evaporates

Methane deposited
in rivers and streams

Subsurface methane source?

Methane / ethane
lake

Methane flows into
Titan lakes

Figure 18.20   *A schematic showing the methane and hydrocarbon cycle on Titan (Source: Charles Cockell).*

hundred kilometres deep, with up to 30% ammonia dissolved in it, acting as an antifreeze. The liquid ammonia, mixed into the composition of Titan when it first formed from the ices and volatiles of the early protoplanetary disc, would also be the source of nitrogen in the atmosphere of the moon. Ammonia is thought to be broken down in the atmosphere by UV radiation into NH_2 radicals, which combine to form hydrazine (N_2H_4), itself broken down by UV radiation to form nitrogen gas and hydrogen.

Beneath the ocean layer may be a layer of high-pressure ice. The interior of Titan may be the source of methane found in its surface features and atmosphere, perhaps exhaled in **cryovolcanoes**, cold volcanic structures that erupt volatiles, although these features have yet to be definitively identified on the surface. The moon probably has a silicate core.

Titan raises a list of astrobiologically important questions such as:

Could it be host to a type of life which uses liquid methane instead of water? There has been some speculation that life could exist on Titan using liquid methane

or ethane and coupling compounds such as ethylene to hydrogen to generate energy. These ideas are difficult to empirically test and there is no evidence for these organisms on the Earth, although of course these low temperature biochemistries would not necessarily be expected on the Earth.

Could this organic surface material feed a biosphere in a deep sub-surface ocean? The possibility of liquid water beneath the surface raises the possibility of deep sub-surface habitable conditions in Titan. To be able to advance this investigation, however, we need to know what this sub-surface water contains and whether it meets the requirements for life. If it contains ammonia, as is supposed, then it would have extreme alkaline pH conditions. We do not know the concentrations of different elements within the ocean or whether it contains biologically available energy supplies.

Could the complex organic surface give us clues about prebiotic chemistry? Even if Titan is too cold for life, its rich hydrocarbon chemistry might well provide important clues to the formation of prebiotic molecules.

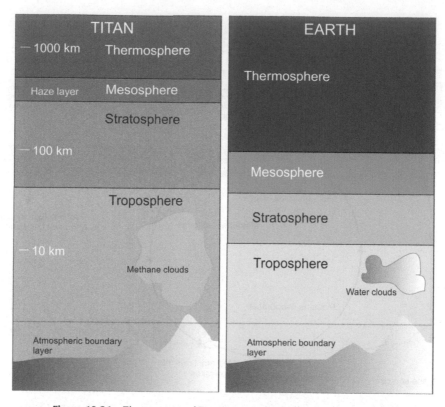

Figure 18.21   *The structure of Titan's atmosphere compared to the Earth.*

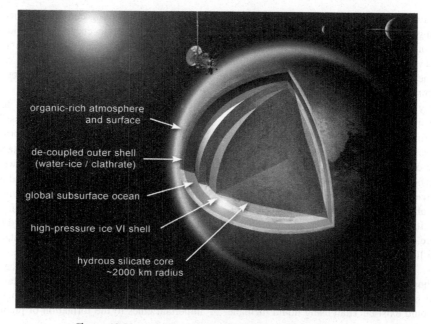

Figure 18.22   *A proposed structure of Titan (Source: NASA).*

We can think of the moon as a giant organic chemistry laboratory for understanding processes such as the UV radiation-induced production of organic materials. They might tell us something about how UV radiation would have interacted with organic compounds on the early Earth before the formation of the ozone shield. In particular, if there is a water layer beneath Titan's surface, there exists the possibility, through processes such as impact events, of short-lived interactions between water and complex organic compounds, leading to the formation of prebiotic compounds such as amino acids through hydrolysis reactions (Chapter 11). A thermal pulse, delivered by an impacting body, would contribute to higher localised temperatures and accelerated reaction rates.

18.7 Other Icy Worlds

Amazingly, the interest of astrobiologists does not stop at Jupiter and Saturn. Even further out in the Solar System other planetary bodies may hold interest for our investigations of habitability.

Some of these distant planetary bodies may contain liquid water. Even if they do not contain life, they offer us yet more geochemical environments in which to study aqueous chemistry and gain insights into the processing of carbon or prebiotic molecules. Some moons, such as Uranus's icy moons, Ariel and Miranda, could be astrobiological targets, although their characteristics are not well understood. Even Pluto (about the same size as Europa) and its moon, Charon, which synchronously rotate about their common centre of mass (facing towards each other), may host geochemically active sub-surfaces in which we could gain significant insights into the interactions of water and planetary bodies. All of these possibilities offer us the chance to increase the statistical robustness of the sample size used to make conclusions about the distribution of water, carbon chemistry and habitability in the Universe. Let's complete this chapter by looking at just two of these more unusual targets.

18.7.1 Triton

Neptune's largest moon Triton (with a diameter of 2700 km) has a retrograde orbit, indicating that it was captured by Neptune about three or four billion years ago and was not formed with Neptune out of the protoplanetary disc. It is likely to be a captured Kuiper Belt object.

Neptune's tidal forces are causing Triton to spiral inwards (this is different to our Moon, which is moving outwards), so that in about 3.6 billion years, the entire

Figure 18.23 *Triton's surface imaged by Voyager 2. The black ovals on the surface are the sites of cold geyser eruptions with material brought up from the sub-surface (Source: NASA).*

moon will be demolished by tidal forces as the moon passes within the **Roche limit**, creating a new ring system for Neptune.

Triton has a surface containing frozen nitrogen (Figure 18.23). The surface temperature, measured by Voyager 2, was 36 K. The moon possesses frozen methane at its poles. The crust is mostly water ice. Models of its interior suggest that it is differentiated. Like Europa, radioactive heating is likely to be insufficient to cause the formation of a deep sub-surface ocean over geological time periods, but tidal heating, if it occurs, could allow for the formation of deep sub-surface liquid water. The moon is thought to have an icy mantle and a core of rock and metal. The core makes up about two-thirds of its mass.

The moon is known to be geologically active. As a consequence, its surface is relatively young, with a complex geological history revealed in intricate cryovolcanic terrains. Chosen as a *Scientific American* 'Wonder of the Solar System', Triton has many cryogeysers, probably erupting nitrogen and possibly dark organic compounds seen on its surface. The geysers contribute to its tenuous nitrogen atmosphere.

18.7.2 Ceres

Another intriguing prospect for water is the dwarf planet, Ceres. It is the largest asteroid in the Solar System, 950 km in diameter. It contains about one-third of the mass of the asteroid belt. It was discovered in 1801 by Giuseppe Piazzi.

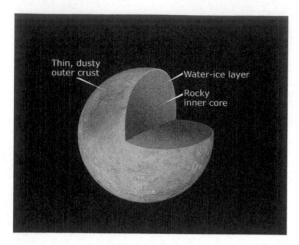

Figure 18.24 *A putative internal structure of Ceres (Source: NASA).*

The surface of Ceres is probably a mixture of water ice and various hydrated minerals such as carbonates and clays. It has a density of $2.09\,g\,cm^{-3}$. Temperature estimates of the surface suggest that the maximum value is $\sim$235 K. It appears to be differentiated into a rocky core and icy mantle, and may have an ocean of liquid water or water ice under its surface (Figure 18.24). Ceres was the focus of the Dawn mission.

18.7.3   Pluto

Geologically active environments even exist in the far reaches of our Solar System. The craterless plains (Figure 18.25) of Pluto suggest recent geological activity involving nitrogen and methane ices. Although these ices are likely too cold to be of astrobiological interest, they yield insights into the different conditions that allow for geological activity in the Solar System and expand our understanding of what types of geological activity may or may not be associated with habitable conditions.

18.8   Planetary Protection

The possibility of life or habitable environments on other planetary bodies, including Mars and the icy moons, motivates concerns about contaminating those worlds with microorganisms on spacecraft sent from Earth (forward contamination), or even contaminating the Earth with maleficent microbes inadvertently returned from alien biospheres (back contamination). These concerns broadly fit under the title of **planetary protection**. Planetary protection is taken seriously and the Committee for Space Research (COSPAR) has categories of missions that correspond to the likelihood of contamination occurring. These protocols were developed in response to the United Nations Outer Space Treaty, which requires nations that

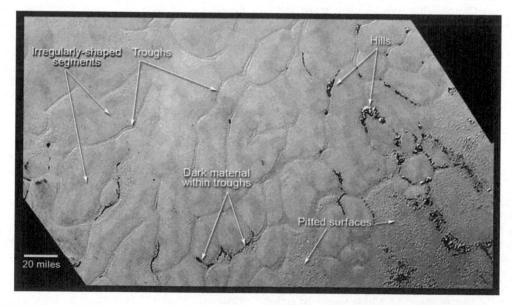

Figure 18.25   *An annotated image of the plains of Pluto showing what appear to be recent geological features (Source: NASA).*

Figure 18.26 *A Viking lander being heat sterilised prior to being dispatched to Mars in the 1970s* (Source: NASA).

carry out exploration of other planetary bodies to 'avoid their harmful contamination'. This statement is a reference to the destruction of scientific information, not an intrinsic concern for the planet itself or hypothetical alien biospheres, although ethical concerns are now playing a role in planetary protection. Planetary protection regulations are not international law, but instead they are regulations suggested by COSPAR that nations abide by.

Planetary protection is not a new endeavour. The Viking landers sent to Mars (Chapter 17) were heated to 111 °C for 40 h in a giant oven to minimise the chances that they could contaminate Mars (Figure 18.26). However, the COSPAR categories have formalised these procedures.

The categories of planetary protection are shown below and you can see that they generally start with the places least likely to have interesting organic chemistry or life (Type I) and move upwards from there to planetary bodies of greater concern.

Category I: A mission to a location not of direct interest for chemical evolution or the origin of life. An example would be the Sun or Mercury.

Category II: A mission to a location of significant interest for chemical evolution and the origin of life, but only a remote chance that spacecraft-borne contamination could compromise the investigations. Examples would be the Moon, Venus, and comets, where there could be interesting investigations to be carried out on organics, but there is no strong evidence that they support life.

Category III: Fly-by and orbiter missions to a location of significant interest for chemical evolution and the origin of life, and with a currently assessed significant chance that contamination could compromise investigations. Examples would include Mars, Europa and Enceladus.

Category IV: Lander or probe missions to the same locations as Category III.

Missions to Mars in category IV are sub-classified further:
Category IVa. Landers that do not search for Martian life. These landers should have the same bioburden as the Viking lander at pre-sterilisation. Prior to being heat sterilised, the Viking Landers had a certain number of organisms on them (like any spacecraft). To achieve this number the lander should not have more than 300 000 spores per spacecraft and 300 spores m^{-2}.

Debate Point: Is planetary protection necessary?

Planetary protection was motivated by a desire to stop the contamination, and therefore destruction, of potential areas of scientific study on other planetary bodies. In recent years there has been interest in expanding this motivation into the ethical sphere, in particular motivated by a concern about interfering with, or causing damage to, an extraterrestrial biosphere, beyond merely the destruction of scientific value. However, some people have said that all of this concern is over exaggerated. They assert, for example, that the chances of contaminating an alien biosphere and causing any sort of destruction are so low that planetary protection requirements are a costly cottage industry slowing down space exploration by imposing difficult constraints on planetary missions. What do you think? Is planetary protection justified from a scientific position or just a position of prudence? How do we go about deciding what is enough cleanliness and what isn't given that a single spore could be enough to contaminate a habitable region? Finally, consider the difference between robotic and human missions. In the former case we can sterilise a robot, in the latter case we cannot sterilise a human, making contamination very likely. What regulations, if any, should accompany the human exploration of other planets?

Fairén, A.G., Schulze-Makuch, D. (2013) The overprotection of Mars. *Nature Geoscience* **6**, 510–511.

Kminek, G., Rummel, J.D., Cockell, C.S. *et al.* (2010) Report of the COSPAR Mars Special Regions colloquium. *Advances in Space Research* **46**, 811–829.

Category IVb. Landers that search for Martian life.

Category IVc. Any component that accesses a Mars Special Region, which is a region where life could propagate, must be sterilised to at least to the Viking post-sterilisation biological burden levels (30 spores in total per spacecraft).

Category V: This category is for sample return missions. It is divided into 'unrestricted' and 'restricted' sample return missions.

Unrestricted Category V. Samples from locations judged by scientific opinion to have no indigenous lifeforms. No special requirements.

Restricted Category V. Samples for which scientific opinion about indigenous life is uncertain. The requirements include: prohibition of destructive impact upon return, containment of all returned hardware which directly contacted the target body, and containment of any unsterilised samples returned to the Earth.

18.9   Conclusions

Several decades ago the idea that the moons of the outer Solar System would be of interest to astrobiologists would have been considered as a fringe suggestion. Today, many of these moons offer important opportunities. The discovery of sub-surface oceans under some of them raises the question of whether they are habitable or even contain life. However, we should not get too focused on the search for life, since these environments will reveal much more general and important insights into the geochemistry of aqueous environments in the Universe and the organic chemistry that might occur there. Other moons, such as Titan and even the surface of Triton, have provided new insights into the processing of organic molecules under planetary conditions, for example with UV radiation, which may advance our understanding of prebiotic conditions on the early Earth. The exploration of these moons and other planetary bodies in the outer Solar System is set to provide us with much comparative data that will enhance our knowledge of the conditions in early solar systems, how organics and prebiotic molecules are produced and whether life is limited to terrestrial-type rocky planets.

Further Reading

Books

Greenberg, R. (2008) *Unmasking Europa: The Search for Life on Jupiter's Moon, Copernicus.* Springer, New York.

Lorenz, R., Mitton, J. (2008) *Titan Revealed.* Cambridge University Press, Cambridge.

Papers

Broadfoot, A.L., Atreya, S.K., Bertaux, J.L., *et al.* (1989) Ultraviolet spectrometer observations of Neptune and Triton. *Science* **246**, 1459–1466.

Collins, G.C., Head, J.W., Pappalardo, R.T., Spaun, N.A. (2000) Evaluation of models for the formation of chaotic terrain on Europa. *Journal of Geophysical Research* **105**, 1709–1716.

Greenberg, R., Geissler, P., Hoppa, G., Tufts, B.R., Durda, D.D., Pappalardo, R., Head, J.W., Greeley, R., Sullivan, R., Carr, M.H. (1998) Tectonic processes on Europa: Tidal stresses, mechanical response, and visible features. *Icarus* **135**, 64–78.

Hand, K.P., Carlson, R.W., Chyba, C.F. (2007) Energy, chemical disequilibrium, and geological constraints on Europa. *Astrobiology* **7**, 1–18.

Lorenz, R.D., Stiles, B.W., Kirk, R., Allison, M.D., del Marmo, P.P., Less, L., Lunine, J.I., Ostro, S.J., Hensley, S. (2008) Titan's rotation reveals an internal ocean and changing zonal winds. *Science* **319**, 1649–1651.

Lorenz, R.D., Mitchell, K.L., Kirk, R.L., *et al.* (2008) Titan's inventory of organic surface materials. *Geophysical Research Letters* **35**, L02206.

McKay, C.P., Porco, C.C., Altheide, T., Davis, W.L., Kral, T.A. (2008) The possible origin and persistence of life on Enceladus and detection of biomarkers in plumes. *Astrobiology* **8**, 909–919.

Paranicas, C., Mauk, B.H., Ratliff, J.M., Cohen, C., Johnson, R.E. (2002) The ion environment near Europa and its role in surface energetics. *Geophysical Research Letters* **29**, doi: 10.1029/2001GL014127.

Raulin, F., Owen, T. (2002) Organic chemistry and exobiology on Titan. *Space Science Reviews* **104**, 377–394.

Roberts, J.H., Nimmo, F. (2008) Tidal heating and the long-term stability of a subsurface ocean on Enceladus. *Icarus* **194**, 675–689.

Sagan, C., Khare, B.N. (1979) Tholins: organic chemistry of interstellar grains and gas. *Nature* **277**, 102–107.

Schmidt, B., Blankenship, D., Patterson, W., Schenk, P. (2011) Active formation of 'chaos terrain' over shallow subsurface water on Europa. *Nature* **479**, 502–505.

Waite, J.W., Lewis, W.S., Magee1, B.A., *et al.* (2009) Liquid water on Enceladus from observations of ammonia and ^{40}Ar in the plume. *Nature* **460**, 487–490.

19

Exoplanets: The Search for Other Habitable Worlds

Learning Outcomes

➤ Understand what exoplanets are.
➤ Understand the different methods used to detect exoplanets.
➤ Be able to apply the equations associated with the Transit and Radial Velocity methods of exoplanet detection to measure a planet's mass, radius and density.
➤ Describe some of the diversity of exoplanets that have been found.
➤ Understand which gases might be potential biosignatures of life and how they might be detected.

19.1 Exoplanets and Life

It would not be an exaggeration to say that one of the most exciting developments in astronomy of relevance to astrobiologists was the discovery of planets orbiting other stars – exoplanets. An extrasolar planet (or exoplanet) is a planet orbiting a star other than the Sun.

Exoplanets have great significance to astrobiologists for many reasons. Just a few of these reasons include:

• Understanding how typical the planetary 'architecture' of our Solar System is.
• Quantifying how common terrestrial-type rocky planets are.

• Cataloguing the diversity of rocky planet types.
• Investigating how many rocky planets orbit stars other than solar-type G stars (including M, K and F stars).
• Finding out whether there are habitable planets that orbit binary stars.
• Searching for evidence of life on distant planets.

This is an impressive and heady list! There can be little hesitation in saying that the last one in the list is the most interesting to astrobiologists. The search for a 'second Earth', as it has sometimes been colloquially described, is an exciting possibility, although one should be quick to point out that there is no reason why other inhabited planets, if they exist, need be a mere replica of the Earth.

There is a significant experimental reason for searching for an inhabited exoplanet. It would finally tell us something about the probability of life in the Universe and whether the origin of life is common. Let's assume a speculative scenario where we find evidence for a second independent origin of life in our Solar System. Although this would be remarkable, it would still be what scientists cause pseudo-replication. We could never know that there wasn't something unusual about our particular protoplanetary disc. Perhaps it contained some rare carbon compound, synthesised in the primordial gas, which is very infrequently distributed in the Universe and was dispersed amongst the planets, causing an origin of life on more than one. Regardless of what we think about the probability of such a scenario, the point of principle is that finding life within our own Solar System

Astrobiology: Understanding Life in the Universe, First Edition. Charles S. Cockell.
© 2015 John Wiley & Sons, Ltd. Published 2015 by John Wiley & Sons, Ltd.
Companion Website : www.wiley.com/go/cockell/astrobiology.

Figure 19.1 *An artist impression of 51 Pegasi b, the first exoplanet to be discovered around a main sequence star* (Source: wikicommons, debivort).

is sampling from the same protoplanetary experiment. Extrasolar planets offer us the possibility of overcoming this limitation and searching for life on planets formed out of independent protoplanetary discs, achieving robust experimental replication and statistical independence of sample alien biospheres, if they exist.

The first exoplanet discovered around a main sequence star was 51 Pegasi b (Figure 19.1). Exoplanet names begin with the star name followed by letters. Michel Mayor and Didier Queloz at the Geneva Observatory discovered the planet in 1995, orbiting the Sun-like 51 Pegasi in the constellation Pegasus, 51 light years away. This object was found to orbit the star in just 4.2 days, an astonishingly short time, and with a minimum mass half that of Jupiter. This remarkable new world defined a new class of objects, now known as **Hot Jupiters**, so named because of their massive size and close-in orbits.

19.2   Detecting Exoplanets

For stars that are close by, modern technology and telescopes could in principle be used to separate and detect a distant planet. The **spatial resolution** of a telescope (not taking into account the effects of the atmosphere) is given by the formula λ/D, where D is the aperture of the telescope (which is the diameter of its main mirror or lens) and λ is the wavelength of light being used. The spatial resolution is given in **radians** where a radian $= 180°/\pi = 57.3°$. Consider a 10 m diameter telescope. Its spatial resolution at 650 nm in the red region of the spectrum would be $6.5 \times 10^{-7} / 10 = 6.5 \times 10^{-8}$ radians. This is equivalent to

0.013″ (**arcseconds**). A gas giant-like planet, let's say orbiting at the distance of Saturn (1.43×10^9 km) around a star 10 light years away (9.4×10^{13} km away), would have an angular separation of $\sim$3.14 arcseconds (using simple trigonometry), higher than the resolution of the telescope, making it detectable. However, this only works for close stars within a few tens of light years. Most stars are much further away.

The problem is much more challenging than merely acquiring a high resolution. Another major problem is that the amount of light reflected by the planet is tiny compared to the light from the star. The light given off by an Earth-sized planet, for example, is about 10 million to 10 billion times less than the parent star, depending on wavelength. Although the direct imaging of exoplanets is possible, there are other ways in which one can indirectly detect these worlds. Let's survey these methods.

19.3   Transit Method for Detecting Exoplanets

When a planet passes in front of its star it will block some of the starlight, causing a dip in the brightness of the star behind it, which can be seen from the Earth. The dip in the brightness will be proportional to the radius of the planet. This approach to finding a planet is called the **transit method**. Figure 19.2 shows the principle of the transit method and a typical light curve for a planet.

The fraction of light blocked when a planet passes in front of a star will be a function of the ratio of the cross-sectional area of the planet to the star, given by:

$$\text{Fraction of light blocked} = \pi R_{\text{planet}}^2 / \pi R_{\text{star}}^2 = R_{\text{planet}}^2 / R_{\text{star}}^2$$
(19.1)

We can calculate the radius of the star, which is deduced from its luminosity (Chapter 8). The fraction of light blocked is calculated from the measured light curve (Figure 19.2), allowing us to use this equation to calculate the radius of the planet.

The method has a disadvantage. Planetary transits are only observed for planets that are almost exactly in the line of sight of the observer. The probability of a planetary orbital plane being directly in the line of sight to a star is the ratio of the diameter of the star to the diameter of the orbit. For a planet orbiting a Sun-like star at 1 AU, the probability of a transit is 0.47%. However, by scanning large areas of the galaxy containing hundreds of thousands of stars, transit surveys can find exoplanets at a high rate. The space-based CoRoT (COnvection ROtation et Transits

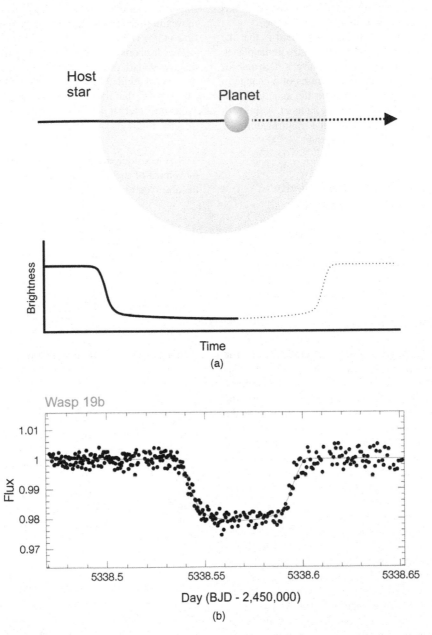

Figure 19.2 *(a) Schematic showing the principle of the transit method. As the planet passes in front of its host star, it blocks out some of the light, causing a dimming in the star light (Source: Charles Cockell). (b) A light curve of WASP-19b, an exoplanet possessing one of the shortest orbital periods of any known planetary body of approximately 18.932 h (the light curve is shown in days, specifically the* **Barycentric Julian Date***; Source: ESO). It has 1.15 Jupiter masses, but has a much larger radius (1.31 times that of Jupiter) making it nearly the size of a low-mass star. It orbits the star WASP-19a in the Vela constellation about 815 light years away.*

planétaires) and Kepler telescope missions used this approach as did the ground-based WASP search.

The main advantage of the transit method is that the light curve can be used to determine the size of the planet. However, the method also tends to have a high rate of false detections because any other causes of a change in brightness (for example **sunspots** on the star) could be falsely interpreted as being caused by a planet. A star with a transit detection requires additional confirmation, for example from the radial-velocity method.

19.4 Doppler Shift/Radial Velocity Method of Detecting Exoplanets

Strictly speaking, a planet does not orbit a star. The planet and star orbit their common centre of mass (the **barycentre**). As the planet generally has much less mass than the star, its orbit is more pronounced, but the star orbits the common centre of mass, which is often within the star itself.

The wobble induced in a star by planets can be detected indirectly and one of the most successful methods of doing this is by studying a star's **Doppler shift**. The wobble of a star causes a Doppler shift in the star light as it moves towards and away from us around the common centre of mass. The wavelength is shifted towards the blue as the star moves towards the observer, towards the red as it moves away. This can be measured as a shift in the spectral lines and from this, the radial velocity, or speed at which the star moves towards us and away from us, can be determined (Figure 19.3). The first exoplanet discovered around a main sequence star, 51 Pegasi b, was discovered using this method.

The radial velocity curves can be used to determine the eccentricity of the orbit of a planet. If the orbit is circular, the radial velocity curves will be perfect sinusoid. If the orbit is eccentric, the planet will travel fast when close to the star and slower when it is further from the star. The radial velocity curve of the star will then not be a perfect sinusoidal curve. It will change more quickly when the planet is close to the star and more slowly when the planet is further away, allowing for the determination of the eccentricity of the orbit by Kepler's laws.

The Doppler shift can be quantified. Consider a light source (a star) moving towards an observer at a velocity v (the radial velocity). As the source moves towards the observer so each time a wavelength of light is emitted, the star is moving a tiny bit closer to the observer, causing the light to become 'squashed' to shorter wavelengths

(blue wavelengths). The distance (d) that the source moves in time, t, is simply given by speed multiplied by time or $d = vt$.

We can choose the time, t, to be the time required to emit one wavelength of light, which is given by the wavelength of light (λ) divided by the speed of light, c, or $t = \lambda/c$. We can then say that the distance travelled by the star in this short time interval is:

$$d = v\lambda/c \qquad (19.2)$$

Now consider the wavelength change induced by the movement of the star. If λ_o is the wavelength as seen by the observer, then the shift in wavelength is given by:

$$\Delta\lambda = \lambda - \lambda_o \qquad (19.3)$$

As the wavelength from the source is shortened by the distance, d, this can be re-written as:

$$\Delta\lambda = \lambda - (\lambda - d) \qquad (19.4)$$

Therefore, substituting, d, as above:

$$\Delta\lambda = \lambda - (\lambda - v\lambda/c) \qquad (19.5)$$

and the Doppler equation is derived:

$$\Delta\lambda c \,/\, \lambda = v \qquad (19.6)$$

where v is the radial velocity of the star.

This equation applies to a moving source (a star) and a stationary observer and so is the appropriate equation for use in searching for exoplanets, where the observer is either fixed on the Earth or using a telescope above the Earth.

For the sake of completeness in understanding the Doppler Effect, we should realise that the equation is different for a stationary source and a moving observer and is given by:

$$\Delta\lambda = \lambda/(1 - \ c/v_o), \qquad (19.7)$$

where v_o is the velocity of the observer.

Furthermore, we can take this equation and the equation derived for the stationary observer (Equation (19.6)) and we can derive the general Doppler Shift equation, which is given by:

$$\Delta\lambda = \lambda_s \ (v_s - v_o)/(c - v_o), \qquad (19.8)$$

where λ_s is the wavelength of the source, v_s is the velocity of the source and v_o is the velocity of the observer. You can see that, if the observer is fixed (i.e. $v_o = 0$), then we recover Equation (19.6).

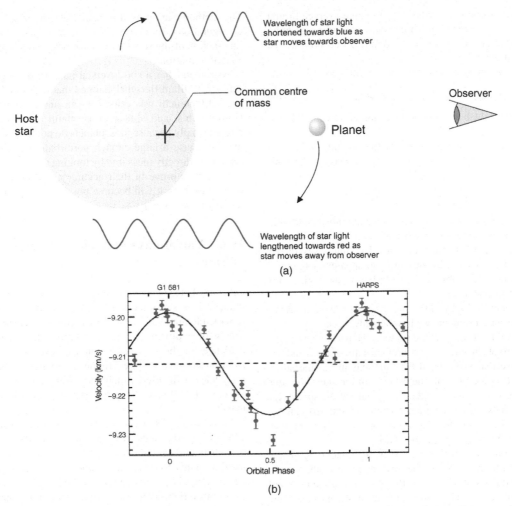

Figure 19.3 *The Doppler method of exoplanet detection. (a) The Doppler method relies on the change of the wavelength of the light to detect the influence of the planet on the movement of the star. The physical movement of the position of the star is also the basis of the astrometry method (Source: Charles Cockell). The lower image (b) shows radial velocity data from Gliese 581, a main sequence M star located about 22 light years away from Earth in the constellation Libra (Source: ESO). The data shows the presence of planet(s).*

The Doppler method lends itself particularly to low mass stars, since these will be influenced more by a gravitational effect of its planet(s). Low mass stars rotate slowly, which reduces the confusion in interpreting the signal caused by the rotation of the star itself. Furthermore, low mass stars have observable spectral lines that can be used to measure Doppler shift in contrast to high mass stars which have spectral lines that are sparser. The method is also good at detecting very massive planets orbiting close to their stars. Even with good targets though, the method still requires great sensitivity. Typical

Doppler shifts can be on the order of 10^{-8} of a wavelength, requiring very high quality and accurate equipment.

Once we have the radial velocity is it possible to calculate the mass of the planet.

The mass of the planet is obtained by using the equation for the conservation of linear momentum:

$$m_{star} v_{star} = m_{planet} v_{planet} \qquad (19.9)$$

We know that the velocity of the planet, v_{planet}, is given by calculating the distance divided by time, which is given by $2\pi a_{planet} / p_{planet}$, where $2\pi a_{planet}$ is the total distance of one orbit, where a_{planet} is the average orbital distance of the planet from the star (its **semimajor axis**) and the time taken for it to complete one orbit is the **sidereal period**, p_{planet}).

Thus the mass of a planet can be written as:

$$m_{planet} = m_{star} v_{star} p_{planet} / 2\pi a_{planet} \qquad (19.10)$$

On the right-hand side of the equation we can fill in the values:

We can calculate the mass of the star (m_{star}) from its spectral type or luminosity by using the mass–luminosity relationship of stars given by:

$$L_{star}/L_{Sun} = (M_{star}/M_{Sun})^a \qquad (19.11)$$

as shown in Chapter 8, where L is the luminosity and M is the mass, a is some value between 1 and 6, generally taken to be 3.5 for main sequence stars. The luminosity of the star can be recovered by measuring its apparent magnitude (its brightness as observed on the Earth) and taking into account its distance.

The radial velocity of the star, v_{star}, is obtained from the Doppler measurements we have already discussed.

The value of p_{planet} is the same as the period of the star as they rotate around their common centre of mass and we also have that value from the time axis in our measurements. It is the time of one rotation of the star about the common centre of mass (the wavelength of the sinusoid).

The distance of the planet from the star (a) can be derived by using the equation for Kepler's third law (Chapter 8): $p_{planet}^2 = 4\pi^2 a^3 / GM_{star}$.

One limitation of this method is that it can only give a planet's minimum mass. The reason is that as the planetary orbit is more inclined to the observer, so the Doppler shift will be less. The observed mass of the planet will be the true mass multiplied by $\sin i$, where i is the angle of inclination of its orbit from the line of sight. It is typically unknown.

You will also recall that if we have the volume of the planet, which we can derive from knowing the radius of the planet obtained from the Transit method, then we can combine the Transit and Doppler method to work out the density of the planet (mass/volume). The density tells us a great deal about the composition of the planet. The best characterised exoplanets so far have been ones on which both of these methods have been applied.

19.5   Astrometry

The wobble in the star as it rotates about the common centre of mass with planets(s) can in principle be measured by **astrometry**, whereby the star's position in the night sky is determined by direct measurement of the star's location.

The method can be used to detect planets in any orbital configuration (from an edge-on orbit, which produces a line motion of the star, to a face-on orbit, which produces a circular motion).

Astrometry has a controversial past. In the eighteenth century, William Herschel claimed that the wobble of the star 70 Ophiuchi was caused by an unseen companion. However, his, and almost every claim since, has been refuted simply because most planets do not have sufficient mass to cause a large enough perturbation in the star's orbit to be directly measured by looking at its position. As telescopes improve in their accuracy, particularly space telescopes, then it will become possible to detect planets reliably by astrometry.

19.6   Variations in Other Attributes of Stars

Other attributes of stars can vary on account of the wobble induced in a star's motion by planets. One particularly remarkable application is the detection of changes in the timing of a **pulsar** (a rapidly rotating neutron star) as evidence for the presence of planets around such stars. Neutron stars are not likely to make very good candidates for planets with life as their radiation environments are extreme, but this method has led to the discovery of planets.

In 1992, the discovery of planets orbiting pulsar PSR 1257+12 by Aleksander Wolszczan was the first discovery of planets outside our Solar System (as discussed above, the first planet to be found around a main sequence star was 51 Pegasi b). The approach has been applied to other stars with variability in their luminosity (something called variable star timing).

19.7   Orbital Brightness Changes

As with the waxing and waning of the Moon, planets reflect light. This changing contribution of light to the total light given off by a star system can be used to detect the presence of a planet. Changes can also be induced by distortions in the shape of the star. The method lends itself particularly well to planets that are close in to their stars since it is easier to detect the light changes over realistic time periods.

19.8   Gravitational Lensing

Einstein's Theory of General Relativity predicts that massive objects should be able to distort light, since massive

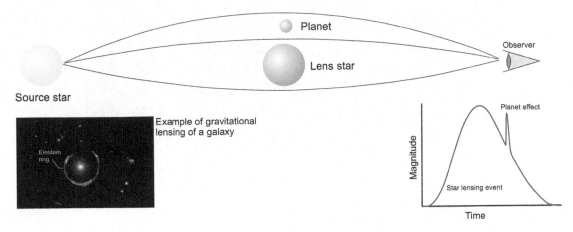

Figure 19.4 *Gravitational lensing showing the effect of a planet on the image obtained during the lensing event* (Source: Charles Cockell). *Inset left is an example of the principle of gravitational lensing showing how an Einstein ring is formed where the gravity of a luminous red galaxy has gravitationally distorted the light from a much more distant blue galaxy* (Source: Hubble Space Telescope).

objects curve spacetime. If two stars are aligned to the observer (Figure 19.4), then the more distant star light will be bent around the nearer star like a lens, creating a near-circular pattern observed by the astronomer and known as an **Einstein ring**. This effect is also seen with distant galaxies. A brightening of the light as it is focused into the ring (Figure 19.4) occurs, but only if it is perfectly aligned. The presence of a planet contributes to the effect and creates a further focussing (and brightening) in the light received (Figure 19.4).

These events are very brief (events last from tens of seconds to days), which means that lots of stars must be continuously monitored. One of its major problems is that it usually cannot be repeated because it depends on the alignment of the stars. An obvious advantage is that it can be used to find planets that are in orbits that are face on to the observer, unlike the transit method, for instance. It is most sensitive to planets orbiting about 1–10 AU from the star. Despite these complications, the method has been successful at detecting planets.

19.9   Direct Detection

Exoplanets can be imaged directly. The method is currently best applied to hot massive planets that are emitting lots of infra-red radiation, have a good separation from their parent star and are quite near to the Earth (Figure 19.5).

Direct detection is not very good at putting mass estimates on a planet. However, it has been successfully used to directly detect objects such as **Brown dwarfs** and

even to directly map variations in the atmospheric characteristics, for example variations caused by temperature differences in the atmosphere.

The method can be achieved using a **coronagraph** which blocks out the light being given off from the parent star (which mainly comes from the corona, the aura of plasma surrounding a star and the primary source of its light).

If direct detection is done from the ground, it is very sensitive to the problem of perturbations in atmospheric conditions. **Adaptive optics** can be added to ground-based telescopes that in real time correct for atmospheric distortions caused by perturbations in the atmosphere in the line of sight. This is achieved by firing lasers into the sky, whose perturbations are used to modify the optics within a telescope to divide out the atmospheric changes. By removing these subtle changes, stability in the image can be achieved that is sufficient to detect distant planets. Yet another approach is **nulling interferometry** (Figure 19.6). This method splits the light (or collects it from two telescopes) and generates a half-wavelength shift in the wavelength so that when combined, the signals cancel each other out, removing the star light and leaving the planet light to be examined.

19.10   Using Direct Detection to Study Protoplanetary Discs

Exoplanet search methods are not only used to find planets. They can also be used to study protoplanetary discs or dust discs in systems with formed planets around early stars. Dust discs can be very diffuse, but their large surface

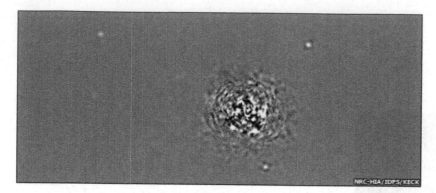

Figure 19.5 *Direct imaging of three planets. Three exoplanets orbiting a young star (HR 8799) 129 light years away are captured using Keck Observatory near infra-red adaptive optics (Source: NASA).*

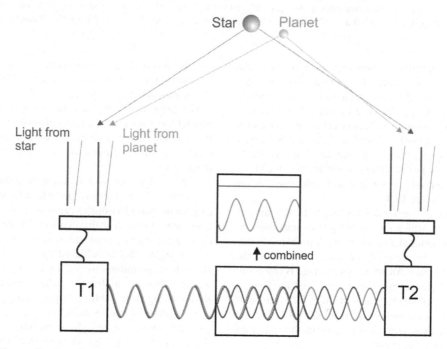

Figure 19.6 *How nulling interferometry works. Two telescopes are set up so that light from a distant star (red beam) hits each telescope, labelled T1 and T2, simultaneously. Before the light beams are combined, the beam from one telescope is delayed by half a wavelength using electronics or optics. When the starlight beams are brought together, peaks from one telescope line up with troughs from the other and so are cancelled out (represented by the straight red line), leaving no starlight. However, light from a nearby object such as a planet (green beam) comes in at two different angles. This introduces a delay in the planet light hitting one of the telescopes. So, even after the half-wavelength change in one of the beams, when they are combined they do not cancel out perfectly, but are reinforced (represented by the large green waves) (Source: Charles Cockell).*

area means that they emit sufficiently large quantities of infra-red radiation to be detected. These discs have been observed in 15% of stars near the Sun and give us important information about the conditions for planet formation and the material left over when this process occurs.

19.11 Exoplanet Properties

19.11.1 General Properties

The search for exoplanets is a rapidly moving area. It would not be worthwhile to try to present the discoveries as an up to date set of knowledge. By the time you buy and read this textbook there will have been many more planets discovered and their properties deduced. There may even be new families of planetary types discovered. You are encouraged to seek out the latest information. Instead of trying to summarise all our knowledge, here I will review some of the major types of planets that have been found and make some general observations about the diversity of objects detected.

If there are any general lessons to be learned that will probably stand the test of time we might include: (i) That our Solar System is not typical. Planets of almost every size, variety and orbit and eccentricity have been detected. We have much to learn about how the variety of solar systems are formed and (ii) Rocky planets are common and we will have many candidates with which to try to detect biosignatures.

Despite the rapidly changing field, it is worth showing a diagram of the types of planets that have been detected (shown for the year 2014). This is shown in Figure 19.7. It is worth noting that the fewer planets with a large semi-major axis is not merely a bias caused by the transit and radial velocity methods, but since exoplanets investigation began in the 1990s there has been little time to detect the large orbital periods of true Jupiter analogue planets orbiting other stars.

Some exoplanets have much higher eccentricities than would be expected based on what we know of planets in our Solar System. The high eccentricities may be caused by dynamical interactions between multiple planets in a star system when it first formed. If some planets are ejected from the system, the remaining planets are thought to be left with higher eccentricities than when they formed.

19.11.2 Hot Jupiters and Neptunes

Many Jupiter-mass exoplanets orbit closer to their parent stars than Mercury is to the Sun. These are known as

Hot Jupiters. In broad terms, a Hot Jupiter is an object with a mass roughly similar to that of Jupiter and an orbit inside 0.1 AU. Hot Jupiters were the first exoplanets to be discovered as they are close to their star and so are amenable to detection by the radial velocity method (they are also more easily detected by the transit method). They surprised a great many people because, until their discovery, it was taken as a paradigm that giant planets orbit in the outer reaches of a star system and rocky planets orbit closer in, much like our own solar system. Modelling suggests that Jupiter-like planets cannot form close to their parent stars as the high temperatures would prevent ices from forming on the dust grains, stripping them of much of their volatiles.

Jupiter-like planets are thought to form beyond the **snowline**. They can migrate in by three possible mechanisms: (i) Gas disc migration, involving tidal interactions between a Jupiter mass planet and the protoplanetary disc, (ii) Planetesimal disc migration, where the planet interacts with remnant planetesimals and (iii) Planet scattering, caused by interactions with other orbiting planets. In some cases they can migrate to the inner disc to become Hot Jupiters, but of course there must be a halting process or they would simply be devoured by the host star. This may well be the fate of some planets, but there are mechanisms by which halting could be achieved. For example, intense winds during the early **T-Tauri** phase of a star might clear a region around the star of dust and create an effective barrier to further inward migration.

There is a smaller class of hot gas planets, known as Hot Neptunes. A **Hot Neptune** is an exoplanet in an orbit close to its star with a mass similar to that of Uranus or Neptune (Figure 19.8). The first hot Neptune discovered was Mu Arae c (or HD 160691 c) at 50.6 light years away.

19.11.3 Super-Earths and Ocean Worlds

If a planet has a radius and/or mass between that of Earth and Neptune, then there is a question as to whether the planet is rocky like Earth or a mixture of volatiles and gas like Neptune. Above a certain radius, planets seem to be gaseous. A radius of about 1.5 times that of Earth is a possible dividing line between the two types of planet.

The discovery of the low-density Earth-mass planet KOI-314c, 200 light years away, shows that there is an overlapping range of masses in which both rocky planets and low-density planets occur. These planets could be **ocean planets** or **super-Earths** with a remnant hydrogen atmosphere, or hot planets with a steam atmosphere, or mini-Neptunes with a hydrogen–helium atmosphere. Other possibilities for low-mass low-density planets are

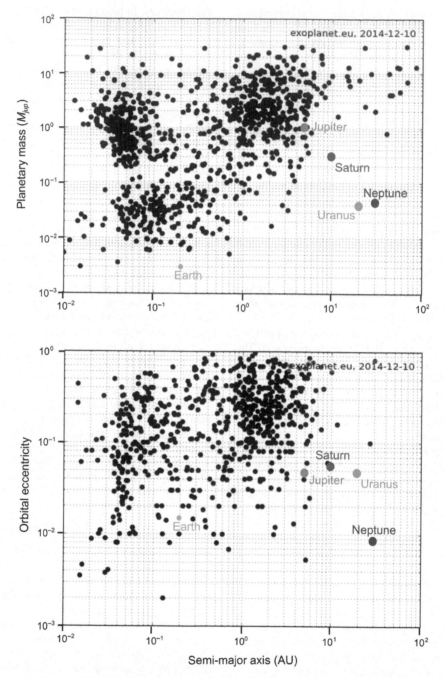

Figure 19.7 *Diagrams illustrating the diversity of exoplanet properties. Planetary semimajor axis against mass measured in multiples of the mass of Jupiter (top) and semimajor axis against eccentricity of known exoplanets (bottom) as of December 2014. These figures, although constantly being updated (and therefore outdated by the time you read this), illustrate the general observation that exoplanet masses and eccentricities have a wide variety of values, in many cases very different from planets in our own Solar System. These are general points that will not go out of date.*

Focus: Astrobiologists. Debra Fischer.

What do you study? My job is to detect planets orbiting other stars and to understand the nature of exoplanetary systems.

Where do you work? I am a professor at Yale University in New Haven, United States.

What was your first degree subject? My first degree was in Nursing. When I returned to school as a pre-med major, I discovered that I loved math and physics so my next degree was in Physics.

What science questions do you address? Which nearby stars have analogs of our Earth? What types of stars are the best hosts for habitable worlds? Do most stars have systems of many planets? What biosignatures could be detected in exoplanetary atmospheres? Over the past decade, my team has been working to improve the instruments that are used to detect exoplanets. Our Doppler measurements allow us to derive the mass of a planet. For planets that transit their host stars we can determine the size of a planet and derive the bulk composition.

What interests you about astrobiology? I think that the discovery of potential biosignatures on other worlds will happen in the next 20 years. In the same way that we have transitioned from wondering if planets were rare around other stars to knowing that almost every star forms with planets, humanity will learn whether life elsewhere is rare or whether many stars have habitable and inhabited worlds.

planets with large atmospheres of carbon monoxide, carbon dioxide, methane or nitrogen. It is difficult to explain how a low mass object could retain a hydrogen atmosphere.

An **ocean planet** (sometimes called a waterworld) is a type of planet which has a substantial fraction of its mass made of water (Figure 19.9). Two candidate ocean planets are the exoplanets GJ 1214 b (about 42 light years away), and Kepler-22b (620 light years away). The formation of these hypothetical worlds could occur by migration. Icy planets might move to orbits where their ice melts into liquid, turning them into ocean planets. The surface of such planets would be completely covered by an ocean of water hundreds of kilometres deep, much deeper than the oceans of Earth. If the atmosphere is thick, then pressures and temperatures might be sufficient to create a supercritical fluid of water (Chapter 2) or an atmosphere of steam if the greenhouse effect is sufficient. The pressures at the bottom of these oceans could lead to the formation of a mantle of high-pressure forms of ice.

Another type of rocky world is a **super-Earth**. Such a world has a mass between 1 and 10 Earth masses. The

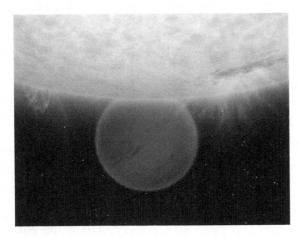

Figure 19.8 *An artist impression of a Hot Neptune (Source: NASA JPL).*

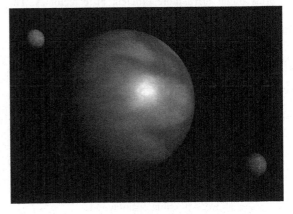

Figure 19.9 *An artist impression of an ocean world with two satellites (Source: wikicommons, Anynobody).*

first super-Earths were discovered around the pulsar PSR B1257+12, about 980 light years away, in 1992. The two outer planets of the system have masses approximately four times that of the Earth, which is too low for them to be gas giants.

The first super-Earth around a main sequence star was Gliese 876 d, discovered in 2005. It orbits Gliese 876, 15.3 light years away, and has an estimated mass of 7.5 Earth masses and an orbital period of about two days. Its host star is a red dwarf, giving it a possible surface temperature of ~450–650 K.

In 2007, two super-Earths were reported on the edge of the habitable zone for the star Gliese 581, at a distance of 20.3 light years. Gliese 581 c is at least five Earth masses and orbits at a small distance from Gliese 581 of 0.073 AU, putting it near the inner edge of the habitable zone. It may be subject to a runaway greenhouse effect like Venus. There was a reported planet, Gliese 581 d, apparently within the star's habitable zone, with an orbit at 0.22 AU and 7.7 Earth masses. However, subsequent analysis of the data suggests that it does not exist and that instead magnetic regions, similar to sunspots rotating with the star, gave spurious radial velocity measurements. Given the excitement that surrounded Gliese 581 d after its discovery, these observations highlight the need for caution in examining and re-affirming exoplanet data.

Super-Earths could potentially be habitable worlds with geochemical and hydrological cycles similar to the Earth. An unresolved matter is the extent to which these planets can sustain plate tectonics. Very massive super-Earths would have high pressures in their centres which would make molten rock more viscous, implying less convection and less efficient plate tectonics. By contrast, however, more water-rich lithospheres such as ocean worlds might drive more efficient plate tectonics. The regime for plate tectonics on super-Earths may involve a complex interplay of a variety of factors including mantle viscosity, the quantity of water and the thickness of the crust, analogously to the Earth (Chapter 16).

19.11.4 Rocky Planets in the Habitable Zone

Of obvious interest to astrobiologists are planets in the habitable zone. Already there are a variety of candidates that will almost certainly increase in number (Figure 19.10). Confirmed planet discoveries in the habitable zone include Kepler-22b, the first super-Earth to be found located in the habitable zone of a Sun-like star, and as we saw above, a possible ocean world.

A planet orbiting the red dwarf Gliese 163 (Gliese 163 c), at a distance of 49 light years from the Earth, is

Figure 19.10 *An artist impression of a rocky world orbiting another star* (Source: ESO).

about 6.9 times the mass of Earth and is considered to be within the habitable zone, although hotter than the Earth. The Kepler-62 (about 1200 light years away) and Kepler-69 (about 2700 light years away) systems host potentially habitable planets (Kepler-62 e, Kepler-62 f, and Kepler-69 c) which are super-Earths possibly covered by oceans.

The sheer number of these candidates has allowed for a statistical assessment of the likelihood that stars can host an Earth-size planet in the habitable zone. Based on Kepler data, it has been calculated that 5.7% of Sun-like stars in our Galaxy have an Earth-sized planet in the habitable zone with an orbital period of 200–400 days. Half of habitable zones around red dwarfs may have Earth-size planets. These numbers will be refined as the statistical occurrence of rocky planets is better understood.

Kepler-186f, about 490 light years away, was the first truly Earth-sized planet (Figure 19.11) in a habitable zone to have been discovered. It is a 1.1 Earth radius planet in the habitable zone of the red dwarf, Kepler-186. Kepler 452b was the first Earth-sized planet to be found orbiting a G star similar to the Sun. In the constellation Cygnus, the planet is 1400 light years away. It has a diameter 60% greater than Earth and is 6 billion years old.

19.11.5 Planets in Binary and Multiple Star Systems

Most known planets orbit single stars. Until recently it was assumed that binary star systems would not be very good places for planets to be. However, planets are now being discovered that orbit one member or both members of a

Figure 19.11 *Schematic showing the size of Kepler-186f, the first habitable zone exoplanet to be observed compared with the Earth (Source: NASA).*

binary star system. The Kepler results show binary star planetary systems are quite common. The study of these planets can put constraints on what might determine the orbital dynamics required for stable orbits to exist.

Potentially habitable planets have even been identified in triple star systems. The red dwarf, Gliese 667C, at a distance of 22.1 light years (Figure 19.12), contains at least two super-Earths in its habitable zone (Gliese 667 Cb, Gliese 667 Cc). The star is part of a remarkable triple star system in which it is orbited by a binary system containing two K stars (Gliese 667A and B).

Planets have even been found in quadruple star systems. Kepler-64, about 5000 light years from the Earth, is such a system and contains Kepler-64b, a Neptune-sized planet of about 20–55 Earth masses that orbits one of the binary pairs (Figure 19.13).

All of these results are significant because it is estimated that about 60% of stars are binaries and most of the remainder in multiple systems such as triple star systems. Stars, like our own Sun, may become solitary after being ejected from multiple star systems, making them the exception rather than the rule.

19.11.6 Strange Worlds

There are a variety of strange worlds out there that have no analogies to any planet in our Solar System. They illustrate the enormous diversity in paths taken in the formation of planets.

Some of the gas giants observed are very different from even the Hot Jupiters already known. Gas giants with a large radius and very low density are called **puffy planets**. Puffy planets usually orbit close to their stars, where the intense heat from the star and internal heating within the planet inflate the planet's atmosphere. Several large-radius low-density planets have been detected by the transit method including WASP-12b (871 light years away), WASP-17b (about 1000 light years away), and Kepler-7b (about 1000–1400 light years away). The mechanism of their formation is not completely certain, but one plausible pathway involves the stellar wind. Expansion would be caused by the interaction between the stellar wind and the planet's magnetosphere generating an electric current through the planet that heats it up, causing the atmosphere to expand. The more magnetically active a star is the greater the stellar wind and the larger the electric current, leading to more heating and expansion of the planetary atmosphere. TrES-4 is a puffy planet about 1.7 times the size of Jupiter located about 1400 light years away from Earth. It orbits its parent star in only three and a half days.

Remarkable in this class of planets is HAT-P-1. Located 453 light years away, it is 52% of the mass of Jupiter but 23% greater diameter and orbits its star in 4.47 days. The density of the object is lower than water. At the time of

Figure 19.12 *At the top, an artist's impression of Gliese 667 Cb with two stars of the three-star system in the background. Below is an artist impression of what it might be like on the surface of 667 Cc which orbits the red dwarf component of the system. In the sky you can see the triple sunset! (Source: ESO/L. Calçada).*

Figure 19.13 *Kepler 64b is a Neptune-size planet that orbits binary stars that themselves are part of a quadruple star system (Source: NASA/JPL-Caltech/T. Pyle, SSC).*

its discovery it was the lowest density exoplanet found (Figure 19.14).

Brown dwarfs constitute yet another type of object, a large number of which have been discovered. These are objects with a mass range between about 15 to 80 times that of Jupiter. Above about 80 Jupiter masses hydrogen fusion is initiated and objects become stars.

Unusual types of worlds are not confined to the gas giants. Rocky planetary bodies include classes of worlds very different to the ones we are familiar with in our Solar System. One of the strangest of these is carbon planets. A carbon planet is a type of planet that contains more carbon than oxygen. Carbon planets (Figure 19.15) would form if the protoplanetary disc is carbon-rich and oxygen-poor.

Such a planet would probably have an iron or steel-rich core like the terrestrial planets of our Solar System. Surrounding that would be molten silicon carbide and titanium carbide. Above that might be a layer of carbon in the form of graphite, possibly containing a thick layer of diamond if there is sufficient pressure. During volcanic eruptions, it is possible that diamonds from the interior could erupt to the surface, forming mountains of diamonds and silicon carbides. The surface might contain frozen or liquid hydrocarbons in analogy to Titan.

An iron planet is a type of planet that has an iron-rich core with little or no mantle. These planets are also known as cannonball planets. They would be the remnants of normal metal and silicate rocky planets like the Earth whose rocky mantles were stripped away by giant impacts. They would form in close-in orbits or around massive stars where the protoplanetary disc contains iron-rich material.

Finally, there is the possibility of habitable exomoons. Chapter 18 explored the possibility of habitable conditions in icy moons of the gas giant planets in our own Solar System. Exoplanets may harbour moons that are similar, or larger, than the icy moons of our own Solar System. They will be detected using the methods described in this chapter, but small moons will require particularly sensitive methods.

19.12   Detecting Life

19.12.1   Biosignature Gases

The long-term objective of exoplanet surveys is to measure the spectrum of light emitted from Earth-like exoplanets.

The spectroscopic detection of life in an exoplanet atmosphere depends on the sufficient build-up of gases associated with metabolic processes (Chapter 5) carried

Figure 19.14 *Schematic showing the size of the HAT-P-1 puffy planet compared to Jupiter (Source: NASA).*

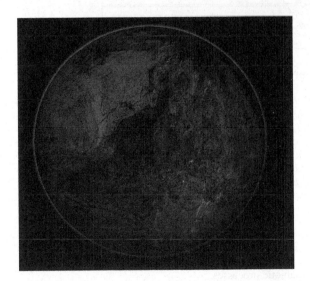

Figure 19.15 *An artist impression of a carbon planet (Source: NASA).*

out by life to yield an absorption spectrum (Chapter 2) in which those gases can be detected. Necessarily this approach is limited to what we know about life on Earth, but the terrestrial atmosphere provides insights into a number of candidate biosignature gases.

If we examine a spectrum of the Earth covering the visible and infra-red regions we notice some major absorptions attributable to components of the terrestrial atmosphere. Figure 19.16 shows some of these features and their detail.

These features were detected by the Galileo spacecraft during its journey to Jupiter in 1990, when, in an ingenious experiment, the spacecraft was turned towards the Earth to see if it could detect life. The near infra-red spectrometer on the spacecraft successfully picked up the presence of the gas, ozone. Ozone (O_3) is formed from the **photolysis** of O_2 in the stratosphere by UV radiation. The gas has a strong and distinctive absorption at 9.6 μm in the infra-red. The presence of the O_3 signature in our atmosphere is clearly shown in Figure 19.16. Our O_3 signature is also accompanied by a strong H_2O signature, which would suggest to alien astronomers the presence of water clouds and possibly a hydrological cycle, suggesting habitable conditions.

The lifetime of biosignature gases is strongly influenced by the radiation environment. Planets around M stars, for example, despite the occasional flares, generally are subjected to lower extremely short wavelength UV radiation. This can be expected to reduce the production of **free radicals** in the atmosphere and enhance the lifetime and potential concentration of biosignature gases.

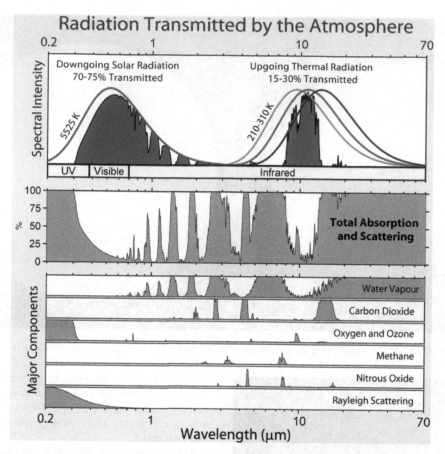

Figure 19.16   *The absorption spectrum of Earth. At the top the radiation received by the Earth from the Sun is shown (downgoing solar radiation) in the wavelength range 0.2–70.0 μm (from the ultraviolet into the infra-red). Superimposed on it is a solid red curve, which shows the radiation from a blackbody radiator at 5525 K. The energy emitted by the planet (upgoing thermal radiation) is shown on the right and is equivalent to a much lower temperature black body (Chapter 16). The different curves (purple, blue and black) show that the equivalent blackbody temperature depends on where on the planet the radiation is emitted from. Beneath the blackbody curves is the total absorption (including scattering) spectrum of the Earth. In the lower part of the graph the absorption of some of the major different atmospheric components is shown as well as the loss of radiation caused by scattering in the atmosphere (Rayleigh Scattering) (Source: wikicommons, Robert A. Rohde).*

Numerous attempts have been made to model the formation of O_2 in an abiotic atmosphere. High concentrations can be formed by **photolytic** reactions in an atmosphere with abundant water, particularly if the O_2 is not efficiently removed by geological processes. However, it is thought that the only mechanism enabling the production of high concentrations of O_2 in a terrestrial-like atmosphere is photosynthesis. This is a strong signature of life. Nevertheless, defining conditions that allow for the so-called 'false positive' detection of life is crucial to the search for biosignatures. Scientists are making considerable efforts to understand how different gases can

accumulate in exoplanet atmospheres and the different planetary atmospheric and chemical processes that might allow for the build-up of gases of biological interest by abiotic processes.

Atmospheric disequilibria are a strong indication of biological activity. The presence of O_2 and CH_4 at significant concentrations in an atmosphere suggests biological processes generating non-equilibrium conditions. Additional evidence of life from Galileo came from the detection of CH_4, which would generally not be expected at high concentrations in the presence of O_2. The use of disequilibria as a means of detecting

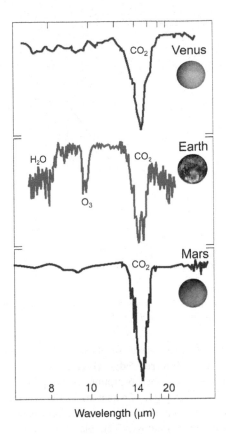

Figure 19.17 *The absorption spectra of Venus, Mars and Earth from the visible into the infrared, showing in particular the detail of the ozone biosignature at 9.6 μm in Earth's atmosphere, caused by high concentrations of oxygen in the atmosphere.*

biosignatures in alien atmospheres was first recognised by environmental scientist, James Lovelock.

Another very strong feature in our atmospheric absorption spectrum is CO_2. This gas is also observed in the atmosphere of Venus and Mars (Figure 19.17). The concentrations are different in each of these planetary atmospheres, but the signature is strong even for terrestrial concentrations of ~400 ppm. You will also notice that alongside CO_2 on Venus and Mars, the O_3 biosignature associated with the Earth is missing.

It is important to realise that the lack of biosignatures does not necessarily prove that a planet is lifeless since it might have extremely low biomass that does not produce a detectable signature, particularly one that can be detected by an instrument with limited spectral resolution.

The Galileo experiment was a simple demonstration of the principle of the remote detectability of life. In 1992,

when the spacecraft performed a similar analysis of the Moon, this control experiment unsurprisingly did not yield any biosignatures.

There are alternative candidates to O_2 that have been suggested as plausible targets for biosignatures. N_2O (nitrous oxide), a product of denitrification, has a short residence time in the atmosphere (~110 years) and its presence in an atmosphere would suggest a biological mechanism for its production.

Sulfur gases produced by organisms have been considered as another class of possible biosignatures. Dimethyl sulfide (CH_3SCH_3, or DMS) and dimethyl disulfide ($CH_3S_2CH_3$, or DMDS) are examples of biogenic sulfur gases. Modelling studies suggest that concentrations of these gases could increase to remotely detectable levels, but only under low UV radiation fluxes, which tend to destroy them. This could occur in the habitable zone of a low activity M star. Organic sulfur gases are thought to increase the concentration of other gases such as ethane (C_2H_6), which could be indirect signatures of the presence of biology.

Technologically-produced gases have even been suggested as a biosignature. Chlorofluorocarbons (CFCs) are used as solvents and refrigerants and were internationally banned to prevent their continued contribution to ozone depletion (the so-called 'ozone hole'). They have strong spectroscopic absorption and could be detected by alien astronomers, although their concentrations would have to be high. As for human civilization, it seems likely that intelligent aliens would attempt to limit these compounds accumulating at very high concentrations in their atmosphere.

N_2O is already a technologically-produced gas. On the Earth, its concentration is over 320 ppb. The gas is produced anthropogenically by agriculture, industrial activity and fossil fuel burning (Figure 19.18). It is about 300 times more effective than CO_2 as a greenhouse gas, so its concentration in the atmosphere is of great interest.

We should bear in mind that the gases present in our atmosphere have changed over geological time periods. Prior to the Great Oxidation Event (Chapter 14), there was insignificant O_2 in the atmosphere. Although the presence of CH_4 from biological production could have been detectable, this signature is more equivocal than O_2 when it is on its own since it can be produced by geological processes. Depending on the concentrations achieved, it may not have been possible for an alien astronomer to definitively tell whether this was biological or geological. Given that the Great Oxidation Event was ~2.4 Ga ago, this means that approximately half of our planetary history was without an O_2 biosignature.

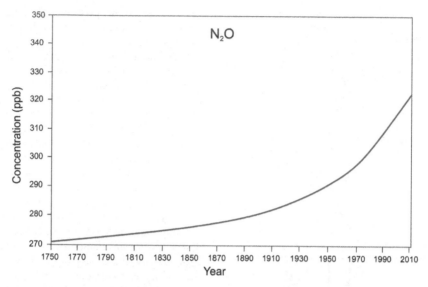

Figure 19.18   *Concentrations in parts per billion (ppb) of N_2O in the atmosphere in modern times. As well as being a general biosignature, N_2O in the atmosphere has also become a biosignature of a technological civilisation.*

A number of missions and approaches are being considered as ways to collect sufficient light to resolve the spectra of Earth mass planets. The transit method that was discussed earlier in the chapter is one way that can be used to look at the light from a planet and thus obtain spectroscopic data about the composition of the atmosphere. When the planet is in front of the star we will have the light spectrum from both the star and the planet. However, when the planet is behind the star (the so-called **secondary transit**), we will only have the spectrum of the star. By subtracting these spectra we can gather information on the atmospheric spectrum of the planet.

These transit methods, proposed for detecting biosignatures, have already been very successful at investigating the gases within the atmospheres of Hot Jupiters, including gases such as CO_2, H_2O, CO and CH_4, showing that the method has great potential for studying the atmospheres of distant Earth-like worlds.

Other technologies being considered include coronagraphs and sunshades (Figure 19.19), which are methods to physically block out the light from the parent star and detect the light of orbiting planets.

19.12.2   Surface Biosignatures

Life produces not merely gases, but also surface pigments which are characteristic of certain adaptive and metabolic activities. These too have been suggested as a means to find life on exoplanets.

The best investigated of these possible signatures is the **vegetation red edge** (Figure 19.20). The red edge results from the absorption by the photosynthetic molecule, chlorophyll (Chapter 5) in the red region of the spectrum contrasted with a strong reflectance in the infra-red region. Photosynthetic plants in particular have a strong reflectance in the infra-red region, possibly as an evolved mechanism that prevents metabolically dangerous overheating from absorbed infra-red radiation. This raises the possibility that high infrared reflectance is an emergent property of structures in photosynthetic organs that evolved in photosynthesis.

The vegetation red edge on Earth is so distinctive in satellite measurements of radiation reflected from the surface that it can be used to map vegetation.

Several groups have measured the Earth spectrum with the technique of **Earthshine**, which uses sunlight reflected from the non-illuminated, or dark side of the moon to get a whole-disc spectrum of the Earth's light, analogously to what would be obtained from an exoplanet. Earthshine measurements have shown that detection of Earth's vegetationred edge is feasible but made difficult owing to its broad, essentially featureless spectrum and the interfering effects of cloud coverage. Averaged over a spatially unresolved hemisphere of Earth, the additional reflectivity of this spectral feature is a small percentage.

Figure 19.19 *Artist's impression of a coronagraph (the star-shaped object in the centre), which is used to block out starlight, aiding in the detection of orbiting planets. The planets are not to scale but shown to illustrate the concept (Source: NASA).*

Debate Point: What happens after a biosignature detection?

It is easy to become focused on the search for a biosignature on an exoplanet and certainly such an event would be a momentous event for astrobiology and science in general. However, an obvious question to debate is what we would do if we found such a signature? Could we seriously discuss a sample return mission? Assuming that such a planet was many hundreds, possibly thousands of light years away, such a mission would be a multigenerational mission even with a probe traveling at light speed. One follow-up priority could be to search for yet more planets with biosignatures to derive a statistically robust estimate of the number of planets in the habitable zone that host life-producing unequivocal biosignatures. However, once even this task was achieved, what next? Does exoplanet science have a future beyond the detection of biosignatures?

The exact wavelength and strength of the spectroscopic edge depends on the plant species and the environment. The spectral signatures of any hypothetical photosynthetic life form elsewhere could shift according to the spectral output of the parent star. Quite different spectra have been suggested for other stars. For example, around M stars, where the spectrum is shifted to the red and the intensity is low, plants may have to harvest as much energy across the whole wavelength range as possible. Under these conditions plants could be black. Plants on planets orbiting F stars, which emit bluer radiation, might have absorbances shifted towards the blue region making them appear green, yellow or even red. These are wild speculations, but they underline the importance of being cautious in making assumptions about the sort of biosignatures one might look for when using terrestrial life as a standard.

Not all pigments are associated with photosynthesis. Carotenoids, which are ubiquitous coloured pigments involved in a range of functions such as quenching **oxygen free radicals**, also produce distinctive absorption peaks and edges, many of which are different from abiotic materials. These too might give a tell-tale signature of biological activity.

19.12.3   How Likely are These Signatures?

The search for extrasolar oxygenic photosynthesis and other biosignatures, including gaseous signatures and surface reflectance spectra involves assumptions, perhaps better referred to as presumptions, on a breath-taking scale, such as the ubiquity of the origin of life on other planets, the likelihood of the evolution of photosynthesis elsewhere (when on the Earth oxygenic photosynthesis

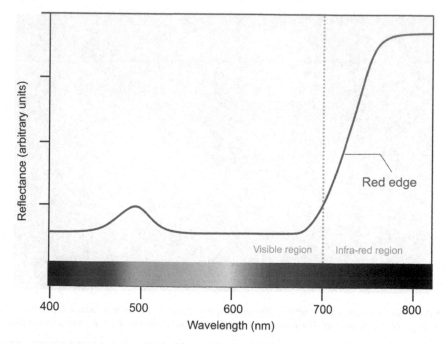

Figure 19.20 *Schematic reflectance spectrum showing the vegetation red edge. The exact spectrum varies for different organisms.* (Source: Charles Cockell).

has only evolved in one known bacterial lineage) and the subsequent alteration of the atmosphere by photosynthetic life sufficient to cause the build-up of a gaseous biosignature that is remotely detectable. However, the search for biosignatures is driven by the desire to test the hypothesis of whether this sequence of events is common or rare. The discovery that they are rare would itself be a remarkably important conclusion. The ability to arrive at these conclusions does, however, require that we understand the abiotic conditions that allow for the production of different gases, thus potentially causing a false-positive detection of life.

19.13   Conclusions

The detection of planets around other stars has brought astrobiology firmly into the empirical realms of science. A variety of methods have been applied individually and in combination to find candidate planets. The success of these methods has allowed for the detection of Earth mass planets. With a catalogue of such worlds we are now in a position to build telescopes to gather spectra from light reflected from these planets and thereby infer their

atmospheric compositions. The objective of astrobiology is to investigate these planets for signs of biosignatures, particularly oxygen. Ultimately, this task will allow us to test the hypothesis that there are other life-bearing planets distributed throughout the Universe.

Further Reading

Books

Casoli, F., Encrenaz, T. (2007) *The New Worlds: Extrasolar Planets*. Springer Praxis, New York.
Seager, S. (2010) *Exoplanets*. University of Arizona Press, Tucson.

Papers

Arnold, L., Bréon, F.-M., Brewer, S. (2009) The Earth as an extrasolar planet: The vegetation spectral signature today and during the last Quaternary climatic extrema. *International Journal of Astrobiology* **8**, 81–94.
Borucki, W.J., Koch, D.G., Batalha, N. *et al.* (2012) Kepler 22b: a 2.4 Earth-radius planet in the habitable zone of a Sun-like star. *Astrophysical Journal* **745**, doi 10.1088/0004-637X/745/1/120.

Burrow, A.S. (2014) Spectra as windows into exoplanet atmospheres. *Proceedings of the National Academy of Sciences* **111**, 12601–12609.

Cockell, C.S. (2014) Habitable worlds with no signs of life. *Philosophical Transactions of the Royal Society* **372**, 20130082.

Des Marais, D.J., Harwit, M.O., Jucks, K.W., Kasting, J.F., Lin, D.N.C., Lunine, J.I., Schneider, J., Seager, S., Traub, W.A., Woolf, N.J. (2002) Remote sensing of planetary properties and biosignatures on extrasolar terrestrial planets. *Astrobiology* **2**, 153–181.

Elkins-Tanton, L.T. (2010) Formation of early water oceans on rocky planets. *Astrophysics and Space Science* **332**, 359–364.

Kaltenegger, L., Jucks, K., Traub, W. (2007) Spectral evolution of an Earth-like planet. *Astrophysical Journal* **658**, 598–616.

Kiang, N.Y., Siefert, J., Govindjee, B., Blankenship, R.E. (2007a) Spectral signatures of photosynthesis. I. Review of Earth organisms. *Astrobiology* **7**, 222–251.

Kiang, N.Y., Segura, A., Tinetti, G., Govindjee, B., Blankenship, R.E., Cohen, M., Siefert, J., Crisp, D., Meadows, V.S. (2007b) Spectral signatures of photosynthesis. II. Coevolution with other stars and the atmosphere on extrasolar worlds. *Astrobiology* **7**, 252–274.

Léger, A., Fontecave, M., Labeyrie, A., Samuel, B., Demangeon, O., Valencia, D. (2011) Is the presence of oxygen on an exoplanet a reliable biosignature? *Astrobiology* **4**, 335–341.

Lovelock, J.E. (1975) Thermodynamics and the recognition of alien biospheres. *Proceedings of the Royal Society of London B* **189**, 167–181.

Petigura, E.A., Howard, A.W., Marcy, G.W. (2013) Prevalence of Earth-size planets orbiting Sun-like stars. *Proceedings of the National Academy of Sciences* **110**, 19273–19278.

Sagan, C., Thompson, W.R., Carlson, R., Gurnett, D., Hord, C. (1993) A search for life on Earth from the Galileo spacecraft. *Nature* **365**, 715–721.

Seager, S., Turner, E.L., Schafer, J., Ford, E.B. (2005) Vegetation's red edge – possible spectroscopic biosignature of extraterrestrial plants. *Astrobiology* **5**, 372–390.

Seager, S., Deming, D. (2010) Exoplanet atmospheres. *Annual Reviews of Astronomy and Astrophysics* **48**, 631–672.

Segura, A., Meadows, V.S., Kasting, J.F., Crisp, D., Cohen, M. (2007) Abiotic formation of O_2 and O_3 in high-CO_2 terrestrial atmospheres. *Astronomy and Astrophysics* **472**, 665–679.

Tinetti, G., Meadows, V.S., Crisp, D., Fong, W., Fishbein, E., Turnbull, M., Bibring, J.-P. (2006) Detectability of planetary characteristics in disk-averaged spectra. I: The Earth model. *Astrobiology* **6**, 34–47.

Tsiganis, K., Gomes, R., Morbidelli, A., Levison, H.F. (2005) Origin of the orbital architecture of the giant planets of the Solar System. *Nature* **435**, 459–461.

Wolstencroft, R.D., Raven, J.A. (2002) Photosynthesis – likelihood of occurrence and possibility of detection on Earth-like planets. *Astrobiology* **157**, 535–548.

Woolf, N.J., Smith, P.S., Traub, W.A., Jucks, K.W. (2002) The spectrum of Earthshine: A pale blue dot observed from the ground. *Astrophysical Journal* **574**, 430–433.

20

The Search for Extraterrestrial Intelligence

Learning Outcomes

➤ Understand the Drake equation and its component factors.

➤ Describe some of the methods used in radio and optical SETI.

➤ Understand some of the history of SETI and the methods used.

➤ Know about the Fermi Paradox and some of its possible solutions.

➤ Understand some of the proposed classification schemes for a civilisation.

➤ Understand some of the policy implications of SETI.

20.1 The Search for Extraterrestrial Intelligence

It was inevitable that, as soon as people began to wonder about life beyond the Earth, they also began to question whether that life might be intelligent. Like all science, this question can only become empirical if testable hypotheses can be constructed. Otherwise it remains within the realms of interesting, but unconstrained speculation.

One hypothesis could be: 'There is intelligent life beyond the Earth'. However, unless we are lucky enough to find some aliens on our cosmic backdoor waiting for us to receive their communications, the hypothesis is liable to become untestable. If the nearest communicative civilisation is in a far-away galaxy, then they could be millions of light years away. For all practical purposes, at least with the technology we have access to and are likely to have available in the foreseeable future, this would render the hypothesis untestable. If the speed of communication or travel is ineluctably set at the speed of light, c, as we think it is, then communication over realistic time periods compatible with the lifetime of different human civilisations may never be possible.

A more tractable hypothesis is: 'There are communications from intelligent civilisations that are currently accessible to us'. This hypothesis is a perfectly reasonable scientific hypothesis. It is readily testable and falsifiable. We can turn telescopes and other apparatus towards the sky and we can see if we get any signals replete with lots of interesting alien information. If we find them then we accept the hypothesis and if we do not, we reject the hypothesis (we accept the null hypothesis).

In hindsight, given that to date the search for such signals has been fruitless, it is easy to laugh at this hypothesis. But sixty years ago, who knew? Maybe the radio and optical frequencies of space were full of signals from frustrated aliens trying to make contact. Only by at least seeking such signals can the hypothesis be tested (Figure 20.1).

The Search for Extraterrestrial Intelligence (SETI) can have Earth-based spin-offs. For example, complex algorithms developed to search for signals in large amounts of data have wider use in developing computer technology and communications technology, for example processing information in many narrow waveband channels.

Astrobiology: Understanding Life in the Universe, First Edition. Charles S. Cockell.
© 2015 John Wiley & Sons, Ltd. Published 2015 by John Wiley & Sons, Ltd.
Companion Website : www.wiley.com/go/cockell/astrobiology.

Figure 20.1 *The Search for Extraterrestrial Intelligence seeks to find out whether there are other communicating intelligences in the Universe. This image shows the Green Bank radio telescope, West Virginia which has been used in SETI programs* (Source: NRAO/AUI).

However, we need no excuses for this endeavour, as it addresses one of the fundamental questions of the human scientific enterprise: are we alone in this vast Universe?

20.2   The Drake Equation

How many intelligent communicative civilisations are there in the Universe? We have no data to answer this question, but we can try to break down the question in such a way as to find out exactly what we need to know to be able to answer it. The first attempt to do this was made by Frank Drake in 1961. His famous mathematical relationship is a simple expression designed to specify the number of such civilisations in the galaxy (although it can be applied to scales beyond the galaxy). The equation is shown below:

$$N = R^* \times f_p \times n_e \times f_l \times f_i \times f_c \times L$$

where the terms are:

N = the number of civilisations in our galaxy with which communication might be possible;

R^* = the average rate of star formation in our galaxy;

f_p = the fraction of stars that have planets;

n_e = the average number of planets that can potentially support life per star;

f_l = the fraction of planets that could support life and that develop life;

f_i = the fraction of planets with life that go on to develop a civilisation;

f_c = the fraction of civilisations that develop a technology that produces detectable signs of their existence;

L = the length of time for which such civilisations release these signals into space.

Of course, it's easy to criticise this equation. Consider f_i. Can we ever know the number of planets with life that will develop a civilisation? This is not predictable and probably can only be determined by empirical observation of many life-bearing planets. The equation does not take into account that another relevant factor is whether the other civilisations happen to be communicating with technology that is compatible with our current civilisation. Our own technology has dramatically changed over time. Two hundred years ago we did not have radio communication and we do not know whether there is scope for radical communication methods that might emerge in the next 200 years. In other words, the contact cross-section is important. Of course, we could simply add a term into the equation that takes into account the fraction of civilisations that have communication methods compatible with our civilisation.

However, the equation has good uses. It allows us to break up the problem and think about what we do and don't know. For example we can, using data on extrasolar planets, as we saw in the last chapter, estimate f_p, a factor that 40 years ago would have been impossible to estimate (Figure 20.2). The equation allows us to track

---

Debate Point: The Drake Equation

When Frank Drake first proposed the Drake Equation, he used it to estimate N as 10 000 civilisations in the galaxy. Since his estimate, we now have Kepler Space Telescope data that allows us to estimate the number of stars with planets and the number of planets in habitable zones. Use the Drake Equation to have an informed discussion about the value of N, identifying those factors that are contingent on the sociology of civilisations and might never be estimated and those factors that are susceptible to experimental determination. How might future advances in the study of exoplanets and the search for spectroscopic signatures of life improve the estimate of N, if indeed they can? You might try to estimate your own value of N, defining your arguments and quantifying the uncertainties in your estimate.

Figure 20.2   *The search for rocky extrasolar planets allows us to constrain f_p in the Drake equation* (Source: NASA).

Figure 20.3   *Frank Drake, a pioneer in SETI searches and originator of Project Ozma* (Source: Science Picture Library).

astronomical progress in our quest to quantify the number of life-bearing planets and civilisations beyond the Earth. It provides a framework for astrobiologists to contemplate whether we can or cannot estimate the number of civilisations in the galaxy and which parts of this question are tractable. Quite apart from that, even if you wonder about its real uses, it has enormous educational use.

20.3   Methods in the Search for Extraterrestrial Intelligence

People have considered the possibility of alien civilisations for a long time. Even in recent centuries we need go no further back than the early twentieth century to find Percival Lowell staring down his telescope, mapping and writing profusely on the Martian canals, constructed by a dying and desiccated civilisation desperately saving themselves as they sought to channel water from the Martian polar ice caps.

However the first serious attempt to pick up a signal using radiowaves from intelligences was Project Ozma in 1960, pioneered by Frank Drake (Figure 20.3) and run from the Greenbank Telescope (Figure 20.1). About 150 h of total observing time were spent studying the Sun-like stars Tau Ceti and Epsilon Eridani. At nearly the same time a paper published by Cocconi and Morrison in 1959 advocated the possibility of finding alien signals. During the 1960s, Soviet scientists also launched a variety of projects. Since that time, other efforts have been devised, including SERENDIP (Search for Extraterrestrial Radio Emissions

from Nearby Developed Intelligent Populations), a project that was initiated by the University of California, Berkeley. None of these projects found signals.

Today most SETI projects either use spare time on radio-telescopes or, like SERENDIP, they piggy-back – collecting data whilst other more conventional radio-astronomy is being performed. Clever approaches to data analysis have included SETI@Home and SETILive, in which personal computers in people's private homes are linked across the internet to collectively analyse the huge quantity of information gathered by radio-telescopes.

New telescopes offer other possibilities for SETI. The Square Kilometer Array (SKA) is proposed as a large number of dishes in South Africa and Australia with an effective total radio dish size of one square kilometre. The facility might be used for SETI, searching within 300 light years for radio leakage similar to that given off by our civilisation.

A question that faced the SETI community from the early days is at which frequency should they search? The electromagnetic spectrum is rather broad and it is difficult to listen across the whole spectrum all at once.

A favoured search location has been at 1420 MHz (21 cm wavelength). This was the wavelength used in Project Ozma. Hydrogen atoms have one proton and one electron. Crudely, in analogy to two planets, their spins can be in the same direction or opposed. When they flip from one state to another a photon is emitted at 1420 MHz. Hydrogen is common in the Universe, so we suppose that alien astronomers might want to look in that region. Near to this frequency is the frequency (1640 MHz) associated with hydroxyl radical (OH) emission. Hydrogen combined with OH produces water (H_2O). Just as animals collect at the water hole in the desert, so in analogy we might expect that communicating civilisations might collect at this cosmic radiowave 'water hole' (Figure 20.4). The atmosphere is relatively transparent to these frequencies, making them quite detectable. Although space telescopes eliminate the theoretical significance of this advantage, they are expensive. This wavelength has been used in many SETI searches, but it is by no means the only sensible place to search. One permutation is to search at frequencies obtained by dividing or multiplying the hydrogen frequency by constants such as π, since π is a universal number that could be known by other intelligences.

The SETI search has one problem in that it assumes that civilisations are communicating regularly and with powerful signatures. As a civilisation, our intentional transmissions to alien civilisations have been very limited, so why would we expect any different from other societies? The signals that we have unintentionally sent out, such as radio and TV, have not been very strong as they are

not intended for successful transmission across interstellar distances. The first major global television transmission was the Berlin Olympic games in 1936, so even if alien civilisations had immensely powerful receivers, we had still only reached a distance of 79 light years in 2015.

More recent searches have turned to 'optical SETI', attempting to pick up signals that could have been transmitted in laser pulses in our direction. Harvard University initiated such a programme. Human civilisation has built immensely powerful lasers that can achieve a brightness about 5000 times greater than the Sun. Perhaps other civilisations are using such technologies to send messages. A problem that optical SETI faces is that there is a great deal of noise in the optical region of the spectrum. More recent ideas have included infra-red SETI, which is less prone to dust absorption in the interstellar medium.

Up until very recently, all of these searches have been done on stars that might have plausible environments for planets, regardless of any knowledge about their planetary

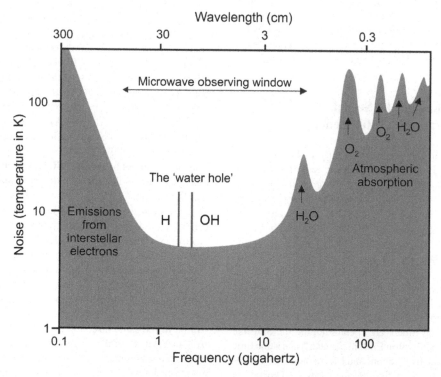

Figure 20.4   *The hydrogen and hydroxyl frequencies provide one location in the electromagnetic spectrum (the 'water hole') where attempts have been made to search for alien signals. This region is also relatively transparent for observation on Earth. The orange denotes the 'noise' which can be considered here to be the absorption caused by the atmosphere. Axes are logarithmic scales.*

systems. There is convergence between exoplanet research and SETI, since exoplanet searches (Chapter 19) enable us to constrain the number of planets worthy of study in SETI. If we make the assumption that oxygen is required for the rise of intelligence, then the search for Earth-mass rocky planets with oxygen biosignatures will provide a further narrowing of potential SETI targets.

It has occurred to some scientists that perhaps we should not assume that all aliens are on their planets attempting to communicate across vast interstellar distances. Perhaps they have sent a probe to our Solar System to observe or communicate with us. The Search for Extraterrestrial Artefacts (SETA) is another ball game, but no less intriguing. The notion behind this approach is that we should go about searching for space probes within our own Solar System. Such objects would be expected to be extremely small and probably very dark across much of the electromagnetic spectrum, making their detection an enormous challenge. Optical and infra-red searches, and very intense ones at that, within the Kuiper Belt for example, could be implemented.

20.4   Communication with Extraterrestrial Intelligence

It would remiss of a civilisation to expect all other civilisations to send messages, but make no effort to do so itself. Although our society has not been very chatty, we have made tentative efforts to send messages to any potential recipients. Our efforts in Communication with Extraterrestrial Intelligence (**CETI**) can be broadly split into two categories: physical materials and messages in the electromagnetic spectrum.

There have been two notable attempts to send physical materials as messages. The Pioneer 10 and 11 spacecraft, sent to Jupiter and Saturn, both took with them a gold anodised aluminium plaque 23 cm wide and 15 cm high (Figure 20.5). These craft are now greater than 80 AU from the Earth.

The plaque depicts several items. At the top left of the plaque is the schematic representation of the spin transition of hydrogen discussed earlier. Below this symbol is a small vertical line to represent the binary digit 1. This spin-flip transition of a hydrogen atom from electron state spin up to electron state spin down can specify a unit of length (wavelength 21 cm) as well as a unit of time (frequency 1420 MHz). Both units are used as measurements in the other symbols.

To the right is an image of a man and woman superposed on a silhouette of the spacecraft to give scale and information about our species. The radial pattern on the left of the plaque shows 15 lines emanating from the same origin and this provides the location from which the probe was sent. The lengths of the lines show the relative distances of pulsars to the Sun. A tick mark at the end of each line gives the Z coordinate perpendicular to the galactic plane. The 15th line extends to the far right, behind the humans. The line indicates the Sun's relative distance to the galactic centre. Giving the location of 14 pulsars provides redundancy so that the location of the origin can be worked out, even if only some of the pulsars are known. Fourteen of the lines have corresponding binary numbers, which stand for the time of the pulses (the periods) of the **pulsars**, using the hydrogen spin-flip transition frequency as the unit. Since these periods will change over time (the pulsar rotations will reduce), the time of the launch of the spacecraft can also be calculated.

A map of the Solar System is shown at the bottom of the plaque indicating from which planet in our Solar System the craft originated.

A second attempt to send a material message was made in 1977, when two 30 cm diameter gold-plated copper records were attached to the Voyager 1 and 2 probes, both of which were destined to explore the outer Solar System and eventually to leave the Solar System altogether (Figure 20.6). The Voyager spacecraft were launched in 1977. Voyager 1 passed the orbit of Pluto in 1990, and left the Solar System in 2013. In about 40 000 years, Voyager 1 will have approached star Gliese 445, located in the constellation Camelopardalis and Voyager 2 will have approached star Ross 248, located in the constellation of Andromeda. Both will come within about 1.5 light years of these stars.

On both records is an ultra-pure sample of uranium-238 to allow for radiometric dating (Chapter 13). A civilisation that encounters the record will be able to use the ratio of remaining uranium to daughter elements to determine the age of the record. The 116 images on the record are encoded in analogue form and composed of 512 vertical lines. The rest of the record is in audio and is designed to be played at 16.66 revolutions min^{-1}. The record even has some instructions on its front for how to play it.

The contents of the record were designed for NASA by Carl Sagan (1934–1996). Sagan and his colleagues assembled the images for the record. The images included ones in black and white and colour. They included mathematical and physical quantities, the Solar System and its planets, the structure of DNA and information on human anatomy. Pictures of landscapes, plants, insects and other animals were included. Images of humanity show people going about their lives and there are images of food and

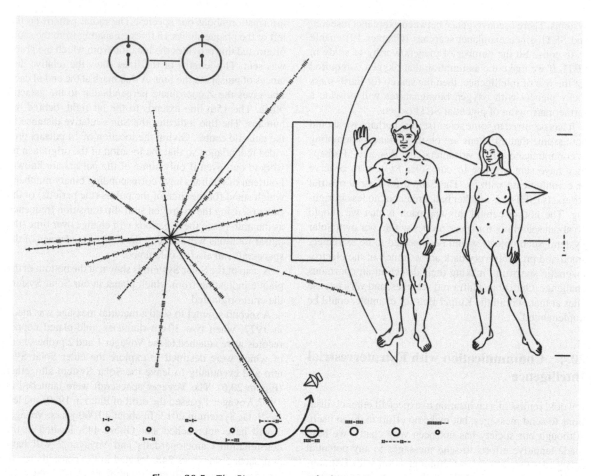

Figure 20.5   *The Pioneer spacecraft plaques* (Source: NASA).

architecture. The record contained a variety of sounds, such as those made by wind, surf, thunder and the sounds of animals, including whales and birds. The musical selection included Mozart, Bach, Beethoven, Stravinsky, Chuck Berry and other musical selections from around the world. There were greetings in 55 ancient and modern languages and printed messages from United States President Jimmy Carter and United Nations Secretary-General Kurt Waldheim. The hydrogen molecule diagram and pulsar map sent on the Pioneer spacecraft were included on the front of the record so that any recipient could fathom from where it had been sent.

The most distinctive attempt to transmit an electromagnetic message was the Arecibo message. On 16 November 1974, a 1000 kW message was sent to the globular star cluster M13, 23 000 light years away and transmitted at

2380 MHz. The message was sent for a duration of three minutes and was made of 1679 binary digits. This number was chosen because it is the unique product of the two prime numbers 23 and 73. Binary notation is used as the simplest way of communicating information. When the message is split in this way (Figure 20.7) it provides a set of information about us and our civilisation. The key components of the message in order from the top are: (i) the numbers one to ten, (ii) the atomic numbers of the elements hydrogen, carbon, nitrogen, oxygen and phosphorus, which make up DNA, (iii) the formulas for the sugars and bases in the nucleotides of DNA, (iv) the number of nucleotides in DNA, and a graphic of the double helix structure of DNA, (v) the height of an average human, a graphic figure of a human and the human population of Earth, (vi) a graphic of the Solar System

Focus: Astrobiologists: Jill Tarter

What do you study? My team uses radio telescopes to attempt to detect evidence of other technological civilisations beyond Earth. We are looking for ETI.

Where do you work? I am now retired, but I spent the majority of my career working for the non-profit SETI Institute in Mountain View, California.

What was your first degree subject? Engineering Physics.

What science questions do you address? We are trying to answer the old question 'Are we alone?'. In particular, we use technology as a proxy for intelligence and then attempt to discover other distant, intelligent civilisations. Using our own twenty-first century technologies that include optical and radio telescopes, we search for signals that have characteristics suggesting they were engineered rather than naturally emitted. This means that we would probably miss evidence of very advanced technologies that we ourselves haven't yet invented. Although we cannot deliberately search for what we cannot conceive, perhaps we might stumble across some evidence while looking for something else.

What excites you about astrobiology? We have wondered about our place in the cosmos throughout recorded history. It is exciting to be participating now in a new science that is developing the necessary tools and using them to launch a grand exploration. Did the laws of chemistry and physics as we now understand them actually lead to life beyond the Earth? Is life as we know it unique, rare, or quite common in our galaxy? Scientists and engineers can now lead the searches for what is actually out there.

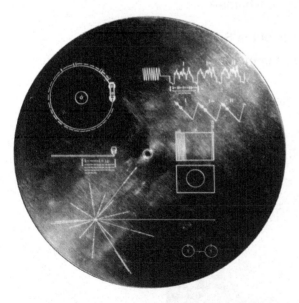

Figure 20.6 *The Voyager record* (Source: NASA).

indicating which of the planets the message is coming from and (vii) a graphic of the Arecibo radio telescope and the diameter of the transmitting antenna dish.

The distance that the message has to travel makes it impractical as a genuine attempt to start a conversation, however it did demonstrate the capability for using a radio telescope to send such messages.

Communication with extraterrestrial intelligence always runs into an obvious problem, which is what language to use. What is universal information? Many people have suggested that mathematics is one obvious way to communicate. Presumably any radio-telescope or laser-building civilisation does mathematics. A message containing clearly non-random information related to constants such as π would suggest alien intelligence. Again, however, how is the information to be structured to make it separable from the interstellar noise?

20.5   The Fermi Paradox

The search for signals from other civilisations has not yet yielded a positive response. It is easy to simply dismiss this result, but it does raise a very perplexing problem that was recognised by physicist Enrico Fermi (1901–1954). We live in a Universe of maybe over 100 billion galaxies. Our own galaxy has 200 billion stars. There are many stars, both in our own galaxy and elsewhere, that are much older than our Sun. So from a purely qualitative consideration of these statistics one might suppose that, assuming the origin of life has occurred elsewhere at a reasonable frequency, there is a high likelihood that not only would there

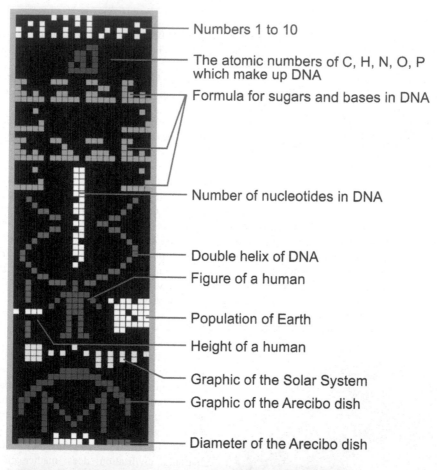

Numbers 1 to 10

The atomic numbers of C, H, N, O, P which make up DNA

Formula for sugars and bases in DNA

Number of nucleotides in DNA

Double helix of DNA

Figure of a human

Population of Earth

Height of a human

Graphic of the Solar System

Graphic of the Arecibo dish

Diameter of the Arecibo dish

Figure 20.7   *Above: The Arecibo message sent in 1974. Below: The Arecibo dish observatory, Puerto Rico, which sent the message* (Source: NAIC – Arecibo Observatory, a facility of the National Science Foundation, USA).

be other civilisations, but that some of them could be very considerably older than ours. In less than half a million years we have come from the Great Rift Valley in Africa to building spaceships. If some of these civilisations were a few million years older still, surely they would be travelling the galaxy? Why, then, is alien contact not an everyday and rather mundane event?

There have been a very large number to responses to the so-called **Fermi Paradox**. It is not useful here to try to list them all. Some of them seem plausible. Some suggestions have been outlandish to say the least. Let's look at a few of these. You can decide what you think the solution to the paradox is.

20.5.1 Civilisations are too Far Apart in Space

I start with this because I have no compunction in saying that this seems to be the most plausible answer. We know from Kepler Space Telescope data that the nearest habitable planets are tens to thousands of light years away. Even if one was to assume that they host intelligent civilisations, efforts to communicate with them, let alone physically visit them, and they visit us, would be multi-century to millennia projects. If the speed of light is the limit of our rate of travel (there is no good physics at the moment that convincingly shows us that we can make a macroscopic spaceship leap across space–time through worm holes and other hypothetical means) then no amount of optimism and technology will make these time scales more sensible. As we approach the speed of light, time dilation for the individuals on the spacecraft will shorten the apparent duration of the journey for them, but it is not clear that we can get sufficiently close to the speed of light to make this effect practically useful. Furthermore, it is not clear that even if we could develop appropriate technology, we would be able to assemble groups of people or societies that could endure these journeys and remain viable over long time scales in the vast emptiness and isolation of interstellar space. The ability to visit aliens and for them to visit us may simply represent a technological and psychological ceiling that is universal.

The problem is compounded by the fact that to even consider such transfer of individuals and messages, the civilisations must be at comparable stages of technological development, which may itself be very rare. If aliens had arrived only five million years ago they would have met a rather uncommunicative bunch of apes (although if you have the capability for rapid interstellar flight it would seem easy enough to come back again, even if that takes you a few thousand years).

There are variations on this theme, which include that travelling between stars is just too expensive for alien civilisations, just as it isn't at the top of our own political and economic agendas.

20.5.2 No Other, or Very Few, Civilisations have Arisen

If the origin of life is a rare event, then either there are no other civilisations in existence or they are very rare and very widely spaced apart. Most of them are so far away they cannot, in practical terms, visit us. Alternatively this rarity might arise from rare events anywhere between the origin of life and the emergence of intelligence. Perhaps the rise of oxygen is unusual. Maybe the transition from oceans to land is not inevitable. Perhaps the emergence of multi-cellular life is infrequent. Conceivably the rise of intelligence itself is very rare (remember our thought experiment about intelligent dinosaurs in Chapter 15). Even if intelligence does emerge, does it always develop a desire to build interstellar spaceships? There have been many evolutionary branch points in the emergence of our own civilisation that we have discussed throughout this book. We do not know which of these are inevitable and which, if any, depend on rare biological contingencies.

Debate Point: Is the Fermi Paradox a scientific hypothesis?

For a problem to fit within our idea of 'good science' it must be experimentally tractable and it must be possible to examine it by constructing testable, falsifiable hypotheses. Are any of the apparent 'solutions' to the Fermi Paradox experimentally tractable? You might like to list the ones presented here, think up others or look them up. Systematically devise a list of those solutions that could be testable and those that are not, listing the types of controlled experiments that could be done to test these ideas. Based on this analysis, develop a ranked set of solutions according to their scientific testability. Consider the question – are the most testable ideas also the most plausible based on our existing knowledge?

Webb, S. (2002) *Where is everybody? Fifty Solutions to the Fermi Paradox and the Problem of Extraterrestrial Life.* Copernicus, New York.

20.5.3   Intelligent Life Destroys Itself

As we witnessed in the Cold War between the 1950s and early 1990s, when there were sufficient nuclear weapons to destroy every major city on Earth several times, civilisations can reach breath-taking scales of self-destructive capability (Figure 20.8). Maybe it is inevitable that civilisations destroy themselves. It is somewhat of an irony that rocket technology to take people into space also allows you to hurl bombs across a planetary surface. It could always be that civilisations can never reach space-faring capacity without also developing planetary-scale self-destructive tendencies. War is not the only challenge. Resource depletion, mass extinction, over-population and disease are other challenges. There are a number of possible ways, either separately or in combination, that stresses may destroy a civilisation or periodically set it back in such a way that it fails to develop an advanced space-faring capability.

From a Darwinian perspective, self-destruction may be an unfortunate outcome of evolution. The psychology that developed during the competition for scarce resources

Figure 20.8 *Are civilisations destined to destroy themselves?* (Source: wikicommons).

over the course of human evolution may have left the species subject to aggressive, instinctual drives. These compel humanity to consume resources and to reproduce, the very factors that led to the expansion and development of society. However, for this to provide an adequate response to Fermi's paradox, it would have to work near universally.

20.5.4   Life is Periodically Destroyed by Natural Events

As a parallel to the previous rejoinder, another possibility is that most civilisations are destroyed by natural events such as asteroid and comet impacts. Chapter 15 should provide plenty of evidence to show that this is plausible. It might be a matter of luck that a civilisation reaches the capacity of interstellar colonisation before being destroyed. That luck could be extremely rare.

20.5.5   It is the Nature of Intelligent Life to Destroy Other Civilisations

This is the darkest response to Fermi's problem and it posits that civilisations, given their own aggressive tendencies, would destroy other civilisations (Figure 20.9). Alternatively, perhaps there is something out there on the rampage. The result is that either civilisations are destroyed, or, if they keep very quiet, they survive. A cosmic selection pressure exists for silence. Yet another chilling possibility is that rather than do this nasty deed themselves, which is rather time-consuming and tedious work, civilisations would dispatch automated probes to destroy other civilisations. These probes might be busy doing this work long after the originating civilisation has gone. We just haven't yet made it to the top of the waiting list. A relevant question to ask is whether these motives are realistic.

An alien civilisation might think it is too sociologically dangerous to visit, either for us or for them. When very different, and previously isolated, civilisations have met on Earth, the results have usually been disastrous for one side or the other. The same may well apply to interstellar contact.

20.5.6   They Exist, But We See No Evidence of Them

This is a response that is difficult to test. The aliens might be out there, but we just can't communicate with them or

Figure 20.9 *Denizens of Mars attack the Earth in HG Wells'* War of the Worlds. *An illustration by Alvim Corréa, from the 1906 French edition. Perhaps all civilisations are destructive* (Source: wikicommons).

recognise them in the same way that ants are probably not very aware of our existence.

20.5.7 They are in the Local Area, But Observing us Rather Than Attempting to Make Contact

Sometimes called the zoo hypothesis, the idea is that they are in our Solar System, or nearby, and are watching the Earth, fascinated by this particular evolutionary experiment (Figure 20.10). Variations on this idea include that they don't come any nearer out of ethical prudence – the principle of non-interference. Alternatively, perhaps we terrify them. If you met a species that dropped bombs on each other's cities, would you be enthusiastic for afternoon tea with them? A criticism of this hypothesis is that in a Universe without a hegemonic power, random civilisations with independent principles would result in one that in all likelihood would make contact.

20.5.8 They are too Busy Online

Other civilisations simply lose an interest in exploration. Tired, cynical and busy looking at photos of cats or their alien equivalent online, old civilisations simply fail to explore. Perhaps any advanced society will develop engaging entertainment before the capacity for advanced space travel. The appeal of these social possibilities is destined to overcome any desire for expensive endeavours such as space exploration. This idea isn't entirely outlandish. We ourselves argue about the value of human exploration and prefer to send robotic emissaries to Mars and other planets as a priority before human missions. The danger and loss of life that might be associated with interstellar adventures just may not appeal to other civilisations. It is not clear that we have sufficient motive to do this ourselves.

20.5.9 They are Here

The aliens are here and they either hide themselves very well or manage to look like us. This is a popular science fiction trope, but very difficult to formulate a testable hypothesis as, by definition, the aliens make themselves undetectable.

20.5.10 The Evidence is Being Suppressed

The aliens are here, the government knows about it, but they are not telling us. Government agencies can do amazing work and they can hide some astonishing things, but hiding aliens and alien spaceships in hangars without this information ever leaking out is not usually their forte.

We could go on with a few more of these ideas, but the intellectual return gets incrementally less. The main point to understand is that there are a variety of responses to the Fermi Paradox. Some of them, unfortunately, are not amenable to experimental testing, making those interesting ideas, but little more. Nevertheless, the overarching discussion is not in vain. It remains the reality that we see no evidence for aliens and we need to find the definitive reason for that fact.

Figure 20.10   *Perhaps, as we observe animals in safari parks, our own planet is being observed but, unlike these giraffes in the West Midland Safari Park (Worcestershire, UK), we are not aware of the prying eyes (Source: wikicommmons, Robek).*

20.6   Classifying Civilisations

Let us say that, regardless of the Fermi Paradox, civilisations are out there. How might they differ from us? There have been a number of attempts to classify civilisations. The Kardashev scale is a method of measuring a civilisation's level of technological advancement, based on the amount of energy it can use. The scale has three categories that are designated Type I, II, and III civilisations. The scale was first proposed in 1964 by the Russian astronomer, Nikolai Kardashev.

A Type I civilisation uses all available resources on its home planet with an energy accessible to it of about 10^{12} W. Human society is still within the bracket of an emerging Type I civilisation.

A Type II civilisation harnesses all the energy of its star, giving it the capacity to harness energies equivalent to the luminosity of our Sun, about 10^{26} W. One method by which a civilisation might achieve Type II status is by building a **Dyson sphere**. A Dyson sphere is a hypothetical giant structure originally described by physicist, Freeman Dyson, that completely encloses a star and captures most or all of its energy output. Other more exotic ideas include 'star lifting', a process where an advanced civilisation would remove a substantial portion of a star's matter in a controlled manner.

A Type III civilisation has managed to grapple with all the available energy in its galaxy – about 10^{37} W. It would achieve this capacity by building Dyson spheres around many stars in a galaxy or more dramatically, it would be able to tap into the energy released from supermassive black holes at the centre of most galaxies.

There have been many modifications to the Kardashev scale. At least one has proposed to extend the scale to even more hypothetical Type IV civilisations that control or use the entire Universe or even Type V civilisations that control collections of universes (assuming that there are other universes).

The Kardashev scale was exclusively focused on energy use. However, other people have suggested that although this is a reasonable metric, it is not the only way to assess a civilisation. Carl Sagan suggested adding the information available to a civilisation as one way to measure the level of advancement. He assigned the letter A to represent 10^6 unique bits of information (which is less than any recorded human culture). Each successive letter was applied to represent an order of magnitude increase in information, so that a level Z civilisation would have 10^{31} bits. The Earth in 1973, when Sagan formulated the scale, was an H civilisation, with access to 10^{13} bits of information.

Robert Zubrin adapted the Kardashev scale to refer to how widespread a civilisation is. A Type I civilisation has colonised its planet, a Type II has extensive colonies in its stellar system and a Type III has colonised its galaxy.

20.7 Policy Implications

Even the remote possibility of contact with an alien intelligence raises a whole variety of questions. One such question is: Who should represent the Earth in the event of communication?

You might like to consider and debate this question. Should it be a powerful nation? Should it be a block of nations (such as the EU)? Should it be a group of scientists and, if so, who? Should it be a non-governmental organisation? Or should it be some randomly selected members of the public? All of these suggestions have their merits and demerits (Figure 20.11). But should contact occur, one of them will have to be chosen.

It is also unclear what our first messages or signals should be. It would be unfortunate to send them some well-meaning greeting, only to discover that we have told them we intend to eat them all. Cultural misunderstandings are rife on the Earth. What are universal messages? We have already seen that we have some experience of inventing messages for the Pioneer plaques and Voyager records, but what about direct contact?

It has occurred to people that a protocol agreed by everyone might be a good idea to prevent a 'loose cannon' of an organisation from transmitting something that could be regrettable to all of us. So far, messages that have been transmitted have been quite ad hoc, with no wider consultation. There have been no ill consequences so far, but it does raise the question of whether there should be international agreements on constructing and transmitting messages to alien civilisations.

It might be easy, in the absence of evidence of alien intelligences, to dismiss these discussions as extravagant childishness, but for the sake of a few days of discussion, it might be worth at least having a protocol, just in case.

The International Academy of Astronautics has drafted a proposed protocol for what should happen following detection. It stipulates that the first step is, unsurprisingly,

Figure 20.11 *Who should coordinate response to alien contact? The United Nations or other human representatives?* (Source: wikicommons, Yann Forget).

to rigorously test to make sure the signal is real and then inform as many scientific groups as possible that are capable of following up the signal to verify and examine it. All national authorities would then be informed (mainly governments). The United Nations would be informed through the Secretary General and then the public would be informed.

20.8 Conclusions

The Search for Extraterrestrial Intelligence is now an old endeavour. It has so far yielded no signal or evidence of alien civilisations. Nevertheless, the possibility of alien civilisations should interest any enquiring human mind. Like all other science, SETI must be subjected to critical analysis and the parts of it that are experimentally testable must be a priority. The Fermi Paradox and its various solutions illustrate well the complexities in disentangling hypotheses that are testable and those that are not. Quite apart from being a tractable scientific endeavour, one should have no embarrassment in pursuing and openly supporting the search for other intelligences as an expression of the inquisitive nature of the human species and its desire to fully comprehend its place in the Universe.

Further Reading

Books

Davies, P. (2011) *The Eerie Silence: Renewing Our Search for Alien Intelligence*. Mariner Books, Chicago.
Shklovskii, I.S., Sagan, C. (1966) *Intelligent Life in the Universe*. Holden Day, San Francisco.

Papers

Almár, I., Race, M.S. (2011) Discovery of extra-terrestrial life: assessment by scales of its importance and associated risks. *Philosophical Transactions of the Royal Society* **369**, 679–692.
Baum, S.D., Haqq-Misra, J.D., Domagal-Goldman, S.D. (2011) Would contact with extraterrestrials benefit or harm humanity? A scenario analysis. *Acta Astronautica* **68**, 2114–2129.
Brin, G.D. (1983) The great silence – the controversy concerning extraterrestrial intelligent life. *Quarterly Journal of the Royal Astronomical Society* **24**, 283–309.
Cocconi, G., Morrison, P. (1959) Searching for interstellar communications. *Nature* **184**, 844–846.
Drake, F.D. (1961) Project Ozma. *Physics Today* **14**, 140–143.
Drake, F.D., Sagan, C. (1973) Interstellar radio communication and the frequency selection problem. *Nature* **245**, 257–258.
Dyson, F.J. (1960) Search for artificial stellar sources of infra-red radiation. *Science* **131**, 1667–1668.
Freitas, R.A., Valdes, F. (1980) A search for natural or artificial objects located at the Earth–Moon libration points. *Icarus* **42**, 442–447.
Galloway, J.F. (1996) An international relations perspective on the consequences of SETI. *Space Policy* **12**, 135–137.
Harrison, A.A. (2011) Fear, pandemonium, equanimity and delight: human responses to extraterrestrial life. *Philosophical Transactions of the Royal Society* **369**, 656–668.
Othman, M. (2011) Supra-Earth affairs. *Philosophical Transactions of the Royal Society* **369**, 693–699.
Race, M.S., Randolph, R. (2002) The need for operating guidelines and a decision-making framework applicable to the discovery of non-intelligent extraterrestrial life. *Advances in Space Research* **30**, 1583–1591.
Tartar, J.C. (2001) The search for extraterrestrial intelligence (SETI). *Annual Reviews of Astronomy and Astrophysics* **39**, 511–548.
Tartar, J.C., Agrawal, A., Ackermans, R., Backus, P., Blair, S.K., Bradford, M.T., Harp, G.R., Jordan, J., Kilsdonk, T., Smolek, K.E., Richards, J., Ross, J., Shostak, G.S., Vakoch, D. (2010) SETI turns 50: five decades of progress in the search for extraterrestrial intelligence. *Proceedings of SPIE* **7819**, 781902.
Turbull, M.C., Tartar, J.C. (2003) Target selection for SETI. I. A catalog of nearby habitable stellar systems. *Astrophysical Journal Supplement Series* **145**, 181–198.
Vakoch, D.A. (2011) The art and science of interstellar message composition: A report on international workshops to encourage multidisciplinary discussion. *Acta Astronautica* **68**, 451–458.

21

Our Civilisation

Learning Outcomes

➤ Understand the history of human evolution.
➤ Understand some of the threats to an intelligent civilisation.
➤ Understand the ideal rocket equation, how it can be used in calculating orbital dynamics and its significance to space exploration and settlement.
➤ Be able to elaborate some of the challenges to establishing a permanent human presence on other planetary bodies.
➤ Be able to describe some of the requirements for life support systems.
➤ Understand why becoming a multiplanet species is a significant step in the evolutionary biology of a species.

21.1 Astrobiology and Human Civilisation

Astrobiology is concerned with the origin, evolution and future of life. We have already seen in this textbook how mass extinctions have afflicted the biosphere throughout its existence. We have also discussed the very far future, when the greenhouse effect will doom all life on the Earth in over a billion years.

Therefore, there remains a question we have unanswered: what is the future of human civilisation? We end our investigation of astrobiology with a foray into this more scientifically uncertain question.

This area of investigation may seem remarkably parochial, **anthropocentric** and even egocentric. After all, humans are just one species of the many millions that have existed on the Earth throughout its history. However, there are three reasons why the question of our future should interest us from the point of view of astrobiology:

1. We are the only intelligence that has emerged on the Earth that is capable of leaving its home planet (at least deliberately; as we saw in Chapter 17, microbes may already have left the Earth in rocks ejected in impacts). This is a significant development in terms of the potential spread of life in the cosmos.
2. Human civilisation provides us with a datum point to consider how an intelligence arises on a planet and influences its geology and biology. We might learn general lessons about intelligence in the Universe, if it exists elsewhere, by examining ourselves.
3. If we want to survive and buck the general trend of extinction, we should use our intelligence to consider what might threaten our future.

21.2 The Emergence of Human Society

In Chapter 13 we explored the emergence of the first land-based animals and the origin of ancestral mammals. We could continue a detailed ancestral journey through all the intermediate stages to our own species. Instead, for our purposes here, we can pick the story up about 15–20 million years ago, when the primate Hominidae

Astrobiology: Understanding Life in the Universe, First Edition. Charles S. Cockell.
© 2015 John Wiley & Sons, Ltd. Published 2015 by John Wiley & Sons, Ltd.
Companion Website : www.wiley.com/go/cockell/astrobiology.

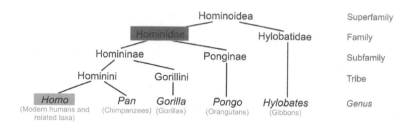

Figure 21.1 *A simple family phylogenetic tree – the emergence of the Hominidae (red) and the genus* Homo *(blue). Our closest non-*Homo *relatives are the chimpanzees* (Source: wikicommons).

family branched off from the gibbons (Figure 21.1). The Hominidae, or the great apes, are a group of large tailless primates. From this family a new group emerged about 14 million years ago, the Orangutans (Ponginae). They were followed about 2–3 million years ago by the gorilla and chimpanzees (*Pan*), also members of the Hominidae.

The beginning of our modern lineage is controversial. The australopithecines seem to have emerged about 4 million years ago and may have been the ancestors of our genus, *Homo*. The australopithecine best represented in the fossil record is *Australopithecus afarensis*. There are more than 100 fossil individuals, found from Northern Ethiopia, including the famous 'Lucy' and fossils in South Africa and Kenya. They show evidence for advanced bipedalism, but these ancestors still had the brain size of apes (~410 cubic centimeters).

The first member of our genus is thought to have emerged about 2.3 million years ago – *Homo habilis*. It had a brain about the same size as a chimpanzee (~630–700 cc), but during the next million years this doubled and the two species that emerged, *Homo erectus* (with a brain size of ~820–1100 cc) and *Homo ergaster*, are thought to have begun the journey out of the Rift Plains of Africa into Asia and Europe about 1.8 million years ago.

The exact nature of these migrations is not fully understood. Some researchers suggest that modern human populations of *Homo sapiens* left Africa more recently, from 100 000 to 50 000 years ago (Figure 21.2) and are derived from species such as *Homo heidelbergensis*. *Homo sapiens* represented yet another leap in brain size to an average of 1360 cc. These modern populations then displaced the existing populations of *Homo erectus* and species including *Homo neanderthalensis* (Figure 21.3). Remarkably, non-African present-day human DNA includes up to 6% Neanderthal DNA, determined from

the sequencing of Neanderthal DNA remains recovered from the bones of this species. This suggests that our early ancestors may have been up to more than just displacing their forerunner species. Archaeological evidence is thought to show that our ancestors may have coexisted with the Neanderthals for about 5000 years.

The evolutionary reasons for the rapid increase in brain size, cognitive capacity and tool-building ability are not fully understood. The retreat of forests, forcing early members of the species into the plains, may have provided a selection pressure for more mentally adept capacities for survival in this new environment. It is likely that any evolutionary innovation, once it is 'discovered' and provided it confers a selection advantage, will tend to be propagated. Developed cognitive capacities would seem to have many potential survival advantages. Amongst them we might list the better ability to evade predators, use tool building to hunt prey and escape environmental extremes, as well as the more complex social arrangements that might ensue from improved communicative abilities and language that confer survival to the individual as part of a well-coordinated group. However, if intelligence so obviously brings advantages, then another question is raised – why doesn't it emerge many times in animals and why didn't it emerge in the dinosaurs who ruled the Earth for over a hundred million years, orders of magnitude longer than our species?

The modern human cognitive capacity has led to impressive demonstrations of tool-building (Figure 21.4). It is not known when the human primate lineage first used tools. Evidence suggests that *Homo habilis* used pebbles as tools ~2.3 million years ago and this marks the beginning of the Paleolithic (the 'Stone Age'), which ran through until the beginning of the last ice age 10 000 years ago.

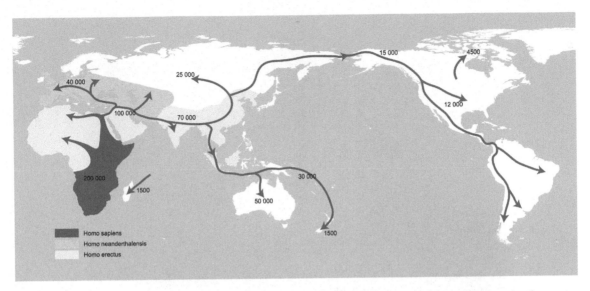

Figure 21.2 *Principal migratory routes of modern humans showing some estimated dates of major migrations.*

Figure 21.3 *A Neanderthal surveys its surroundings* (Source: wikicommons).

'Archaic' *Homo sapiens*, the forerunner of modern humans, is thought to have evolved between 400 000 and 250 000 years ago and was using stone tools.

About 50 000 years ago (although some put the date earlier), *Homo sapiens* began to engage in distinctly more complex and modern behaviour, including cave painting, burying dead, making specialist tools and jewellery. It is not known whether this behavioural change marked something of a revolution of cognitive capacities or a gradual change. As human tool-building abilities reached a more complex zenith, so the stage was set for the dramatic expansion, geographically and in numbers, of this new species of primate.

The birth of what we might think of as civilisation depends on what we define as a civilisation. If we consider the emergence of agriculture (when plants and animals were tended and harvested by communities) to be the first time fixed settlements came into existence, then this seems a reasonable point to place the transition of hunter-gathering bands of early humans to social organisation that we associate with a coordinated 'civilisation'.

The Neolithic Revolution, sometimes called the Agricultural Revolution, was the world's first revolution

Figure 21.4 *Tool-building ability: one of the distinguishing traits of human society is the ability to construct large artefacts.*

in agriculture and it allowed for an increasingly large population. Archaeological data indicates that the domestication of various types of plants and animals began in separate locations worldwide. Early agriculture is believed to have originated and become widespread in Southwest Asia around 10 000–9000 BP, though earlier individual sites have been identified. The **Fertile Crescent** region of Southwest Asia is the centre of domestication for three cereals (einkorn wheat, emmer wheat and barley) and four legumes (lentil, pea, bitter vetch and chickpea) and is thought to be one of the most important regions in which this early agricultural transition occurred. Some of the earliest domesticated animals included dogs (about 15 000 years ago), sheep, goats, cows and pigs.

The Neolithic Revolution instigated much more than the use of new food-producing techniques. It transformed the small and essentially nomadic groups of hunter-gatherers that dominated human pre-history into sedentary societies based in villages and, later, towns. These societies radically modified their natural environment by means of crop cultivation such as irrigation and deforestation and they introduced a range of related technologies including food storage. This technology allowed the development of extensive surplus food production, introducing robustness into the survival of societies and ultimately allowing them to

survive cold winter months in northern latitudes. These developments provided the basis for high population density, specialisation and the division of labour, the development of more elaborate forms of art, including fixed works of architecture, the first economies, centralised administration and political structures, systems of knowledge such as writing and eventually property ownership.

Despite the rapid expansion of our species across the globe, at least in terms of geological time spans, aided by its growing technological prowess at subjugating its environment, tool-building capacity does not secure our civilisation from the possibility of extinction or catastrophic collapse. What are these threats?

21.3   Threats to a Civilisation

A civilisation's tenure on a planet is not assured. As with all species, it faces a number of threats. Some of these are natural threats. In Chapter 15 we investigated just a few of these and how they have decreased species diversity during periods of mass extinction in the Phanerozoic. Our own civilisation faces a variety of possible threats; only a few of these could potentially destroy civilisations as a whole,

Figure 21.5 *Porcelain Basin, Yellowstone National Park. The park is the location of a dormant supervolcano* (Source: wikicommons).

but a large variety might incapacitate or seriously set back society.

Supervolcanoes are one potential threat. Yellowstone National Park is a volcanically active remnant of a supervolcano (Figure 21.5). About two million years ago, again at about 1.2 million years ago and yet again 640 000 years ago, this volcano erupted probably over many years. Its eruption two million years ago was sufficient to create an 80-km diameter crater. Such an event today would cripple the world economy and could plunge the Earth into a short-term **volcanic winter**, similar to the effects of an **impact winter**. More recent large volcanic eruptions have been recorded, albeit smaller than the Yellowstone eruptions. In 1815, the eruption of Tambora volcano in Indonesia injected about 200 million tonnes of sulfur-rich ash into the atmosphere. A global temperature drop averaging 0.7 °C occurred and 1816 was widely known as 'the year without a summer'.

Scientists compare the eruption magnitude (the mass of material ejected) and the eruption intensity (the rate at which it is erupted) using the **Volcanic Explosivity Index (VEI)**. This logarithmic scale goes from 0, which would be small eruptions on the islands of Hawaii to VEI of 3 or 4, which correspond to large eruptions that destroy houses and cause local devastation. The Tambora eruption, capable of causing global effects, reached a VEI

of 7. The historical Yellowstone eruptions would have VEI values of 8.

Even more devastating are flood basalts that we encountered as possible causes of extinction in Chapter 15. Capable of perturbing the global system over thousands of years, such events are thought to be capable of causing mass extinction. The onset of such an event during the prime of development of a civilisation would surely be a disaster for its progress. They cannot be easily measured using the VEI scale because they are not comprised of a single explosive eruption, but rather erupt over long time periods which could include explosive episodes.

Asteroid and comet impacts pose a realistic threat and there is good evidence for an extinction event 65.5 million years ago caused by these objects. In Chapter 15, the deleterious environmental effects of large asteroid impacts were considered. The implications for human civilisation are clear. Asteroids could potentially be diverted with enough warning. Comets, which travel faster and have more eccentric, unpredictable orbits, pose a more undefined threat.

Disease: it is not impossible that humanity could succumb to a catastrophic disease. The emergence of new viral strains such as Severe Acute Respiratory Syndrome (SARS) virus in 2009, the re-emergence of old viral strains such as the appearance of West Nile Virus in the United

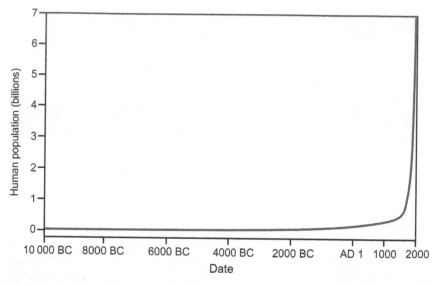

Figure 21.6   *The massive rise in the human population poses one serious threat to the sustainability of the species.*

States in 1999 and the spread of Ebola in 2014 reminds us that we are still very much at the mercy of microorganisms. Although genetic sequencing efforts have greatly enhanced our ability to unravel the genetics and biochemistry of our worst microbial foes, this information does not automatically give us the tools to control it. Pandemics are still a very real threat. Disease is unlikely to destroy civilisation completely, but it could cause a serious disruption and global turmoil.

If there are civilisations elsewhere, they may face similar perturbations from within their planet and from the astronomical environment, and perhaps even their own biological foes. Other civilisations may even face threats that are not considered a priority for our own civilisation. Supernovae explosions might pose a danger to civilisations whose parent star is close to other old stars nearing the end of their lives (such as in a dense star cluster). In our own specific case, we know of no close high mass star that is likely to explode within the next few million years and so this is not regarded as a near-term threat.

These natural threats on a planet draw us inexorably to the conclusion that the best way for a civilisation to ensure its long-term survival is not merely to find ways to combat these threats and minimise their effects, but also to spread itself in the Universe in such a way that it is not confined only to the surface of one planet. We will return to this possibility later.

As well as natural threats, a civilisation can also face threats of its own making. In our own case, there is the challenge of massive population growth (Figure 21.6). The detail of policies designed to deal with this problem need not detain us here, but from an astrobiological perspective an obvious question concerns the future trend of this population growth. It could stabilise in something akin to microbial stationary phase. It could crash in an event that might include disease, environmental deterioration and other stresses. It could continue to increase for much longer, but at a less extreme rate, reaching some sort of carrying capacity before stabilising. However, it is clear that a species cannot reproduce in a near exponential fashion indefinitely. It should also be noted, that barring some massive environmental disaster, a crash in population might be unlikely to make the species extinct. Many human-induced problems, such as overpopulation and resource depletion, threaten the size and scale of a civilisation, but probably do not fundamentally threaten its very existence.

21.4   Climate Change and the Challenge to Seven Billion Apes

An important phenomenon that has accompanied the emergence of humanity, and probably would be the case for any intelligence, is the capacity to change atmospheric composition. The atmosphere has a mass of 5.15×10^{18} kg, of which about three-quarters is within 11 km of the surface. The bulk of the atmosphere is therefore an extremely thin veneer on the Earth. As a civilisation emerges, extracts resources and generates waste gases from various processes, it is likely that this activity will become extensive enough to change the atmosphere. A crucial question is whether the civilisation becomes cognisant of this problem before it causes long-term environmental change in ways that may be extremely deleterious.

The discovery of the destruction of the UV-screening ozone layer (the so-called 'ozone hole') in the stratosphere by chlorofluorocarbons (CFCs) used in spray cans and refrigerators, led to the internationally agreed Montreal Convention, signed in 1987, to ban the use of these substances. Since that time, concentrations of CFCs in the atmosphere have begun to decline (Figure 21.7). I emphasise that here I am not going to discuss policy or government agreements. That is a discussion for other books. However, from an astrobiological perspective, the Montreal Convention illustrates one important point – that an intelligent species can develop a planetary-scale awareness of its impact on the home world and implement policies to reverse the damage. It is not the case that technological species with incredible capacity to manipulate their environment are doomed to cause irreversible destruction to their planet.

Unfortunately, there are many gases and waste products that a technological civilisation can produce. In our own case, increases in the concentrations of CO_2 (Figure 21.8) is attributed to the burning of fossil fuels, releasing the carbon for a long time stored in buried organic matter. These anthropogenic increases in CO_2 are associated with enhancing the greenhouse effect with all of its attendant

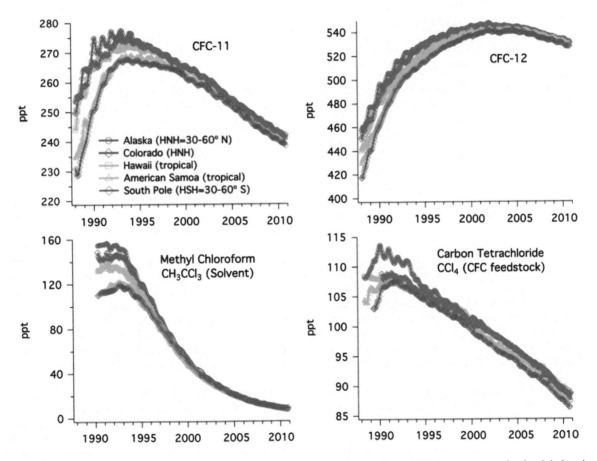

Figure 21.7 *The Montreal Convention, which banned the use of chlorofluorocarbons (CFCs), is one example of a global-scale protocol to change atmospheric composition that has been successful. Here the concentrations of some key CFCs (in parts per trillion; ppt) in different regions of the world are shown since the implementation of the ban* (Source: NOAA).

Monthly Carbon Dioxide Concentration

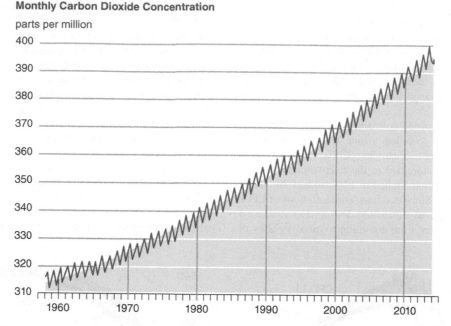

Figure 21.8 *The rise in carbon dioxide in the atmosphere measured at Mauna Loa, Hawaii since records began. The concentrations are shown in parts per million (ppm)* (Source: CO_2: Sémhur/Wikimedia Commons/CC-BY-SA-3.0).

knock-on effects, such as an increase in sea levels caused by the melting of ice sheets and glaciers and changes in the ranges of organisms as they begin to respond to changing temperatures. International attempts to curb CO_2 emissions again challenge a civilisation to generate planetary-scale agreement on how to manage its planetary atmosphere.

At the current time, these efforts are being overseen by the Intergovernmental Panel on Climate Change (IPCC) that assimilates the scientific data being gathered around the world and tries to achieve consensus on what the overwhelming weight of evidence is showing us. Global mean surface temperatures are estimated to rise by between 1.1 and 6.4 °C by 2100. The models predict an increase in sea level of between 18 and 59 cm. If the ice sheets of Greenland and Antarctica melt, this rise could be much higher, between 28 and 79 cm.

The effects of climate change are uncertain, as they depend on our efforts to cut emissions and uncertainties in the climate system. However, more chaotic weather systems, water shortages, major impacts on agricultural productivity and other knock-on effects have been predicted.

The response of the Earth to a gradual warming seems unlikely to be a simple linear response. Instead there

may be **tipping points** at which the system flips from one state into another. These could be precipitated by a diversity of mechanisms including the sudden melting of buried **methane gas hydrates** and rapid changes in ocean circulation patterns affecting heat transport across the planet in a short space of time. For example, the collapse of the North Atlantic Ocean Circulation, which brings warm water to northern Europe, would plunge northern Europe into temperatures more consistent with other parts of the Earth at the same latitude (e.g. Alaska). The very great concern with tipping points is that they are unpredictable and could cause sudden and very serious climatic perturbation in short time periods (decades). Past mass extinctions raise chilling questions about whether such sudden changes could set in motion other interlinked changes that are capable of causing collapse within the biosphere.

We cannot unravel the whole story of the human impact on the Earth here, but while we ponder the effect of our species on the Earth, there is another question about our long-term plan for our civilisation: will we move beyond the Earth? We will explore this question next and then return to consider whether these two objectives, living successfully on the Earth and settling space, are really so different.

21.5   The Human Future Beyond the Earth

In Chapter 20 we explored the Fermi paradox, which questions why we have not been visited by alien intelligences. Regardless of what we think the answer to the paradox might be, it raises the matter of whether humans themselves will ever leave the home planet and travel elsewhere.

Humans have long had aspirations to leave the Earth (Figure 21.9), ambitions that were first realised when Yuri Gagarin (1934–1968) made the first orbit of the Earth in the Vostok spacecraft in 1961 and fully implemented in 1969 when Neil Armstrong (1930–2012) became the first human to step onto the surface of another planetary body during the Apollo 11 mission to the Moon.

These efforts to leave the Earth were based on fundamental understandings derived from Newton's laws that allowed scientists to calculate the trajectories and energies required to reach other planetary bodies. The most important of these equations, and one of the earliest attempts to apply physics to space exploration, is the famous rocket equation.

21.5.1   The Rocket Equation

That simple physics could be used to predict how to get rockets to other planetary bodies was recognised long before the beginning of space travel. In 1903, Konstantin Tsiolkovsky (1857–1935) derived the **ideal rocket equation**. The equation relates the 'delta-v' (Δv) with the effective exhaust velocity of the rocket and the initial and final mass of a rocket (or other reaction engine). Delta-v is the maximum change of speed of the rocket if no other external forces act on it. It can be considered colloquially to be the 'effort' needed to change a rocket's direction. The equation is given as:

$$\Delta v = v_e \ln(m_0/m_1) \qquad (21.1)$$

where v_e is the rocket exhaust velocity, m_0 is the initial mass and m_1 is the final mass of the rocket (which is different from the initial mass because the rocket uses propellant). Note that if we are launching a rocket from a planetary body then Δv is the **escape velocity**.

The velocity of the rocket exhaust can also be expressed as a function of the specific impulse. The specific impulse (usually abbreviated I_{sp}) is the force of the rocket with respect to the amount of propellant used per unit time. It tells you the efficiency of the rocket engine. The higher the specific impulse, the lower the propellant flow rate required for a given thrust. It can be expressed as:

$$I_{sp} = v_e/g_0 \qquad (21.2)$$

where g_0 is the nominal gravitational acceleration of an object in a vacuum near the surface of the Earth (defined as 9.81 m s^{-2}).

Figure 21.9   *A station on the Moon* (Source: NASA).

The ideal rocket equation also allows a person to calculate the energy that must be applied to move a rocket from one trajectory to another, for example in moving from a trajectory towards a planet to insertion into orbit around the planet, perhaps to land on it.

From this simple equation, the science of orbital mechanics was born and with it the possibility of planning human voyages beyond the Earth.

21.6   Settling the Solar System

The human exploration of the Moon during the Apollo Program (1961–1972) was a relatively brief sojourn beyond the Earth. However, if humanity is to achieve a presence beyond its home world permanently, then it must build bases capable of self-sufficiency (Figure 21.9).

The political and scientific motives for travelling to other worlds vary. Mars has a history (Chapter 17) that bears similarities to the Earth in its early history. This fact motivates the exploration of the planet, both to search for life and to understand better the evolution of terrestrial rocky planets. However, the Moon is closer than Mars (with transit times on the order of two to three days in contrast to months for Mars), which makes it a more appealing destination to optimise life support technology for the exploration of other planetary bodies (Figure 21.10). On the Moon, errors can be more easily rectified and the failure of systems is less likely to result in a stranded population facing death.

As we have seen in Chapter 17, Mars may crudely be viewed as the most Earth-like of the planetary bodies of the inner Solar System and considerable efforts have gone into plans for the human exploration and settlement of the planet. Many vital resources can be extracted from its environment. For example, water is present in the surface as permafrost, which could be dug from the soil, melted and cleaned. Although the atmosphere has a low water content (~0.03%), this could be extracted in large volumes in a fan-like extractor rig, condensed and collected. The situation is more challenging when establishing a station on the Moon. In this location, water ice might be obtained from ice in permanently shadowed craters at the polar regions or it must be shipped in from external sources.

On Mars, fuel for rovers, heating and other systems could be made from methane in the Sabatier reaction operated optimally at 300–400 °C using a nickel or ruthenium catalyst:

$$CO_2 + 4 \ H_2 \rightarrow CH_4 + 2 \ H_2O + energy$$

Figure 21.10   *The exploration of Mars* (Source: NASA).

The H_2 would be acquired from the electrolysis of water and the CO_2 extracted from the atmosphere (which contains 95.3%). The methane could be liquefied and used for a large variety of fuel-requiring processes.

Another useful resource to gather on Mars is carbon monoxide. CO can be produced by a number of pathways. CO_2 can be reacted with carbon at high temperatures (~800°C) in the Boudouard reaction:

$$CO_2 + C \rightarrow 2CO$$

Another source is 'water gas', a mixture of hydrogen and carbon monoxide produced via the endothermic reaction of steam (produced from Martian water) and carbon:

$$H_2O + C \rightarrow H_2 + CO$$

CO also exists in the Martian atmosphere, although at a low concentration of 0.08%. Once CO is produced, 'syngas' (CO and H_2) can be used for producing diesel, petroleum products, lubricants, plastics and so on. A whole variety of industrial processes can be envisaged for Mars.

Debate Point: Terraforming

Ambitious plans for settling other planets have been proposed that involve the alteration of planetary atmospheric conditions to better suit life and eventually humans. Given our own impact on the terrestrial atmosphere discussed earlier, it may not be so far-fetched to think that an atmosphere could be intentionally altered. In the case of Mars, for instance, warming the planet by injecting CFCs into the atmosphere (synthesised from rocks on the surface) is proposed as a means to initiate a greenhouse effect, to release CO_2 into the atmosphere, thus increasing atmosphere pressure and enabling liquid water to be sustained on the surface. Under such conditions cyanobacteria and algae might be introduced to produce biomass and oxygen, with increasing complexity of organisms possible as the oxygen concentrations of the atmosphere increase. The time it would take to pull off such schemes vary wildly between different calculations and proponents, from centuries to many millennia. Discuss both the technical and ethical aspects of terraforming. What are the major technical uncertainties in such an enterprise based on the feedback processes that you understand are important for habitability? Consider the case of Mars. What effect would the lack of plate tectonics have on the very long-term sustainability of a terraformed atmosphere? Even if we could do it, would it be ethically acceptable to alter a planetary body if we had not definitively ruled out life there?

Birch, P. (1992) Terraforming Mars quickly. *Journal of the British Interplanetary Society* **45**, 331–340.

Fogg, M.J. (1995) *Terraforming: Engineering Planetary Environments*, SAE International, Warrendale, PA.

McKay, C.P., Toon, O.B., Kasting, J.F. (1991) Making Mars habitable. *Nature* **352**, 489–496.

Debate Point: What does it take to live in self-sufficiency on the Moon and Mars?

Each planetary body in the Solar System has unique combinations of physical extremes and available resources. The Moon has water in polar craters, but lacks an atmosphere. In contrast, on Mars the CO_2-dominated atmosphere could be a source of the gas to generate methane for fuel. The lack of an atmosphere on the Moon makes it more exposed to ionising radiation. The Moon, however, is closer to the Earth and so would provide the opportunity for more rapid emergency return to the Earth. Draw up a list of pros and cons of selecting the Moon and Mars as a target for a human settlement, discussing physical differences, resource differences and logistical differences for human explorers. Use this information to decide which planetary body makes a better choice for a first human settlement.

Seedhouse, E. (2008) *Lunar Outpost: The Challenges of Establishing a Human Settlement on the Moon*. Springer, Heidelberg.

Zubrin, R. (2011) *The Case for Mars. The Plan to Settle the Red Planet and Why We Must*. Free Press, New York.

Although Mars is further away from the Sun than the Earth and light levels are on average ~43% of those on Earth (varying depending on its location in its orbit), these are high enough to grow plants. Martian **regolith**, which is mainly ground volcanic rock, could be used as a substrate for growth. Some regions of Mars have high concentrations of salts such as sulfates. Perchlorates are thought to be ubiquitous at ~0.4–0.6% weight in Martian soils. Thus, some regolith materials may need washing of these contaminants. Lacking in significant ozone screening, the surface is exposed to damaging UV radiation down to 200 nm, but this can be screened by glass (itself fabricated from silica-rich rocks) or plastics. Thus, UV screening greenhouse structures for crop growth are possible on the surface.

As with most planetary bodies, Mars is exposed to higher levels of ionising radiation than the Earth. Its atmosphere provides an equivalent screening of $16\,g\,cm^{-3}$ of aluminium (compared to the Earth's of $\sim1000\,g\,cm^{-3}$). The two orders of magnitude higher ionising flux and the greater intensity of particles fluxes from **Solar Particle Events** (SPE), which are high-intensity particle (mainly proton) storms, mandate shielding, such as covering a station in Martian regolith.

All factors considered, it is plausible to image the establishment of a human settlement on Mars using locally sourced materials to provide much of what people need in terms of basic survival materials. A self-sustaining branch of human civilisation on that planet seems possible.

Beyond the inner Solar System, there are a variety of choices for building human settlements. The asteroid belt offers vast resources. The low gravity of the objects there means that comparatively little energy is required to escape the **gravity well** of asteroids compared to planets. For example, the escape velocity of Mars is $5.0\,km\,s^{-1}$. By contrast, the escape velocity of Ceres (950 km in diameter), the largest asteroid, is 0.51 km s^{-1}. All other asteroids have even lower escape velocities, making it easy to leave them and move among them.

Unlike the Earth, where heavier metals are differentiated close to the core, apart from in surface ores, metals in many asteroids are distributed throughout them, making them easier to extract. Asteroids contain valuable and useful materials such as iron, nickel and rare platinum group metals, often in significantly higher concentration than found in mines on Earth. As many crucial elements such as gold, cobalt, manganese, iron, osmium, platinum, rhodium and so on are being extracted and used from the Earth's crust, it is becoming increasingly difficult to find high concentrations of the elements, making space-based resources more economically compelling. S-type asteroids, which make up about 17% of asteroids, are some of the most industrially promising targets. A small 10-m S-type asteroid contains about 650 000 kg of metal with 50 kg in the form of metals such as platinum and gold. M-type asteroids are rarer, but contain about ten times as much metal.

The economics of such enterprises cannot be readily calculated. The successful collection of such material might bring down market prices on the Earth by enhancing availability, so there is an economic feedback which is not readily predictable. Certainly the quantity of the materials available is not in doubt. The asteroid 16

Psyche, a 200-km diameter asteroid, and one of the most massive M-type asteroids in the asteroid belt, contains about 1.5×10^{19} kg of nickel–iron, which could supply the world production requirement for several million years. A small metallic asteroid of ~2 km diameter contains more than US$ 20 trillion worth of industrial and precious metals at current market prices.

Asteroids not only provide a source of metals of interest to Earth industry, but they are also rich sources of volatiles such as water to sustain a long-term human presence in space, with the water being used to make oxygen and other commodities such as fuel.

Going further afield and establishing settlements beyond the asteroid belt requires identifying locations that are relatively stable and free of the enormous radiation fields associated with the giant gas planets. In the Jovian system, Callisto has been suggested as a potential location for a station (Figure 21.11).

As the furthest out of the Galilean satellites, Callisto experiences the lowest radiation field and the low tidal forces mean that the surface is ancient and therefore geologically relatively stable compared to the inner moons. The moon is subject to 0.036 **sieverts** (Sv) year^{-1}. On the Earth, a typical person receives about 0.006 Sv year^{-1} from natural background sources. Radiation shielding on Callisto would be necessary to bring doses down to acceptable levels, but the engineering challenge is

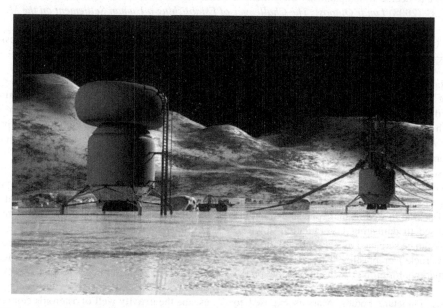

Figure 21.11 *The geological stability and relatively low radiation (compared to the other Jovian moons) has made Callisto a suggested target as a base and refuelling station for deep space missions to the outer Solar System* (Source: NASA).

Figure 21.12 *A toroidal spaceship houses many thousands of humans* (Source: NASA).

considerably less than the surface of Europa, for example, where the *daily* dose is 5.4 Sv, enough to kill a human (Chapter 18).

Ultimately, for the human settlement of the Solar System to be significant, it would be necessary to build large spaceships capable of housing many tens of thousands of people.

Gerard O'Neill (1927–1992) was an American physicist who considered designs for large toroidal and cylindrical ships in which artificial gravity, created by rotation, would allow for large populations to be sustained (Figure 21.12) in conditions approximating to the Earth (Figure 21.13). These ideas are currently beyond our engineering capabilities, but in principle, if resources could be mined and extracted from asteroids, thus avoiding the enormous energetic costs of transporting them from the Earth, we could envisage the assembly of these types of human settlements.

21.7 Avoiding Extinction or Collapse: A Multiplanet Species

There are many reasons to build permanent settlements in space. Earlier, I alluded to the vast resources in the asteroid belt. However, one very significant reason is the possibility of avoiding extinction or at least providing robustness and redundancy against a catastrophic collapse of civilisation. There are a variety of disasters that could befall a civilisation on the Earth, as we have seen. Some of these we might be able to mitigate, such as asteroid impacts and climate change. However, supervolcanoes,

Figure 21.13 *The interior of a large-scale spaceship houses entire communities of people* (Source: NASA).

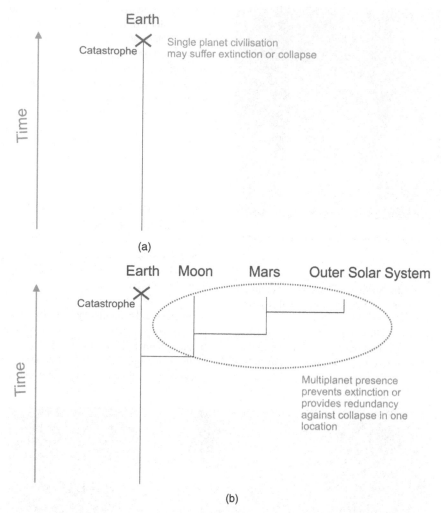

Figure 21.14   *Becoming a multi-planet species. (a) A civilisation located on one planet (represented as a single black line) has a chance that a major catastrophe will destroy or cripple it. (b) A civilisation that has independent, self-sustaining branches in different parts of the Solar System (represented as branching black lines) has a chance of surviving a catastrophe that destroys or temporarily incapacitates one of those branches* (Source: Charles Cockell).

fast-moving comets and other unknown disasters mean that we can never ensure that civilisation is free of the possibility of extinction or catastrophic collapse.

However, if we build a civilisation that is dispersed on a variety of planetary bodies, we increase the chances of avoiding extinction. The destruction of civilisation would require a Solar System-wide catastrophe. In the less extreme case, other branches would provide redundancy and robustness to the collapse (even if not extinction) of civilisation in one location. If the presence in space becomes expansive enough and its economic scale sufficiently large, it might even be possible that space branches of humanity could help Earth civilisation achieve a faster reconstruction if disaster befell it. This is the concept of a **multiplanet species** and it is illustrated in Figure 21.14.

When we consider humanity's future from this point of view, we realise that space technology is more than just an exploratory flutter from rich nations. It may eventually be essential to our survival. Space technology and exploration

Figure 21.15 *A dramatic depiction of dinosaur extinction. Unless we build asteroid and comet deflection programmes and establish permanent settlements beyond the Earth as insurance, our civilisation will be prone to the same disaster depicted here. Intelligence on its own does not secure the future of a species* (Source: wikicommons).

has the potential to allow us to avoid extinction. Other technology improves our quality of life and our capacities, but does not fundamentally change our long-term prospects for survival as a species.

Merely being intelligent is not sufficient to avoid extinction. A species must also use its intelligence to build self-sustaining settlements beyond the Earth. The dinosaurs and many other organisms met their fate 65.5 million years ago (Figure 21.15). At the time of writing, human civilisation, despite its superior intelligence and technology, is still, like the dinosaurs, at the mercy of asteroid impact.

21.8   Environmentalism and Space Exploration as a Single Goal?

There is a tendency to view the exploration and settlement of space and the environmental protection of the Earth as an either/or choice. Some environmentalists view space exploration as distracting from the urgent problems we face on Earth, such as climate change. Some proponents of space settlement see environmentalism as an inward-looking perspective that distracts from the longer-term vision of establishing a civilisation beyond the Earth. It is therefore not unusual to find the view that we are at a branch-point where we must decide which of these objectives is a priority.

However, we could view environmentalism and space settlement as one and the same goal. Both of them seek to achieve the sustainable continuity of humans in the cosmos. One of them happens to focus on the Earth, the other on space. Both groups are concerned with the maintenance of a life support system. In the case of the Earth, this life support system is the planetary-scale biosphere, in the case of space it is life support systems on the Moon, Mars or elsewhere.

There are many synergies between the two objectives. As we have already seen in this book, analogue environments on Earth have similarities to extraterrestrial environments and can be used to gain insights into the geology, geochemistry and potential habitability of extraterrestrial environments (Chapter 17). Extraterrestrial environments can yield new information with which to understand the history of the Earth. Therefore, the environmental study of the Earth and other planetary bodies are inextricably linked.

There are technological synergies. Satellite technology and imaging sensors used to study the Earth can be used to study other planets and vice versa. Technologies used to survive in extraterrestrial environments where water and food must be grown in conditions without naturally occurring liquid water may improve technologies for recycling and waste reclamation on the Earth. In short, the technology we need to live on the Earth whilst minimising destruction to the environment is, in principle, the same sort of technology we need to live efficiently elsewhere.

Seen from this perspective, perhaps an intelligent species that seeks to optimise its persistence on a planetary surface and in outer space is actually engaging in exactly the same activity – the creation of sustainable populations in the cosmos.

21.9   Sociology: The Overview Effect

Is the exploration and settlement of space and the development of a planetary-scale perspective of life on Earth merely technological or can a species begin to change its view of its place in the Universe? With this question astrobiology enters into distinctly philosophical and sociological realms, but it is a question that we must answer to better understand how intelligence develops on a planet.

In the 1980s, an interesting perspective was offered by Frank White in his book, *The Overview Effect*. Here he proposed that once a large number of humanity directly sees for itself the Earth from space, in its serene isolation and insignificant scale compared to the cosmos, there may occur a change in perspective as a greater proportion of the species begin to realise the futile pettiness of national borders and conflict. He referred to this planetary-scale perspective as The Overview Effect and used evidence from astronauts to bolster this idea of a change in perspective offered by journeying into space.

Whatever we think about the capacity for change in basic human outlook, White's book raises a more fundamental question of whether the experience of space exploration can change the perspective of a species. Are we forever bound to the competitive and sometimes destructive behaviours that we presume arise from our evolutionary history or is it possible for a species to develop a more connected planetary perspective? We have already seen that international accords such as the Montreal convention on CFCs can yield real results, but are these planetary-scale actions merely examples of how a species can act on a large scale when it is forced to by expediency? Is it possible for a cultural perspective on a planetary scale to become inwrought into the way in which a species behaves?

21.10   Will We Become Interstellar?

Finally, we could ask the question – will humans not merely leave the Earth, but eventually leave the Solar System? In the very long-term, the Earth, at least in its current location (and assuming it is not moved using technology!) will be doomed by the Sun's transition off the main sequence into Red Giant status. However, the Earth will become uninhabitable long before this time when a moist greenhouse removes the surface water (Chapter 16). On those time scales, if we want to survive, we will be forced to move out.

To move beyond the Earth, we would first have to find a convincing second home. New technological capacities that allow us to search for Earth-like planets around other stars offer us the possibility of detecting potentially habitable worlds, even those with oxygen and colonisation potential. Whether we progress to considering these as potential homes drives us into ethical and philosophical problems – would it be unethical to colonise a planet that already had life, even if only single-celled?

However, the ethical challenge may well be foreshadowed by the technological problem of making human journeys beyond our star system. It is possible to build small robotic spacecraft that can achieve interstellar flight. In 2013, Voyager 1 (Figure 21.16), a 722-kg spacecraft launched in 1977 and originally sent to the outer Solar System to explore the giant gas planets, left the heliopause and entered interstellar space. Voyager 1 will reach the Oort cloud in about 300 years and take about 30 000 years to pass through it. In about 40 000 years, it will pass within about 1.5 light years of the star Gliese 445, which is in the constellation Camelopardalis.

To reach the nearest stars, travelling at the speeds of Voyager, would be multi-millennia trips. A variety of propulsion technologies have been proposed for robotic and human interstellar craft to achieve faster speeds up to fractions of the speed of light, including nuclear fission and fusion and matter–antimatter collisions.

Faster than light travel remains in the realms of technological speculation. Einstein's Theory of General Relativity predicts a curved space–time. Wormholes, which are hypothetical links between different areas of this curved space–time, have been a favoured suggestion for how one could move from one point to another in the Universe without having to violate faster than light speed travel. However, there is no evidence that such features exist or, even if they do exist, that it would be possible to pass a macroscopic object through them, such as a spaceship.

Even if we cannot travel between the stars in short time periods, we could conceivably accept that very long journey durations are a necessary reality and build spaceships, or worldships, that would allow large populations of people to travel over multi-generational time scales to distant stars. Would such societies survive?

I have often been challenged with the observation that after multi-millennia year journeys across interstellar

Figure 21.16   *Human civilisation became an interstellar spacefaring civilisation in 2013. The Voyager 1 spacecraft, launched in 1977, left the Solar System after exploring the giant gas planets of the Solar System. Will humans follow it to other stars? (Source: NASA).*

space towards a new star, societies of humans would forget what their original purpose was and merely exist for the sake of existing. The mission would be pointless. But we might ask in return: is this any different to all of us on Spaceship Earth? Orbiting round and round a star that is heading to no particular destination, what is our civilisation's long-term objective? What do you think your role is in this future?

21.11   Conclusions

In a relatively short time, at least in geological terms, a species has developed that has the ability to manipulate its environment in ways that have allowed it to build spaceships and contemplate establishing a permanent human presence beyond the Earth. The future of this species is uncertain. It has the potential to mitigate certain natural disasters, such as asteroid impacts, but it faces other challenges, from population control to resource depletion. Outer space cannot be regarded as a panacea for solving many of the challenges of an emerging civilisation, but access to resources, energy, protecting the Earth from impact and avoiding extinction by becoming

a multiplanet species are just a few of the benefits that could be derived from establishing a permanent presence beyond the Earth. Whilst it does these things, this species must also learn how to sustainably live on one planet that can support its presence – the Earth. Eventually this species may contemplate the challenge of leaving the Solar System and travelling to other stars.

In all these endeavours and challenges on Earth and in space, astrobiology is the science that will drive us forwards.

Further Reading

Books

Cockell, C.S. (2006) *Space on Earth: Saving Our World by Seeking Others.* MacMillan, London.

Lewis, J.S. (1997) *Mining the Sky: Untold Riches from the Asteroids, Comets and Planets.* Basic Books, New York.

O'Neill, G. (1977) *The High Frontier: Human Colonies in Space.* Morrow, New York.

Rees, M. (2004) *Our Final Century: Will Civilisation Survive the Twenty-first Century?: Will the Human Race Survive the Twenty-first Century?* Arrow Books, London.

White, F. (1987) *The Overview Effect: Space Exploration and Human Evolution*. Houghton–Mifflin, New York.

Zubrin, R. (2012) *The Case for Mars. The Plan to Settle the Red Planet and Why we Must*. Simon and Shuster, New York.

Papers

Bostrum, N. (2013) Existential risk prevention as global priority. *Global Policy* **4**, 15–31.

Cirkovic, M.M. (2004) Forecast for the next eon: applied cosmology and the long-term fate of intelligent beings. *Foundations of Physics* **34**, 239–261.

Crawford, I.A. (2009) The astronomical, astrobiological and planetary science case for interstellar spaceflight. *Journal of the British Interplanetary Society* **62**, 415–421.

Hopfenberg, R., Pimentel, D. (2001) Human population numbers as a function of food supply. *Environment, Development and Sustainability* **3**, 1–15.

Krauss, L.M., Starkman, G.D. (2000) Life, the universe, and nothing: life and death in an ever-expanding universe. *Astrophysical Journal* **531**, 22–30.

Landis, G.A. (1995) Footsteps to Mars: an incremental approach to Mars exploration. *Journal of the British Interplanetary Society* **48**, 367–342.

Mellars, P. (2006) Why did modern human populations disperse from Africa ca. 60,000 years ago? *Proceedings of the National Academy of Sciences* **103**, 9381–9386.

Miller, C.F., Wark, D.A. (2008) Supervolcanoes and their explosive supereruptions. *Elements* **4**, 11–16.

Troutman, P.A., Bethke, K., Stillwagen, F., Caldwell, D.L., Manvi, R., Strickland, C., Krizan, S.A. (2003) Revolutionary Concepts for Human Outer Planet Exploration (HOPE). *American Institute of Physics Conference Proceedings* **654**, 821–828.

Wood, B., Collard, M. (1999) The changing face of Genus *Homo*. *Evolutionary Anthropology* **8**, 195–207.

Appendix

A.1 The Astrobiological Periodic Table

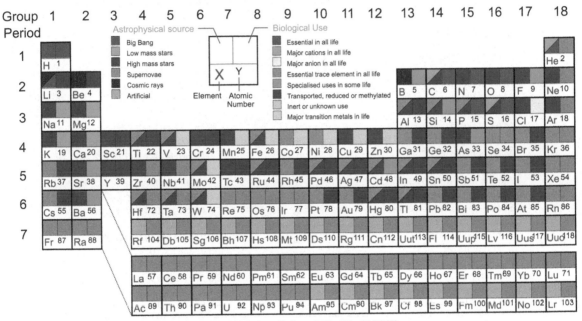

The Astrobiological Periodic Table
© Charles S Cockell, v. 1.0 [June 2015]: The Astrobiological Periodic Table

Biological data from Wackett, L.P., Dodge, A.G., Ellis, L.B.M. (2004) *Applied and Environmental Microbiology* **70**, 647-655.

A.2 Units and Scales

A.2.1 Standard International Base Units

Quantity	Name	Symbol
length	metre	m
mass	kilogram	kg
time	second	s
electric current	ampere	A
thermodynamic temperature	kelvin	K
amount of substance	mole	mol
luminous intensity	candela	cd

A.2.2 Basic Physical Constants

Name	Symbol	Value
Speed of light	c	$2.99792458 \times 10^8 \ m/s$
Planck constant	h	$6.6260755 \times 10^{-34} \ J \cdot s$
Planck constant	h	$4.1356692 \times 10^{-15} \ eV \cdot s$
Planck hbar	$\hbar$	$1.0545727 \times 10^{-34} \ J \cdot s$
Planck hbar	$\hbar$	$6.582121 \times 10^{-16} \ eV \cdot s$
Gravitation constant	G	$6.67259 \times 10^{-11} \ m^3 \cdot kg^{-1} \cdot s^{-2}$
Boltzmann constant	k	$1.380658 \times 10^{-23} \ J/K$
Boltzmann constant	k	$8.617385 \times 10^{-5} \ eV/K$
Universal gas constant	R	$8.314510 \ J/mol \cdot K$
Charge of electron	e	$1.60217733 \times 10^{-19} \ C$
Permeability of vacuum	μ_0	$4\pi \times 10^{-7} \ N/A^2$
Permittivity of vacuum	ε_0	$8.854187817 \times 10^{-12} \ F/m$
Coulomb constant	$1/4\pi\varepsilon_0 = K$	$8.987552 \times 10^9 \ N \cdot m^2/C^2$
Faraday constant	F	$96485.309 \ C/mol$
Mass of electron	m_e	$9.1093897 \times 10^{-31} \ kg$
Mass of electron	m_e	$0.51099906 \ MeV/c^2$
Mass of proton	m_p	$1.6726231 \times 10^{-27} \ kg$
Mass of proton	m_p	$938.27231 \ MeV/c^2$
Mass of neutron	m_n	$1.6749286 \times 10^{-27} \ kg$
Mass of neutron	m_n	$939.56563 \ MeV/c^2$
Atomic mass unit	u	$1.6605402 \times 10^{-27} \ kg$
Atomic mass unit	u	$931.49432 \ MeV/c^2$
Avogadro's number	N_A	$6.0221367 \times 10^{23} \ /mol$
Stefan–Boltzmann constant	σ	$5.67051 \times 10^{-8} \ W/m^2 \cdot K^4$
Rydberg constant	R_∞	$10973731.534 \ m^{-1}$
Bohr magneton	μ_B	$9.2740154 \times 10^{-24} \ J/T$
Bohr magneton	μ_B	$5.788382 \times 10^{-5} \ eV/T$
Flux quantum	Φ_0	$2.067834 \times 10^{-15} T/m^2$
Bohr radius	a_0	$0.529177249 \times 10^{-10} m$
Standard atmosphere	atm	$101325 \ Pa$
Wien displacement constant	b	$2.897756 \times 10^{-3} \ m \cdot K$

A.3 Temperature Scale Conversion

	from Celsius	to Celsius
Fahrenheit	$[°F] = [°C] \times {}^9/_5 + 32$	$[°C] = ([°F] - 32) \times {}^5/_9$
Kelvin	$[K] = [°C] + 273.15$	$[°C] = [K] - 273.15$

A.4   Composition of the Sun

Gas	Abundance (% of total number of atoms)	Abundance (% of total mass)
Hydrogen	91.2	71.0
Helium	8.7	27.1
Oxygen	0.078	0.97
Carbon	0.043	0.40
Nitrogen	0.0088	0.096
Silicon	0.0045	0.099
Magnesium	0.0038	0.076
Neon	0.0035	0.058
Iron	0.0030	0.14
Sulfur	0.0015	0.040

A.5   Some of the Major Star Types and Temperatures and Colour

Stellar class	Effective temperature	Conventional colour description	Actual apparent colour	Mass (solar masses)	Radius (solar radii)	Luminosity (bolometric)	Fraction of all main-sequence stars
O	$\geq 30\,000$ K	Blue	Blue	$\geq 16\,M_\odot$	$\geq 6.6\,R_\odot$	$\geq 30\,000\,L_\odot$	~0.00003%
B	$10\,000–30\,000$ K	Blue white	Deep blue white	$2.1–16\,M_\odot$	$1.8–6.6\,R_\odot$	$25–30\,000\,L_\odot$	~0.1%
A	$7500–10\,000$ K	White	Blue white	$1.4–2.1\,M_\odot$	$1.4–1.8\,R_\odot$	$5–25\,L_\odot$	0.6%
F	$6000–7500$ K	Yellow white	White	$1.04–1.4\,M_\odot$	$1.15–1.4\,R_\odot$	$1.5–5.0\,L_\odot$	3%
G	$5200–6000$ K	Yellow	Yellowish white	$0.8–1.04\,M_\odot$	$0.96–1.15\,R_\odot$	$0.6–1.5\,L_\odot$	7.5%
K	$3700–5200$ K	Orange	Pale yellow orange	$0.45–0.8\,M_\odot$	$0.7–0.96\,R_\odot$	$0.08–0.6\,L_\odot$	12%
M	$2400–3700$ K	Red	Light orange red	$0.08–0.45\,M_\odot$	$\leq 0.7\,R_\odot$	$\leq 0.08\,L_\odot$	76%

A.6   Three- and One-letter Designations of Amino Acids

Amino Acid	3-Letter Code	1-Letter Code
Alanine	Ala	A
Cysteine	Cys	C
Aspartic acid or aspartate	Asp	D
Glutamic acid glutamate	Glu	E
Phenylalanine	Phe	F
Glycine	Gly	G
Histidine	His	H
Isoleucine	Ile	I
Lysine	Lys	K
Leucine	Leu	L

Amino Acid	3-Letter Code	1-Letter Code
Methionine	Met	M
Asparagine	Asn	N
Proline	Pro	P
Glutamine	Gln	Q
Arginine	Arg	R
Serine	Ser	S
Threonine	Thr	T
Valine	Val	V
Tryptophan	Trp	W
Tyrosine	Tyr	Y

A.7   Codon Table for the Genetic Code (also shown in Chapter 4; Figure 4.12)

A.8 Planetary Data

	Mercury	Venus	Earth	Moon	Mars	Jupiter	Saturn	Uranus	Neptune	Pluto
Mass (10^{24} kg)	0.330	4.87	5.97	0.073	0.642	1898	568	86.8	102	0.0131
Diameter (km)	4879	12,104	12,756	3475	6792	142 984	120 536	51 118	49 528	2390
Density (kg m^{-3})	5427	5243	5514	3340	3933	1326	687	1271	1638	1830
Gravity (m s^{-2})	3.7	8.9	9.8	1.6	3.7	23.1	9.0	8.7	11.0	0.6
Escape velocity (km s^{-1})	4.3	10.4	11.2	2.4	5.0	59.5	35.5	21.3	23.5	1.1
Rotation period (h)	1407.6	−5832.5	23.9	655.7	24.6	9.9	10.7	−17.2	16.1	−153.3
Length of day (h)	4222.6	2802.0	24.0	708.7	24.7	9.9	10.7	17.2	16.1	153.3
Distance from Sun (10^6 km)	57.9	108.2	149.6	0.384	227.9	778.6	1433.5	2872.5	4495.1	5870.0
Perihelion (10^6 km)	46.0	107.5	147.1	0.363	206.6	740.5	1352.6	2741.3	4444.5	4435.0
Aphelion (10^6 km)	69.8	108.9	152.1	0.406	249.2	816.6	1514.5	3003.6	4545.7	7304.3
Orbital period (days)	88.0	224.7	365.2	27.3	687.0	4331	10 747	30 589	59 800	90 588
Orbital velocity (km s^{-1})	47.4	35.0	29.8	1.0	24.1	13.1	9.7	6.8	5.4	4.7
Orbital inclination (degrees)	7.0	3.4	0.0	5.1	1.9	1.3	2.5	0.8	1.8	17.2
Orbital eccentricity	0.205	0.007	0.017	0.055	0.094	0.049	0.057	0.046	0.011	0.244
Obliquity (degrees)	0.01	177.4	23.4	6.7	25.2	3.1	26.7	97.8	28.3	122.5
Mean temperature (°C)	167	464	15	−20	−65	−110	−140	−195	−200	−225
Surface pressure (bar)	0	92	1	0	0.01	Unknown	Unknown	Unknown	Unknown	0
Number of moons	0	0	1	0	2	67	62	27	14	5
Ring system?	No	No	No	No	No	Yes	Yes	Yes	Yes	No
Global magnetic field?	Yes	No	Yes	No	No	Yes	Yes	Yes	Yes	Unknown

A.9 Geological Time Scale

From: Cohen, K.M., Finney, S.C., Gibbard, P.L., Fan, J.X. (2013) The ICS International Chronostratigraphic Chart. *Episodes* **36**, 199–204.

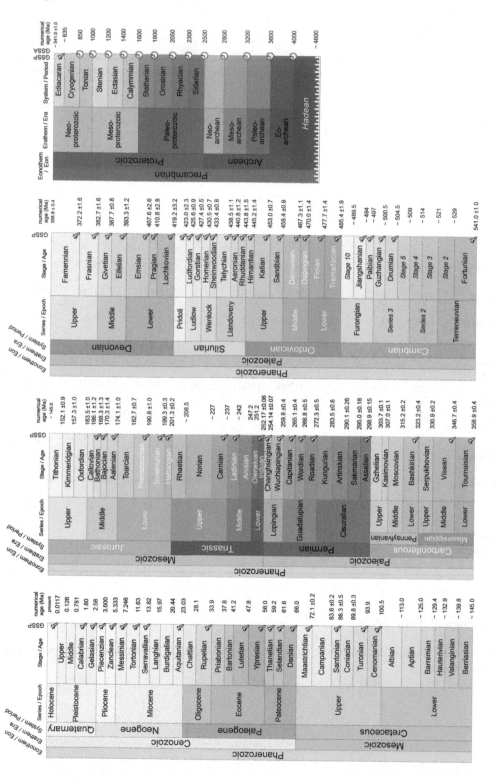

Glossary

16S rRNA. A specific type of bacterial and archaeal ribosomal RNA. The DNA sequence that codes for 16S rRNA is used to build *phylogenetic trees* and analyse microbial diversity.

Abiotic. Non-biological matter.

Absorption spectrum. Absorption spectroscopy refers to techniques that measure the absorption of radiation, as a function of frequency or wavelength, due to its interaction with a sample such as a gas.

Accretion. The gravitational accumulation of matter to form larger objects.

Achondrites. A stony meteorite lacking *chondrules* with a similar composition to terrestrial *mafic* rocks. Achondrites are inferred to have once been part of planetary bodies or asteroids that have undergone *differentiation* or partial differentiation and have undergone melting and recrystallisation.

Acetate. A simple organic ion with the formula $C_2H_3O_2^-$.

Acidophile. An organism that needs to live in environments with low pH.

Action spectrum. A plot of wavelength of light against biological effect for a specific biological process (e.g. photosynthesis, DNA damage), particularly used for specifying effects of UV radiation.

Active site. The active site is a small region in an *enzyme* where substrate molecules bind and undergo a chemical reaction.

Adaptive optics. A method of correcting for atmospheric perturbations in astronomical observations by changing telescopic optics.

Adaptive radiation. In evolutionary biology, adaptive radiation is a process in which organisms diversify rapidly into new forms, particularly when a change in the environment makes new resources available, creates new challenges or opens environmental *niches*, such as a mass extinction.

Adenosine triphosphate (ATP). The energy molecule of life with energy stored in its phosphate bonds.

AGB star. The asymptotic giant branch (AGB) is the region of the *Hertzsprung–Russell diagram* populated by evolving low- to medium-mass stars (0.6–10.0 solar masses). The stars are characterised by a carbon core and double shell burning of helium and hydrogen.

Acetogenesis. A process through which acetate ($C_2H_3O_2^-$) is produced from CO_2 and an electron source (e.g., H_2, CO, formate, etc.) by *anaerobic* bacteria.

Alpha particle. Alpha particles consist of two protons and two neutrons (a helium nucleus).

Alpha decay. Radioactive decay involving the loss of an alpha particle (helium nucleus).

Alkalinophile. An organism that needs to live in environments with high pH.

Alkane. A saturated hydrocarbon consisting only of hydrogen and carbon atoms and all bonds are single bonds. Alkanes have the general chemical formula C_nH_{2n+2}.

Amazonian. The period of Mars history from about 3 Ga ago to the present day.

Amino acid. Amino acids are biologically important organic compounds composed of amine ($-NH_2$) and carboxylic acid ($-COOH$) functional groups, along with a side chain specific to each amino acid.

Amorphous. Material without a well-defined structure.

Amphiboles. An important group of generally dark-coloured minerals, forming prism or needlelike crystals, composed of double-chain silica tetrahedra.

Amphiphilic. The property of a molecular compound with both *hydrophilic* (water-loving) and *hydrophobic* (water-hating) regions.

Anaerobic. Oxygen-free (anoxic) conditions.

Analogy (*in phylogenetics*). Two organisms that share similar characteristics, but are not genetically closely related (similarities caused by *convergent evolution*).

Angiosperms. The taxonomic class that includes flowering plants.

Anions. Negatively charged ions.

Astrobiology: Understanding Life in the Universe, First Edition. Charles S. Cockell.
© 2015 John Wiley & Sons, Ltd. Published 2015 by John Wiley & Sons, Ltd.
Companion Website : www.wiley.com/go/cockell/astrobiology.

Anisotropy. The property of being directionally dependent, as opposed to isotropy, which implies identical properties in all directions.

Anoxia. Oxygen-free (*anaerobic*) conditions.

Anoxygenic photosynthesis. Use of light to gain energy using compounds other than water as the electron donor (e.g. sulfide, ferrous iron, hydrogen, hydrogen sulfide or organics).

Anthropocene. The proposed name of a new geological epoch defined by the activities of humans on the planet.

Anthropocentric. The view that human beings are the central or most significant entities.

Anthropic principle. The philosophical consideration that observations of the physical Universe must be compatible with the conscious life that observes it.

Antifreeze proteins. Proteins that inhibit the formation and growth of ice crystals.

Anticodon. A three-nucleotide sequence on a tRNA molecule that is complementary to a *codon*.

Antimatter. Material composed of antiparticles, which have the same mass as particles of ordinary matter but have opposite charge.

Apatite. A group of phosphate minerals, usually referring to hydroxyapatite, fluorapatite and chlorapatite, containing high concentrations of OH^-, F^- and Cl^- ions, respectively, in the crystal.

Aphelion. The point in the orbit of a planet furthest from the Sun.

Apoptosis. The process of programmed cell death (PCD) that can occur in multicellular organisms.

Arcseconds. A second of arc (arcsecond, arcsec) is 1/60 of an arc minute and 1/3600 of a degree.

Archaea. A domain of single celled microorganisms, genetically distinct from bacteria. Archaea, which comes from the Greek for 'ancient things', are often found in extreme environments.

Archean. The period of Earth history from 4.0 to 2.5 Ga ago.

Arrhenius relationship. A formula for expressing the temperature dependence of reaction rates.

Aromatic. Aromatic compounds, or aromatics, are compounds that contain planar ring systems.

Asteroid. Remnants of planetesimals. Most of them orbit between Mars and Jupiter.

Astrobiology. The study of the origin, evolution and distribution of life in the Universe.

Astrometry. The measurement of the position of a star.

Astronomical Unit (AU). An astronomical distance where 1 AU is defined as exactly 149 597 870 700 m, roughly equal to the average Sun–Earth distance (which is how it used to be defined).

Atomic mass number. The total number of protons and neutrons (together known as nucleons) in an atomic nucleus.

Atomic number. The atomic number of a chemical element (also known as its proton number) is the number of protons found in the nucleus of an atom.

Autotroph. An organism that uses simple compounds to make organic compounds, generally using CO_2 as the source of carbon.

Bacilli. Bacteria that possess a cylindrical or rod-like cell shape.

Bacteria. A domain of single celled microorganisms that has spread to inhabit almost every habitat on Earth. Bacteria lack a cell nucleus.

Bacteriochlorophyll. Light harvesting pigments in certain photosynthetic bacteria.

Bacteriorhodopsin. A simple light driven proton pump found in some archaea, particularly the Halobacteria (which are archaea, despite their name).

Bacteriophage. A virus that infects bacteria.

Barophile (or piezophile). An organism that requires high pressures to grow.

Barred spiral galaxy. A type of spiral galaxy with a central bar-shaped structure composed of stars. Bars are found in approximately two-thirds of all spiral galaxies.

Barycentre. Common centre of mass of a system, particularly in astronomy.

Barycentric Julian Date. The Julian Date corrected for differences in the Earth's position with respect to the barycentre (centre of mass) of the Solar System. Used by astronomers.

Baryonic matter. Matter made up of baryons, which are particles themselves made of the subatomic particles, quarks. The most familiar baryons are the protons and neutrons that make up most of the mass of the visible matter in the universe. Electrons (the other major component of the atom) are leptons.

Basalt. A common *extrusive igneous* (volcanic) rock formed from the rapid cooling of basaltic lava exposed at or very near the surface of a planet or moon.

Base pairs. Base pairs, which form between specific *nucleobases* (also termed nitrogenous bases), are the building blocks of the DNA double helix and contribute to the folded structure of both DNA and RNA. The nucleobases are thymine, adenine, cytosine and guanine in DNA, with thymine replaced by uracil in RNA.

Beta decay. Radioactive decay involving the loss of an electron as a neutron decays into a proton.

Binary fission. The process by which cells divide to form two genetically identical 'daughter' cells.

Binary star. A star system containing two stars.

Biodiversity. A measure of the degree of the variation of life. It can refer to genetic variation, ecosystem variation, or species variation within an area, biome, or planet.

Biofilm. A group of microorganisms that grow stuck together on a surface, either in a single layer or multiple layers.

Biologically effective irradiance. The integrated value of a light spectrum between two wavelengths multiplied by an *action spectrum* for a given biological process.

Biomarker. A biomarker, or biological marker, generally refers to a measurable indicator of some biological state or condition.

Biominerals. Minerals produced by biological processes.

Biomineralisation. The precipitation of minerals on a biological surface.

Binomial. The two-part name for a species.

Bioremediation. A method using organisms to remove or control toxic substances in land.

Biosignature gas. A gas derived from biological activity.

Biospace. The physical and chemical boundaries within which life can survive, grow or reproduce (a biospace must be defined with reference to one of these requirements).

Biosynthesis. The biological production or building of new molecules.

Black smoker. A black smoker or sea vent is a type of hydrothermal vent found on the ocean floor. They produce sulfide-rich (black) fluids.

Blackbody. An idealised physical body that absorbs all incident electromagnetic radiation, regardless of frequency or angle of incidence.

Blackbody radiation. Electromagnetic radiation within or surrounding a body in thermodynamic equilibrium with its environment, or emitted by a blackbody (an opaque and non-reflective body) held at constant, uniform temperature.

Boltzmann constant. The Boltzmann constant (kB or k), named after Ludwig Boltzmann (1844–1906), is a physical constant relating energy at the individual particle level with temperature.

Bottleneck *(evolutionary biology)*. A rapid population reduction in a given species or group due to an external environmental (e.g. volcanic eruption) or anthropogenic (e.g. hunting) event.

Boundary *(in geology/extinction)*. The transition between geological times, sometimes when an extinction is recorded.

Brown dwarf. Substellar objects too low in mass to sustain hydrogen fusion reactions in their cores, unlike main-sequence stars. They occupy the mass range between the heaviest gas giants and the lightest stars, with an upper limit around 75 to 80 Jupiter masses.

Calvin cycle. A metabolic pathway found in autotrophic bacteria and in the chloroplasts of phototrophs in which carbon enters in the form of CO_2 and leaves in the form of sugar.

Cambrian. The first geological period of the Phanerozoic (541.0 ± 1.0 to 485.4 ± 1.9 million years ago). It includes the rise of multicellular animals.

Cap carbonates. Layers of distinctively textured carbonate rocks which typically form the uppermost layer of sedimentary sequences reflecting major glaciations in the geological record.

Capsid. The protein shell of a virus particle.

Carbohydrates. A molecule, or macromolecule, consisting of carbon (C), hydrogen (H) and oxygen (O) atoms, usually with a hydrogen:oxygen atom ratio of 2:1 (as in water).

Carbonaceous chondrite. A stony meteorite containing *chondrules* and organic compounds. The group includes some of the most ancient known meteorites which are believed to be similar in composition to that of the original solar nebula.

Carbonate–silicate cycle. The feedback process that regulates planetary temperature.

Carbon dating. The use of the decay of carbon-14 to nitrogen-14 (with a half-life of 5730 years) to accurately date objects.

Carbonatisation. The formation of a variety of carbonate minerals through water-rock interactions involving fluids with high concentrations of dissolved CO_2.

Carboxydotrophy. The use of carbon monoxide as an electron donor by microorganisms.

Carboxylic acid. An organic compound that contains a carboxyl group (C(O)OH). Long-chained carboxylic acids are also called *fatty acids*.

Carotenoids. A class of long chain molecules involved in quenching *reactive oxygen species*, UV screening and other functions. Many of them confer colour on microorganisms.

Catastrophic outflow channels. Large features on Mars which are suggested to form from catastrophic release of sub-surface water onto the surface of the planet.

Cations. Positively charged ions.

Cell differentiation. The process by which groups of cells become specialised during the growth of a multicellular organism or biofilm.

Cellulose. A polysaccharide containing thousands of glucose subunits linked in linear chains and an important structural component of plant cell walls. It has the general formula $(C_6H_{10}O_5)_n$ and is the most abundant polymer on Earth.

Central peak. The rebounded central uplift within an impact crater.

CETI. Communication with *ExtraTerrestrial Intelligence*.

Chandrasekhar limit. The Chandrasekhar limit is the maximum mass of a stable *white dwarf* star.

Chaos terrain. Chaos terrain (or chaotic terrain) is a planetary surface area where features such as ridges, cracks, and plains appear jumbled and enmeshed.

Chaotropicity. The tendency of a substance to induce disorder in a molecule.

Chaperonins. Molecules involved in enabling correct folding in proteins.

Chemiosmosis. Chemiosmosis is the movement of ions across a selectively permeable membrane, down their electrochemical gradient. More specifically, it relates to the generation of ATP by the movement of hydrogen ions across a membrane during cellular respiration.

Chemoautotrophs. A chemoautotroph is an organism that uses inorganic electron donors to make energy.

Chemolithotroph. See chemoautotroph.

Chemotaxis. Movement of single celled organisms in response to chemical stimuli.

Chert. A fine-grained silica-rich sedimentary rock.

Chiral. A type of molecule that has a non-superimposable mirror image (more technically it has two optically active *isomers*).

Chlorophyll. A green pigment found in cyanobacteria and the chloroplasts of algae and plants. Involved in capturing sunlight in photosynthesis.

Chondrules. Round grains found in chondrites. Chondrules form as molten or partially molten droplets in space before being attached to their parent asteroids in *accretion*.

Chromosomes. A single long piece of DNA that contains many genes. Chromosomes can be packaged into coiled shapes, or remain unpackaged and loosely circular.

Clade. A clade (monophylum, i.e. 'monophyletic') is a group consisting of an ancestor and all its descendants.

Clathrates. A chemical substance consisting of a lattice that traps or contains molecules. See *methane gas hydrates*.

Clay. A generic term for a fine-grained material that combines one or more clay minerals with traces of metal oxides and organic matter. Clays are a type of *phyllosilicates*.

Coacervates. A spherical droplet of organic molecules which is held together by hydrophobic forces from a surrounding liquid. Sometimes used to describe early cells.

Coccus. Microorganisms that possess a spherical cell shape.

Codon. A three-nucleotide sequence in *messenger RNA* that codes for a specific amino acid.

Coenzyme. A substance that enhances the action of an enzyme.

Comet. Small icy Solar System bodies. Comets are distinguished from asteroids by the presence of an extended, gravitationally unbound atmosphere surrounding their central nucleus. Short-period comets originate in the *Kuiper Belt*. Long-period comets are thought to originate in the *Oort Cloud*.

Compatible solute. Small molecules that act as osmolytes and help organisms survive extreme osmotic stress, particularly found in *halophiles*.

Competence. The ability of a cell to take up DNA from its environment.

Conjugation. The transfer of genetic material (typically a plasmid) between bacterial cells by direct cell to cell contact or by a pilus connection between two cells.

Continental shelf. An underwater landmass which extends from a continent, resulting in an area of relatively shallow water known as a shelf sea.

Continuously habitable zone. The zone around a star where surface liquid water is stable for geological time periods.

Convergent Boundary. Also known as a destructive plate boundary (because of *subduction*), is an actively deforming region where two (or more) tectonic plates or fragments of the lithosphere move toward one another and collide.

Convergent evolution. The independent evolution of similar features in species of different lineages. Usually caused by exposure to similar environmental conditions or selection pressures.

Coronal mass ejection. A massive burst of solar wind and magnetic fields rising above the solar corona or being released into space.

Coronagraph. A method of blocking out sunlight using a shade or similar device to acquire light from planets without the confounding effects of starlight.

Cosmic Background Radiation. The cosmic microwave background (CMB) is the thermal radiation thought to be left over from the Big Bang of cosmology.

Cosmic rays. High-energy radiation originating outside the Solar System. They produce showers of secondary particles when they collide with the atmosphere and sometimes reach the surface. They are composed primarily of high-energy protons and atomic nuclei with highly variable energies typically between 10^8 and 10^{15} eV.

Covalent bonds. A chemical bond that involves the sharing of electron pairs between atoms.

Crater age dating. A method of estimating the age of planetary surface features by counting the number and distribution of craters, predicated on the assumption that impacts occur at a constant rate over geological time.

Craton. An old and stable part of the continental *lithosphere*. Having survived cycles of merging and rifting of continents, they are generally found in the interiors of tectonic plates.

CRISPR (Clustered regularly interspaced short palindromic repeats). Short sequences in the bacterial or archeal *genome* that act as anti-viral agents.

Critical point. The critical point, also known as a critical state, is the point with conditions (such as specific values of temperature, pressure or composition) at which no phase boundaries exist.

Cross-over *(genetics)*. The exchange of genetic material between two complementary chromosomes during meiosis.

Crust. The solid shell of a rocky planetary body that has undergone *differentiation*.

Crustal dichotomy *(Mars)*. The split in the geology of Mars between the northern lowlands and the more ancient southern cratered terrains.

Cryoconite holes. Indentations or holes within the surface of ice sheets or glaciers that usually contain water and microbial communities.

Cryosphere. The frozen region of a planet.

Cryovolcano. A cryovolcano (colloquially known as an ice volcano) is a volcano that erupts volatiles such as water, ammonia or methane, instead of molten rock.

C-value paradox. The unexplained lack of correlation between eukaryotic genome size and organism complexity. A large genome does not imply a complex organism.

Cytoplasm. The fluid-filled interior of a cell, enclosed by the cell membrane. The cytoplasm contains proteins, salts and water, as well as organelles and other structures.

Dark Energy. A hypothetical form of energy that permeates all of space and tends to accelerate the expansion of the universe.

Dark Matter. A type of matter hypothesised to account for astrophysical effects that appear to be the result of mass where such mass cannot be seen.

Death phase. The gradual death of a microbial population if all nutrients have been depleted and no fresh sources are available.

Degenerate matter. Degenerate matter is a collection of free, non-interacting particles with a pressure and other physical characteristics determined by quantum mechanical effects. The degenerate state of matter arises at extraordinarily high density (in compact stars) or at extremely low temperatures in laboratories. It occurs for matter particles such as electrons, neutrons, protons, and is referred to as electron-degenerate matter, neutron-degenerate matter and so on.

Dendrogram. A *phylogenetic tree* diagram.

Denitrification. A microbially facilitated process of nitrate reduction performed by a large group of heterotrophic facultative *anaerobic* bacteria.

Detrital. Minerals deposited in sediments, such as in rivers.

Deoxyribonucleic acid (DNA). The information coding molecule of all cellular life, formed of long chains of a sugar-phosphate backbone with nitrogen containing *nucleobases*.

Diagenesis. The change of sediments or existing sedimentary rocks into a different sedimentary rock during and after rock formation (lithification), at temperatures and pressures less than that required for the formation of *metamorphic* rocks.

Differentiation. The separation under melting of the different constituents of a planetary body where the denser material moves towards the centre creating a core and mantle.

Diploid. Cells possessing two copies of each chromosome, as in many animals and plants.

Dipole. An electric dipole is a separation of positive and negative charges.

Dipole moment. The measure of net molecular polarity, which is the magnitude of the charge at either end of a molecular dipole.

Disaster taxa. Taxa that survive extinction and proliferate.

Dispersal filter. A physical barrier that prevents migration of an organism from one place to another.

Disulfide bridge. A covalent bond, usually derived by the coupling of two thiol (–C-SH) groups. The linkage is also called an SS bond. Formed between cysteine amino acids in proteins.

Divergent Boundary. A linear feature that exists between two tectonic plates that are moving away from each other. Divergent boundaries within continents produce rift valleys. Most active divergent plate boundaries occur between oceanic plates and exist as mid-oceanic ridges.

DNA. *Deoxyribonucleic acid*, the molecule that stores genetic information.

DNA gyrase. An enzyme that relieves strain as DNA is unwound by a *helicase*.

DNA polymerase. An enzyme that synthesises DNA by assembling individual nucleotides.

Domain. The highest, or broadest, level of classification of organisms. All cellular life on Earth is classified into one of three 'domains': Bacteria, Archaea or Eukarya.

Doppler Shift (or Doppler Effect). The change in frequency of a wave for an observer moving relative to its source.

Drake equation. The Drake equation is a probabilistic argument used to estimate the number of active, communicative extraterrestrial civilisations in the galaxy.

Dwarf planet. An object the size of a planet (a planetary-mass object) but that is neither a planet nor a moon or other natural satellite.

Dyson sphere. A hypothetical structure that completely encompasses a star and hence captures most or all of its power output.

Earthshine. The reflected earthlight visible on the Moon's night side.

Eccentricity. The orbital eccentricity of an astronomical object is a parameter that determines the amount by which its orbit around another body deviates from a perfect circle. The eccentricity, e, can be calculated by the relationships $r_p/r_a = (1-e)/(1+e)$, where r_a is the radius at apoapsis (the furthest distance of the orbit to the centre of mass of the system, which is a focus of the ellipse) and r_p, the periapsis, is the closest distance to the centre of mass.

Einstein ring. The ring seen around an object caused by the gravitational lensing of a more distant object.

Electron. The electron (symbol: e^-) is a subatomic particle with a negative electric charge.

Electron tower. A series that depicts the relative electron potential of half reactions involved in reduction–oxidation reactions (*redox reactions*).

Electron volt. The electron volt (symbol eV; also written electronvolt) is a unit of energy equal to approximately 1.6×10^{-19} J. By definition, it is the amount of energy gained (or lost) by the charge of a single electron moved across an electric potential difference of 1 V.

Electrode potential. A measure of the tendency of a chemical species to acquire electrons and thereby be reduced. It is measured in volts (V), or millivolts (mV). Each species has its own intrinsic reduction potential.

Electron transport chain. A series of compounds that transfer electrons from electron donors to electron acceptors via redox reactions.

Emission spectrum. The spectrum of frequencies of electromagnetic radiation of a chemical element or chemical compound emitted due to an atom's electrons making a transition from a high energy state to a lower energy state.

Enantiomer. One of two isomers that are mirror images of each other (also see *chiral*).

Endospore. A dormant, tough, and non-reproductive structure produced by certain bacteria.

Endosymbiosis. The mutually beneficial living together of two partner organisms, one of which lives inside the cell of the other.

Entropy. Entropy (usual symbol S) is a measure of the number of ways in which a thermodynamic system may be arranged. According to the second law of thermodynamics the entropy of an isolated system never decreases. Such systems spontaneously move towards thermodynamic equilibrium.

Enzymes. Biological catalysts (e.g. proteins and ribozymes) responsible for the thousands of metabolic processes that sustain life.

Eon. A unit of geological time, for example Archean eon.

Epoch. A unit of geological time, for example Holocene epoch.

Equation of state. A thermodynamic equation describing the state of matter under a given set of physical conditions.

Era. A unit of geological time, for example Cenozoic era.

Escape velocity. The speed needed to 'break free' from the gravitational attraction of a massive body.

Eukaryote. An organism whose cells contain a nucleus and other organelles enclosed within membranes.

Eutectic point. A eutectic system is a mixture of chemical compounds or elements that has a single chemical composition that solidifies at a lower temperature than any other composition. This composition is known as the eutectic composition and the temperature is known as the eutectic temperature or eutectic point.

Exobiology. A term generally referring to the study of the possibility of life beyond the Earth (now part of the wider term, *astrobiology*).

Exons. Portions of a gene that remain in the mRNA molecule after *introns* are removed, and are hence translated into protein.

Exoplanet. A planet orbiting a star other than the Sun.

Exponential phase. A phase of rapid growth in a microbial population, often characterised by an exponential increase in cell numbers.

EXPOSE Facility. An apparatus deployed on the outside of the International Space Station by the European Space Agency to study the response of organics and organisms to the space environment, particularly the UV radiation environment.

Extinction event. A period of pronounced extinction of species (when the rate of extinction increases with respect to the rate of speciation).

Extinct radionuclides. Radioactive elements that have short half-lives (on a geological time scale), were produced in the early Solar System, but now no longer exist in the Solar System.

Extracellular polysaccharide (EPS). Long-chained sugars produced by organisms in or outside their cell wall.

Extremophile. An organism that requires certain extreme physical or chemical conditions to grow.

Extremotolerant. An organism that can tolerate certain extreme conditions.

Extrusive. A rock that is formed by crystallisation at the surface of the Earth. They usually have fine-grained textures and small crystals. An example is basalt, formed from lava.

Facies. Specific rock characteristics formed under certain conditions, reflecting a particular process or environment.

Faint young Sun paradox. An apparent paradox describing the conflict between the inferred presence of liquid water on the early Earth and the lower solar output at this time.

Facultative. Used in reference to microorganisms to mean that an organism can grow with a certain set of conditions, but does not require them. For example a facultative *anaerobic* microorganism that can grow in the presence or absence of oxygen. Compare to *obligate*, which means a requirement for certain conditions.

Fatty acid. A carboxylic acid with a long aliphatic tail (carbon atoms arranged in a long chain), which is either saturated (containing no double or triple bonds) or unsaturated (containing double or triple bonds). They have a chain of an even number of carbon atoms, from 4 to 28.

Feldspar. Feldspars ($KAlSi_3O_8 - NaAlSi_3O_8 - CaAl_2Si_2O_8$) are a group of rock-forming minerals that make up as much as 60% of the Earth's crust and form three-dimensional 'framework' structures. The composition of many feldspars can be expressed in terms of the composition of three end-members; sodium-rich or albite ($NaAlSi_3O_8$), calcium-rich or anorthite ($CaAl_2Si2O_8$) and potassium-rich or K-feldspar ($KAlSi_3O_8$).

Fermentation. A metabolic process that converts sugar to acids or alcohols. It occurs in a wide diversity of organisms including bacteria and yeast, but also in the oxygen-starved muscle cells of animals.

Fermi Paradox. The problem raised by physicist Enrico Fermi, which is why if there are so many stars in the Universe, we are not visited by aliens from older civilisations.

Ferrodoxins. A family of ancient iron-sulfur proteins involved in electron transfer.

Fertile Crescent. A crescent-shaped region containing comparatively fertile and moist land from the Persian Gulf, through modern-day southern Iraq, Syria, Lebanon, Jordan, Israel and northern Egypt.

Fischer–Tropsch reactions. A set of chemical reactions that converts a mixture of carbon monoxide and hydrogen into hydrocarbons.

Flagellum (plural flagella). Microbial appendages that allow for movement.

Flood basalt. A flood basalt is the result of a giant volcanic eruption or series of eruptions that coats large stretches of land or the ocean floor with basalt lava.

Fluid inclusions. Tiny bodies of fluid and gas contained within a crystal.

Foraminifera. Foraminifera (or forams for short) are single-celled mainly aquatic microorganisms with shells or tests (a technical term for internal shells) commonly made of calcium carbonate.

Formose reaction. Chemical reaction involving the formation of sugars from formaldehyde.

Fractional crystallisation. The process by which, as magma cools, different minerals crystallise from the solution at different temperatures.

Free radical. A free radical is an atom, molecule, or ion that has unpaired valence electrons or an open electron shell.

Functional extinction. Functional extinction occurs when the population of an organism is no longer viable to maintain the species.

Fusion crust. The melted outside of a meteorite caused by heating and ablation of the outside of the object during transit through the atmosphere.

Ga. Giga annum (one billion years). Sometimes written as Gyr.

Gabbro. A large group of dark, often coarse-grained, mafic *intrusive igneous* rocks. Chemically identical to basalt.

Gamma radiation. Radiation of extremely high frequency and therefore high energy per photon. Gamma rays are *ionising radiation*, and thus biologically hazardous.

Gamma radiation decay. The decay of a radioactive element involving the release of gamma rays.

Gas gun. A method for accelerating small projectiles to km s^{-1} velocities. Used in impact studies. A gas gun uses explosives to accelerate the projectile along a barrel and into a solid target at the end.

Gas hydrates. Clathrate hydrates (or gas clathrates, gas hydrates, clathrates, hydrates etc.) are crystalline water-based solids physically resembling ice, in which small non-polar molecules (typically gases such as methane) or *polar* molecules with large *hydrophobic* groups are trapped inside 'cages' of hydrogen-bonded water molecules.

Gene. A DNA sequence which contains the information for building one protein.

Genetic degeneracy. The redundancy of the genetic code; the fact that multiple possible *codons* may code for the same *amino acid.*

Genome. The full complement of genetic information in an organism.

Geological proxy. A measured geological variable used to infer a variable of interest.

Genotype. The genetic make-up of an organism.

Geothermal. Heat from the Earth's interior.

Geothermal gradient. The change (increase) in heat as one goes to greater depths into the Earth's interior.

Genera. See *genus.*

Genus (plural genera). A taxonomic rank used in the biological classification of living and fossil organisms. In biological classification hierarchy it comes above species and below family.

Germ Theory. The early idea that diseases can be caused by microorganisms now accepted as fact, but only convincingly demonstrated with the new discoveries of microbiology in the eighteenth century.

Giant impact model. A model which attempts to explain the formation of Earth's Moon through an impact event with a Mars-sized rocky body.

Gibbs free energy. The energy associated with a chemical reaction that can be used to do work.

Glass. An *amorphous* solid (non-crystalline) material.

Glycolysis. The metabolic pathway that converts glucose into pyruvate.

Glycoproteins. Proteins with sugars covalently bonded to them.

Goldilocks Zone. A colloquial term used for the *habitable zone.*

Gram stain. A method for classifying bacteria and archaea based on their tendency to be stained with a specific dye. Used to distinguish Gram negative and positive bacteria.

Granite. A common type of *felsic intrusive igneous* rock which is coarsely granular in texture. It is rich in silica and has low concentrations of Fe and Mg. It is one of the major constituents of the continents.

Graphite. The most stable form of carbon, with a layered planar structure made entirely of carbon atoms.

Gravitational focussing. The way in which the gravity of a *planetesimal* attracts other planetesimals during planet formation.

Gravity well. A term used to refer to a surface of a planetary body on which one must use energy to escape its gravitational attraction.

Gray. A unit of ionising radiation dose in the International System of Units (SI). It is a measure of the absorbed dose and is defined as the absorption of one joule of radiation energy by 1 kg of matter.

Great Oxidation Event (GOE). The biologically-induced appearance of oxygen in the Earth's atmosphere. Geological, isotopic, and chemical evidence suggest that this major environmental change happened around 2.4 billion years ago.

Greenhouse effect. The trapping of infra-red radiation by a planetary atmosphere causing the surface to warm. A runaway greenhouse effect refers to a greenhouse effect so strong that the surface is too hot for liquid water to persist. Any water that is present boils away.

Greenhouse gas. A gas that traps infra-red radiation.

Gymnosperm. A taxonomic class that includes plants whose seeds are not enclosed in an ovule but in other structures (e.g. a pine cone).

Habitable zone. The zone around a star where surface liquid water is stable on a planetary surface.

Habitat. A physical space where an organism lives.

Halophile. An organism that requires high salt concentrations to grow.

Half-life. The time it takes for the concentration of a radioactive material to decay to half its original concentration.

Haploid. Cells possessing one copy of each chromosome.

Heat shock proteins. Proteins involved in stress response, particularly to high temperatures.

Helicase. An enzyme that unwinds the DNA double helix.

Hertzsprung–Russell diagram. The Hertzsprung–Russell diagram is a scatter graph of stars showing the relationship between the stars' absolute magnitudes or luminosities versus their spectral types or classifications and effective temperatures.

Hesperian. The period of Mars history from about 3.7 Ga ago to about 3.0 Ga ago.

Heterocysts. Specialised cells capable of fixing nitrogen from the air. Heterocysts are formed by some cyanobacteria during periods of nitrogen starvation.

Heterotrophs. Organisms that use organic carbon as a source of carbon.

Hold-over taxa. Taxa that survive an extinction.

Homology (*in phylogenetics*). Similarities in organisms caused by common descent.

Homoplasies. A trait (genetic, morphological etc.) that is shared by two or more taxa because of *convergent evolution*. An alternative term is *analogy*.

Hopanoids. Natural compounds based on the five-ringed (pentacyclic) chemical structure of hopane.

Horizontal gene transfer (HGT). The transfer of genes between organisms in a manner other than traditional reproduction. Also termed lateral gene transfer (LGT).

Hormones. A class of biochemicals in multicellular organisms that are capable of regulating many aspects of growth, physiology and organ function.

Hot Jupiters. Extrasolar gas giants with high surface temperatures induced by close orbital trajectories around their parent star and similar to the mass of Jupiter.

Hotspot volcanism. Where volcanoes often form one after another in the middle of a tectonic plate, as the plate moves over the hotspot. The volcanoes progress in age from one end of the chain to the other.

Hot Neptune. Extrasolar gas giants with high surface temperatures induced by close orbital trajectories around their parent star and similar to the mass of Uranus or Neptune.

Huronian. The Huronian glaciation (or Makganyene glaciation) lasted from 2.4 to 2.1 Ga, during the Siderian and Rhyacian periods of the Paleoproterozoic era. The Huronian glaciation followed the *Great Oxidation Event*.

Hydrodynamic escape. Hydrodynamic escape refers to a thermal atmospheric escape mechanism that can lead to the escape of heavier atoms of a planetary atmosphere through numerous collisions with lighter atoms.

Hydrogen bonding. A hydrogen bond is the attractive interaction between *polar molecules*, in which hydrogen is bound to a highly electronegative atom (one that has a high tendency to attract electrons), such as oxygen, nitrogen or fluorine.

Hydrophilic. Having a strong affinity for water.

Hydrophobic. Having a tendency to be repelled from water.

Hydrothermal. The action of heated aqueous solutions.

Hyperthermophile. Organisms that require a temperature of greater than 85 °C to grow.

Ideal rocket equation. The relationship that describes the motion of vehicles that behave according to the basic principle of a rocket.

Igneous. Igneous rock is formed through the cooling and solidification of magma or lava.

Impact erosion. The gradual removal of a planetary atmosphere by impact events.

Impact winter. A term used to refer to a period of catastrophic cold and dark resulting from the injection of large amounts of dust into the atmosphere by an asteroid impact.

Incompatible elements. During the crystallisation of magma and magma generation by the *partial melting* of the Earth's mantle and crust, elements that have difficulty in entering *cation* sites of the minerals are concentrated in the melt phase of magma. These are incompatible elements. They form economically important metal ores.

Insolation. The cumulative solar radiation input to a specific surface area over a given time period.

Interstellar grains. Grains of siliceous and carbonaceous materials covered in ices.

Interstellar medium (ISM). The matter that exists in the space between stars in a galaxy.

Introns. Portions of a gene that are transcribed, but removed from the mRNA before protein synthesis.

Intrusive. A rock that crystallises at depth in the Earth (and usually has large crystals). An example is *granite*.

Ion. An atom or molecule in which the total number of electrons is not equal to the total number of protons, giving the atom a net positive or negative electrical charge.

Ionic bond. A type of chemical bond that involves the electrostatic attraction between oppositely charged ions. The bond involves the transfer of electrons.

Ionising radiation. Radiation that has enough energy to liberate electrons from atoms or molecules, thereby ionising them.

Island biogeography. The branch of ecology that investigates how discrete populations of organisms interact and cross-migrate.

Isochron dating. Isochron dating is a common technique of *radiometric dating* and is applied to date certain events. This technique can be applied if the daughter element has at least one stable isotope other than the daughter isotope into which the parent nuclide decays.

Isomer. Molecules with the same molecular formula but different chemical structures.

Isomerisation. The process by which one molecule is transformed into another molecule which has exactly the same atoms, but the atoms have a different arrangement.

Isotope. Isotopes are variants of a particular chemical element such that, while all isotopes of a given element have the same number of protons in each atom, they differ in the number of neutrons.

Isotopic fractionation. The preferential enrichment in a particular *isotope*.

Jarosite. A basic sulfate of potassium and iron with a chemical formula of $KFe^{3+}_3(OH)_6(SO_4)_2$. This sulfate mineral is formed in ore deposits by the oxidation of iron sulfides.

Jeans mass. The critical mass of a gas cloud when it just lacks sufficient gaseous pressure to balance the force of gravity and collapses.

Julian Date. The Julian Day Number (JDN) is the integer assigned to a whole solar day in the Julian day count, with Julian day number 0 assigned to the day starting at noon (Greenwich Mean Time) on 1 January, 4713 BC. The Julian Date (JD) of any instant is the Julian day number for the preceding noon in Greenwich Mean Time plus the fraction of the day since that instant. Julian Dates are expressed as a Julian day number with a decimal fraction added. For example, the Julian Date for 00:30:00.0 UT 1 January, 2013, is 2 456 293.520833.

Junk DNA. An outdated term to describe portions of an organism's genome that appear to have no function.

Keratin. A family of fibrous proteins.

Kerogen. A generic term for undefined complex organic carbon compounds such as in sedimentary rocks or meteorites.

Kardashev scale. A method of measuring a civilisation's level of technological advancement, based on the amount of energy a civilisation is able to use.

Komatiites. Rare *ultramafic igneous* rocks. More abundant early in Earth's history when the mantle was presumed to be hotter.

Krebs cycle. The Krebs cycle (or tricarboxylic acid cycle or citric acid cycle) is a part of cellular respiration involved in producing CO_2 from the metabolism of carbohydrates, fats and proteins. It is a series of chemical reactions used to generate energy.

Kuiper Belt. The Kuiper belt, sometimes called the Edgeworth–Kuiper belt, is a region of the Solar System beyond the planets, extending from the orbit of Neptune (at 30 AU) to approximately 50 AU from the Sun. It is composed of many small bodies from the early Solar System formation.

Lag phase. The initial growth phase of a cell population immediately following colonisation. The lag phase is often characterised by very low or undetectable growth rates.

Last Universal Common Ancestor (LUCA). The hypothetical organism from which all life on Earth evolved.

Late heavy bombardment. The Late Heavy Bombardment (commonly referred to as the lunar cataclysm, or the LHB) is a period of intense bombardment thought to have occurred approximately 4.1–3.8 billion years ago.

Lichen. An organism consisting of a fungus (the mycobiont) and a photosynthetic partner (the photobiont or phycobiont) growing together in a *symbiotic* (interdependent) relationship.

Light year. A unit of length used in astronomy equal to the distance that light travels in one year.

Lignin. A complex polymer of aromatic alcohols known as monolignols. It is most commonly found in wood, and is an important part of the secondary cell walls of plants and some algae.

Lipopolysaccharides. Large molecules composed of a lipid and a polysaccharide. They are found in the cell walls of *Gram*-negative bacteria.

Lipoproteins. A biochemical assembly that contains both proteins and lipids.

Lipoteichoic acid. Teichoic acids that are covalently bound to lipids. They are found within the cell walls of *Gram*-positive bacteria.

Lithified. Turned into rock.

Lithosphere. The rigid, outermost shell of a rocky planet.

Lithostatic pressure. The pressure exerted by rock.

Low Earth Orbit. An orbit around Earth with an altitude between 160 km (with an orbital period of about 88 min) and 2000 km (with an orbital period of about 127 min).

LUCA. Last Universal Common Ancestor, the progenitor of all life on Earth.

Lyman-α. Sometimes written as Ly-α line, is a spectral line of hydrogen, or more generally of one-electron ions, emitted when the electron falls from the $n = 2$ orbital to the $n = 1$ orbital. In hydrogen, its wavelength is 1215.67 Å (121.567 nm), corresponding to the ultraviolet region of the electromagnetic spectrum. Such radiation is observed to be emitted from certain galaxies, for example.

Lysis. The breakdown and bursting of a cell, resulting in cell death.

Lysogeny. Reproduction of foreign DNA by its incorporation into the host DNA.

Ma. Mega-annum (one million years ago). Sometimes written as Myr.

Macromolecules. Large molecules.

Mafic. A silicate mineral or rock that is rich in magnesium and iron. The term is a portmanteau of the words 'magnesium' and 'ferric'.

Magma. A mixture of molten or semi-molten rock, volatiles and solids that is found beneath the surface of the Earth.

Magnetite. A magnetic mineral iron oxide with chemical formula Fe_3O_4.

Magnetosphere. The area of space near an astronomical object in which paths of charged particles are controlled by that object's magnetic field.

Main sequence. Stars that are fusing hydrogen atoms to form helium atoms in their cores.

Mantle. The layer between the crust and outer core of a rocky planet or natural satellite large enough to have undergone *differentiation*.

Mass dependent fractionation. Any chemical or physical process that acts to separate isotopes, where the amount of separation scales in proportion with the difference in the masses of the isotopes.

Mass extinction. A substantial increase in the extinction rate over a short geological period of time over a large geographical area.

Mass independent fractionation. Any chemical or physical process that acts to separate isotopes, where the amount of separation does not scale in proportion with the difference in the masses of the isotopes.

Mass spectrometry. An analytical technique that produces spectra of the masses of the atoms or molecules constituting a sample of material.

Maximum likelihood. Maximum likelihood, also called the maximum likelihood method, is the procedure of finding the value of one or more parameters for a given statistic which makes the known likelihood distribution a maximum (used in *phylogenetics*).

Maximum parsimony. A character-based method that infers a phylogenetic tree by minimising the total number of evolutionary steps required.

Meiosis. A process of cell division unique to sexual reproduction in eukaryotes. Unlike mitosis, meiosis produces two 'gametes' which each contain half of the full complement of genetic material, that is one set of chromosomes, instead of two sets (*diploid*) for most animal cells. These are called *haploid* cells.

Mesophile. An organism that needs to grow at temperatures between 20 and 45 °C.

Messenger RNA (mRNA). RNA molecules that carry a complementary copy of genetic information from DNA to the ribosome, where proteins are synthesised.

Metabolism. The set of life-sustaining chemical reactions within the cells of living organisms.

Metallicity. In astronomy, the metallicity of an object is the proportion of its matter made up of chemical elements other than hydrogen and helium.

Metamorphic. Metamorphic rocks arise from the transformation of existing rock types, in a process called *metamorphism.*

Metamorphism. Metamorphism is the change of minerals or geologic texture.

Metasomatism. The process by which a mineral is chemically altered, for example by hydrothermal fluids.

Meteorite. A meteorite is a solid piece of debris, from such sources as asteroids or comets, that originates in outer space and survives its impact with the Earth's surface.

Methane gas hydrates. Clathrate hydrates (or gas clathrates, *gas hydrates*, clathrates, hydrates etc.) are crystalline water-based solids physically resembling ice, in which methane is trapped inside 'cages' of hydrogen-bonded water molecules.

Methanogenesis. The process of biological methane production, for example during microbial energy production using CO_2 and H_2.

Methanogens. Microorganisms that produce methane.

Methanotrophs. Organisms that metabolise methane.

Methylotrophs. Organisms that metabolise carbon compounds with a single carbon atom (e.g. methane, methanol).

Microbial mat. A thick assemblage of microorganisms producing a mat-like macroscopic structure.

Microfossils. Microscopic fossils.

Micron. One micrometre. Equivalent to one-thousandth of a millimetre, one-millionth of a metre. Sometimes written µm.

Mineral. A naturally occurring substance that is solid and stable at room temperature, representable by a chemical formula, usually abiogenic, and has a distinct ordered atomic structure.

Mitochondrion. An organelle found in eukaryotic cells. It is responsible, among other functions, for the production of *adenosine triphosphate* (ATP) through aerobic respiration.

Mitosis. The process by which eukaryotic cells divide to form two, genetically identical 'daughter' cells.

Mole. A unit of measurement which expresses the amount of a chemical substance that contains as many elementary entities (e.g. atoms, molecules, ions, electrons) as there are atoms in 12 g of pure ^{12}C. This corresponds to the Avogadro constant, which has a value of 6.022×10^{23} elementary entities of the substance.

Molecular clock. The use of changes in biomolecules to infer times of divergence in evolutionary lineages.

Molecular systematics. The study of the diversification of living forms, both past and present, and the relationships among living things through time using genetic or other molecular information.

Monophyletic. See *clade*.

mRNA. See *messenger RNA*.

Multimeric. A protein made up of more than one subunit.

Multiplanet species. A species that has a presence on more than one planetary body.

Multiverse. The hypothetical set of infinite or finite possible universes (including the universe we experience) that together comprise everything that exists.

Mutation. A change in the nucleotide sequence of the genome of an organism caused, for example, by errors in the reading of the genetic code, ionising radiation or chemicals.

NAD$^+$. See *NAD(P)H*.

NADH. See *NAD(P)H*.

NAD(P)H. The acronym that refers to both nicotinamide adenine dinucleotide phosphate (NADPH) and nicotinamide adenine dinucleotide (NADH), two molecules used in organisms to carry and donate electrons in metabolic pathways. NADP$^+$ and NAD$^+$ are both electron acceptors that become reduced to NADPH and NADH. These two molecules can then go on to donate electrons ('reducing power' as electron donors) to other reactions (e.g. CO_2 fixation in the Calvin cycle).

Near-Earth Object. A Solar System object whose orbit brings it into proximity with Earth.

Neutrino. An electrically neutral, weakly interacting elementary subatomic particle.

Neutron. A subatomic particle. Neutrons have no net electric charge and a mass slightly larger than that of a proton.

Neutron star. A type of stellar remnant that can result from the gravitational collapse of a massive star during a supernova event.

Niche. The 'niche' of an organism usually means its place in the biotic environment, its relation to food and nutrients and its functional role in an ecosystem.

Nitrogenase. The enzymes involved in fixing atmospheric nitrogen gas.

Noachian. The period of Mars history prior to 3.7 Ga ago (often including the Pre-Noachian; 4.5–4.1 Ga ago).

Node *(in phylogenetics)*. The time at which a formerly cohesive population diverged into two genetically isolated descendant populations.

Non-coding DNA. DNA sequences that do not code for protein.

Nucleic acid. Macromolecules including *DNA* and *RNA* made of chains of *nucleotides*.

Nuclear fission. A nuclear reaction or a radioactive decay process in which the nucleus of an atom splits into smaller parts.

Nuclear fusion. A nuclear reaction in which two or more atomic nuclei collide at a very high speed and join to form a new type of atomic nucleus.

Nucleases. An enzyme capable of cleaving the phosphodiester bonds between the nucleotide subunits of nucleic acids. Essentially an enzyme that breaks down nucleic acids.

Nucleobases. Nitrogen-containing molecules encoding genetic information in DNA (adenine, thymine, guanine and cytosine) and RNA (containing uracil instead of thymine).

Nucleosynthesis. The process that creates new atomic nuclei from pre-existing sub-atomic particles.

Nucleotide. Organic molecules that serve as the subunits of nucleic acids like DNA and RNA.

Nulling interferometry. A type of interferometry in which two or more signals are mixed to produce observational regions in which the incoming signals cancel themselves out, leaving signals of interest.

Obligate. Used in reference to microorganisms to mean a requirement for a condition (e.g. obligate anaerobe that requires the absence of oxygen to grow) – in contrast to *facultative*, which means it does not have the requirement (e.g. facultative anaerobe that can grow in the presence or absence of oxygen).

Obliquity. The angle between an object's rotational axis and its orbital axis (sometimes called axial tilt).

Ocean planet (waterworld). A hypothetical type of planet which has a substantial fraction of its mass accounted for by water.

Olivine. A magnesium iron silicate with the formula $(Mg^{2+}, Fe^{2+})_2 SiO_4$.

Oort Cloud. A spherical cloud of predominantly icy planetesimals believed to surround the Sun at up to 50 000 AU.

Ophiolite. A section of the Earth's oceanic crust and the underlying upper mantle that has been uplifted and exposed above sea level and often emplaced onto continental crust.

Organic carbon. Carbon containing bonds to H, N, O, S and other carbon compounds.

Origins of replication. Specific sequences in the *genome* where DNA replication is initiated.

Outgroup. In cladistics or *phylogenetics*, an outgroup is an organism or group of organisms that serve as a (usually distantly related) reference group when determining the evolutionary relationships of other organisms of interest.

Oxidation/reduction potential (ORP). A measure of the tendency of a chemical species to acquire electrons and thereby be reduced. It is measured in volts (V) or millivolts (mV). Each species has its own intrinsic reduction potential.

Oxidation state (of an element). The degree of oxidation (loss of electrons) of an atom in a chemical compound. For example, iron can be found as Fe^{2+}, where it has lost two electrons, or in a more oxidised state, Fe^{3+}, in which it has lost three electrons. A greater oxidation state means an element is in a form where it has lost a greater number of electrons (as a result the ion has a more positive charge since it has more protons than electrons).

Oxygen free radicals or Reactive Oxygen Species (ROS). *Radicals* of oxygen and oxygen-containing molecules that are produced by radiation and chemical agents. They are reactive and can damage biological macromolecules. Also see *free radical*.

Oxygenic photosynthesis. Oxygenic photosynthesis utilises light energy to generate 'excited' electrons from water and thus drive the energy-gathering reaction: carbon dioxide + water + light energy → carbohydrate + oxygen.

PAHs. See *Polycyclic aromatic hydrocarbons*.

Paleosols. Ancient soils.

Pangaea. A supercontinent that existed during the late Paleozoic and early Mesozoic eras. It formed approximately 300 million years ago and began to break apart after about 100 million years.

Panspermia. The theory that life can be transferred between planets.

Parsec. The distance from the Sun to an astronomical object which has a parallax angle of one arcsecond.

Partial melting. Partial melting occurs when only a portion of a solid is melted. For mixed substances, such as a rock containing several different minerals, this melt can be different from the bulk composition of the solid. In rocks made up of several different minerals, some minerals will melt at lower temperatures than others, causing this partial melting.

Partial pressure. The pressure exerted by an individual gas in a mixture of gases.

Pelagic. The zone in a water body neither close to the bottom nor near the shore.

Peptide bond. The bond formed between two *amino acids* in a dehydration reaction.

Peptidoglycan. A molecular compound composed of sugars and *amino acids* that forms a mesh-like covering on the outside of bacterial cell membranes.

Per mil (per thousand). Usually used in notation of *isotopic fractionation* and given the symbol ‰.

Peridotite. A dense, coarse-grained *igneous* rock, consisting mostly of the minerals *olivine* and *pyroxene*.

Periglacial. Geomorphic processes that result from seasonal thawing of snow or ice in areas of permafrost, the runoff from which refreezes in ice wedges and other structures.

Perihelion. The point in the orbit of a planet closest to the Sun.

Period. A unit of geological time, for example Cambrian period.

Periplasm. The fluid filled space between two cellular membranes, particularly between the inner and outer membranes of *Gram*-negative bacteria.

Permineralisation. The process whereby an organism's remains are replaced by minerals.

Phase diagram. A phase diagram in physical chemistry, engineering, mineralogy, and materials science is a type of chart used to show conditions at which thermodynamically distinct phases can occur at equilibrium.

Phenotype. The physical observable characteristics of an organism.

Phosphorylation. The addition of a phosphate (PO_4^{3-}) group to a protein or other organic molecule.

Photic zone. The photic zone, euphotic zone, or sunlight zone is the depth of the water in a water body (for example a lake or ocean) that is exposed to sufficient sunlight for photosynthesis to occur.

Photodissociation. The process by which a chemical compound is broken down by photons.

Photolysis. A chemical reaction in which a chemical compound is broken down by photons (see *photodissociation*).

Photooxidised. The oxidation of elements by light.

Phyllosilicate. Compounds with a structure in which silicate tetrahedrons are arranged in sheets. Examples are mica and clays.

Phylogenetic bracketing. A method for inferring the characteristics of extinct organisms using the characteristics of their extant relatives.

Phylogenetics. The field within biology that reconstructs evolutionary history and studies the patterns of relationships among organisms.

Phylogenetic tree. A diagram showing the evolutionary relationships or distances between organisms.

Phylogeny. The history of the evolution of a species or group, especially in reference to lines of descent and relationships among broad groups.

Phylum. A taxonomic rank below kingdom and above class.

Piezophile (barophile). An organism that requires high pressures to grow.

Pili. Hair-like appendages found on the surface of many bacteria.

Pillow lava. Lavas with distinct pillow shapes formed when *basalt* lava erupts under water.

Planet. An astronomical object orbiting a star or stellar remnant that is massive enough to be rounded by its own gravity, is not massive enough to initiate thermonuclear fusion, and has cleared its neighbouring region of *planetesimals*.

Planetary embryos. Large accumulations of material in a protoplanetary disc that assemble to form planets (see also *planetesimals*).

Planetary island biogeography. The area of planetary sciences that investigates whether planets are biogeographical isolated islands with respect to the interplanetary transfer of life (sometimes called *panspermia*).

Planetary protection. The protocols involved in preventing the transfer of life to another planetary body (forward contamination) or the contamination of the Earth with extraterrestrial organisms (backward contamination).

Planet migration. The modification of a planet's or other stellar satellite's orbit by interactions with *planetesimals* or a gas disc.

Planetesimals. Kilometer-sized bodies of gas and dust existing in the *protoplanetary disc* that *accrete* under gravitational attraction to form protoplanets.

Plankton. A diverse group of marine organisms including drifting animals, protists, archaea, algae, or bacteria that inhabit the *pelagic* zone of oceans, seas, or bodies of fresh water. Plankton are therefore defined by their ecological *niche* rather than phylogenetic or taxonomic classification.

Plasma. One of the four fundamental states of matter (the others being solid, liquid, and gas). When air or gas is ionised, plasma forms.

Plasmid. A small DNA molecule that is separate from, and can replicate independently of, chromosomal DNA within a cell.

Plate flyer. A method of detonating explosive above a small plate to achieve shock pressures to study impact effects.

Plate tectonics. A theory describing the large scale movement of the Earth's lithosphere.

Polar molecule. A molecule that has an unequal electron distribution leading to a charged (tiny bar-magnet-like) property.

Polycyclic aromatic hydrocarbons (PAHs). Hydrocarbons – organic compounds containing only carbon and hydrogen – that are composed of multiple aromatic rings.

Polyextremophile. An organism that is adapted to growing in environments with multiple extremes.

Polymerase Chain Reaction (PCR). A method for amplifying single or multiple strands of DNA.

Polypeptides. Chains of amino acids (generally proteins).

Polysaccharide. An alternative name for long-chained sugars or carbohydrates.

Porins. Hollow cylindrical proteins which allow substances to pass through the cell membrane.

Positron (or antielectron). A positron is the antiparticle or the antimatter counterpart of the electron. It has a positive change of +1.

Prebiotic. Compounds or molecules that are involved in assembling life or were the precursors of biological molecules.

Precambrian. The Supereon of Earth history from the formation of the Earth 4.56 *Ga* ago to 541 *Ma* ago. Covers the Hadean, Archean and Proterozoic eons.

Primase. An enzyme involved in initiating the replication of DNA.

Prion. An infectious agent composed of protein only. Prions replicate by acting as a template for the production of new prion proteins.

Progenitor taxa. Taxa that survive an extinction and act as the phylogenetic ancestors of new dominant groups.

Prograde direction. A planetary body having a direction of rotation or movement in its orbit that is anticlockwise as viewed from the north pole of the sky or a planet.

Promoters. A region of DNA that initiates *transcription* of a particular gene. Transcription enzymes bind to promoters.

Protease. An enzyme that degrades proteins.

Protein. Biological molecules, or macromolecules, consisting of one or more long chains of amino acid residues. Many of them are *enzymes*.

Proton. A subatomic particle with the symbol p or p^+ and a positive electric charge.

Proton–proton chain. One of several fusion reactions by which stars convert hydrogen to helium.

Proton motive force. Energy that is generated by the transfer of protons or electrons across an energy-transducing membrane and can be used for chemical, osmotic or mechanical work.

Protoplanetary disc. A rotating circumstellar disc of dense gas surrounding a young newly-formed star.

Protoplanets. Bodies formed from the accretion of planetesimals which collide to eventually form planets.

Protostar. An early star in a protoplanetary disc.

Protozoa. A group of unicellular, heterotrophic eukaryotes.

Provirus. A virus genome that is integrated into the DNA of a host cell.

Psychrophile. An organism that requires a temperature below 15 °C to grow.

Puffy planet. An exoplanet with very low density.

Pulsar. A pulsar (the name is a portmanteau of pulsating star) is a highly magnetised, rotating neutron star that emits a beam of electromagnetic radiation.

Pyroxene. A group of important rock-forming minerals found in many *igneous* and *metamorphic* rocks. They share a common structure consisting of single chains of silica tetrahedra.

Quanta. Plural of quantum. The minimum amount of any physical entity involved in an interaction, particularly used in reference to light.

Quark. An elementary particle and a fundamental constituent of matter.

Quasar. Distant objects in the universe. They are the furthest objects away from our galaxy that can be seen. Quasars are extremely bright masses of energy and light. The name quasar is short for quasi-stellar radio source or quasi-stellar object.

Quartz. The second most abundant mineral in the Earth's continental crust, after *feldspar*. It is made up of a continuous framework of SiO_4 tetrahedra, with each oxygen being shared between two tetrahedra, giving an overall formula SiO_2 (*silica*).

Quinone. A class of organic compounds that are derived from aromatic compounds by conversion of an even number of –CH= groups into –C(=O)– groups with any necessary rearrangement of double bonds.

Quorum sensing. A system of communication in populations of microbes. Signals are passed through the group by signalling molecules, influencing the behaviour of all cells.

Racemic. A mixture that has equal amounts of left- and right-handed *enantiomers* of a chiral molecule.

Radial velocity method. The radial velocity method to detect exoplanets is based on the detection of variations in the velocity of the central star (by the *Doppler Shift*), due to the changing direction of the gravitational pull from an (unseen) exoplanet as it orbits the star.

Radians. A standard unit of angular measure. An arc of a circle with the same length as the radius of that circle corresponds to an angle of 1 radian.

Radical (sometimes free radicals, such as oxygen free radical). An atom, molecule, or ion that has unpaired electrons or an open electron shell. These unpaired electrons make free radicals highly chemically reactive towards other substances.

Radiogenic. Produced from radioactive decay.

Radioisotope. An element that decays into new atoms over time by radioactive decay.

Radiolysis. Break down of a compound by radiation. For example, the radiolytic break down of water by radiation in minerals.

Radiometric dating. A technique used to date materials such as rocks or carbon based on a comparison between the observed abundance of a naturally occurring radioactive isotope and its decay products.

Reactive Oxygen Species. See *Oxygen Free Radicals*.

Recovery interval. The period after an extinction during which diversity recovers.

Red edge. The region of a large increase in reflectance of vegetation or other *chlorophyll*-containing organisms in the near infrared range of the electromagnetic spectrum.

Red Giant. A luminous giant star of low or intermediate mass in a late phase of stellar evolution characterised by a helium core and a hydrogen burning shell. Also called Red Giant Branch stars (*RGB stars*).

Redshift. Redshift happens when light or other electromagnetic radiation from an object is increased in wavelength, or shifted to the red end of the spectrum.

Reducing (*atmospheric chemistry*). An atmospheric condition in which oxidation is prevented by the lack of oxygen and other oxidising gases.

Regolith. A layer of loose, heterogeneous material covering solid rock.

Redox couple or reaction. A Reduction–Oxidation reaction involving an electron donor and acceptor.

Replication fork. A structure formed during DNA replication as the double-stranded DNA helix splits apart, allowing the formation of complementary strands.

Replicons. A continuous region of DNA or RNA produced during replication. Several replicons can originate from a single chromosome.

Restriction endonuclease. An enzyme that cuts DNA at specific sequence sites.

Retrograde motion. The opposite of *prograde* motion.

RGB stars. *Red Giant Branch* stars. See *Red Giant*.

Ribonucleic acid (RNA). A molecule capable of carrying genetic information and making complementary copies of the DNA code. Compared to DNA, contains the uracil nucleobase instead of thymine and a hydroxyl group on the ribose, which is not present in DNA.

Ribose. An organic compound with the formula $C_5H_{10}O_5$, specifically, a pentose monosaccharide with linear form $H(C=O)–(CHOH)_4–H$.

Ribosomal RNA (rRNA). RNA molecules that form part of the structure of ribosomes.

Ribosome. A complex structure of protein and RNA where proteins are assembled.

Ribozyme. A ribozyme (ribonucleic acid enzyme) is an RNA molecule that is capable of catalysing specific biochemical reactions.

RNA. *Ribonucleic acid.*

RNA World. The hypothetical period in the origin of life when RNA dominated early biochemistry.

Roche limit. The Roche limit, sometimes referred to as the Roche radius, is the distance within which a celestial body, held together only by its own gravity, will disintegrate due to a second celestial body's tidal forces.

Rock. A rock is a naturally occurring solid aggregate of one or more minerals or a glass.

Rock cycle. A basic concept in geology that describes the dynamic transitions through geologic time among the three main rock types: *sedimentary*, *metamorphic* and *igneous*.

R-process. A nucleosynthesis process that occurs in core-collapse supernovae, and is responsible for the creation of approximately half of the neutron-rich atomic nuclei heavier than iron.

Rubisco. Ribulose bisphosphate carboxylase/oxygenase. An enzyme involved in the first major step of carbon fixation.

Runaway greenhouse effect. See *greenhouse effect.*

Rydberg Constant. The Rydberg constant (R_∞) represents the limiting value of the highest wavenumber (the inverse wavelength) of any photon that can be emitted from the hydrogen atom. It is equal to 1.097×10^7 m^{-1}. The Rydberg unit of energy, symbol Ry, is closely related to the Rydberg constant (1 Ry is equivalent to $hc\,R_\infty$). It corresponds to the energy of the photon whose wavenumber is the Rydberg constant, that is the ionisation energy of the hydrogen atom. Its value is 13.6 eV.

Rydberg unit of energy. See *Rydberg Constant.*

Secondary transit (or secondary eclipse). When the observation of a planet's reflected light is blocked by the star (one means of achieving the subtraction of starlight from the signal that includes both star and planet light).

Sedimentary. Types of rock that are formed by the deposition of material at the Earth's surface and within bodies of water.

Semimajor axis. Half the distance across a planetary ellipse along its long principal (major) axis.

Septation. Segmentation, such as in cyanobacteria.

Serpentine. A group of common rock-forming hydrous magnesium iron *phyllosilicates* with the general formula $(Mg, Fe)_3Si_2O_5(OH)_4$.

Serpentinisation. Serpentinite is a rock composed of one or more serpentine group minerals. Minerals in this group are formed by serpentinisation, a hydration and metamorphic transformation of *ultramafic* rock from the Earth's mantle.

SETI. *S*earch for *ExtraT*errestrial *I*ntelligence

Sheet silicates. Silicates forming two-dimensional sheet structures.

Shocked quartz. Quartz that has been subjected to high pressures during asteroid or comet impact.

Sidereal period. The time that it takes an object to make a full orbit.

Sievert. A derived unit of ionising radiation dose in the International System of Units (SI).

Sigma factor. A bacterial protein that initiates *transcription* of a specific gene. They are used to regulate gene expression by controlling mRNA formation.

Silica. Silicon dioxide (SiO_2). In minerals formed from silica tetrahedral (SiO_4) where the oxygen atoms are shared between silicon atoms leading to the overall formula SiO_2. Found in many rocks such as *quartz* and *granite*.

Silicates. A major component of terrestrial crustal rocks defined as a compound containing a silicon-bearing anion. Most of them contain *silica* (oxygen as the anion), but some contain fluorine as the anion.

Silicification. The process whereby silica is deposited on a material, such as an organism.

Singularity *(cosmology)*. The initial singularity was the gravitational singularity of infinite density thought to have contained all of the mass and spacetime of the Universe before fluctuations caused it to rapidly explode in the Big Bang and subsequent inflation, creating the present-day Universe.

Sister taxa. The closest relatives of a group in a *phylogenetic tree*.

Smectite. A type of clay mineral that includes minerals such as montmorillonite, nontronite and saponite.

Snowball Earth. Periods when the entire, or substantial parts of, the Earth became covered in ice sheets.

Snowline. In astronomy or planetary science, the snowline, also known as the frostline or iceline, refers to a particular distance in the protoplanetary disc from the central protostar where it is cold enough for hydrogen compounds such as water, ammonia and methane to condense into solid ice grains.

Sol. One Martian day.

Solar constant. The solar constant, a measure of flux density, is the amount of incoming solar electromagnetic radiation per unit area that is incident on a plane perpendicular to the rays, at a distance of one *astronomical unit*.

Solar Particle Events. Occurs when particles (mostly protons) emitted by the Sun become accelerated to very high energies either close to the Sun during a solar flare or in interplanetary space by the shocks associated with *coronal mass ejections*.

Solar Wind. A stream of plasma released from the upper atmosphere of the Sun, typically containing mainly protons and helium nuclei with energies between 10^6 and 10^8 eV.

Spaceguard. The term Spaceguard loosely refers to a number of efforts to discover and study *Near-Earth Objects* (NEO).

Spatial resolution. The resolving power of a telescope, given by λ/D, where D is the aperture of the telescope (which is the diameter of its main mirror or lens) and λ is the wavelength being used for observation.

Species-Area Relationship (or Species-Area Curve). The ecological relationship whereby the number of species in a physical area correlates with the size of that area.

Specific epithet. The second part of the species *binomial* that defines the particular species within its genus.

Spectral class. The classification of stars based on their spectral characteristics. The spectral class of a star is a short code summarising the ionisation state, giving an objective measure of the photosphere's temperature and density.

Spirochaetes. A *phylum* of bacteria with distinctive, helical (corkscrew-shaped) cells.

Spontaneous generation. The hypothetical process by which living organisms develop from non-living matter. Generally regarded as obsolete, although it must have happened at least once when life first originated.

Sputtering. The erosion of planetary atmospheric gases by, for example, the *solar wind*.

Stable isotope. Chemical isotopes that may or may not be radioactive, but if radioactive, have half-lives too long to be measured.

Standard Mean Ocean Seawater (SMOW). An isotopic standard for seawater. A mean concentration of present-day ocean water, maintained by the International Atomic Energy Agency in Vienna, Austria. SMOW oxygen atoms are about 99.76% ^{16}O and 0.20% ^{18}O.

Stratigraphy. A branch of geology which studies rock layers (strata) and layering (stratification). Can also be used to describe a layered sequence of rocks.

Stationary phase. A period of little or no microbial growth, but high numbers of cells. The stationary phase usually follows a stage of *exponential growth* and is caused by a depletion of nutrients.

Stoichiometric. A compound (such as a mineral) having its component elements present in the exact proportions indicated by its formula.

Strecker synthesis. A series of chemical reactions that synthesise an amino acid from an aldehyde or ketone.

Stromatolites. Stromatolites or stromatoliths are layered structures formed in shallow water by the trapping, binding and cementation of sedimentary grains by biofilms of microorganisms, especially cyanobacteria.

Subduction. The process that takes place at *convergent boundaries* by which one tectonic plate moves under another tectonic plate.

Sublimation. The transition of a substance directly from the solid to the gas phase without passing through an intermediate liquid phase.

Sunspot. Temporary phenomena on the photosphere of the Sun or other stars that appear visibly as dark spots compared to surrounding regions.

Supercritical fluid. Any substance at a temperature and pressure above its *critical point*, where distinct liquid and gas phases do not exist.

Super-Earth. An exoplanet of between about one and ten Earth masses.

Supernova. A stellar explosion that briefly outshines an entire galaxy, radiating as much energy as the Sun is expected to emit over its entire life span, before fading from view over several weeks or months. The fate of high mass stars.

Symbiosis. A close and often long-term interaction between two or more different species.

Synodic period. The time required for a body within the Solar System, such as a planet, the Moon or an artificial Earth satellite, to return to the same or approximately the same position relative to the Sun as seen by an observer on the Earth.

Syntrophy. A process by which the waste products of one organism provide energy for another organism.

Systematics. The study of the diversification of living forms, both past and present, and the relationships among living things through time.

Taq polymerase. A DNA replicating polymerase enzyme isolated from the thermophile *Thermus aquaticus*, used in processes such as *Polymerase Chain Reaction*.

Taxa. See *taxon*.

Taxon (plural taxa). Any unit used in the science of biological classification, or taxonomy.

Taxonomy. The science of defining groups of organisms on the basis of shared characteristics and giving names to those groups.

TCA cycle (tricarboxylic acid cycle). See *Krebs cycle*.

Teichoic acid. Polysaccharides that provide rigidification to *Gram*-positive bacterial cell walls by attracting positively charged ions.

Tektites. Small gravel-size objects that are composed of black, green, brown or grey natural glass that are formed from terrestrial debris ejected during extraterrestrial impacts.

Ter protein. The 'DNA replication terminus site-binding protein'. An enzyme which stops DNA replication in bacteria.

Tetrapods. Four-legged animals.

Thermophile. An organism that requires high temperatures, between 45 and 80 °C, to grow.

Thermotaxis. Movement in response to heat.

Tholins. Complex organic molecules formed by solar ultraviolet irradiation of simple organic compounds such as methane or ethane.

Tidal locking. When a planetary body orbits another body such that the same face always faces the body around which it orbits. Sometimes it is referred to as synchronous rotation.

Tiger stripes. The tiger stripes of Enceladus consist of near parallel, linear depressions in the south polar region of the Saturnian moon.

Tipping point. A sudden change in climatic conditions caused by transition across a critical threshold in certain systems.

Tolman-Oppenheimer-Volkoff limit. An upper bound to the mass of stars composed of neutron-*degenerate matter* (i.e. *neutron stars*).

Topology *(in phylogenetics).* The shape of a *phylogenetic tree*.

Transcription. The process by which a complementary copy of RNA is made from DNA.

Transcription bubble. A small section of double-stranded DNA that is unwound to allow *transcription* of a gene. A transcription bubble moves along the DNA strand as *mRNA* is synthesised.

Transduction. The process by which DNA is transferred from one microorganism to another by a virus.

Transfer RNA (tRNA). RNA molecules that transfer amino acids to the ribosome to be assembled into proteins. Each tRNA molecule carries a single amino acid with its corresponding *anticodon* sequence.

Transform boundary. A transform fault or transform boundary, also known as conservative plate boundary since these faults neither create nor destroy lithosphere, is a type of fault whose relative motion is predominantly horizontal. Most common where two plates are moving past one another.

Transformation. The genetic modification of an organism resulting from the uptake of foreign genetic material for the environment.

Transit method. The transit method is based on the observation of a star's small drop in brightness that occurs when a planet passes ('transits') in front of the star.

Translation. The process by which genetic information carried by *mRNA* is converted to protein at the *ribosome*.

Transmissible spongiform encephalopathies. The neural tissue disease caused by *prions*, symptoms of which include the development of microscopic holes in tissues.

Tricarboxylic acid cycle. See *Krebs cycle*.

Triple alpha reaction. A set of nuclear fusion reactions by which three helium-4 nuclei are transformed into carbon.

Triple point. The temperature and pressure at which the three phases of that substance coexist in thermodynamic equilibrium.

tRNA. See *transfer RNA*.

Trophic level. A level within a food web from the bottom (primary producers) to higher level predators.

T-Tauri. A phase of stellar development between the accretion of material from the slowing rotating material of a solar nebula and the ignition of the hydrogen that has agglomerated into the *protostar*. The 'T-Tauri' wind is named after the young star (a variable star in the constellation Taurus) that is currently in this stage.

Ultramafic (sometimes called ultrabasic rocks). *Igneous* rocks with a very low *silica* content (less than 45%), generally greater than 18% MgO, high FeO, low K and composed of usually greater than 90% *mafic* minerals.

Uniformitarianism. The assumption that the same natural laws and processes that operate in the Universe now have always operated in the Universe in the past and apply everywhere in the Universe.

Uninhabitable environment. An environment in which no known organism can be active.

Uninhabited habitat. An environment capable of supporting the activity of at least one known organism, but containing no such organism.

Unsaturated fatty acids. *Fatty acids* with one or more double bonds.

Van der Waals' interactions. The sum of the attractive or repulsive forces between molecules (or between parts of the same molecule) other than those due to *covalent bonds* or the electrostatic interaction of ions with one another or with neutral molecules or charged molecules (including *ionic bonds*).

Vector. A DNA molecule used in molecular biology laboratories to carry foreign genes into a cell.

Vegetation red edge. See *Red Edge*.

Virus. A small infectious agent that replicates only inside the cells of other organisms.

Viscosity. A measure of the resistance of a liquid to gradual deformation. It corresponds to the informal concept of 'thickness' of a liquid.

Volatiles. A generic term, particularly in planetary sciences, referring to the group of chemical elements and compounds with low boiling points that are associated with a planetary body's crust or atmosphere. Examples include water, carbon dioxide, nitrogen and methane.

Volcanic arc. A chain of volcanoes positioned in an arc shape as seen from above. Caused by plate *subduction*. This mechanism for forming volcanoes is different from *hotspot volcanism*.

Volcanic Explosivity Index. An index that scales the magnitude and intensity of a volcanic eruption.

Volcanic winter. A term used to describe a catastrophic cooling period following a large volcanic eruption and the injection of ash and aerosols into the upper atmosphere.

Water activity. Water activity or a_w is the partial vapour pressure of water in a substance divided by the standard state partial vapour pressure of water.

Weathering. The *in situ* degradation of natural or anthropic materials through physical/chemical/biological processes such as wind, rain and microbial interactions.

White dwarf. A white dwarf, also called a degenerate dwarf, is a stellar remnant composed mostly of electron-*degenerate matter*.

Widmanstätten patterns. Lineations of long nickel-iron crystals, found in iron meteorites.

Zircon. A tough *silicate* mineral containing thorium and uranium. Zircons are able to survive many geological processes and can be used to date ancient geological features.

Index

Astrobiology: Understanding Life in the Universe, First Edition. Charles S. Cockell.
© 2015 John Wiley & Sons, Ltd. Published 2015 by John Wiley & Sons, Ltd.
Companion Website : www.wiley.com/go/cockell/astrobiology.

CPSIA information can be obtained
at www.ICGtesting.com
Printed in the USA
BVOW07s0023241217
503487BV00001B/1/P